FUNDAMENTALS OF GEOGRAPHIC INFORMATION SYSTEMS

FUNDAMENTALS OF GEOGRAPHIC INFORMATION SYSTEMS

Michael N. DeMers

New Mexico State University

JOHN WILEY & SONS, INC.
New York · Chichester · Brisbane
Toronto · Singapore · Weinheim

Acquisitions Editor	Nanette Kauffman
Executive Marketing Manager	Catherine Faduska
Production Editor	Deborah Herbert
Text Designer	Lynn Rogan
Cover Designer	Laura Nicholls
Cover Art	Lynn Rogan
Manufacturing Manager	Mark Cirillo
Photo Researcher	Lisa Passmore
Illustration	Edward Starr

This book was set in 10 on 12 Cheltenham by Ruttle Shaw & Wetherill, Inc., and printed and bound by Hamilton Printing Co. The cover was printed by Phoenix Color Co.

Recognizing the importance of preserving what has been written, it is a policy of John Wiley & Sons, Inc. to have books of enduring value published in the United States printed on acid-free paper, and we exert our best efforts to that end.

Library of Congress Cataloging in Publication Data:
DeMers, Michael N.
    Fundamentals of geographic information systems / Michael N.
DeMers.
        p.   cm.
    Includes bibliographical references and index.
    ISBN 0-471-14284-0 (cloth : alk. paper)
    1. Geographic information systems.   I. Title
G70.212.D46   1997
910′ .285—dc20   96-34504
                                                                            CIP

Printed in the United States of America

10 9 8 7 6 5 4 3 2

PREFACE

Geographic information systems (GIS), in their automated form, originated in the 1960s. During the early days of the technology their use was limited to a small group of practitioners. GIS instruction was even more limited to a select few universities fortunate enough to have faculty who were aware of the potential of these systems. Since then the numbers and types of systems, the potential users, and the types of technical training and conceptual education have all increased at a rate that must amaze even the originators. In fact, the letters GIS now also stand for geographic information science—a recognition that this discipline has gone well beyond a set of techniques and tools to a field of scientific investigation in its own right.

My purpose in writing this book is to serve two educational missions, both reflecting the current growth of GIS technology and its intellectual content. The challenge is to provide a conceptual level of understanding of spatial analysis through the implementation of modern GIS software while presenting useful material for technical training. Thus, I have provided, wherever possible, functional links between the technician and the academic. In this way, I help technicians understand that a knowledge of fundamental geographic concepts will enhance their ability to perform the day-to-day tasks required of the software. At the same time, I show academics how the tasks they request of technicians are made easier if they can formulate their models with an understanding of the technical limitations of the software and the hardware.

My approach is notably a geographic one, as one might expect given my background. Accordingly, in my estimation, anyone who studies and employs the techniques of GIS is a geographer, at least, by avocation if not by vocation. The applications for which GIS are currently applied are likely far fewer than they will be in the future. As these applications grow so will an understanding of the role that automated geography can play in spatial decision making.

For this reason, I have written this book from the perspective of geographic exploration. My intent is to share the enjoyment I have experienced by discovering new lands and new settings within the geographic databases I see in my own GIS work. My objective throughout the book is to show that a well-prepared GIS professional is a precious resource, capable of providing a set of decision-making tools far beyond what was available only a few short decades ago. This book is designed to provide the basics of those GIS skills, somewhat independent of the technology.

Although this book is intended to be an introduction to GIS, it is important for instructors and users to supplement any course lecture material with hands-on practical exercises using some form of GIS. Some will want to use high-powered professional software, especially those who are learning GIS as a set of techniques. Others, especially those whose education in GIS will extend to multiple courses at different levels, may be more comfortable with software that is easily accessible to the student but that maintains the essential operational capabilities of a professional package. Depending on the particular situation, laboratory exercises should be adaptable to the general concepts and ideas in this book. The text is purposely designed to be software (and data) independent, trusting that course instructors will use whatever software and hardware are most applicable to the course mission and available to them. I have provided some sources for software, data, and GIS output in Appendix A, which should prove useful in either a classroom or a self-learning situation.

Finally, I recognize that a comprehensive book such as this is likely to contain errors, minor inconsistencies, and misunderstandings. I have provided a point of contact through the World Wide Web in Appendix B that allows instructors or students to ask questions or to clarify information. My hope is that readers will share information, methods, laboratories, and learning techniques with others so that we all gain from our experiences. In so doing, users of this book will play a role in enhancing the preparations of all future GIS professionals.

Michael N. DeMers

ACKNOWLEDGMENTS

It is no small irony that, after having written nearly 200,000 words for this textbook, I find myself struggling with the acknowledgments. Perhaps it is because I am painfully aware that I owe so many people an enormous debt. I hope that this work will, in some way, show them the measure of the gratitude I have for them.

I offer this work first to William A. Dando, who put my feet on the path of geographical knowledge and who has been an unwavering supporter of my work. He not only shared his knowledge and skills with me but, more importantly, he shared his time, his family and, above all, his love. No student ever received so much from an adviser. His has been the pattern that I have tried to emulate.

Much of my early understanding of GIS came from two of my former teachers as well as from two who were unaware that they taught me. Kang Tsung Chang, who taught me both manual and automated cartography, was a model of patience and understanding. His insights are still with me today. T. H. Lee Williams provided a structured, incremental, well-designed educational framework for both remote sensing and GIS. He also provided many unique opportunities to learn about the techniques of geography through seminars, directed readings, and mixes of topical and regional geography. More importantly, he allowed me the freedom to pursue my own interests in my own way. His faith in my abilities kept me going when the clouds of self-doubt threatened. Some of the material he used to teach me about GIS was derived from seminars and short courses taught by C. Dana Tomlin and Joseph K. Berry. Many of their ideas are scattered throughout this book because they are part of the way I view GIS. So, in a sense, they, too, have been my teachers. I hope that all of these fine educators continue to share their approaches through their classes, seminars, and writings. The field is better for having them.

Isaac Newton once wrote, "If I have seen further it is by standing upon the shoulders of giants."* This is certainly true of the textbook you now hold in your hands. During the many long days I spent throwing words at the computer, I pondered who the muse was who provided me with the material that you will soon read. Perhaps it was not a single muse but rather all those whose research and ideas continue to expand the field of automated geography. Because this discipline is relatively new, I am fortunate that many of these key players are still expanding our knowledge. I have been blessed by knowing some of them and by being able to count them among my friends.

Those who have befriended me include Duane Marble with whom I had the great good fortune to work for six years and whom I consider to be one of my

*Letter to Robert Hooke, February 5, 1675.

best friends. I doubt whether my knowledge of GIS would have grown enough to consider writing this book without his continued support, his demand for excellence, and his willingness to share his ideas, experience, time, and writings. Much of the last chapter of this book can be attributed to his seminal work on GIS design. Among my good friends and most ardent supporters I am also fortunate to include Vince Robinson, who shared his ideas and writings, and who has probably written more letters in support of my efforts than anyone. The efforts of Michael Goodchild for recommending me as a possible candidate for academic posts are gratefully acknowledged—perhaps, even more than some others because he was unaware that I knew of this kindness. Jerome Dobson also requires many thanks for suggesting me for the science advisory panel of the NOAA C-CAP program, for writing letters in support of my career, and for his continued encouragement of my efforts to link traditional geography to automated geography. He also deserves credit for the term automated geography, which I believe is the correct term for what I do.

Thanks are also due to Nicholas Chrisman, who encouraged me during some very difficult times and from whose work I have drawn many insights over the years. I am also grateful to him for having given me the opportunity to share my own musings on the first draft of his upcoming GIS textbook, *Exploring Geographic Information* (Wiley, 1997). My hope is that, as our two books share space on the publisher's display, we too, may share in our mutual goal of producing the next generation of GIS professionals.

Another friend who has supported my work and who has gained my professional and personal respect is Peter Fisher. I have learned much in my short professional association with him and perhaps even more by my personal association. Many other friends are responsible for this book through their academic research efforts and their personal encouragement. Robert Aangenbrug, David Cowen, Lee De Cola, Earl Epstein, John Jensen, David Mark, and Donna Peuquet are those who readily come to mind. To them and to all who have supported me, I offer this book as a way of thanking them for their continued efforts on behalf of the field of geography.

I thank all of my GIS students, many of whom have waded through the larval stages of this work in the form of fragmented and often incomplete course notes. My thanks also to the graduate teaching assistants who produced new laboratory exercises and improved on some existing ones for my GIS courses. Without their efforts this work would be an empty shell of facts. The College of Arts and Sciences and the Department of Geography at New Mexico State University provided me with an environment conducive to producing this work. I am grateful to Dean René Casillias, Robert Czerniak, Jerry Mueller, Albert Peters, and Jack Wright for their support and continued encouragement.

I thank the following reviewers of my manuscript for their diligence and professionalism: D. G. Barber, Bryan Baker, Alan Forsberg, Michael Hodgson and Keith Rice. Reviewing someone else's manuscript can be a tedious task that rewards the author more than the reviewer. The reviewers should be reminded that any success that is attendant upon this book is in large part the result of their efforts. The folks at Wiley have been nothing short of spectacular in their efforts to keep this manuscript moving forward. Thanks to my geography editors—Frank Lyman for believing in my book proposal and Nanette Kauffman for supporting that belief—and to the production and manufacturing staffs for their efforts at turning this manuscript into a bound volume.

Finally, a special thank-you is due to my loving wife Dolores, who fervently believed that this was a work that needed to be performed and that I was the right person to do it. While resisting every effort to become interested in automated geography, she has remained steadfast in her support of my writing. Without her the daunting task of completing this project would have loomed ever larger. She gave up many hours of her time to take care of things that might have distracted me from writing, gave up vacations so there would be no interruption of the work, and has had to endure an endless flood of "what I did on my book today" stories. While I spent fewer and fewer hours with her while I worked on the manuscript, my love grew stronger because I knew she was always there. This one's for you, but you don't have to read it.

M. N. D.

CONTENTS

Chapter 4   Cartographic and GIS Data Structures   83

UNIT 3   INPUT, STORAGE, AND EDITING   123

Chapter 5   GIS Data Input   125

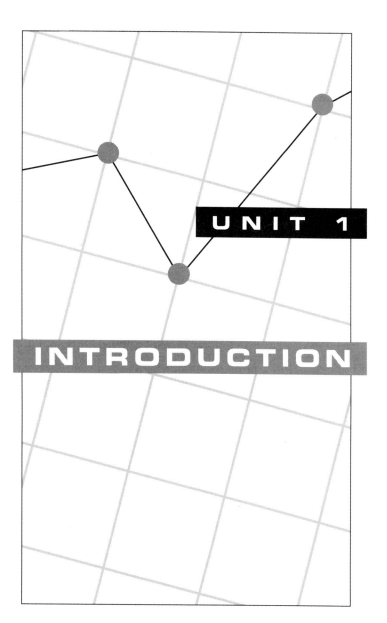

UNIT 1

INTRODUCTION

CHAPTER 1

Introduction to Automated Geography

This is a book about geography. It is also about **Geographic Information Systems (GIS)**—a fundamental set of automated ideas and concepts rooted in over 2500 years of exploration and geographic research (Dobson, 1995) and designed to provide answers to questions based on mapped data. As a practicing geographer I have long been intrigued by the idea of geographer as explorer. Visions of some Indiana Jones-like figure chopping his way through the tropical vegetation in search of some ancient ruins have filled me with wonder about the world in which we all live. I've been thrilled to picture myself as Professor Challenger, the Arthur Conan Doyle (1989) character from the book *The Lost World,* seeking out proof of the existence of live dinosaurs. The early explorers shared this wonder as they searched for new lands, new people, and new resources. As more of these lands were discovered, geographers began to use new tools to investigate the **spatial** distribution of people, plants, animals, and natural resources. They employed new methods of mapping and more efficient ways of examining the maps that others had produced. Geographic exploration became more than going to new places and describing what explorers saw. Instead, it became a way of trying to decide why the patterns explorers saw existed and what impacts those patterns might have on the health and well-being of both the people and fragile environments in which they dwelled.

Today, with very few exceptions, much of the earth has been explored by conventional means. The machete and the pith helmet have been replaced by satellites and computers. Long, hazardous journeys through deserts and tropical rainforests, have been supplanted by computerized maps and statistical data—quantitative measures of unexplored terrain. They provide a different window into new worlds as well as old, much as the microscope and the telescope provided new eyes for the biologist and the astronomer. We can now see deeper and farther than we could before, allowing us to map more of what is present on the landscape and to ask questions that could not have been imagined. Questions of where things were on the earth have been replaced by those that ask why they are there and how that knowledge could be applied to predict future distributions and patterns. These predictions allow us to plan for the future, to design our natural and human world for the mutual benefit of both.

We are still explorers. But our mission has an even greater significance than our predecessors. The ideas and tools you are about to learn are the supplies you will need to be effective in your travels through uncharted lands.

Before we begin our introduction to the tools of the new explorer a note of caution is necessary. Exploration of geographic data, like exploration of unknown regions of the earth, is exciting. It is also filled with much potential danger; with quicksand and rockslides, dangerous precipices and great chasms. The greatest risks arise from lack of knowledge of what these hazards might be and how to avoid them. Good planning is essential to success of any journey and having the right tools is a necessary first step. And, like the explorers of old, having access to the tools is not enough. Nor is it enough to know how to use the tools. You will also need to know why and when the tools should be employed to greatest effect. That is what this book is about. It is about concepts. It is about ideas. It is meant to prepare you for a lifetime of exploration. When the tools change you will be able to employ them because you are on familiar ground.

So, I invite you now to prepare for your trip into the modern world of geographic exploration. No matter what field of science or social science you pursue, the ideas you will learn will both fortify you and enhance your own understanding of your chosen field. You will learn to think both spatially and quantitatively. My hope is that each of you will feel more confident in an increasingly technological world and that this confidence will enable you to be successful and to enjoy the contributions you can make to your society and to your natural world.

LEARNING OBJECTIVES

When you are finished with this chapter you should be able to:

1. Define what a GIS is.

2. Explain the initial impetus for its development and describe some of the difficulties encountered during its early development.

3. Describe the relationships among GIS, CAC, and CAD.

4. Describe the relationship between the traditional analog map and GIS based on the four-subsystem definition.

5. Describe some basic analytical capabilities of a modern GIS.

6. Suggest possible users of GIS and how it might benefit them.

WHY GEOGRAPHIC INFORMATION SYSTEMS?

Imagine that it is the early 1960s. You are part of a team working for the national department of natural resources and development for a large country. Among your many duties in managing the resources in your area is to inventory all the available forest and mineral resources, wildlife habitat requirements, and water availability and quality. Beyond just a simple inventory, however, you are also tasked with evaluating how these resources are currently being exploited, which are in short supply and which are readily available for exploitation. In addition, you are expected to predict how the availability and quality of these resources will change in the next 10, 20, or even 100 years.

All of these tasks must be performed so that you or your superiors will be able to develop a plan to manage this resource base so that both the renewable and nonrenewable resources remain available in sufficient supply for future generations without seriously damaging the environment. You have to keep in mind, however, that exploitation of resources frequently conflicts with the quality of life for people living in or near the areas to be exploited, because noise, dust, and scenic disturbances may be produced that impact the physical or emotional health of local residents, or devalue their property by creating nearby eyesores. And, of course, you must comply with local, regional, and national legislation to ensure that those using the resource base also comply with the applicable regulations.

It takes little foresight to realize that the mission just outlined is a daunting one, requiring enormous amounts of data gathering, compilation, evaluation, analysis, and modeling. Yet this is often the very task that faces natural resource managers. Immediately, it becomes obvious that a map of the resources would permit extent, quality, and current rate of use to be viewed at a single glance. The manual production of such very helpful maps covering an entire country, especially a large one, would call for the employment of perhaps hundreds of cartographers and considerable amounts of time and money. Many of you are painfully aware of the limited coverage and variable quality of topographic, vegetation, and soil maps for your own region or nation. Depending on the size of your area it is entirely possible that long before such a task was completed the resources themselves will have disappeared, leaving the environment despoiled and the local residents up in arms.

Wishing to avoid these negative consequences, you turn immediately to the available computer technology for solutions. Obviously you need a faster, more economical method of producing maps of large regions. It would be wonderful if the computer could be used to input and store these large volumes of data and then produce a map of any region of interest or of any of the resources under your domain. But remember, it is the early 1960s, and computers are in their infancy. The largest computers of that day required rooms just to house them. In addition, their costs often relegated their ownership to large, very well-funded organizations such as the military. But let's assume that you were able to borrow time on such a machine—your problems would still not be over. Computers were not particularly adept at much beyond simple algebra and trigonometry. In fact, their memory, data storage, and analysis capabilities were not substantially beyond those currently available in small, inexpensive, hand-held calculators. The production of graphics was not a task originally envisioned for the computer. The traditional character line printer output did not give fancy graphics. The usual product was calendars, with the occasional accompanyment of cartoon characters. Even if it were possible to produce quality graphics, you would have yet another problem—there is, in the 1960s, no way to input the graphical data to the computer. In fact, output devices, including video and hard copy devices, were generally not available. Nor was there any way to link the graphical data (what we will later call **entities**) to what they were meant to represent on the ground (what will later be called **attributes**). And as yet you have not addressed the problem of how to get the computer to do analysis with the cartographic data, assuming you could input them and get the computer to store them. In addition, the predominant computer language was the cumbersome and often quirky PL/1 that was designed more for business applications than for graphics. University research in this area is relatively lacking, as well, forcing you to develop your own computerized solutions to very large and very difficult problems. To complete your assigned mission, you are going to have to develop a computerized system for the management and analysis of geographic information—a geographic information system (GIS).

For those who are only beginning their study of geographic information systems, the scenario above seems highly unlikely, even ludicrous. Certainly, by today's standards, it is. However, it is not a fabrication. This was precisely the situation that existed for the Canadian Department of Forestry and Rural Development at the beginning of the 1960s (Tomlinson, 1984). Canada, with one of the world's largest landmasses and among the largest collections of natural resources anywhere, realized that its knowledge of the extent, quality, and longevity of the national resource base was limited. Government planners also recognized that just to map such a large area would require more trained cartographers than were then available. And, of course, recognizing that such a massive task would require far longer than they could allow if they wanted to develop successful management plans for their resources, they arrived at essentially the same conclusions. What they needed was a GIS for Canada. Thus the newly instituted Regional Planning Information Systems Division, funded by the federal government, was assigned to produce what was to become the first geographic information system ever built—the Canada Geographic Information System (CGIS). Its initial task was to classify and map the land resources of Canada.

Today there is an ever increasing recognition of the need to perform large-scale mapping and map analysis operations for a wide variety of traditionally manual tasks. Foresters wanting to keep an up-to-date inventory of their timber resources see GIS as an efficient management tool for their day-to-day operations. Fire departments have a need for GIS to enhance their routing capabilities to ensure rapid response in emergencies. The military could use GIS to determine appropriate battle plans and to organize troop movements. Cellular phone companies, wanting to provide the best service for a mobile customer base, must site their transmission towers to avoid conflicts with neighbors and still allow clear line of sight for signal transmission. Local governments employ GIS to develop growth and development plans and to modify zoning regulations to account for increasing population pressures. Businesses are using GIS to market products, and even to develop mailing lists based on selected spatial criteria. Real estate companies are beginning to use GIS to isolate available housing based on customer criteria such as proximity to schools, type of neighborhood, or access to highways. Police departments are currently using GIS to compile information to characterize the movements and operational settings of suspected serial killers. Academic disciplines such as geography, biology, geology, landscape architecture, range science, and wildlife management now have the capability to employ the technology to develop and test hypotheses concerning patterns of natural phenomena on the earth. The potential users of GIS are nearly limitless, and the types and numbers of users are growing at a logarithmic pace.

This growth is indicative of the nature of GIS as an empowering technology. It is not unlike the development of the printing press, the creation of the first telephone, the replacement of the horse and buggy with the automobile, or the introduction of the first computer. All these innovations had a profound impact on the way in which we communicated, the way we moved from place to place, and the way we solved problems—even on the nature of the problems we solved. The modern geographic information system has enhanced the utility of the map by replacing it with a large number of mapped coverages, each with an interrelated theme. These coverages can be automatically analyzed, and their themes combined to give meaningful answers for decision makers. The GIS is changing the way we do things with maps, the way we think about geographic information, even the way in which geographic data are collected and compiled. Tasks that were impossible with traditional maps are now commonplace.

Current market trends for the technology indicate that it is a major growth industry, far outstripping many others, even during recession years (*Newsweek,* 1993). As more organizations become familiar with GIS, the need to become familiar with its basic principles will grow as well. There will also be an increase in the demand for people knowledgeable about the concepts behind the technology. We will examine those concepts here, with an aim toward understanding how spatial phenomena can be manipulated and how the technology can help us in an increasingly complex world.

WHAT ARE GEOGRAPHIC INFORMATION SYSTEMS?

In the broadest possible terms, geographic information systems are tools that allow for the processing of spatial data into information, generally information tied explicitly to, and used to make decisions about, some portion of the earth. This working definition is neither comprehensive nor particularly precise. Like the field of geography itself, the term is difficult to define and represents the integration of many subject areas. As a result, there is no absolutely agreed upon definition of a geographic information system. The term itself is becoming hybridized and modified to conform to intellectual, cultural, economic and even political objectives (Table 1.1). This terminology has, in fact, become extremely elastic, resulting in an increasingly confusing jargon due to new definitions that constantly creep into both the scientific and the popular literature.

This lack of accepted definition has resulted in many gross misconceptions about what a GIS is, what its capabilities are, and what such a system might be used for. It has lead some people to believe, for example, that there is no difference between computer assisted cartography, computer assisted drafting, and GIS. Because the graphic display from these three systems can look identical to both the casual and the trained observer, it is easy to assume that they are, with minor differences, the same thing. Anyone attempting to analyze maps will soon discover, however, that **computer assisted cartographic (CAC) systems,** computer systems designed to create maps from graphical objects com-

TABLE 1.1 Examples of Synonymous Terms for Geographic Information System and the Source or Motivation Behind Their Derivation

Terminology	Source
Geographic information system	United States terminology
Geographical information system	European terminology
Geomatique	Canadian terminology
Georelational information system	Technology-based terminology
Natural resources information system	Discipline-based terminology
Geoscience or geological information system	Discipline-based terminology
Spatial information system	Nongeographical Derivative
Spatial data analysis system	Terminology based on what system does

bined with descriptive attributes, are excellent for display, but generally lack the analytical capabilities of a GIS. Likewise, for pure mapping purposes it is highly desirable to use a computer assisted cartographic system developed specifically for the input, design, and output of mappable data, rather than working through the myriad analytics of the GIS to produce a simple map. **Computer assisted drafting (CAD)**—(a computer system developed to produce graphic images but not normally tied to external descriptive data files)—is excellent software for the architect, speeding the process of producing architectural drawings, and simplifying the editing process. It would not be as easy to use for producing maps as would CAC, nor would it be capable of analyzing maps—generally the primary task assigned to the GIS (Cowen, 1988).

For the experienced user of GIS technology, there is no need for a definition. The complex geographical queries that demand its use normally could not be addressed by CAC and CAD. But for those who have only heard of these tools a definition might prove useful. A preliminary definition for consideration might be that of David Rhind, who defined GIS as "a computer system for collecting, checking, integrating and analyzing information related to the surface of the earth" (Rhind 1988). This definition has some highly worthwhile elements that should be examined. First, it indicates that the GIS deals with the surface of the earth. Although this is not an absolute requirement, the vast majority of GIS applications do deal with portions of the earth. Moreover, the statement that the GIS is used to collect, check, integrate, and analyze information enumerates a large number of the necessary groups of operations for any geographic information system.

Many additional definitions of GIS have been proposed. Some have shown the strong linkage between manual and computer-based methods of map analysis (Dickinson and Calkins 1988, Aronoff 1989, Star and Estes 1990). Most others have explicitly stated among its primary objectives, to act as a tool for analyzing data about the earth (some examples are Aronoff 1989, Parker 1988, Dueker 1979, Smith et al. 1987, Cowen 1988, and Koshkarion, Tikunov and Trifimov 1989). As we will see at the end of this text, one can also extend the definition to include the organizations and people involved in working with spatial data as well (Carter 1989). Like any technology that changes as quickly as does GIS, the definitions themselves will likely change as well.

For this text I have chosen to use a definition that more closely resembles the way the GIS operates as a series of subsystems within a larger system. That definition proposed as a standard by Marble and Peuquet (1983), and used in some form by others in their own definitions (Parker 1988, Ozemoy, Smith and Sicherman 1981 and Burrough 1986), pretty much sums up what it is we do with a GIS and how we do it. It states that GIS deals with space-time data, and often, but not necessarily, employs computer hardware and software. More importantly, perhaps, is the subsystem nature of his definition that provides an easily understandable framework for the study of GIS. The GIS, according to this definition, has the following subsystems:

1. A data input subsystem that collects and preprocesses spatial data from various sources. This subsystem is also largely responsible for the transformation of different types of spatial data (i.e., from isoline symbols on a topographic map to point elevations inside the GIS).

2. A data storage and retrieval subsystem that organizes the spatial data in a manner that allows retrieval, updating, and editing.

3. A data manipulation and analysis subsystem that performs tasks on the data, aggregates and disaggregates, estimates parameters and constraints, and performs modeling functions.

4. A reporting subsystem that displays all or part of the database in tabular, graphic, or map form.

The subsystem definition allows for easy comparison between the modern automated GIS and its analog counterpart, particularly when considering the steps in the cartographic process (Table 1.2). The first GIS subsystem, the data input subsystem, is roughly equivalent to the first and second steps in the **cartographic process**—data collection and map compilation (Robinson et al. 1995) (Table 1.3). In traditional cartography the cartographer compiles or records a map made up of points, lines, and areas on a physical medium such as paper or Mylar. The data are collected from such sources as aerial photography,

TABLE 1.2 Comparison of the Cartographic Process as Applied to Traditional Cartography (Map) and Geographic Information Systems (GIS)

Map	GIS
Data collection: aerial photos, surveys, etc.	Data collection: aerial photos, surveys, etc.
Data processing: aggregation, classing, etc.; linear process	Data processing: aggregation, classing, plus analysis; circular process
Map production: final step except for reproduction and dissemination	Map production: not always final step; normally one map used to produce still more
Map reproduction	Map reproduction

TABLE 1.3   Analog Versus Digital GIS: A Comparison of Input Subsystem Functions

Map	GIS
Input: recorded (compiled) on paper from a collected source	Input: "encoded" into the computer from a collected source
• Points • Lines • Areas	• Points • Lines • Areas
Sources	Sources
• Aerial photography • Digital remote sensing • Surveying • Visual descriptions • Census data • Statistical data, etc.	• Same as map data • Digital line graph (DLG) • Digital elevation models (DEM) • Digital orthophotoquads • Other digital databases

digital remote sensing, surveying, visual descriptions, and census and statistical data. The automated counterpart uses electronic devices to record or **encode** points, lines, and areas into a computer system. Data collection sources are often the same as those used for traditional mapping, but now include a wide variety of digital sources: **digital line graphs, digital elevation models, digital orthophotoquads,** and many more. Although the mechanics differ between the two technologies, the actual methods are strikingly similar.

This is also the case for the second subsystem, the storage and retrieval subsystem (Table 1.4). Although there is no actual counterpart in the cartographic method, the map itself is the storage and retrieval tool. Points, lines, and areas that have been placed on the cartographic document are stored there for retrieval by the map reader. It has been said that the map is the most compact medium for the storage of spatially related information and may be the most complex form of graphic device available. In fact, the compactness of the map and its complexity frequently hamper the map reader's ability to extract information.

The GIS storage and retrieval subsystem has some advantages over the

TABLE 1.4   Analog Versus Digital GIS: A Comparison of Storage and Retrieval Subsystem Functions

Map	GIS
Points, lines, and areas are drawn on paper with symbols.	Points, lines, and areas are stored as grid cells or coordinate pairs and pointers in computer.
Retrieval is simply a matter of map reading.	Attribute tables are associated with coordinate pairs.
	Retrieval requires efficient computer search techniques.

graphic map in that queries can be made of the data and only the appropriate, context-specific information recalled (Table 1.4). This format places more emphasis on formulating queries and asking the appropriate questions and less on overall map interpretation. In general terms this subsystem stores, either explicitly or implicitly, the graphic locations of point, line and area objects (entities), and their associated characteristics (attributes). Computer search methods are inherent in the GIS programs themselves to allow questions to be asked and for appropriate answers to be given.

In the analysis subsystem, once again there is no exact cartographic method counterpart, except that the map is a fundamental tool for the analysis of spatially related data (Table 1.5). The analog map requires rulers to measure distances, compasses to find directions and dot grids or **planimeters** to measure areas (Marble, 1990). Furthermore the map analyst is restricted to the graphic methods used to present the data on the piece of paper or Mylar. Still, these map analysis tools have been used for a great many years because of the known utility of comparing spatially related phenomena in a quantitative manner.

The analysis subsystem is the heart of the GIS. The need to analyze maps to compare and contrast patterns of earth-related phenomena, exemplified by the long-standing tradition of doing so with traditional maps, provides an impetus to find more convenient, faster, and more powerful methods. GIS analysis uses the power of the modern digital computer to measure, compare, and describe the contents of the databases. It allows ready access to the raw data and allows aggregation and reclassification for further analysis. Not only is it not limited in the types of data it can retrieve but it can combine selected data sets in unique and useful ways far beyond what the traditional map could provide on a single sheet (DeMers, 1991).

Of course, once an analysis has been performed, there is generally a need to report these results. In cartography, whether it be traditional analog cartography or its digital equivalent, computer assisted cartography, the output is generally the same—a map. The most common purpose of cartography, at least from the user perspective, is to produce a map product, usually in copies for multiple recipients. In fact, production and reproduction are the final two steps in the cartographic method (Robinson et al., 1995).

A major difference between GIS and cartography, beyond the emphasis on analysis in GIS, is the method of reporting the results of analysis (Table 1.6). Although many users, perhaps even most, will still require mapped output,

TABLE 1.5 Analog Versus Digital GIS: A Comparison of Analysis Subsystem Functions

Map	GIS
Requires rulers, planimeters, compass, and other tools all used by the human analyst	Uses the power of the computer to measure, compare, and describe contents of the database
Restricted to the data as they are aggregated and represented on the paper map	Allows ready access to the raw data and allows aggregation and reclassification for further analysis

TABLE 1.6 Analog Versus Digital GIS: A Comparison of Reporting and Output Functions

Map Output	GIS Output
Graphic device only	The map is only one type of GIS output
Many forms of maps	With minor exceptions, GIS offers same options as traditional hand-drawn maps
Modifications can include cartograms, etc.	
	Also includes tables, charts, diagrams, photographs, etc.

there are many options available in modern GIS. Some typical noncartographic output could include tables listing, for example, the anticipated crop yields per hectare by soil type or predicted changes in population densities by census district. Alternatively, either of these results also could be output as a series of histograms or line graphs. Supplementally, digitally encoded photographs of selected sites could be placed on the map margins or within the tables or charts.

More advanced GIS features are available, as well. Examples include output in the form of printed mailing labels for a search of a database of potential customers to facilitate the distribution of advertising. A 911 emergency system database could be connected to a police or fire department, so that when a caller reports an emergency, the information can be directly routed to the nearest emergency service. This output could also be in the form of a route map showing the fastest path from the emergency branch to the site of the emergency. In fact, the types of output are often dictated more by the use for which the GIS is employed than by the software. And, like the users of maps, the outputs are many and varied.

Among the more interesting phenomena arising from the wide range of users is a new set of terms defining the system on the basis of what it does. For example, one could have a police information system, a natural resources information system, a census information system, a rangeland evaluation system, a land information system, a **cadastral** information system, and so on. Although these terms are generally descriptive of the use for which the GIS is being employed, they do little to clarify the exact nature of the system. In fact, they generally add considerably to the confusion. Perhaps a more structured approach to classification of GIS in the form of a taxonomy would prove useful (Figure 1.1).

This taxonomy diagram clearly shows the separation between spatial and nonspatial information systems. The GIS appropriately fits under the spatial information systems category. Two general classes of spatial information systems are identified: geographic and nongeographic. Nongeographic information systems, although they frequently deal with some portion of geographic space, seldom have strong locational links to the earth itself. In other words, they are not generally geocoded. Thus such systems as computer assisted drafting and computer aided manufacturing, come under the nongeographic spatial information systems heading.

Within geographic information systems there is yet another bifurcation. GIS are divided into **land information systems (LIS)** and nonland information sys-

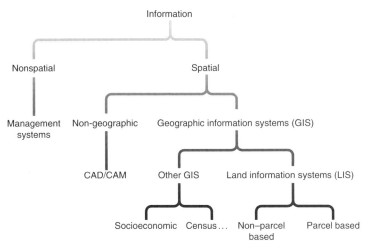

Figure 1.1   A taxonomy of information systems. The illustration shows how GIS and LIS fit in.

tems, or other geographic information systems. Although the division is somewhat artificial, it is important because it separates the applications of GIS technology into those that are primarily focused on the land itself and those that, although being geocoded, are more focused on information that might either affect or be affected by land-related factors. These uses include census information systems whose primary focus is on populations and their housing and economic activities, rather than on the land on which they reside or even on their use of the land. Another (non-land-related) GIS application might include applications surrounding political redistricting. Although political redistricting, by its very nature subdivides or apportions the land into discrete portions, such activities generally have little or no direct and immediate impact on the land itself. Rather, political redistricting affects the voting patterns of those living on the land surface. A common non-land-related use of GIS is market analysis, which may include a determination of the amount of market within reasonable reach of a business **(allocation)** or might involve an analysis of existing facilities to determine where best to put a competing or complementary facility **(location).** Locating fire stations, schools, and other facilities falls into this category. In general, non-land related GIS activities tend to entail social, economic, transportation, and political types of activities.

   Land-related activities provide the framework for the second, and possibly the most often used type of GIS, the land information system (LIS). Such systems are based most often on the ownership, management, and analysis of portions of the earth most frequently of interest to humans primarily because of their condition of ownership. Land information systems are further subdivided into parcel based and non–parcel based. Non–parcel-based LIS include natural resources information systems, such as those used by national park services, forest services, land management agencies, and the like. Activities within the non-parcel-based LIS could include habitat evaluation, conservation easement procurement, wildlife evaluation, earthquake and landslide prediction, flood hazard abatement, chemical contamination evaluation, forest and range management, and scientific investigation.

Parcel-based LIS applications are generally focused on landownership and other cadastral investigations. The defining criterion is that the land be divided into surveyed parcels having legal descriptions. Although this terminology could also apply to such portions of land as national forests, it generally assumes that the parcels are smaller than this (National Academy of Science 1980, 1983). Fundamental to applications of these types is a highly accurate **geodetic framework** upon which the parcels can be precisely described. LIS applications involve traditional survey methods and are among the largest users of NAV-STAR's **Global Positioning System (GPS)** for acquiring this locational information. Once an accurate geodetic framework and cadastral system have been developed, many analyses of land-tenure change can be performed with the assurance of a high degree of measurement accuracy. Included in such studies are those attempting to arrive at compatible multiple land uses within selected parcels of land. Some of these studies may require the incorporation of a multipurpose cadastre—a parcelization framework that allows analysis of multiple land-parcel-related phenomena.

Whether they are land related or human related, the applications of GIS technology are many and varied, offering enormous possibilities for both simple and extremely sophisticated analyses. Most of today's applications are quite limited in sophistication, however. Generally, this under use of system capabilities seems to be related more to a lack of understanding of the existing potential of GIS, rather than to actual software limitations. Before we can ask software to perform a particular task, we must be aware of what that task might be. Then we can see whether the software is capable of accomplishing it. People using today's GIS software are frequently heard saying, "Hey, I didn't know we could do that with the computer!" The exclamation is one of discovery, not unlike the reaction of geographers of old as they ventured into the jungle with pith helmet and machete. For the person newly introduced to GIS software, the journey into new dimensions of geographic exploration has just begun.

WHERE DO I BEGIN?

GIS is an exciting, even glamorous, field with rapidly expanding opportunities for those who are familiar with the concepts and the technology. Because it is exciting, and because of the potential for a rewarding career, we want to get started as soon as possible. But any long journey is best begun through careful planning, and undertaken a step at a time. It is a common but fallacious notion that because GIS has become readily available and is showing up in a wide variety of organizations, it should be possible to just sit down and start using the software. GIS software, however, is not like the personal computer word processing software so many of us have grown so accustomed to. Although most of us know some basics about writing and are perhaps very familiar with computer word processing, few of us are as comfortable with the analytical operations necessary to make decisions with maps. Just as word processing software assumes that you can organize your thoughts and ideas into coherent sentences and paragraphs, however, GIS assumes you are familiar with the vocabulary of maps.

When asked, most of us will say that we are fairly comfortable with maps.

We use road maps routinely, and when necessary, we consult a world atlas, with its political, physical, and economic boundaries, and associated colors, graphic symbols, text and, of course, north arrows. Most of us, however, don't think much about how much information a map contains. Nor do we give much thought to the generalization processes that take place to decide which detail gets included and which does not. Nobody wants to think about the problem of representing an essentially spherical surface onto a flat piece of paper. Because the map is such an elegantly designed document—so well thought out—we simply accept it at face value.

On occasion, however, the limitations of the cartographic art begin to show through. How often have you wondered why a road that looks straight on a map really curves all over the place? The graphic limits imposed on the cartographer by available data quality, pen size, size of paper, and other conditions all require him or her to make conscious decisions about how much detail can and should be placed on a given map document. Much of this generalization is imposed by the map scale. The smaller the map scale (the larger the mapped area), the greater is the required generalization to produce the cartographic model.

This concept of the map as a model of reality is perhaps the most important concept that must be learned by the future GIS professional. Because the map has such a strong visual appeal, the viewer tends to accept it as reality. Those who work with maps, and especially those who work with the interactions of many maps, must constantly remind themselves of the limitations of the cartographic product. Here are just a few simple exercises you can do to become familiar with the cartographic model and some of its limitations.

Take a look at a number of world maps from different atlases you might find in your library. Pick out a country familiar to you. Notice how maps of it differ with respect to sizes, shapes, boundary configurations, numbers of cities, and the like on these maps. You might be surprised at the wide variation from one map to another. Consider, then, what you would have to do if you were to digitize a map of this country into a GIS. Which one would you select? Why? How does focusing on the purpose of your GIS project help you decide which map you want?

Obtain two or three adjacent topographic maps for your area. What are the dates for each map? Are they the same? Different? Now the fun begins. With clear plastic tape (preferably removable), tape the maps together so that all the lines match. Be sure to turn on some relaxing music while you do this. What do you discover? The lines don't match exactly. Imagine how you are going to input 20 or 30 of these documents to a GIS if the lines don't line up with any pair of them.

Soil data might be nice to include in your GIS. Try the last experiment with your local soil map sheets. The match between sheets is even worse. If you are using soil maps from the U.S. Department of Agriculture Soil Conservation Service, you might be quite taken with the use of the aerial photography in the background. Admittedly, this feature is nice to have. But the addition comes at a cost. If you are going to input this map with other maps inside a geographic information system, you will have to coregister it with all the others so that the features match. This requires that the locational coordinates be specified on all maps. Try to find these on the soil survey maps. How do you solve this little dilemma?

If the foregoing examples haven't convinced you of the importance of understanding the vocabulary of maps before you begin speaking GIS, perhaps this one will. You need to create a map of presettlement vegetation for your state or region. It turns out that three very well known vegetation mappers have compiled such maps for portions of your area of interest. Taking a trip to the library to obtain these maps, you discover that the first shows vegetation classified by its structural components (herbaceous, grasses, trees, shrubs, etc.), and, the second map, which intersects with the first map, shows vegetation classed by floristics (based on species). You also note, to your annoyance, that the two systems seem to have only limited map areas that correspond. Hoping for help from the third map, you discover that although it is classed based on a combination of floristics and structure, its area does not overlap either of the other two maps; in fact, it is well separated from them.

Classification problems of the type just described are common ones requiring the student of GIS to become more than a student of the technology. Before you can master the technology you should first master its concepts. We will begin this first step in the journey in the next chapter, where we will look closely at the nature of geographic data and the methods by which they can be represented on map documents. This first step will give us a better appreciation of the fundamental building blocks of GIS and will ensure a more cautious approach when we begin implementing geographic analysis and cartographic modeling.

Terms

spatial

computer assisted cartography

allocate

digital line graphs

digital orthophotoquads

cadastral

digital line graphs

entities

locate

encode

land information systems

global positioning system

attributes

computer assisted drafting

cartographic process

digital elevation models

planimeters

geodetic framework

Review Questions

1. What was the initial impetus for the development of the first GIS? Why was it so difficult to develop in the first place?

2. What is a geographic information system? How does your definition differ from that offered by David Rhind? Why are there so many names for geographic information systems?

3. What is the relationship between the traditional map and its automated counterpart? What are the relationships among the four subsystems in GIS and the map?

4. What is the difference between GIS and computer assisted cartography? Between GIS and CAD?

5. What are the basic analytical capabilities that one would normally find in a modern GIS?

6. Who would normally use a GIS? What accounts for its popularity?

References

Aronoff, S., 1989. *Geographic Information Systems: A Management Perspective.* Ottawa, Canada: WDL Publications.

Burrough, P.A., 1986. *Principles of Geographic Information Systems for Land Resources Assessment.* Oxford: Clarendon Press.

Carter, J.R., 1989. "On Defining the Geographic Information System." In *Fundamentals of Geographic Information Systems: A Compendium,* W. J. Ripple, Ed. Falls Church, VA: ASPRS/ACSM, pp. 3–7.

Cowen, D.J., 1988. "GIS Versus CAD Versus DBMS: What Are the Differences?" *Photogrammetric Engineering and Remote Sensing,* 54(11):1551–1554.

DeMers, Michael N., 1991 "Classification and Purpose in Automated Vegetation Maps." *Geographical Review* 81(3):267–280.

Dickinson, H., and H. Calkins, 1988. "The Economic Evaluation of Implementing a GIS." *International Journal of Geographical Information Systems,* 2:307–327.

Dobson, J.E.,1995. "Defining the University Consortium for Geographic Information Science," *GIS World* 8(3):44–46.

Doyle, A.C. 1989. *The Complete Professor Challenger Stories.* Ware, Hertfordshire: Wordsworth Editions.

Dueker, K.J., 1979. "Land Resource Information Systems: A Review of Fifteen Years Experience." *Geo-Processing,* I:105–128.

Koshkariov, A.V., V.S. Tikunov, and A.M. Trifimov, 1989. "The Current State and the Main Trends in the Development of Geographical Information Systems in the USSR." *International Journal of Geographical Information Systems,* 3(3):257–272.

Marble, D.F., and D.J. Peuquet, 1983. "Geographic Information Systems in Remote Sensing." In Manual of Remote Sensing, Vol. 1, 2nd ed., R.N. Colwell, Ed. Falls Church, VA: American Society of Photogrammetry, pp. 923–957.

Marble, D.F.,1990. "The Potential Methodological Impact of Geographic Information Systems on the Social Sciences." In *Interpreting Space: GIS and Archaeology,* Allen, Zubrow, and Green, Eds. Washington, D.C.: Taylor and Francis.

National Academy of Sciences, 1980. Need for a Multipurpose Cadastre. Panel on a Multipurpose Cadastre, Committee on Geodesy, Assembly of Mathematical and Physical Sciences. Washington, D.C.: National Academy Press.

National Academy of Sciences, 1983. Procedures and Standards for a Multipurpose Cadastre. Panel on a Multipurpose Cadastre, Committee on Geodesy, Commission on Physical Sciences, Mathematics, and Resources, National Research Council. Washington, D.C.: National Academy Press,

Ozemoy, V.M, D.R. Smith, and A. Sicherman, 1981. "Evaluating Computerized Geographic Information Systems Using Decision Analysis." *Interfaces,* 11:92–98.

Parker, H.D., 1988. "The Unique Qualities of a Geographic Information System: A Commentary." *Photogrammetric Engineering and Remote Sensing,* 54(11):1547–1549.

Rhind, D.W., 1988. "A GIS Research Agenda." *International Journal of Geographical Information Systems,* 2:23–28.

Robinson, A.H., J.L. Morrison, P.C. Muehrcke, A.J. Kimerling, and S.C. Guptill, 1995. *Elements of Cartography,* 6th ed. New York: John Wiley & Sons.

Smith, T.R., S. Menon, S. Star, and J.L. Estes, 1987. "Requirements and Principles for the Implementation and Construction of Large-Scale Geographic Information Systems." *International Journal of Geographical Information Systems,* 1:13–31.

Star, J., and J.E. Estes, 1990. *Geographic Information Systems.* Englewood Cliffs, NJ: Prentice Hall.

Tomlinson, R.F., 1984. "Geographic Information Systems: The New Frontier." *The Operational Geographer,* 5:31–35.

U.S. News & World Report, 1995. "20 Hot Job Tracks: 1996 Career Guide," October 30, pp. 98–108.

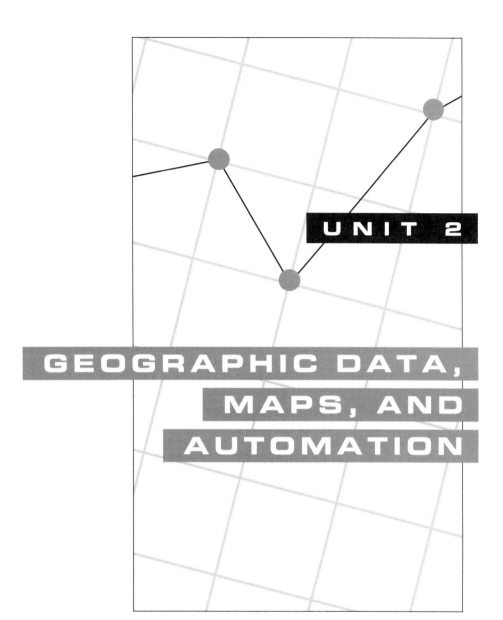

UNIT 2

GEOGRAPHIC DATA, MAPS, AND AUTOMATION

CHAPTER 2

Spatial Analysis: The Foundation of Modern Geography

Before we begin our journey of spatial data exploration, it is important to be aware of the types of environments we will travel. In GIS we travel the same jungles, swamps, plains, deserts, mountains, and valleys of our predecessors. However, these environments are now graphic and numerical representations of the real world. To become familiar with the potential hazards of our travels, we must know the elements that make up our digital environment, just as the traveler of old was well advised to know the types of terrain, the vegetation and the climate, the likely disposition of the natives, and the likelihood of encountering quicksand, cliffs, and other natural features that had to be avoided or searched for.

Unlike the real world, our environments will be composed of cartographic objects representative of the individual components of the earth. These objects will differ in size and shape, in color and pattern, and in scale of measurement and degree of importance. They can be measured directly by instruments on the ground, sensed by satellites hundreds of miles above the surface, collected by census takers, or extracted from the pages of documents and maps produced in ages past. Some will be very important for our journey, others merely useful, and others will need to be discarded as we begin our navigations. Thus to explore the modeled world we encounter, cartographic objects must be collected, organized, and synthesized.

As you read through this chapter, keep in mind that the nature of the data often dictates not only how we will later represent the earth inside a GIS database, but how effectively we will analyze and interpret the results of that analysis. In turn, how we view and experience our environment will affect what features we note and how we will eventually represent them. The points, lines, and areas we encounter are all different. Moreover, their representation and utility will depend, in large part, on our ability to recognize what features are important and to identify those that may be modified by the temporal and spatial scales at which we observe them. This information, in turn, will dictate

21

how the data are stored, retrieved, modeled, and finally output as the results of analysis.

In addition to the temporal scale and the physical sizes of objects stored in a GIS database, we must consider the measurement level we will use to represent their descriptive conditions or attributes. At one scale, for example, points could include whole cities with no areal extent, while at another, objects as small as insects or even microbes will be the important point data under consideration. The cities may also include descriptive attributes, such as their names (nominal measurement scale), whether their viability for placing an industry would be considered major, moderate, or minor (ordinal measurement scale), their average annual temperature (interval measurement scale), or the average annual per capita income (ratio measurement scale). Each of these types of data represents a fundamentally different criterion, measured with a substantially different measuring device, and with a different level of data precision. The same can easily be said for lines, areas, and surfaces.

Always remember that the first step toward better GIS skills is to begin to think spatially. We all operate in a spatial environment, but we are more often than not oblivious to the space around us, paying no attention to how we and other objects occupy, move through, interact with, and even modify our space. The old cliché requesting that you "stop and smell the roses" may be the best advice to begin the transition from spatial insensibility to true spatial cognition. Take the time to notice the arrangements of plants and animals on the landscape; become aware of the differences in urban and suburban neighborhoods; notice the difference in your own pace as you travel up and down hill; keep track of the different routes you take to get places and analyze why you make these decisions; recognize the patterns of agriculture as you drive through or fly over rural landscapes; take note of where pollution flows in waters and in the air. In short, become familiar with all the possible patterns, interconnections, distances, directions, and spatial interactions in your world. As you become more sensitive to the objects themselves, you will find it an easy next step to imagine how the objects and interactions can be measured and what meter stick you will need to record them.

It is important to know these two interconnected concepts of objects and measurements well before you begin producing results from a GIS. As you begin to learn each type of spatial data and its measurement level, try to imagine the types of data you would most often encounter in the work you already do, or would like to do, with maps. A useful exercise is to compile lists of these as you encounter each new type or level of data. This will both enhance what you learn by giving you additional, concrete examples, and allow you to relate the concepts to your own everyday experiences. More importantly, as you progress through the rest of the book you will begin to see how these are represented in map form, how they will be encoded and stored, and some ways in which they could be analyzed. With this firsthand experience at moving from the conceptual data model to the computer representational model, you will be well prepared to deal with the analytical models you are likely to encounter as you use GIS for your own specialty.

When you are finished with this chapter you should be able to:

1. Show how an improved spatial or geographic vocabulary improves your perception of the world.

2. Illustrate the impact of scale on our perceptions of the world.

3. Understand the difference between spatial and aspatial data, and indicate how surrogates can be used when spatial data are lacking.

4. Understand the difference between discrete and continuous data and give examples for point, line, area, and surface features.

5. Understand the relationships among nominal, ordinal, interval, and ratio scales of data measurement and give examples for point, line, area, and surface features.

6. Understand the necessity of a structural framework or grid system for locating yourself and your relationships to other earth features in both absolute and relative terms.

7. Understand the meanings of terms such as orientation, arrangement, diffusion, pattern, dispersion, density, and spatial association and be able to use them when discussing geographical phenomena.

8. Understand the use of ground sampling methods and the advantages and disadvantages of directed ground sampling, probability-based ground sampling, and remote sensing for gathering data about the earth.

DEVELOPING SPATIAL AWARENESS

We live in a complex world. To succeed, we must be aware of this complexity and be able to organize it around a framework that allows us to understand how such a seemingly disordered system continues to function. Throughout your university education you have been presented with a large variety of such frameworks, each designed to help you understand and order particular aspects of your world. Your English professors provided you with an organizational structure to allow you to better communicate your ideas with language. Likewise, your history professors showed you the structural relationships of events through time—how historical events have causes and effects. The study of history, then, places a framework around which you can begin to organize, synthesize, and make decisions concerning temporal phenomena. Your political science professors showed you the structure placed around governments that allows them to operate with a sense of order. In biology you learned how living cells and organisms function and interact. In fact, the word "organism" means an ordered biological creature.

All these disciplines have in common a search for order amid apparent disorder. Our natural curiosity stimulates a search for knowledge that will allow us to structure the aspects of our world whereupon we will be able to determine both how it works now and how it will work if something changes. This is no less true of the discipline of geography and the tools this discipline has developed. The focus on one simple underlying principle—the search for spatial order—separates geography from all the other fields of study you have encountered. Many disciplines you encountered in your studies have, on occasion, asked how people, places, critters, natural features, and the like changed or interacted spatially, but only geography continues to concentrate on spatial relationships as its principal intellectual framework. Because of this focus, geographers have developed a language that reflects the way they think about space. This spatial language, like any language, allows the geographer to think more clearly and communicate more concisely about space, examining only the structures and arrangements of the data at hand that are necessary to explain spatial phenomena. As taxonomists use the Latin and Greek names for biological organisms to more quickly and more precisely communicate the often subtle differences, geographers use their terminology to describe, explain, and analyze spatial arrangements with equal precision.

The spatial language, like any other language, becomes an intellectual filter through which only the necessary information passes (Witthuhn et al., 1974). It modifies the way we think, what we observe as important, and how we make decisions. When you first began to speak, you had difficulty explaining exactly what it was you wanted. When your words were too few to explain those needs, you were often reduced to crying, hoping that someone would take the time to determine what your needs were. In addition, your language skills were not well developed, and the complexity of your ideas was proportionately limited. As your language grew, so did your understanding of the world and your ability to communicate it to others.

Childhood experiences did more than teach verbal communication. In large part you began to experiment with your spatial environment. The excitement of parents when a baby first notices his or her big toe is in part a recognition that the process of spatial exploration has begun (Piaget et al., 1960). Later spatial explorations include development of the concepts of movement and speed (Piaget, 1970) through crawling along the floor, discovering stairs, climbing furniture, reaching out to hot stovetops, opening cabinets, and the like—often to the dismay and concern of others. Still later we began climbing trees, exploring under bridges, wading in streams and ponds, wandering in the woods, finding the locations of new friends, walking to local parks, and so on. Many of these travels required us to learn route finding so that we could work our way back to where we began. Most often these early route-finding excursions were very ribbonlike. Once we found a route, we simply backtracked to return (Muehrcke and Muehrcke, 1992). Eventually we became more sophisticated spatially and we began recognizing that there were shortcuts. In other words, our spatial world took on a two-dimensional structure. We began to comprehend that places and things were either near or far, straight ahead, or at an angle. And we began to take note of obstructions that had to be traveled around, of hills that slowed us down or allowed us to coast on our bicycles. In short we were beginning to think geographically.

Thus we are all geographical creatures. Throughout our lives we continue to expand our knowledge of different places: we will travel to new towns, enjoy new recreational areas, and perhaps visit new countries. But, as with the artistic skills many of us abandoned soon after our crayon days, we often fail to exercise our spatial skills. For the same reason some of us are unable to draw a horse that looks like a horse rather than a brontosaurus with a very short neck, we lose our ability to read a map or find our way in the woods. But as our artistic skills can be relearned by exercising the right side of our brain (the graphic side) (Edwards, 1979), our geographic skills can be enhanced by carefully planned exercise. And as we can, through practice, become better artists, we can become better at viewing, analyzing, and understanding spatial patterns. This will make us better at interpreting space in general and at using GIS in particular.

As we experience space, we will encounter features and objects of many different types. Many disciplines have provided us with additional vocabularies to help us decipher them. Throughout geography's 2500-year history, it has borrowed extensively from numerous disciplines, synthesizing what others have learned and applying the filter of the spatial language. You will need to draw on this background to integrate this knowledge into a spatial framework. If you encounter an urban setting, a knowledge of urban studies will allow you to see patterns of residences, stores, and industries. A background in biology is invaluable when it is necessary to perceive and interpret the locations, patterns, and interdependencies of plants and animals. Geology and geomorphology will provide an awareness of the differences in exposures of rock and the many landforms you encounter. Expertise in economics allows you to suspect underlying forces determining locations of businesses and industries. The list is nearly endless and far beyond the scope of this book.

Many wide-ranging examples will be given throughout the remaining chapters. As you encounter each one, try to envision additional examples more closely related to your own area of expertise. Try also to expand the range of possibilities to include as many different settings, features, and objects as you can. The purpose is to exercise the geographic skills you learned as a child and to increase your spatial vocabulary to allow you to model the widest possible range of spatial phenomena. As a GIS practitioner, you are almost certain to be exposed to far more spatial objects and modeling situations than the disciplinary specialist. This makes your task somewhat more difficult than those who specialize, but it also makes it more rewarding and much more fun.

SPATIAL ELEMENTS

We begin to exercise our geographical skills by examining the types of object and feature we will encounter. Spatial objects in the real world can be thought of as occurring as four easily identifiable types: points, lines, areas, and surfaces (Figure 2.1). Collectively, these four can represent most of the tangible natural and human phenomena that we encounter on an everyday basis. Inside the GIS, real-world objects will be represented explicitly by three of these object types. Points, lines, and areas can be represented by their respective symbols, which we will study in Chapter 3, while surfaces are most often represented either by point elevations or other computer structures, with which we will deal in Chapter 4. What is most important now is that in the GIS, all data are *explicitly* spatial. Phenomena that are by their nature aspatial (ideas, beliefs, etc.) cannot

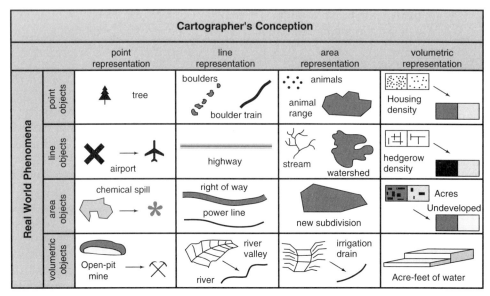

Figure 2.1   Comparison of real word phenomena and the cartographer's conception. Point, line, area, and surface features with examples.   *Source:* P. C. Muehrcke, and J.O. Muehrcke, *Map Use: Reading, Analysis and Interpretation,* 3rd ed., JP Publications, Madison, WI, © 1992, Figure 3.18, page 84. Used with permission.

directly be explored in a GIS unless a way can be found to assign to them a representative spatial character. The process of finding **spatial surrogates** is a difficult one, and we will develop this idea in more detail in later chapters. For the time being we restrict our discussion to the more tangible forms of geographical data.

Point features are spatial phenomena, each of which occurs at only one location in space. From your own experiences you can easily recognize such features as trees, houses, road intersections, and many more. Each feature is said to be **discrete** in that it can occupy only a given point in space at any time. For the sake of conceptual modeling, these objects are assumed to have no spatial dimension—no length or width—although each can be referenced by its locational coordinates. Points then are said to have "0" dimensionality. In reality, of course, all point objects have some spatial dimension, however minute, or we would not be able to observe them in the first place. We assume an absence of length and width, so that in measurements of air pressure, for example, which are characterized by a potentially infinite number of points, the points themselves will always occur at distinct locations with absolutely no overlap. The spatial scale at which we observe these objects or features places a framework on which we can determine whether we should think of them as points. For example, if you are looking at a house from only a few meters away, the structure seems large and occupies a substantial amount of length and width. This concept changes, however, as you move away: the house appears less like an areal object and more like a point object, the farther away you are (Figure 2.2). You choose your spatial scale based on different criteria: whether you want to examine the arrangement of people and furniture in the

Figure 2.2 Effect of scale on spatial dimensions. A house observed in a close-up aerial view appears to have length and width, but as we pull back, its length and width dimensions disappear, leaving as with the impression of a house as a point.

house, for example, or whether you are interested in the house only in relation to other houses, perhaps for an entire city. In the latter case the house would be considered to be a point. Your observation has been filtered according to how you want to view the object.

Linear or line objects are conceptualized as occupying only a single dimension in our coordinate space. These "one-dimensional" objects may be roads, rivers, regional boundaries, fences, hedgerows, or any kind of object that is fundamentally long and very skinny. The scale at which we observe these objects once again places a fundamental limitation on our ability to conceive of them as having no width. As you are fully aware, rivers, roads, fences, and hedgerows all occupy two dimensions when observed at close range. The farther we move away from them, however, the skinnier they become. Eventually they look so skinny that it is impossible to imagine that they are anything but essentially linear objects. It also becomes impossible to represent them as more than lines because we are so far away that we can no longer measure them (Figure 2.2). Other lines, such as political boundaries, have no width dimension to be concerned about. In fact, these lines are not physical entities at all, but rather, a construct of political convention and agreement. Despite their lack of tangibility, however, they can be thought of as explicitly spatial because their existence separates two portions of geographic space.

Linear objects, unlike point objects, allow us to measure their spatial extent by simply finding out how long they are. In addition, since they are not sited at a single location in space, we must know at least two points, a beginning and an ending point, to describe the location in space of a given linear object. The more complex the line, the more points we will need to indicate exactly where it is located. If we take a stream as an example of a linear object, the description of its many twists and turns may require a large number of points. In addition, it may be desirable to describe more than the stream's beginning and ending points. Because we have included a geometric dimension, we can also measure the shapes and orientations of linear objects. We will deal later with these additional spatial concepts.

Objects observed closely enough to be clearly seen to occupy both length and width are called areas. Examples of areas of "two-dimensional" objects include the area occupied by a yard, the areal extent of a city, and an area as large as a continent. To describe the locations of areas in space, we recognize that they are composed of a series of lines that begin and end at the same location. Besides merely indicating the locations of areas using lines, we are now able to envision three additional properties: as with lines, we can describe their shapes and orientations; but in addition, we can now describe the amount of territory they occupy.

Adding the dimension of height to our area features allows us to observe and record the existence of surfaces. While we could certainly observe a house at close range and describe it in terms of its overall length and width, we often want to know whether it is a one-story or two-story structure. In this way we need to observe the house not as an area, but rather as a "three-dimensional" object, having length, width, and height. Surfaces occur all around us as natural features. Hills, valleys, ridges, cliffs, and a whole host of other features can be described by citing their locations, the amount of area they occupy, how they are oriented, and now, with the addition of the third dimension, by noting their heights. As it turns out, surface features are composed of an infinite number of possible height values. We say that they are **continuous** because the possible values are distributed without interruption continuously across the surface (Figure 2.3). In fact, because the height of a three-dimensional object varies from one place to another, we can also measure the amount of change in height with a change in distance from one edge to another. With this information at hand we could also determine the volume of material contained in the feature itself. The ability to make such calculations is very useful if it is necessary to learn how much water is contained in a reservoir or how much surface material (overburden) lies on top of a coal seam.

All these features, whether point, line, area, or surface features, occur in space. And all can be located as to exactly where they may be found. But how can we communicate the importance of a circular areal feature located in a

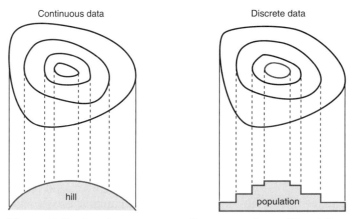

Figure 2.3 Continuous versus discrete surfaces. The difference between discrete and continuous data types.

particular area, occupying 10 hectares of space and oriented north–south? We need a way of classifying these features based on other observable properties and using terminology that others can understand.

SPATIAL MEASUREMENT LEVELS

So far we have examined the types of features and their locations in space. The objects themselves are called entities, and as we have seen, they have associated with them a set of coordinates that allow us to describe where they are located. All these spatial features or entities we observe contain information not only about how they occupy space, but also about what they are and how important they are to what we are studying. A tree, for example, viewed as a point feature, might be classified based on the taxonomist's set of terminology as a pine or an oak. We could also investigate the age of the tree by drilling a core through the trunk and counting its annual rings. With this additional information we not only know that there is a tree located at some point in space, but that it is a 35-year-old oak tree. The additional nonspatial information that helps us describe the objects we observe in space comprises the feature's **attributes**. As we observe our spatial environment, we begin to categorize its attributes and to note where they are. It is our nature as curious creatures to want to classify or pigeonhole the objects we find, so that we can place an additional level of organization on our space. We can now indicate that a particular feature, with a particular name, and with some measurable attributes, exists in a documented location.

But before we assign these properties and attributes, we must know how to measure them. Otherwise we cannot compare the objects at one location with those at another. What meter stick do we use? How accurately can we describe the objects we encounter? What effects will these different levels of measurement have on our ability to compare them?

Fortunately, there is already a well-established measurement framework for nearly all forms of data, including geographic data. These so-called levels of **geographic data measurement** range from simply naming objects, to give ourselves something to call them, to precise measurements that allow us to directly compare the qualities of different objects. The measurement level we use will be partly determined by what we are classifying, partly by what we want to know, and partly by our ability to make measurements at our selected scale of observation. Figure 2.4 illustrates the levels of measurement in terms of three commonly used geographic features.

The first level of measurement is the **nominal** scale. A useful mnemonic device is to think of the word "nominal" as "named." These are named data. The system allows us to make statements about what to call the object, but it does not allow direct comparisons between one named object and another. For example, we could say that at one location we have a maple tree and at another we have an oak. Although this statement certainly separates the objects, we cannot say that one is better than the other because the two are inherently different. In geographical terms, we can no more say that a church is better than a fire station than we can make such a judgment about apples and oranges.

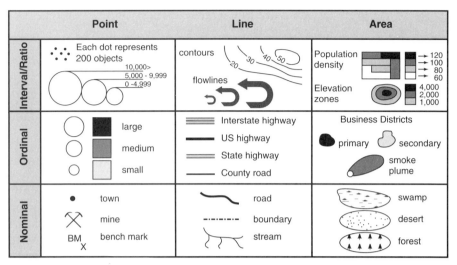

Figure 2.4   Measurement levels for castographic objects. Point, line, area, and surface features at the nominal, ordinal, interval, and ratio levels. *Source:* Robinson et al., *Elements of Cartography,* 6th ed., John Wiley & Sons, Inc., New York, © 1995, modified from Figures 16.1, 16.2, 16.3, pages 272, 273, 274. Used with permission.

If we want to compare two objects, we must be more precise in our level of measurement. As always, this is determined by what it is about the objects that we want to compare. Using our tree example, if we were interested in knowing how well a maple tree, an ash tree, or a pine tree might serve in producing a comfortable setting for a picnic, we could place each one on an **ordinal** scale from best to worst for that particular question. Because pine trees often have low branches and tend to drop sap on the ground, we probably would class pines as the worst of the three choices for our picnic area. While ash trees produce less debris and don't have low branches requiring us to hunker down while we eat, their leaves are small and some sunlight is likely to get through. We might want to class ash trees as moderate for our picnic setting. The maple tree produces very dense shade, however; moreover, it doesn't put a great deal of sap on the ground, and it doesn't have low branches. Therefore we would likely assign a classification of excellent to the maple tree.

In our picnic example we produced a spectrum of values ranging from best to worst. Clearly, however, the spectrum is based entirely on what we intended to use the information for. Because we have classed maple as best, ash as moderate, and pine as worst, the classification does not extend to other uses of the trees. Neither the maple nor the ash would be a very interesting Christmas tree, for example. So our classification is based on a single spectrum representing a single set of circumstances. Clearly, then, ordinal data can give us some insights into logical comparisons of spatial objects, but comparisons are limited to the specific spectrum that serves as the basis of the questions we are asking. Put another way, the grade you receive in your GIS class cannot be compared to the one you receive from your calculus instructor. Two fundamentally different sets of criteria are involved.

If we want to be more precise in our measurements, we move to the **interval** level of data measurement, in which numbers are assigned to the items measured. Data measured at the interval level can be compared, as in the case of ordinal data, but they can be compared with more precise estimates of the differences. A good example of spatial data measured at the interval level consists of soil temperatures across a study area containing widely different soil types. We may find that temperatures of some very dark, organic-rich soils are much higher than those for other parts of our study area that lack organic material and are a lighter color. In fact we can say that the soil types, measured at the same time, exhibit a difference of 8 Fahrenheit degrees. The dark soil, for example, might be 85°F, while the lighter soil is only 77°F. We now have an easily measured, precisely calibrated difference between the soils at two different locations.

One limitation remains in our ability to make comparisons with data measured at the interval level. Take two additional, extremely different soils, one nearly white and the other almost black. When we measure these at the same time, we find very different temperatures: 50°F and 100°F. As before, we can determine the numerical difference between these two soils as 50 Fahrenheit degrees. The numbers 50 and 100 look appealing as we try to read more meaning into the numerical difference. Is it fair to say, for example, that the dark soil is twice as warm as the light soil? At first glance it certainly appears so. However, you should remember that the Fahrenheit temperature scale we have used has an arbitrary starting point. For us to make a ratio of the two numbers, our starting point—in this case 0 degrees—must represent a true starting point for measuring temperature. To make the comparison we have suggested, it would be necessary to convert all the temperatures to the Kelvin scale, where the starting point, or zero, now represents a total lack of the atomic movement commonly associated with heat. Converting our two temperatures of 50°F and 100°F to kelvins, we get 283 and 311 K, respectively. When we form a ratio from these two numbers, we see that the dark soil is not twice as warm as the light soil.

In our conversion of Fahrenheit temperatures to the Kelvin scale, we moved to the final, most useful, level of data measurement—**ratio**. By converting our temperatures, we were able to perform a meaningful ratio operation on the two numbers because we were operating at the ratio scale. As the name implies, this is the only level of data measurement that allows us to make a direct comparison between two spatial variables.

Another example might be useful because it uses a set of numbers with which we are frequently concerned. Suppose you are studying two sets of homes located in different parts of a city: one in an upscale neighborhood in the suburbs, the other in a low-income neighborhood in the inner city. By surveying the residents, you determine that the median annual income for the suburban neighborhood is $50,000, while its counterpart in the inner city is $25,000. Because $0 means no income at all, we can safely say that we are operating at the ratio level of data measurement, hence comparisons are meaningful. By dividing one value by the other, we can now correctly say that the median family income in the suburban neighborhood is twice that of the inner city neighborhood. Such comparisons are valid and will become quite useful for your GIS data analysis. But, as with non-GIS data, you can make them only with ratio level data.

SPATIAL LOCATION AND REFERENCE

Thus far we have seen that we can observe a wide variety of features. We can group them, based on the scale at which we observe them, according to whether they are points, lines, areas, or surfaces. And we can categorize them using measurements at four different levels of measurement—nominal, ordinal, interval, and ratio—depending on how we want to describe them and how we wish to compare them. But we can only begin to describe our world knowing simply where things are, what they are, and what measurements we can assign them. We also need to know how these spatially referenced, classed, and measured objects interact in space to produce the overall sites we observe.

Until now we have indicated that we can **locate** objects in space. Location is the first important spatial concept we need; but to locate objects means that we must have a structured mechanism to communicate the location of each object observed. The first type of location is called **absolute location** and will give us a definitive, measurable, fixed point in space. But first we must have a reference system against which to evaluate such a location. In addition, the reference system must have a fixed relationship to the earth we measure.

The earth is a roughly spherical object, with some large-scale and some small-scale deviations from that shape. If we are observing the earth as a whole, it is generally convenient to consider it to be perfectly spherical. Around that spherical shape we can use simple geometry to create a spherical grid system that corresponds to the rules of geometry. This **grid system**, known as the spherical grid system, places two sets of imaginary lines around our earth (Figure 2.5).

The first set of lines for our spherical grid starts at the middle of the earth, or equator. These lines are called **parallels** because they are parallel to each other, and they circle the globe from east to west. At the equator, the first of these parallel lines, we assign a starting value of 0. As we move both north and south from the equator, we draw additional parallels until we reach the poles. Because each of these lines is a given angular distance, measured between the center of the earth and the intersection of the line, each can be used to measure the angular distance from our starting point at the equator to where the last line would occur at the poles. The angular distance, called **latitude**, from the equator to either pole is equal to one-fourth of a circle, or 90 degrees. So we can measure the angular distance from the equator north or south up to a maximum value of 90 degrees.

This is only half of our spherical grid system, however. To complete the grid we need another set of lines running exactly perpendicular to the first. These lines, called **meridians**, are drawn from pole to pole. The starting point for these lines, called the **prime meridian,** runs through Greenwich, England, then circles the globe, becoming the **international date line** on the opposite side of the earth. The prime meridian is the starting point or zero point for angular measurements east and west, called **longitude**. Longitude is then measured east 180 degrees of the prime meridian, until it reaches the international date line. Likewise it can be measured west of the prime meridian up to 180 degrees, again until it reaches the international date line.

This system of angular measurements allows us to state the absolute location of any point on the earth by simply calculating the degrees of latitude north or

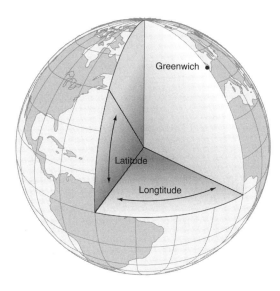

Figure 2.5 The geographic grid. Spherical grid system showing parallels and meridians. Parallels allow us to measure angular distance north and south (latitude) from the equator (0 degrees latitude) up to a maximum of 90 degrees north (North Pole) and 90 degrees south (South Pole). Meridians start at the prime meridian and allow us to measure angular distance east and west (longitude) up to a maximum of 180 degrees where they would meet at the international date line. *Source:* Robinson et al., *Elements of Cartography,* 6th ed., John Wiley & Sons, Inc., New York, © 1995, modified from Figure 4.4, page 47. Used with permission.

south of the equator and the degrees of longitude east and west of the prime meridian. With this system we can then describe with relative ease the locations of any of the objects we wish to describe (Figure 2.6). In addition, these angular measurements can readily be converted to feet or miles, meters or kilometers, thus allowing us to measure short or long distances on the ground using devices designed for calculating them. Although we will need to make adjustments to this system to adapt for the development of flat maps that closely approximate the locations and arrangements of objects on the globe, what we have is a quite elegant system for determining absolute location.

But as we continue to explore our world, we quickly note that it would be very useful to be able to describe not only the absolute locations of objects,

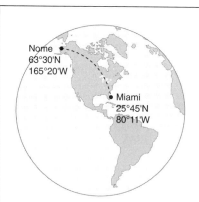

The Great Circle distance *(D)* on the sphere between two points A and B and be calculated using the standard formula in spherical trigonometry:

$$\cos D = (\sin a \sin b) + (\cos a \cos b \cos|\delta\lambda|)$$

where a and b are the geographic latitudes of A and B, and $|\delta\lambda|$ is the absolute value of the difference in longitude between A and B. (The product of the sines will be negative if A and B are on opposite sides of the equator.

Example: Great circle distance between Nome, Alaska, and Miami, Florida.

$$\cos D = (\sin 63.30 \sin 25.45) + [\cos 63.30 \cos 25.45 \cos (165.20 - 80.11)]$$
$$\cos D = (0.89337 \times 0.42972) + (0.44932 \times 0.90296 \times 0.08559)$$
$$\cos D = 0.384 + 0.0347 \qquad \cos D = 0.419$$
$$D = \cos^{-1} 0.419 = 65.25°$$

Because this is approximately 65.41 deg and 1 deg on a great circle = 69 miles, we can multiply these two numbers to get 4513.29 miles from Nome to Miami.

Figure 2.6 Calculating great circle distance. The illustration shows the calculation of spherical or great circle distance between two points on a sphere. *Source:* P.C. Muehrcke, and J.O. Muehrcke, *Map Use: Reading, Anslysis and Interpretation,* 3rd ed., JP Publications, Madison, WI, © 1992, Figure 12.3, page 255. Used with permission.

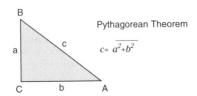

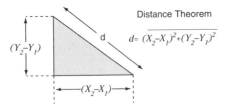

Figure 2.7   Calculating planar distance. The calculation of distance between two points using the Pythagorean theorem and the distance theorem.

but their relationships to other objects in geographic space. In fact this **relative location** becomes quite prominent in our GIS analyses, especially when relative location to other objects affects operation. With our absolute grid system we can determine relative locations by knowing the absolute distances between any two objects by simply subtracting the coordinates of the smaller from the larger. Of course, relative distance is only half of what we might want to know. It would also be useful to know the difference in direction between the two. We could, for example, indicate that a landfill is located 1500 meters southeast of the center of the city. This measure of distance and direction provides us with a framework for describing the precise location of the landfill relative to the city. From the standard Pythagorean theorem (Figure 2.7*a*) for relating the parts of a triangle that has one 90 degree angle, we can determine the distance of the hypotenuse by the so-called distance theorem (Figure 2.7*G*); which is expressed as follows:

$$d = \sqrt{(X_2 - X_1)_2 + (Y_2 - Y_1)^2}$$

where

$$X_2 - X_1 = \text{difference in the } X \text{ direction or longitude}$$

$$Y_2 - Y_1 = \text{difference in the } Y \text{ direction or latitude}$$

$$d = \text{distance between the two points}$$

However, we could provide additional relative information that would be of use to people living in the city. If we know that the prevailing winds are from the northwest, we can say that the landfill is located 1500 meters downwind of the city, thus illustrating that the odor of the dump will not normally be a problem for city dwellers. This latter approach, although less precise in terms of measurement, has more practical utility. This also provides us with a means of interpreting what we have observed—of determining the important relationships between and among features.

SPATIAL PATTERNS

In the preceding section we alluded to the importance of knowing the relationships among objects in space. A primary purpose of GIS is to analyze these relationships. That is why we put so much spatial data into the GIS database in the first place. When we begin to examine the analysis subsystem of the GIS, we will look at the many possibilities available to us. In the meantime, it will be instructive to begin a general development of our spatial language with respect to spatial comparison. This will enable us to begin thinking about what things must be collected during our excursion, because we will have a broad framework for ascertaining the types of spatial relationship that can be examined, and what they can tell us about our environment that we could not find out from a knowledge of features, their locations, and descriptions alone.

We begin by examining our last measure of location, in which we determined the relative location of a landfill to the nearby city. This easy calculation was a measure of **proximity,** the quality of being near something. Proximity can be measured, as we have already seen, by noting the distance between two features or objects. However, we can also set limits of acceptable proximity—a distance within which it is either acceptable or unacceptable for two objects to be found. In our example, if the landfill is too close to the city, even if it is downwind, some people will have to look at the unsightly feature. Beyond 2000 yards, however, we may assume that it is not visible to any city resident. We can then give a minimum acceptable proximity of 2000 yards, to ensure that any landfill will be placed at least 2000 yards beyond the edge of the city.

Our landfill example showed the interaction of two spatial objects. However, many features and objects occur in far greater numbers: cities in a country, houses in cities, animals in natural areas, natural areas in states, trees in forests, roads across nations, tributaries along streams, even plants in a garden all occur as multiples. But they do not all occur uniformly located within these areas. Each set of objects exhibits a particular **spacing** or set of **arrangements**. We begin to notice that these arrangements and spacings seem to have underlying controls or processes that dictate their placement. The instructions on the tomato seedlings may tell us to place the plants 3 feet apart in the garden. By following the directions, we produce a **regular** or uniform pattern of evenly spaced plants (Figure 2.8). This uniform pattern differs considerably from the locations of trees in a forest, which seem to be scattered at **random**, with no apparent underlying design. Alternatively, when we look at the locations of cities across nations, we often see that these population centers are located near lakes, oceans, and streams. Knowing that water bodies provide sources of drinking water and recreation and are useful in commerce, we can easily see that the tendency of cities to **cluster** near such features is driven by these needs. The clustered distribution of cities demonstrates a type of distribution with very high **density** of features, while the distribution of other objects such as farmhouses, scattered about a nation's rural landscape, demonstrates a more **dispersed** pattern.

As we continue our travels we discover that certain features seem to be organized in yet another way. We see, for example, that the vegetation on the north side of a steep slope tends to be decidedly different from that on the south side. In other words, vegetation assemblages have a particular **orienta-**

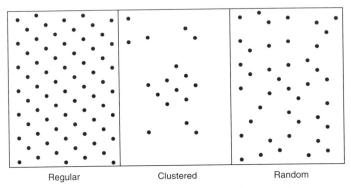

Regular   Clustered   Random

Figure 2.8   Point distribution patterns. Regular clustered, and random distributions of spatial objects.

tion. The property of orientation is also found in planted shelterbelts that are oriented at right angles to the direction of the wind. Some city streets are oriented along a relatively square grid, while others seem to be scattered about with no particular sense of direction. Some of you may have noticed that a great many American Civil War statues are oriented with the figures facing in the direction from which the foe would arrive. In some regions of the world that experienced continental glaciation, we find piles of debris called moraines oriented at right angles to the flow of ice, or lines of giant rocks called boulder trains that give evidence of the direction of retreat of the ice.

If we revisit some locations we've seen before, we find still other spatial processes that give us clues about changing geography. If you return to your hometown or city after an absence of a few years, you may see that it is larger than it was when you left. The town has experienced **diffusion** into the neighboring farmlands. Or you may see that your former downtown area is less important as a center of commerce, and instead there has been a diffusion of stores into malls on the outskirts of town. In fact the process of diffusion can occur daily as people concentrated in the residential areas of large cities travel outward toward their workplaces elsewhere in the city. Diffusion can even be seen in the introduction of GIS concepts and technology, as the once small number of universities offering courses expands to include more and more each year.

But as we view all these spatial configurations throughout our explorations, we also ask for the relationships that might occur between one set of features and another. We touched on this when we mentioned some causes for the development of cities near water bodies. What we discovered is that one spatial pattern may be partially or totally related to some other spatial pattern. The spatial arrangement of steep slopes was seen to have a strong **association** with the vegetation that occurs there. We can now begin to ask questions about the causes, not only of single distributions, but of **spatially correlated** distributions of phenomena (Figure 2.9). This is among the more powerful capabilities of modern GIS: the ability to illustrate, describe, and quantify spatial associations, allowing us to examine clues to the mechanisms that cause these associations. This has long been a major task among geographers, as evidenced by the writings of Carl Sauer (1925). We are beginning to acquire the spatial language of geography and now, with the use of GIS, we have the ability to automate

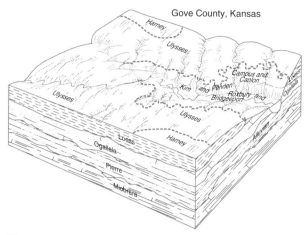

Figure 2.9   Spatial relationships between spatial phenomena. Spatial correlation between geological formations and the types of soils formed from them.   *Source:* USDA county soil survey for Gove County, Kansas.

these geographic concepts. Because we know the language, we begin to think in spatial terms—to filter our thinking to observe the important spatial features, to identify spatial patterns and, finally, to ask the questions that will elicit explanations for the causes of pattern interactions.

GEOGRAPHIC DATA COLLECTION

Thus far, we have taken a rather casual approach to the matter of how we will make and record observations as we continue on our journey. However, even the earlier explorers took along paper and pencil, measuring devices, and even large box cameras when they became available. Times have changed, the nature of the environment has changed, the devices we use have changed and, of course, the very questions we are able to ask have changed as well. These changes call for a quick survey of the instruments available for our journey. Of course, you will become far more familiar with some of these than others, depending on your career and the kinds of data with which you ultimately work. However, because many GIS data are acquired from other sources, it is a good idea to become somewhat familiar with as many as possible.

Many data are still observed through ground survey methods not unlike those used by our predecessors. Tape measures, still fundamentally useful for measuring relatively small objects, have been replaced in some cases by relatively inexpensive range finders that send a beam of light from the observer to an intervening obstacle. Other observations, such as qualitative, or categorical data, are still acquired by means of direct visual observation, or by collecting sample specimens for later identification. These observations should be evaluated carefully before they are accepted as truth, since often their quality is a function of the observer's experience. The traditional method of conducting vegetation surveys illustrates this point. Vegetation surveyors used to base landownership maps on so-called **witness trees,** trees left behind after lands had been cleared for agriculture. The surveyors relied on the assumption that these trees were representative of the commonly available forest types and generalized from these easily identifiable objects to the surrounding areas. While some of the witness trees may have been representative of the presettlement vegetation, however, many were far less typical, having been left because of their prominent size or convenient position.

Absolute positions on the earth were once recorded relative to the locations of celestial bodies such as the North Star. The handheld compass is a simple device that relies on the tendency of the earth as a rotating body to create opposing poles at its north and south extremes. For more local observations, locations are also relatively easily and cheaply obtained using simple handheld Brunton pocket transits, or the more sophisticated **plane-table** and **alidade** survey devices. These devices are still in use today where both the amount of terrain and the budget are limited. Each device assumes also that there are some nearby points of known location against which the object locations can be compared.

From the point of known location, surveyors can measure distances and angles through a process called **dead reckoning,** they can intersect a number of line-of-sight measurements (**triangulation**), or they can create a known baseline against which only the distances between objects and the ends of the baseline need to be measured (**trilateration**) (Figure 2.10).

Differences in elevations at different places on the landscape can be measured with the use of a device called a **dumpy level,** which has a telescopic site much like the alidade but is placed on a taller stand and is capable of rotating up and down to examine elevation differences at a distance. A number of traditional survey devices have been updated to offer much greater accuracy, as

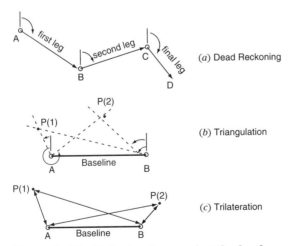

Figure 2.10 Methods of surveying the land. Measurement by dead reckoning (*a*), triangulation (*b*), and trilateration (*c*). *Source:* P.C. Muehrcke and J.O. Muehrcke, *Map Use: Reading, Analysis and Interprelation,* 3rd ed., JP Publications, Madison WI, © 1992, Figure 2.1, page 43. Used with permission.

well as digital readouts and superior ease of use. **Theolodites** are one example. All such improved instruments, like their older counterparts, have somewhat limited utility if it is desired to survey large portions of the earth simultaneously. The time required to survey large areas with these devices, however, is far outweighed by the value of the highly accurate positional data obtainable for applications such as locating the boundaries of property lines. Thus their continued utility is likely to be ensured.

Technological innovations have improved the methods by which we can obtain positional information, especially for large portions of the earth. Circling the globe today are satellites whose positions are known with great accuracy. These satellites receive radio transmissions from field units on earth and return a signal that is processed by a nearby ground station, then sent to the field unit as a set of coordinates. The coordinates give positional location as well as elevational location, providing a very useful set of data to the user. Perhaps the most promising of these devices, and now among the most widely used, was mentioned in Chapter 1: the NAVSTAR Global Positioning System (**GPS**) (Figure 2.11). GPS depends for its accuracy and precision on the number of satellites, the amount of detail or service provided by the military (currently in charge of the program), the locality of the base station providing the baseline data, the age of the field unit in its place, and the sophistication of the field and base stations. Available systems today can give locational accuracies ranging from

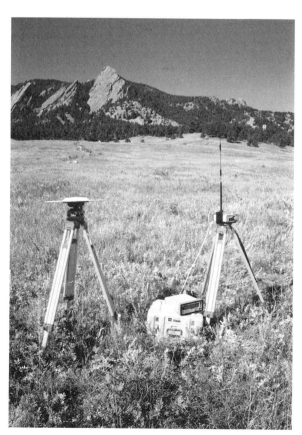

Figure 2.11   GPS field unit.

a relatively coarse 100 meters down to at least decimeter levels. Direct line of sight is not necessary between the base station and the field unit but is required between the field unit and the satellite itself. Thus the usefulness of these devices is limited in regions with dense tree canopies that obscure the line of sight.

For field surveys of animals, there is a system of devices using similar technology to track the mobile locations of deer, birds, small mammals, and even bees (if the units are small enough). Traditionally the units are attached to the creatures, and their locations are determined by monitoring a continuously operating signal at a field unit. The field unit may be located on the ground or carried aloft with aircraft. A recent trend in this form of spatial data collection, called **radiotelemetry,** is to combine it with the GPS unit, allowing much greater areas to be monitored. As these two technologies continue to improve in quality, ease of use, and cost, there will be more and better data available for later input to our GIS.

When our concern is to gather information about the distribution of objects, plants, animals, or even people, rather than individual locations, we employ another form of data collection called a **census**. The census device depends on the nature of the data to be collected. We might, for example, use direct contact to physically note the exact locations and characteristics of individual shrubs in a grassy area. As implied by this example, the purpose of the method is to obtain information about an entire population of objects in space. The most common example, of course, is the census of population conducted by a government. Such survey devices attempt to gather both locational and attribute data about people so that generalizations can be made about populations in selected areas or districts. Data obtained often include marital status, income, housing, age, gender. From these data generalizations it is possible to determine, at least approximately, which areas have the lowest per-capita annual income, whether the overall population is getting older (based on percentage of the total population), whether there are more people living in apartments or in houses, and so on. These results, in turn, allow governmental bodies to suggest ways of adapting to changing circumstances. In other words, governments now have a method of planning based on a knowledge of the spatial distribution of people.

The potential power of the census to describe spatial relationships among a nation's people has prompted a move toward automation of the data itself. The census has become computerized, both gaining from and contributing to the move toward automated geography. We will see in Chapter 4 how the U.S. Bureau of the Census has created and modified its methods of keeping track of data to make census information accessible to most commercial GIS products.

At times, however, neither ground survey nor census is an appropriate means of gathering data and providing contextual spatial information about a large portion of the earth. This statement applies particularly to natural phenomena, but it also holds true for anthropogenic features and cultural data under selected circumstances. Obtaining data about phenomena on the regional or even the continental scale may require the use of secondary or indirect methods of data collection. These indirect methods often rely on sensing devices far removed from the observer and are therefore called collectively **remote sensing**. Although the term most often implies some form of satellite sensing of radiometric data from the surface of the earth, we will use the term more loosely to include the use of aerial photography, which is sometimes separated from the definition because of the strikingly different methods by which the data are interpreted. As a rule, "remote sensing" refers to the use of satellites that obtain data about the surface of large regions of the planet, most often coming in the form of computer-compatible data that can immediately be operated on by computer software.

Point remote sensing devices are normally placed at strategic locations that are meant to be indicative of the general character of the surrounding area. In addition, sensors are scattered throughout the region of interest so that data can be obtained for as much of the region as possible. The weather station is perhaps the best-known point remote sensing device. However, remote sensor networks have also been developed to collect continuous or periodic data on soil temperatures, soil moisture content, and other useful soil factors. These nets are most often connected through a telecommunications network to a base station that receives and stores the information, either as hard-copy displays in the form of paper polygraphs or as digital data inside a remote computer. Even more exotic devices have been developed to record the interception of flying insects, crawling animals, and moving sand grains, or to monitor water or atmospheric pollution, seismic changes and, of course, the prowling jewel thief. Collectively these point-specific sources of information can provide a large regional picture of the phenomena they sense.

The use of aerial photography permits the collection of data describing the continuous change in phenomena from one place to another, making it unnecessary to rely on discrete observations. This type of remote sensing most often relies on the use of aircraft to carry a photographic device designed to sense and record portions of the electromagnetic spectrum. These devices come in various sizes and can use a range of films, from traditional black and white to color to false-color infrared films, depending on the data being collected. In many cases a particular film and filter combination is used to eliminate unwanted portions of the spectrum and enhance the visibility of regions more indicative of the features being sensed.

Aerial photography, a mainstay in a wide range of spatial analyses, has a long tradition of use for forest and range evaluation and management because the photos allow analysts to see substantial portions of areas at a glance. Soil scientists have used aerial photographs to assist in perceiving the subtle changes in soil type over large areas, but also as base data on which soil maps are placed. Urban specialists have used aerial photographs to estimate populations by counting dwellings and interpolating from known averages per household. Geologists have long used aerial photography as a source of information on the spatial distribution of surface features as well as subsurface phenomena such as salt domes and fault zones. The military, of course, has used aerial photography for some time as a means of reconnaissance. In fact, color infrared film, originally called camouflage detection film, was developed largely under military auspices. Thus the use of aerial photography as a means of examining geographical data is well established and still common for areas that are not overly large. For extremely large areas, such as states, regions, or countries, the costs in money and time generally are prohibitive. Similarly, a number of more exotic airborne devices such as **side-looking airborne radar (SLAR), scanning radiometers, color video,** and **digital photography** tend to be ruled out for large-scale applications again, primarily because of the time and effort needed to cover major portions of the earth.

Larger areas can be surveyed by other methods, however, many of which deploy by satellite, hundreds of miles from the surfaces they are viewing, some of the same technology currently being carried in aircraft. The great distances between the sensing platform and its object allow the satellite to view large areas synoptically at each instant. In addition, because satellite remote sensing devices are orbiting the earth, they are able to sense much of the planet in a fraction of the time that would be needed by a conventional airborne sensing device. As indicated, a wide range of sensing devices is available, each with its own spectral, temporal, and spatial characteristics. Some radar satellites are designed primarily for observing weather, but others are able to interpret ground features as well. The *SEASAT* satellite, for example, is designed primarily for examining oceanic features such as waves and icebergs. Thermal scanners on board satellites allow geologists to evaluate the landscape for hot spots, giving them data they need to help predict volcanic activity. Radiometers, such as the advanced very high resolution radiometer (AVHRR) on board the *GOES* satellite, have been used to map very large portions of the earth because these devices can sense the surface of the entire planet in a fraction of the time needed by most other satellite remote sensing systems.

It should be remembered that remote sensing data are not gathered directly. Data from aerial photography, radar, and digital remotely sensed imagery are all more or less representative of what is on the surface. Although they do not, for example, sense types of vegetation, or kinds of human activity, analysts can use the electromagnetic signals received as surrogates of what is actually on the ground. In most cases the signals must be processed by experienced interpretation specialists before the object categories can be properly identified. Most often the classified data are used as input to a GIS database, rather than the raw data themselves. As we will see later, the ability of the human or machine interpreter to correctly identify objects in space has much to do with the utility of such data in decision making.

The field of satellite remote sensing is a vast and ever changing one, both because of the addition of new systems with new sensing devices and because remote sensing scientists are finding new and innovative ways to use them. Each system produces a set of data that provides spatial insights about what it views. Each also sees things through different eyes. Some view the land as a series of lines, while others produce huge checkerboard structures, building up images from the different shades and colors of each cell. Remote sensing scientists will continue to investigate the utility of these devices, and every GIS specialist should be aware of the potential capabilities of this technology for understanding our planet. As GIS scientists, our role is most often to determine how best these remotely sensed data can be incorporated into GIS databases and combined with other forms of data, to provide a more complete view of the lands we explore.

POPULATIONS AND SAMPLING SCHEMES

One problem yet remains before we are ready to explain what we see on our journey. We now have a spatial language at hand that allows us to know where to look and what to look for as we examine our environment. And we know about devices that allow us to collect large volumes of data by means of satellites. But the sheer complexity of our planet can very often become overwhelming simply because there are so many objects and so many possible factors that might be examined. And while remote sensing devices allow us to see large regions at a single glance, the resolution with which they view the earth often limits the sizes of objects that can be observed. Some features—for example, animal burrows in a prairie—are well below the ability of many remote sensing devices to explore. And because such features lie in a large territory, it is prohibitively expensive to obtain detailed aerial photography, and we must do our viewing and evaluating on the ground. This is easy because the features of interest are numerous enough to visually dominate the landscape. But there are far too many for us to count each burrow and put an absolute location for each. To determine why these features have a particular pattern requires us to know that a pattern exists in the first place, which ideally would be done by examining and recording their locations. Since, however, we are unable to obtain a complete census of all the holes, we must sample them to estimate the entire population from a smaller, representative subset.

Sampling can be performed in a number of ways; some are harder than others, and some give us a better ability to make inferences about the larger population. Although sampling is effective for either aspatial or spatial data, we will restrict ourselves to spatial sampling because the GIS deals with explicitly spatial data. Within the spatial domain we sample data in one of two primary ways, directed and nondirected sampling. Each method will be determined both by the limitations of acquiring spatial data and by the inferences we are trying to make about the population of data using the sample.

Directed sampling, as the name implies, involves making decisions about what objects are going to be viewed and later catalogued. In other words, we direct our sampling based on our ability to make samples from and about a **target population** of objects or features. This target population occupies a given **sampled area** within which our samples are taken (McGrew and Monroe, 1993). Together, the target population and the sampled area comprise the **sampling frame,** which includes the types of object of interest bounded by specific and identifiable spatial coordinates. Within directed sampling, sometimes called purposive or judgmental samples (McGrew and Monroe, 1993), we use a combination of experience with the study area and its target population, accessibility to the objects we want to investigate, likelihood that we can obtain pertinent information about each individual (e.g., using only data from survey forms returned by people to whom they were sent), and focused (nonrandomly selected) study areas or case studies designed to demonstrate a particular phenomenon. Although this form of sampling is often discouraged as being "nonscientific," it is often necessary. Take some of the following situations.

You are planning a spatial survey of the types of vegetation that arise in an area that was cleared of vegetation, later farmed, and then abandoned. You know of a large area that has experienced these conditions, and so you plan to study that area, rather than all areas in your state or nation that also have these conditions. Your study is more focused because accessing all the areas that have been cleared and abandoned would be impossible. Upon visiting your potential study area, you find that at least half of it is surrounded by barbed wire and the owners don't want you tramping around on their property. Your study area is now further restricted. In addition, you discover that many of the areas still available for study are inaccessible by vehicle. Now you must direct your investigation to focus only on regions within a reasonable distance from roads. Thus you have had to focus on a small case study area known to have experienced the conditions of interest, and you've had to limit even this area because access to it is restricted. This type of situation is common and will limit the sampling strategy.

Another situation has already been alluded to concerning the use of surveys of populations. If, for example, you want to determine the spatial extent of cable television users and find out the programs they watch, you will use a common survey device that consists of a set of questions about which premium cable television programs the respondents watch. The questionnaires will be sent to the people in your city (the target area). However, you have to remember that some people do not have television, and many set owners do not subscribe to a cable service. You want to survey only people who have cable television (your target population), perhaps by contacting your cable companies to obtain lists of their customers who are willing to be surveyed. Once you have identified cable viewers, you send each one a survey document requesting a response to the questions and indicating how the document should be returned to you. Because most of us have received such surveys, we know that many potential respondents do not return the questionnaires. This will likely be the case with your survey as well. What you are left with is a directed survey, focusing only on cable television users and presenting results based on the responses of people who returned the survey.

Many more scenarios could be included here, but each of you will have to make decisions based on personal circumstances. While a nondirected, probability-based sample generally is preferable because it eliminates bias, at times this approach will be impossible. Before you begin your sampling procedure, try to determine whether the data can be collected by means of a probabilistic sampling procedure. Only if this is not a possibility should you default to the directed approach.

If you are able to use a probabilistic spatial sampling methodology, each object or feature you select from your sampling frame is expected to have a known chance of being selected for study. This known probability of being selected is used to design the procedure for taking the sample. Stated differently, you use the known probability to set up a sampling methodology that will allow all objects the same probability of being selected.

The methodology for probabilistic sampling can easily be divided into four general categories: random sampling, systematic sampling, stratified sampling, and homogeneous sampling (Figure 2.12). Of course these could be combined to create a hybrid sampling design if so desired, but we will examine the basic types and let you modify them later as you see fit. Random spatial sampling is the most basic sampling design. Its purpose is to allow each individual point, line, area, or surface feature to be selected with the same probability as the next. If the spatial data you are sampling are discrete, such as trees, lakes, or people, your purpose is to observe something about a few of these by selecting them, at random, for examination. In such circumstances each object is assigned a unique number, say from 1 to 1000. With the use of random number generators found in nearly all computers and many hand calculators, or a set of random number tables, it is easy enough to select a fraction of these numbers, again randomly. We may, for example, select 100 of the 1000 spatially located objects' numbers for measurement. If the data occur continuously, such

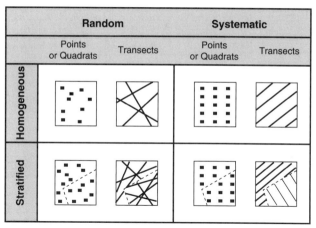

Figure 2.12   Spatial sampling methods. Random, systematic, stratified, and homogeneous sampling designs.   *Source:* P.C. Muehrcke and J.O. Muehrcke, *Map Use: Reading, Analysis and Interpretation,* 3rd ed., JP Publications, Madison, WI, © 1992, Figure 2.6, page 51. Used with permission.

as with topography, barometric pressure, or soil temperature, we will randomly select points at which these properties can be measured and number them, selecting points to be examined as before. In both cases it is possible to select random points, random areas called quadrats (often used to examine the amount of above-ground biomass for grasses), or line transects for use in selecting our objects of study.

Systematic designs operate almost exactly the same as random designs, but now we decide on a repeatable pattern, rather than random numbers, to use as a basis of selection. For point data we could, for example, select every tenth tree, or trees located approximately 20 meters apart. To examine small plots or quadrats, we would select these in the same manner, by examining every nth object or an object every n meters. Likewise, if our sampling involves using line transects, a popular method for examining vegetation assemblages, we would systematically determine where each line transect would fall and take our census of the vegetation along each of these transect lines. Or, if we wanted to completely survey individual plots or quadrats, we would once again select these using a systematic, repeatable pattern for selecting each quadrat for study.

Stratified spatial sampling adds another dimension by selecting small areas within which individual spots, objects, or features are sampled. Stratifying simplifies the process of sampling by dividing the task into small regions that can, for example, be examined by one person, or in one day of sampling. Within each stratified area we can decide whether a random or a systematic design will be used to examine the objects or features. In a modification of this technique called *clustering,* we determine at the onset whether the objects to be observed are clustered or dispersed throughout your entire study area. Each of these groups can then be selected as a separate study subarea, much as we did when we stratified our overall study area. Again, we can use point, quadrat, or transect approaches, and we can choose a systematic or a random method of sampling within each subarea. The advantage of this approach is that homogeneous objects often imply an underlying process. Selecting these areas out for individual study is likely to give us more accurate data relating to that process than will treating it as a part of the whole, in which case we are assuming that the processes operating throughout the whole study area are generally the same. This approach, however, has the disadvantage of requiring us to make decisions about which subareas are more clustered than others that may turn out to be inaccurate.

MAKING INFERENCES FROM SAMPLES

Spatial data taken from spatial sampling yield three major types of manipulation: data at nonsample locations can be predicted from sampled locations; data within regional boundaries can be aggregated (so that a single class of data characteristics can be assigned); data from one set of spatial units can be converted to others with different spatial configurations (Muehrcke and Muehrcke, 1991). As you continue learning about GIS, you will discover many situations involving predictions, and a general understanding of prediction problems will save you time and energy later on.

While sampling shortens the time necessary to collect data about a region, it also leaves gaps in our knowledge about the locations not sampled. Given that most GIS software is based heavily on the idea of areas rather than points, we must be able to determine or predict the missing point values. This need commonly arises when we gather information about surfaces using point samples. For example, to gain an idea of what an entire surface looks like, we might select a number of points to measure elevation. When we view topographic maps, either as contour lines or as three-dimensional representations, the missing data are not measured, but are instead spatially predicted (Muehrcke and Muehrcke, 1992). Under these circumstances there are two general types of predictive model. **Interpolation** is used to predict missing values when we have values bounding, or on both sides of, the gap. If there are values on one side of the missing data, but none on the other side, we call the methodology **extrapolation**. Interpolation can be as simple as assuming a linear relationship between the known values and filling in the sequence. More sophisticated methods are based either on assumptions of a nonlinear relationship between these known values or on a weighted distance or **local operator model,** where nearer values are more likely to be useful in predicting missing values than points farther away. **Surface fitting models** involve fitting the observed values to an equation and then solving the equation for each missing value (Muehrcke and Muehrcke, 1992). Surface fitting models are more likely to be useful for extrapolation than the others, because the equation can easily be extended beyond the known data values. All these methods allow us to predict missing values, but it should be remembered that predictions are not measurements, and each prediction will have its own set of problems and errors. We will look more closely at these methods later on when we study surfaces.

Extrapolation is more appropriate when we are sampling point data as estimates of areas rather than surfaces. From these areas we want to be able to make further predictions about points in other areas that have not been sampled. Let's say that we are sampling the density of trees in a number of small areas, and we want to be able to predict the densities in other surrounding areas. Such a task usually requires us to operate using three assumptions. First, we calculate the average densities, on a region-by-region basis, to prevent region size from affecting the values. Next we assign each density to a single point inside each region (usually some central point). Then, having completed these steps, we can return to the point interpolation methods to predict density values for each of our missing areas.

A final predictive problem based on sampling requires the prediction of area quantities based on a set of quantities sampled from a different set of areas with different sizes and shapes. In the simplest approach, which entails a form of spatial overlay, it is assumed that the data inside each area are essentially uniform and homogeneous (that is certainly true). When these two sets of areas are overlaid, again assuming that the data are homogeneous, the amount of overlap should be directly proportional to the amount of data in each.

One other conversion needs to be considered here that might relate to sampling, although it can also apply to total population census. Suppose you are sampling discrete objects, such as the locations of animals. Once you have located the creatures, and for the sake of simplicity we will say that they don't move around much, you will want to know what portion of the earth they generally occupy. In other words, you need to know their range. This is a common problem for wildlife managers, for example, who use radiotelemetry devices to locate animals and often must operate in a GIS that is poorly adapted for working with point data, but strong on area-based data. Some relatively easy computer techniques can be applied, as well as some biological approaches, and we will consider these in more detail later. For the time being it is enough that you be aware of their existence as you look for ways to explain what you observe.

A final note is in order concerning the case of area prediction using overlay techniques. To perform this nifty feat we are going to need a device that has been mentioned only briefly, but will be the major topic of the next chapter—the map. The map is the fundamental device by which we abstract our environment's space, and within which the GIS will operate to analyze it. In the next chapter, then, we will expand our spatial vocabulary and improve our spatial filter by examining how we can move from spatial data in the conceptual sense to spatial data in a graphic sense. The skills we will be learning, which are called "graphicacy" (Balchin, 1976), will serve us well as we move toward a better understanding of automated geography inside a GIS.

Terms

spatial surrogates	discrete	continuous
attributes	geographic data measurement	ordinal
interval		locate
absolute location	nominal	grid system
parallels	ratio	latitude
longitude	meridians	international date line
proximity spacing	prime meridian	relative location

random	arrangements	regular
dispersed	clustered	density
association	orientation	diffusion
plane-table and alidade	spatially correlated	witness trees
trilateration	dead reckoning	triangulation
GPS	dumpy level	theodolites
remote sensing	side-looking airborne radar (SLAR)	radio telemetry census
scanning radiometers	color video	digital photography
target population	extrapolation	local operator models
interpolation	clustering	sampling frame
surface fitting models	sampled area	

Review Questions

1. Why is it important that we understand the language of geography before we can be effective in GIS analysis? What impact does an increased spatial vocabulary have on our ability to work with spatial phenomena?

2. What impact does scale have on how we experience our world and on how we model it? Give an example, other than the one in the text, of how scale change might permit you to view an object with length and width as if it were a point.

3. Why is it important to develop a spatial vocabulary for GIS? What impact does an increased spatial vocabulary have on how we explore our environment? How does a geographical language help us in filtering the data we view in our explorations?

4. What are spatial surrogates? Can you give some examples of aspatial data that must be given explicit spatial dimensions before they can be incorporated into the explicitly spatial framework of a GIS?

5. What are discrete data? Can you give some examples of them for point, line, area, and surface features?

6. What are continuous data? Give some examples of these, especially with regard to surface data.

7. Why is it important to understand the levels of data measurement when we are observing and recording attribute data for the objects we encounter?

8. Give some concrete examples of nominal, ordinal, interval, and ratio data for each of the following features: point, line, area, and surface.

9. Why do we need a structural framework such as the latitude–longitude grid system? What does it add to the way we view our world?

10. What is the difference between absolute location and relative location? Give some examples of each.

11. Some of the terms added to our vocabulary are orientation, arrangement,

diffusion, pattern, dispersion, and density. How do these terms improve the way we observe our earth?

12. What does spatial association mean? What does it have to do with GIS?

13. What is the impact of modern technology on ground sampling? Give some concrete examples.

14. What has remote sensing added to our geographic tool kit that ground sampling lacks? How will improvements in remote sensing impact on GIS?

15. Why would we choose to use directed sampling, as opposed to probability-based sampling, if the former is considered to be less "scientific"? What are some of the conditions that force us to use directed sampling?

References

Edwards, B., 1979. *Drawing on the Right Side of the Brain: A Course in Enhancing Creativity and Artistic Confidence.* Los Angeles: J.P. Tarcher.

Balchin, W.G.V., 1976. "Graphicacy." *American Cartographer,* 3:33–38.

McGrew, J.C., Jr., and C.B. Monroe, 1993. *An Introduction to Statistical Problem Solving in Geography.* Dubuque, IA: Wm. C. Brown.

Muehrcke, P., and J. Muehrcke, 1992. *Map Use: Reading Analysis and Interpretation,* 3rd ed. Madison, WI: J.P. Publications.

Piaget, J., 1970. *The Child's Perception of Movement and Speed,* translated by G.E.T. Holloway and M.J. Mackenzie. New York: Basic Books.

Piaget, J., B. Inhelder, and A. Szeminska, 1960. *The Child's Perception of Geometry,* translated by E.A. Lunzer. New York: Basic Books.

Sauer, C.O., 1925. *Morphology of Landscapes.* University of California Press, Berkeley.

Witthuhn, B.O., D.P. Brandt, and G.J. Demko, 1974. *Discovery in Geography.* Dubuque, IA: Kendall/Hunt Publishing Company.

CHAPTER 3

The Map as a Model of Geographic Data: The Language of Spatial Thinking

The map is the fundamental language of geography. It is, therefore, the fundamental language of automated geography. This graphic form of spatial data abstraction is composed of different grid systems, projections, symbol libraries, methods of simplification and generalization, and scales. If you are comfortable with the map as a method of modeling your environment, you may be able to skip this chapter and move on to Chapter 4, especially if your background includes coursework in map use and cartography. (Even so, it might prove useful to review this chapter before moving on, as a reminder of the cartographic method and all it implies to correct analysis and interpretation.) If you have not had such courses, or if your experience in mapping or map reading is minimal, you should spend some time focusing on this material. You might also find it useful to select a couple of good books on cartography or map reading to supplement the short descriptions given here. As you evaluate your map familiarity remember that inside a GIS, you are likely to encounter a variety of maps far more abundant than you might have expected on the basis of courses in geology, surveying, or soil science. In addition to the geological, topographic, cadastral, and soil maps used in these disciplines, the thematic content of GIS coverages includes vegetation maps, transportation maps, animal distribution maps, utility maps, urban plans, zoning maps, land use maps, land cover maps, and remotely sensed imagery. These maps will be portrayed as prism maps, choropleth maps, point distribution maps, dasymetric maps, surface maps, graduated circle maps, and a host of other types. If some of these terms are new to you, then read this chapter if for no other reason than to become generally familiar with the myriad possibilities.

Remember, our exploration of the earth through GIS is predicated on our ability to think spatially. Spatial thinking requires us to be able to select, observe, measure, catalog, and characterize what we encounter. But the depiction of objects in cartographic form depends on what questions are being asked, whether we are trying to display the cartographic form or analyze it in a GIS, whether we are observers or users, whether our data are collected in the field or through remote sensing, whether archived maps of previous observations are going to be used in our analysis, and many other factors. The more we know about the possible combinations of graphic elements, and how they are manipulated on the cartographic document, the stronger our geographical language. And, as you have seen, the larger our spatial vocabulary, the better we are at making decisions about spatial phenomena and their distributions in space. A knowledge about cartographic methods will increase a portion of our spatial vocabulary that we have called **graphicacy**.

An improved level of graphicacy will assist us in all four subsystems of the GIS tool kit. When we consider putting existing maps into the GIS, we will be aware of the impacts of different levels of generalization, scales, projections, symbolization, and the like on what is input and how it is done. Once inside the GIS, we will be able to identify potential problems that will require editing: two adjacent input maps originally produced through different projections, for example, or large symbol sizes that permitted us to place data points in incorrect or illogical locations. As we begin to analyze our data, we will be aware of the potential for error of some coverages that were created with very small-scale maps. Finally, as we produce output from our GIS, we will have a better feel for how best to display the results of analysis, because we will be familiar with the cartographic method and its design criteria.

In our journey of spatial data exploration, we will continue to change the way we conceptualize and abstract spatial phenomena. In Chapter 2 we began thinking about how to isolate spatial phenomena into geometrically defined points, lines, areas, and surfaces. We learned the levels of spatial data measurement within which the objects or their attributes can be gauged. Further, we saw how questions of spatial relationships among objects could be formulated, and how all these could be collected and measured directly or indirectly. Now we will learn how spatial features and their relationships can be depicted in a form that allows us to view these relationships more easily and, therefore, how we can begin to formulate methods of analysis.

LEARNING OBJECTIVES

When you are finished with this chapter you should be able to:

1. Understand the role of graphicacy and cartographic communication in improving our understanding of the world around us.

2. Understand the relationship between the analytical and holistic cartographic paradigms as they apply to GIS.

3. Understand how scale is illustrated on a map and the possible impact on the results of analysis of putting differently scaled maps into a single GIS database.

4. Show how the map legend acts to link entities and attributes on a map.

5. Know the different families of map projections, indicate how they modify

the properties of the maps, and tell how this knowledge can be used in decisions regarding selection of appropriate map projections.

6. Be familiar with some basic grid systems and their operation, recognizing their advantages and disadvantages for GIS work.

7. Understand the impact of scale, class interval selection, symbolism, and simplification on the development of cartographic databases.

8. Know the difference between cartographic and geographic databases and understand the potential problems associated with producing a GIS database composed of both.

9. Have a basic feel for the potential problems of some specialty maps such as soil maps, vegetation maps, historical maps, and remotely sensed imagery, as they apply to GIS work.

MAP AS MODEL: THE ABSTRACTION OF REALITY

The map is a model of spatial phenomena. It is an abstraction. It is NOT a miniature version of reality that is meant to show every detail of a study area. While this may sound rather evident, at times we have all either ignored or forgotten this simple fact. Even the most experienced map readers sometimes find themselves exclaiming at road maps that fail to show the constant, annoying curves in the road they are driving. We know enough about maps to recognize that these curves couldn't possibly be drawn on such a small piece of paper, yet we still forget. There are limits to what we can do with cartographic skills. Yet while we would not expect the plastic pilot of a model airplane to fire up the engines and take off down the dining room table, somehow we think the map should be a perfect replication of reality. And where we would not expect dolls representing people to have warts, zits, and facial hair (although they do seem to be getting closer), we expect our hometown of 18 people to be on a map of the country depicted on a sheet of paper measuring 11 inches by 17 inches.

The primary reason for our misinterpretation of the limits of maps to display reality is that they are among the most elegantly designed graphical instruments created to communicate spatial data. Maps have been around for thousands of years, and we have all become more or less accustomed to seeing them. Such familiarity, combined with the compactness of information maps contain, and their powerful visual appearance, all lend a sort of authoritative infallibility that is difficult to discount. This is all the more reason for us to be aware of the basic character, level of abstraction, symbology, and methods of production that comprise the map product.

Maps come in a wide variety of forms and subject matter. The two primary types are the general reference map and the thematic map. As we've seen already, most of what we will deal with in GIS has to do with thematic maps, although at times reference maps are used as input to a GIS, primarily to permit the disassembly from the more complex map of selected (thematic) data. Although we will largely limit our discussion to thematic maps, much of what is covered in this chapter can be readily applied to the reference map.

A PARADIGM SHIFT IN CARTOGRAPHY

As we continue to move to an automated cartographic and GIS environment, we should be aware that the way maps are viewed has changed considerably over the past few decades. These changes have, in some respects, contributed to the enhanced utility of the GIS because of the way we treat the spatial data we input to the system. The traditional approach to mapping, called the **communication paradigm,** assumed that the map itself was a final product designed to communicate a spatial pattern through the use of symbols, class limit selection, and so on. This paradigm is the traditional method of viewing cartography, but it is limited because raw, preclassified data are not readily available to the map user. In other words, the user, being restricted to handling the final product only, is incapable of regrouping the data into forms more useful for changing circumstances or needs.

An alternative approach to cartography, one that maintains the raw data for later reclassification, developed at approximately the same time that map makers began to apply major advances in computer technology. This method, known as either the **analytical paradigm** (Töbler, 1959) or sometimes the **holistic paradigm,** maintains the raw attribute data inside a computer storage device and displays data based on user needs and user classifications. An early precursor to computer-assisted cartography and even GIS itself, this method today is much more flexible in its application than its predecessor. The impetus for its development is the idea that a map, especially with the use of computer technology, should allow for both communication and analysis.

The analytical paradigm was originally designed for use with value by area mapping (covered later in this chapter), where each area would have its own unique color or shading pattern relative to the raw data it represented. These maps, called classless choropleth maps (Töbler, 1973), had the initial limitation of not being easily interpreted by the user. In this respect, they were not unlike unclassed satellite images. With the use of the computer as a storage and data classification device, however, the method expanded in concept to allow multiple classifications of the data, and each set could be viewed readily by the viewer. This enhancement prompted some to change the name from "analytical" to "holistic" because it signals the return to prominence of the map's role as a communication device. We will see this in Chapter 14, where we discuss output from the GIS, a truly holistic approach to cartography as well as to geography in general.

An example, using identical study areas, for different purposes, may be useful for understanding the different cartographic paradigms. Suppose you are creating a map for a state park area. The initial assignment is to create a map that will allow tourists to enjoy the activities and sights offered by the park. You design the map to highlight the lake, boat docks, camping areas, cabins, the park ranger station, geysers or other natural features, walking trails, fishing sites, restricted areas, food vendors, and so on, showing the roads that provide

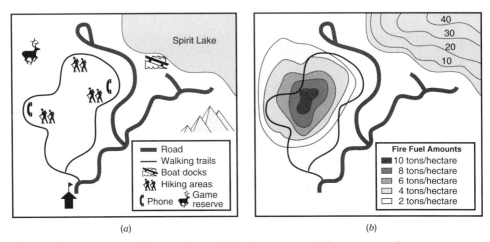

Figure 3.1   Cartographic paradigm shift. Two maps of the same study area—a state park—showing *(a)* the use of the communication paradigm to create a visual display of spatial relationships and *(b)* the added utility of the analytical paradigm for the same region.

access to each feature (Figure 3.1*a*). The features may also be classed to illustrate, for example, the quality of the roads, the types of boat that can use the docks, and whether there are hookup facilities for recreational vehicles available at the campsites. The fundamental reason for providing this map is to show the spatial distribution of the phenomena of interest in a way that indicates how each one may be reached. This is illustrative of the communication paradigm.

However, the park rangers will need information to manage the facilities and resources for which they are responsible. They may need such quantitative information as the average tree size within the forested regions, or the buildup of fire fuel in the forests, the numbers of rare or endangered species in restricted areas, and the sizes and areal distributions of gaps in the forest that provide needed environments for browsing species. With these data, the park rangers can begin to determine whether deer populations are rising or falling, whether some trees may need to be cut to lessen the threat of fire, or where old and dying trees should be thinned to allow smaller trees to grow in. Many of these situations will call for additional attribute information about tagged species, diseased trees, or annual differences in animal or plant species composition.

Eventually the amount of information, particularly the attribute information, becomes so dense that a single map is incapable of illustrating the necessary conditions for the park rangers to plan for future management strategies. The data must first of all be detailed for each location. If, for example, there are 200 different types of vegetation assemblage, data must be recorded for each, but displaying all these areas on a single map will obscure the information visually; moreover, multiple maps would have to be produced for each variable at each area. The primary function of these cartographic representations is analysis, rather than simply viewing spatial distributions (Figure 3.1*b*). The communication paradigm is inadequate for this application, and it must be augmented to include more analytical capabilities such as are now available in the holistic paradigm with a GIS.

MAP SCALE

No matter which paradigm we choose, when we consider converting our spatial concept of space into a map, we must always remember that maps are **reductions** of reality. Although it might be intellectually appealing to envision a map that physically covers our entire study area, such a map would require us, once again, to explore the planet on foot. A primary purpose of any thematic map is to allow us to view important detail, for a large region at a single glance, without the distractions of inconsequential or extraneous detail. The amount of reduction is primarily a function of the level of detail we need to examine our area. If we are looking at a very small area, such as a single field (say of 20 hectares) we are not required to reduce reality as much as we would if we were looking at a study area of 1000 square kilometers.

Scale (Figure 3.2) is the term commonly applied to the amount of reduction found on maps. It can most easily be defined as the ratio of distance on the map to the same distance as it appears on the earth. For example, a map legend might indicate that one inch on the map is equal to 63,360 inches on the ground. In other words if we placed an object one inch long on the ground, we would have to line up 63,359 more equivalent objects to match the amount of ground covered by a single inch on the map. A scale expressed as "one inch equal to 63,360 inches" is called a **verbal scale.** This common method of expressing scale has the advantage of being easily understood by most map users. Another common method is the **representational fraction (RF)** method, in which both the map distance and the ground distance are given in the same units, as a fraction, thereby eliminating the need to include units of measure. The RF method is most often preferred by experienced map users because it reduces confusion. The GIS practitioner is especially warned to find out which of these two methods of expressing scale is being used. Many of you will undoubtedly experience a sense of delight when a county extension agent or other government official offers you maps (or aerial photographs) at a scale of "1:600." You expect to see maps at a representative fraction of 1 to 600—very detailed maps. However, you are really hearing a shorthand version of the verbal scale, and your contact is trying to say that the map scale is 1 inch = 600 feet. This

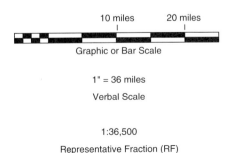

10 miles   20 miles

Graphic or Bar Scale

1" = 36 miles

Verbal Scale

1:36,500

Representative Fraction (RF)

Figure 3.2   Methods of illustrating map scale. Examples of the three most common methods of illustrating scale on a map. The verbal scale, the representative fraction (RF), and the graphic or bar scale all have advantages and disadvantages for analog or digital mapping projects.

translates into a representative fraction of 1:7200, for a much smaller scale map covering a relatively large area.

A **graphic scale,** also shown in Figure 3.2, is another basic method of expressing scale; here measured ground distances appear directly on the map. Actual area measures may be illustrated on the map as well, but this is less common. Although the graphic scale is difficult to communicate orally to the user, it has the advantage of being easily transferred from one scale to another, which far outweighs the limitation. GIS map manipulations are likely to entail many changes in scale of output depending on the format required by the user. A graphic scale device can be placed on the map during input and, as the map scale changes on output, so will the scale bar itself.

As you begin working with GIS, you will find that most software can accomplish changes of scale very easily. And, of course, the scale at which you input the data may differ from the scale at which you display the results. The ability of the software to convert to nearly any scale frequently inparts a feeling of confidence that can translate into problems later on. As you will see in some detail in Chapter 5, reliable analysis is greatly influenced by the quality of the data that are input to the system. This reliability is, in turn, affected to a large degree by the scale of the cartographic documents you input. Keep in mind that on a very small-scale map, say of 1:100,000, a line of 1 millimeter covers over 100,000 millimeters, or 100 meters, on the ground. That's approximately the length of a football field. When the scale is changed in the GIS for later output at, say a scale of 1:1000, the line that is drawn will be a millimeter in width, giving the reader the impression that the line was very accurately located when it was input. In fact its location is far from accurate.

Another example might be quite useful in illustrating the problem of scale as it affects analysis. Someone not familiar with maps presents you with a map of the 48 contiguous states of the United States depicted on a piece of paper about the size of a 3.5-inch floppy diskette. The person needs to have the map blown up to a much larger size, say a meter on a side, so that the areas of each state can be measured and the results used to calculate population density by state. Whether this map is to be enlarged through some xerographic approach, or input to a GIS to have the areas measured, the end result is the same. The values obtained through dividing the number of people for each state by the highly error-filled measures of area are certain to be useless (Figure 3.3). Thus

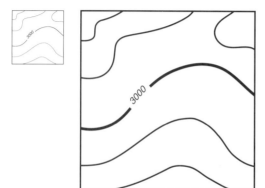

Figure 3.3 Effect of scale on accuracy. A portion of an extremely small scale map blown up to a larger scale. The lines have become thicker and show a high level of generalization; areas are less precise; and measurement and analysis are practically impossible.

the following simple, long-held, rule of thumb: it is always better to reduce a map after analysis than to enlarge it for analysis. This rule applies as much to the automated environment as to the manual one.

MORE MAP CHARACTERISTICS

Maps, as images of the world we model with them, represent the locations of objects in space as well as their qualities or magnitudes. These two related items are called **entities** and **attributes,** respectively, and both are as necessary in the cartographic document as they were when we collected the data in the first place. But regardless of whether entities and objects represent points, lines, areas, or surfaces in the real world, they cannot be displayed as minia-turizations of reality because of the limitations of scale. Instead, we must store them in the computer's memory and then, upon output, to assign a set of symbols to represent them. In turn, the symbols must have an interpretive key called a **map legend,** to be referred to by the user. The legend effectively unites the entities with their attributes, whereupon each displayed entity can be un-derstood to represent a real feature with measurable attributes. In this way the map reader is able to envision what was actually seen as the original data were collected, through the variety of methods we've already discussed.

As you remember from Chapter 2, entities are measurable, and their attrib-utes can be delimited using a number of different levels of data measurement. Associated with each different spatial data type and each level of data meas-urement is a set of symbols. We will take the spatial data types in turn and examine their potential to be mapped, the variety of symbols that can represent each one, and the ways in which the process of map abstraction limits how they can be used as input to the GIS.

MAP PROJECTIONS

Although when viewed at close range the earth appears to be relatively flat except for the surfaces we traverse, we all know that it is relatively spherical. Maps, as we have seen, are reductions of reality, designed to represent not only its features, but its shape and spatial configurations as well. The use of globes is a traditional method of representing the earth's shape. While globes preserve the majority of the earth's shape and illustrate the spatial configuration of continent-sized features, they are very difficult to carry in one's pocket, even at extremely small scales (often as little as 1:100,000,000). Most of the thematic maps encountered in practice, as in GIS analysis, are of considerably larger scales, say on the order of from 1:100,000 to as large as 1:1000, depending on the level of detail. A globe of this size would be difficult and expensive to produce and even more difficult to carry around or to spread out on a digitizing table for input to our GIS. As a result, cartographers have developed a set of techniques called **map projections** designed to depict with reasonable accuracy the spherical earth in two-dimensional media.

As the term implies, the process of creating "map projections" was originally envisioned as positioning a light source inside a transparent globe on which opaque earth features were placed, then projecting the feature outlines onto a two-dimensional surface surrounding the globe. Different ways of projecting could be produced by surrounding the globe in a cylindrical fashion, as a cone, or even as a flat piece of paper. Each of these methods as originally envisioned produces what is called a **projection family**. Thus there is a family of **planar projections,** a family of **cylindrical projections,** and another called **conical projections** (Figure 3.4). A fourth, referred to as **azimuthal projections,** which is based on the idea that the light is projected at a selected azimuth onto a flat material, is discussed later in this section. Today, of course, the process of projecting a spherical surface onto a flat medium is done using the mathematical principles of geometry and trigonometry that recreate the physical projection of light through the globe.

Cartographers have studied, modified, and produced new projections to better represent the earth on a flat medium. Unfortunately, many noncartographers in the geographic community have largely neglected this important aspect of spatial representation. The reasons for this are the stuff of the history and philosophy of geography. But with the advent of satellite remote sensing and geographic information systems, the subject can no longer be ignored. It is even more relevant now than in the past because of the need to minimize the distorting effects of projections on the true representation of spatial phenomena, combined with the wide range of mapped data the automated geographer will encounter every day. These matters will impact on all phases of GIS and should be understood to limit their negative influences.

Projections are not absolutely accurate representations of geographic space. Each will impose its own types and amounts of distortion on the map document. The important characteristics of our maps that must be retained for accurate analytical operations will often dictate which projections should be used. These characteristics or properties include angles (or shapes), distances, directions, and areal sizes. It is impossible to preserve all these properties at the same

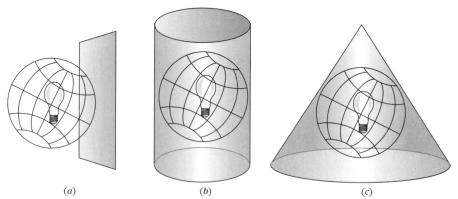

(a)   *(b)*   *(c)*

Figure 3.4   The three families of map projections. They can be represented by *(a)* flat surfaces, *(b)* cylinders, or *(c)* cones.   *Source:* P.C. Muehrcke, and J.O. Muehrcke, *Map Use: Reading, Analysis and Interpretation,* 3rd ed. © 1992, Figure C.9, page 573. Used with permission.

time when performing a map projection. We will look at each one separately, but first we must establish some useful terminology to help us understand the changes in properties as they are altered by the projection process.

The process of map projection as it is mathematically produced is twofold: first an obvious scale change converts the actual globe to a **reference globe** based on the desired scale; then the reference globe is mathematically projected onto the flat surface, where the globe's three-dimensional surface is transformed to a flat surface (Robinson et al., 1995). When we reduce the scale from the actual globe to its reference globe, we have changed the representative fraction to reflect this scale change. The representative fraction for the reference globe, called the **principal scale,** is calculated by dividing the earth's radius by the radius of the globe. We now have a representative fraction that is uniform throughout the reference globe because the shape of the latter is nearly identical to that of the earth. That is, the actual scale will be the same as the principal scale everywhere on the globe.

Before we perform the second step of projecting the reference globe to its flat map counterpart, we should note that the **scale factor (SF),** defined as the actual scale divided by the principal scale, is by definition 1.0 at every location on the reference globe. When we move from a spherical reference globe to the two-dimensional map, however, the scale factor will necessarily change because the flat surface and the spherical surface are not completely compatible. Therefore, the scale factor will differ in different places on the map (Robinson et al., 1995). We will need to keep the idea of a changing scale factor in mind as we examine the types of distortion made through the projection process.

When working with a globe, the cardinal directions of the compass rose found on its side will always occur at 90 degrees of one another. In other words, east will always occur at a 90 degree angle to north. This property of maintaining correct angular correspondence can be preserved on a map projection as well. A projection that retains this property of **angular conformity** is called a **conformal** or **orthomorphic** map projection. As we have seen, the SF is always 1.0 in every direction for every point on the reference globe. Conformal projections permit us to mathematically arrange the stretch and compression so that, within the projected map, the SF is kept the same in every direction. Since, however, the SF will not then be equal to 1.0 at every point, the parallels and meridians in the resulting map will always be at 90 degrees to one another as they were on the globe, but the areas shown on the map will be distorted (Robinson et al., 1995). Keep in mind that maintaining true angles is difficult for large areas and should be attempted only for small portions of the earth.

As we have seen, the conformal type of projection results in distortions of areas, meaning that if area measures are made on the map, they will be incorrect. We can preserve areas through projections called **equal area** or **equivalent projections,** in which the product of scale factors in cardinal directions is equal to 1.0 (Robinson et al., 1995). This procedure, of course, ensures that if you find the areas of, for example, square surface features, their two dimensions multiplied together will result in an area identical to what would be calculated on the reference globe. This is because the product of these two dimensions results in the identical SF. However, once we have achieved this identity, we find that although the amount of area is correctly measurable, the scale factor will vary in every direction about a point except the cardinal directions. In other words, by preserving area we distort angles. Thus it is a fundamental feature of these two properties of projected maps—that areas and angles cannot be preserved at the same time.

If our purpose in projecting a map is to accurately measure distances, we must select a projection that preserves distances. Such projections, called **equidistant projections**, require us to keep the scale of the map constant; it must also be the same as the principal scale on the reference globe (Robinson et al., 1995). There are two ways in which this can be done. The first maintains a scale factor of 1.0 along one or more parallel lines called **standard parallels**. Then distances measured along these lines will be accurate representations of real distances. The second approach is to maintain the scale factor as 1.0 in all directions from either one or two points. Distance measured from those starting points will then be accurate representations of true distance. A distance measured from any other point on the map will be inaccurate. As you might guess, the choice of starting point is vital here. Usually you choose the point at which most of your measurements are going to be made.

When maps are used for navigation, the primary interest is to preserve the directions shown. The preservation of true direction on a map is limited to preservation of great circle arcs, which define the shortest distance from point to point on the earth. Normally our purpose is to depict these great circle routes as straight lines. There are two primary methods of doing this. The first uses very limited areas to show great circles as straight lines between all points on the map. If you intersect the meridians with these great circles, however, the angular intersections will be incorrect. Both the limited area and the angular intersections between meridians and great circles severely limit the use of this projection. An alternative, called the **azimuthal projection**, is more commonly used for preserving directions. As with the equidistant projection, we begin by selecting one or two points of reference from which our directions will be preserved. In this case, straight lines directed from these reference points will illustrate true direction. Again, measures of direction made starting at any other points will be incorrect.

Later on we will discuss the problem of mixing map projections inside a GIS, especially on input, but in the meantime we need some rules of thumb for determining which of the many map projections we might want, depending on the types of analysis to be performed. If an analysis requires us to analyze motion or changing directions of objects—for example, the type of data obtained by a game manager using telemetry to record the location of every member of a herd of reindeer at different times—the conformal projection is best. This type of projection also is best for the production of navigational charts and when angular orientation is important, as it often is with respect to meteorological or topographic data. This group of projections includes the **Mercator, transverse, Lambert's conformal conic,** and **conformal stereographic** projections.

General reference and educational maps most often require the use of equal area projections, but our interest is in analysis. As the name implies, these maps are best used when calculations of area are the dominant calculations you will perform. If, for example, your interest is in the calculation of the changing percentage of a land cover type with time, or if you are trying to analyze a particular land area to determine whether it is large enough to be considered for a shopping mall, equal area projections are best. When considering the use of an equal area projection, you will need to take into account the size of the region involved and also the distribution and amount of angular deformation. Small areas will display far less angular distortion when equal area projections are used, and this property might be relevant if both shape and size are important characteristics for your analysis. Alternatively, the larger the area you are analyzing, the more precise your area measures will be if you use an equal area projection rather than another type. Types of equal area projection for use with medium-scale maps most often encountered in GIS work are **Alber's equal area** and **Lambert's equal area projections.**

GIS projects that require a determination of shortest routes, especially for long distances, call for the use of azimuthal projections because of their ability to show great circle routes as straight lines. These projections are most often used for creating maps for airline traffic, determining ranges of radio signals, keeping track of satellites and satellite data, and mapping other celestial bodies (Robinson et al., 1995). Only recently have these projections become popular, but their use will increase as these uses for GIS increase. The most common azimuthal projections you will encounter include **Lambert's equal area, stereographic, azimuthal equidistant, orthographic,** and **gnomonic** projections. A final note about the use of these projections is that some maintain both direction and area. This property may prove useful for analyzing aerial distributions such as volcanic plumes that are likely to travel great circle routes as they disperse into the atmosphere and follow the general circulation patterns of the earth.

There are many map projections from which to choose—far more than are listed here. Some are special-purpose projections especially useful for depicting the entire earth or very large portions of it. Other projections allow for better coordination of large-area mapping programs such as the topographic mapping of whole continents, which is done in smaller portions. The list is quite large. As you will see later, the selection of projection is a fundamental process of designing a GIS. You should take the time to select a good reference on map projections, with special emphasis on the properties each best maintains. Two sources that might prove particularly useful are those of Nyerges and Jankowski (1989) and a well-known reference work by Snyder (1988).

GRID SYSTEMS FOR MAPPING

We have seen that a grid or coordinate system is necessary to reckon distance and direction on the earth. This geographical coordinate system based on latitude and longitude is useful for locating objects or features when they are confined to the spherical earth or its reference globe. Because we will most often operate with two-dimensional maps projected from this reference globe, however, we also need one or more coordinate systems that correspond to the distortions introduced through that process. These reference systems, called **rectangular coordinates** or **plane coordinates,** allow us to locate objects correctly on these flat maps.

The basic rectangular coordinate system is also the one that most of us learned when working with graphs and number lines. It consists of two lines, an abscissa and an ordinate. The abscissa is a horizontal line that contains

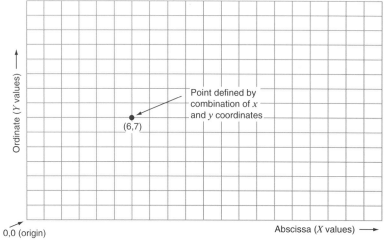

Figure 3.5 A cartesian coordinate system. Classic rectangular grid system used to represent the number system. The abscissa (*X* values) is horizontal and the ordinate (*Y* values) is vertical. Any point can be defined by combinations of these two.

equally spaced numbers starting from 0, called the origin, and extending as far as we wish to measure distance in either of two directions (Figure 3.5). The values are called *X* coordinates; they are positive if we move to the right of 0 and negative if we move to the left. The second line, the ordinate, allows us to move vertically from the same point of origin in a positive or negative *Y* direction. Together the *X* and *Y* coordinates allow us to locate any point or feature by combining values of *X* and *Y*. As you will see later, **digitizers,** the devices we use to input geographic coordinates into a GIS, are based on this simple **Cartesian coordinate system.** It is possible to obtain reasonably accurate results with such a system as long as the map exhibits the property of conformality. Maps of this type are most likely to illustrate only small portions of the earth's surface. Each of these large-scale maps must maintain its own coordinate system to ensure a degree of accuracy of measurement (Robinson et al., 1995).

By tradition, when reading maps using rectangular coordinates, we give the *X* value first and the *Y* value second. When a map is oriented with north at the top, the normal situation, the *X* value is called an **easting** because it measures distances east of the origin or starting point. In like manner, *Y* value is called a **northing** because it measures distance north of the origin. As you can see, there are no westings or southings. Instead, the origin is placed so that all references are positive; or, stated another way, they are placed in the northeast quadrant of our reference system. This allows us to read first right, then up from the origin, a process called reading "right–up." In some cases, the size of the area will require us to construct a number of **false origins** to ensure that each portion of the earth will be accurately represented on a flat surface. Measures from false origins are then called **false eastings** on the abscissa, and **false northings** on the ordinate line.

As indicated, plane coordinates are not normally used on small-scale maps because of the potential for distortion. For small-scale maps, adjustments must be made to compensate for the distortions introduced during projection. Despite the large number of map projections available, the vast majority of plane coordinate systems attempt to adjust for conformality by using only conformal map projections, typically the transverse Mercator, polar stereographic, and Lambert's conformal conic. This will not always be the case, however. If, for example, your study area is located in equatorial locations, a Mercator projection may prove more useful (Robinson et al., 1995).

In the United States, five primary coordinate systems are used, some based on properties of the map projections and others on historical land subdivision methods. When you encounter maps of other nations in your GIS work, you will need to ascertain the projections, coordinate structure, and other properties of the systems of those nations. Many countries will use one or another of the types discussed below, but will require you to become familiar with their points of origin and the zones of the earth they occupy before beginning your GIS input procedures.

Perhaps the most prevalent plane grid system used in GIS operations is the universal transverse Mercator (UTM) (Figure 3.6). It has been adopted for much remote sensing work, topographic map preparation, and natural resource database development because it allows precise measurement using the metric system of measurement, which is accepted by many countries and by the scientific community at large. The basic unit of measure is the meter.

UTM divides the earth from latitude 84 degrees north and 80 degrees south latitude into 60 numbered vertical **zones** that are 6 degrees of longitude wide. To allow for all coordinate locations to be positive, the UTM has two primary ordinate starting points, one at the equator and the other 100,000,000 meters south of the equator (located at latitude 80 degrees south). For reference, these zones are numbered starting at the 180th meridian, in an eastward direction. Each zone in turn is divided into rows or sections of 8 degrees of latitude each, with the exception of the northernmost section, which is 12 degrees, allowing all the land of the northern hemisphere to be covered with the system.

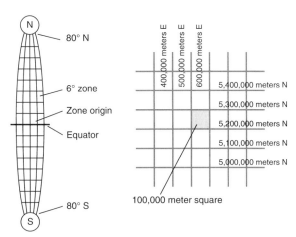

Figure 3.6   The universal transverse Mercator system (UTM).   *Source:* P.C. Muehrcke, and J.O. Muehrcke, *Map Use: Reading Analysis and Interpretation* 3rd ed. JP Publications, Madison, WI © 1992, Figure 10.8, page 215. Used with permission.

Since each section can be located by a combination of the number and letter (read right–up as before), relatively small sections of the globe can be isolated. With the exception of the northernmost grouping, each of these sections will occupy 100,000 meters on a side and can be designated by eastings and northings of up to five-digit accuracy. Thus measurements should be correct to within a one-meter level of resolution.

As the name implies, the UTM system uses a transverse Mercator projection. For each of the 60 longitudinal zones, a separate Mercator projection is applied to reduce distortion. The *Y*-coordinate origin is placed at the exact center of each zone (comprising 6 degrees of latitude), and a false origin is offset 3 degrees west of that. The scale factor does not vary from its 0.99960 in the north–south direction, but it does vary in the east–west direction. Still, at its farthest point from the *Y*-coordinate origin, the scale factor is very nearly the same at 1.00158. This near equivalence illustrates the minimization of distortion possible with UTM that results in accuracies approximating 1 meter variation in every 2500 meters distance (Robinson et al., 1995).

For polar regions that extend beyond the area covered by the UTM grid system, while still maintaining the same level of accuracy, we use a **universal polar stereographic (UPS)** grid. This system divides the polar regions into concentric zones, then splits them into two equal halves east to west at longitude 0 and 180 degrees. These zone halves are assigned different designations for North and South poles. In the northern hemisphere the west half is designated grid zone *Y* and the east half grid zone *Z*. In the southern hemisphere the west half is designated zone *A,* and the east half zone *B*. As in the case of the UTM grid system, measurements are made as eastings and northings of up to 2,000,000 meters, corresponding to 180 degrees in longitude. Also as in the UTM, the zones can be divided into 100,000-meter squares, each with its own projection, resulting again in an accuracy of approximately 1 meter in 2500. Collectively, the UTM and the UPS systems provide a global coverage with only minor distortion and reasonably accurate measures.

Perhaps the premier mapping organization in the United States and among the largest contributors to GIS data, the USGS (United States Geological Survey) has found it convenient to simplify the UTM system for its topographic mapping. Because the United States falls entirely inside the northern hemisphere, you only need a zone number (1–60) and a single pair of easting and northing values to locate a point on the map. The equator is designated as the origin for northing values, and meridian in the center of each meridian zone is given a value of 500,000 meters as a false easting. These simple modifications make the designations simpler, but the locations are identical to what would be found on UTM. Most USGS topographic maps have this coordinate system printed in blue lettering on the margins, together with another grid system we will discuss a little later.

A system devised in the 1930s by the U.S. Coast and Geodetic Survey (now the U.S. Chart and Geodetic Survey), called the state plane coordinate (SPC) system, uses a unique set of coordinates for each of the 50 states (Figure 3.7) (Claire, 1968; Mitchell and Simmons, 1974). This grid system uses either a transverse Mercator or Lambert's conformal conic projection tied to a national geodetic framework. Its original design was to provide a permanent record of land survey monument locations. Like the UTM system, the SPC uses two or

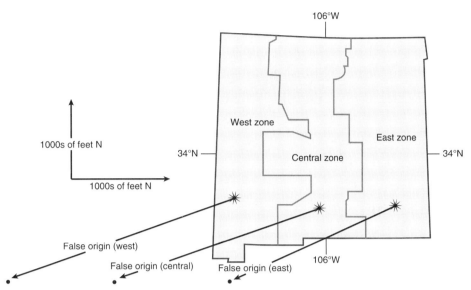

Figure 3.7   The United States state plane coordinate system. Applied to New Mexico, this illustrates the potential for problems at cross-state boundaries when using SPC for GIS work in large regions.

more zones, in this case overlapping, to maintain an accuracy of one part in 10,000. Each zone has its own projection and coordinate grid that is measured in feet rather than meters, although metric equivalents have recently been included. To define a location using this system, you give the state name, the zone name, and the easting and northing values (in feet). The advantage of SPC is in its accuracy, estimated at four times that of the UTM system. For the GIS applications specialist, the disadvantage is a lack of coordination between state borders (Figure 3.7). If your study area extends across one or more state boundaries you will have to spend time making adjustments among the disparate grid systems. As you might guess, this means that the object locations on each side of the boundary are also going to be misaligned.

The final plane or rectangular grid system in common use in the United States is the U.S. **Public Land Survey System (PLSS)**, established in 1785 as a method of land subdivision (Figure 3.8) (Pattison, 1957). As a land subdivision system it was not, and is not today, formally tied to any particular map projection, or to a reference globe at all. It was meant as a tool for recording ownership of land. And its basic unit of measure is an area of 1/640 of a square mile, or one acre. Each square mile of area is called a section, and sections are grouped in larger, 36-square-mile groups that collectively are called townships. The sections are numbered in a serpentine fashion starting in the upper right with section 1 and ending in the lower right with section 36. Each of the 36 sections can be divided and subdivided into halves and quarters, permitting ownership of smaller portions of land than an entire square mile (640 acres). For example, a person can own a quarter of a quarter of a quarter section, meaning that he or she owns 1/4 of 1/4 of 1/4 of 640 acres, or 10 acres. One could also own a quarter of half a section, or 1/4 of 1/2 of 640 acres, or 80 acres. By defining the location of the subdivisions as to north, south, east, or west for halves; or

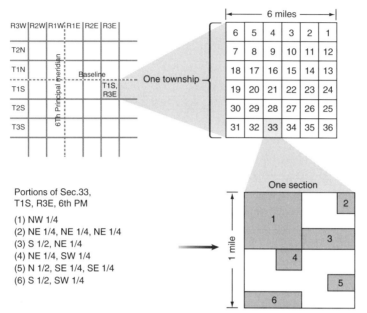

Portions of Sec.33,
T1S, R3E, 6th PM

(1) NW 1/4
(2) NE 1/4, NE 1/4, NE 1/4
(3) S 1/2, NE 1/4
(4) NE 1/4, SW 1/4
(5) N 1/2, SE 1/4, SE 1/4
(6) S 1/2, SW 1/4

Figure 3.8   The public land survey system of the United States (PLSS). Township (T) and range (R) lines intersect at the center of a township. The township occupies 36 square-mile sections (S), each of which is 640 acres in area. When more than one fractional portion of a lot is given, the position in the section map is located by reading the subportions from right to left.   *Source:* P.C. Muehrcke, and J.O. Muehrcke, *Map Use: Reading, Analysis and Interpretation,* 3rd ed. JP Publications, Madison, WI, © 1992, Figure 10.20, page 226. Used with permission.

northeast, northwest, southeast, or southwest for quarters, individual land-owners can determine which lands belonged to them.

Townships and their smaller, square-mile sections are located within a larger grid of established horizontal and vertical lines. The horizontal lines, called baselines allow measurements of township lines (not the same as townships themselves although they bound townships), while the vertical lines, called principal meridians (not the same as prime meridians), measure the longitudinal bounds of land through measurements called ranges. Township and range lines run through the center of each township of land. Together, these lines, chosen as conventions by the U.S. Continental Congress, allow the designation of landownership by stating first the subportion of the numbered section (e.g., NE 1/4 N 1/2 of section 18), and then the township and range numbers that identify the township they intersect.

Although the Public Land Survey System (PLSS) as just described is not particularly difficult to understand, a number of problems have been encountered in its implementation in the GIS environment. First, boundaries with Canada and Mexico interrupt and often distort the system, preventing a designation of land without substantial minute subdivisions, especially along meandering stream borders. Second, past surveyors' errors are compounded when markers are lost, creating many situations in which landownership disputes occur. Finally, because the earth is round and maps are flat, and the PLSS is often portrayed on projected maps together with other rectangular grid systems, the square sections are often offset east and west to account for converging meridians as we move away from the equator. All these conditions, combined with the lack of an earth-bound original geodetic framework, contribute to the limited usefulness of this system for GIS. However, you will encounter the PLSS, and you will be required to provide a projected reference system on which to map this system.

THE CARTOGRAPHIC PROCESS

Beyond the identification of scale, projection, and grid system, the GIS analyst should keep in mind the major steps in the process of making a map, especially in view of the paradigmatic shift from communication alone to communication and analysis. There are four general steps in the cartographic process: data collection, data compilation, map production, and map reproduction. We looked at data collection as our first step in formalizing our conceptual model of space. And in Chapter 14, we will look at production and reproduction as a process fundamental to the output phase of the GIS. However, the data compilation phase of the cartographic process should be scrutinized here, before we consider either data structure in Chapter 4 or data input in Chapter 5.

The process of cartographic data compilation has traditionally involved the selection or development of a base map on which to place the spatial data collection. Symbol sets are used to represent the points, lines, areas, and surfaces, each of which may skew exact locations to make room for their placement on the limited map surface. A preselection and sorting process is invoked to group the data by means of statistical techniques. Under the communication paradigm, grouping has most often been performed as a separate step or set of operations before mapping is begun. With the advent of the analytical or holistic paradigm, grouping can be performed within the GIS with the use of statistical or database management techniques. However the cyclical nature of GIS operation (see Chapter 13) has obscured grouping as a separate and easily identifiable portion of the cartographic process. Many maps that will be included in the GIS database nevertheless will be input from maps created under the traditional communication paradigm. As we have already seen, this compilation process can obscure the raw data by creating categories based on the original design criteria for the map, leaving the GIS analyst without a complete set of attribute data for input.

MAP SYMBOLISM

In a later section we will discuss the problems related to computerized data abstraction, particularly those surrounding input of existing map documents. But first it is useful to briefly review some of the basic concepts of data compilation that produce these map documents. The concepts will prove beneficial, as well, when we discuss GIS output, because at that point we will return to basic cartographic compilation within an automated environment.

As we have already seen, maps are not exact miniaturizations of reality, but rather abstractions. We've seen that geographic objects occur as points, lines, areas, and surfaces, and we've also seen that they can be described by four different levels of measurement. When we move to cartographic abstractions, we need to represent all these objects, no matter what their level of measurement, by careful selection, categorization, and symbolization so that the results will physically fit on the space provided and the reader can understand what is being presented.

A frequently useful diagram (Figure 3.9, p. 70) is an expansion of Figure 2.9, which we used earlier to show the relationships between spatial data types and their levels of measurement (Muehrcke and Muehrcke, 1992). As you can see, we now have some sample sets of symbols that correspond to points, lines, areas, and surfaces at all levels of measurement. Figure 2.4 grouped the last two levels of measurement as interval/ratio, rather than keeping them separate, because the symbol set is generally the same for these measurement levels. It is important to note this because it adds another level of abstraction. If we are not aware of how the data were originally gathered, we may not know whether they are interval or ordinal. The symbols do not tell us. We may find ourselves performing a ratio-type analysis on a set of data that do not have an absolute starting point. Results from such an analysis would be meaningless.

You should also note that lines can be used to illustrate surfaces. This may add some confusion as well, for two reasons. First, there is the tendency of the novice to misinterpret lines as one-dimensional features rather than as symbols of surfaces. This form of confusion is easily overcome by continued familiarity with map features and symbols as more and more maps are encountered. The second problem due to the use of line symbols to represent surfaces is that of interpreting the lines themselves as accurate representations of point elevation values (see Chapter 2). As we have seen, the lines are estimates or predictions of specific elevations made through the process we have called interpolation. Although lines may be good predictors of elevations, the GIS student should keep their predictive nature clearly in mind, especially if they are used as GIS input. The only nonpredictive elevational values are those that were actually measured, as exemplified by bench marks on topographic maps. As we will see later, most commercial GIS software has its own interpolation capabilities. It is up to you to become familiar with both the data and the reliability of the interpolation method before deciding to input known point locations or to sample the lines themselves for input (see Chapter 5).

The use of lines to represent surfaces is one of a large number of **changes of dimensionality** that can occur when the data are abstracted with the use of symbols. Figure 1.1 clearly indicates the wide range of dimension changes that can be symbolized on a map. Good judgment will always be necessary in deciding whether the symbol geometry and dimensionality given are truly representative of an object. For example, if an areal symbol is used to depict a point feature, you must be aware that despite the two-dimensionality of the symbol, only a single point needs to be recorded or input for the feature.

The same can be said regarding levels of measurement. Symbols such as graduated circles, indicative of attributes at point locations, are often manipulated to achieve a particular visual response (Robinson et al., 1995). This change in symbol size is not directly proportional to the changes in the data; rather, it is designed to allow the viewer to visually perceive proportionality. Such manipulations are commonplace in cartography, and the GIS analyst should examine the map carefully before proceeding with data encoding based on symbol manipulation. As always, if the raw data are available, the attributes should be input from them rather than their graphic depictions.

A major difference between the communication and holistic paradigms is the classification-oriented manipulation of data prior to map production. Because a single thematic map is meant to serve a single purpose, this classification process normally takes place only once and the raw data are no longer available. Consider the case of choropleth or value-by-area mapping. When mapping areas under the communication paradigm, we want to group the areas into meaningful as well as visually appealing aggregations. The many common methods of grouping or aggregating these areas are collectively called **class interval selection.**

Among these methods of class interval selection we find several categories that merit examination. The first category is called constant intervals. This group of methods includes having the same number of areas within each category or the same number of data points for each class, or simply dividing the range of values from start to finish into an equal number of classes. Each of these constant interval methods has its own characteristics. Some will produce well-balanced map output (see Chapter 14); some will be more convenient; some will ensure that all classes will have data.

The second group of classing techniques is called variable intervals. These approaches produce maps that are less intuitive but may be highly useful for isolating certain high or low values, or for highlighting variations in value. Variable interval methods can be systematic, including arithmetic, logarithmic, or other mathematical series; or they can be unsystematic, with the clustering of the data into natural groupings used to determine where the class intervals will be designated.

Both general types of class interval selection can be used for point, line, and area symbols to depict all spatial data types. Take, for example, the use of contour intervals on topographic maps. The selection of contour interval is as much a method of class interval selection as grouping areas for choropleth maps. Likewise, creating a discrete set of point symbols to show variation in attribute variable is also a method of class interval selection because the cartographer knows that the human eye is incapable of discerning extremely subtle changes in sizes.

All these interval selection techniques create documents that to varying degrees disguise the original data and, if poorly selected, can obscure the original distributional patterns. We should keep these principles in mind when we prepare to produce GIS data output. If, on the other hand, we intend to use these classified maps as input to a GIS database, we must be very careful about the analyses we perform with the highly manipulated data.

Of course, symbolism and classification are not the only cartographic compilation methods we need to be aware of. Among the most important compilation processes the GIS analyst will encounter is the graphic simplification that takes place on the map as these classified data and their symbols are transferred to paper. This process presents a particular problem during GIS input, but it also affects the results of subsequent measurements and other analyses. Simplification goes further than classifying the data we will find on a map, or determining the types and levels of abstraction for the symbols used. Instead, simplification eliminates some features that are not wanted, or smoothes, aggregates, or further modifies the features on the map. We saw some of this in our earlier discussion of the changes in dimensionality. Ultimately the purpose of map simplification is to provide for readability of the cartographic document once it is produced. Two basic methods are used: feature elimination and feature smoothing. We will discuss them separately.

In feature elimination we perform a function that is very much like the process of spatial data gathering itself (Figure 3.9; Muehrcke and Muehrcke, 1992). When we observe a portion of the earth, we use our geographic filter to make decisions about which features we will note and which we will ignore. The importance of the features during the data gathering process is determined before we begin, and it is largely controlled by our reasons for gathering the data in the first place. In fact, the selection of objects for investigation will, by default, act as a **passive** process of feature elimination on our maps, because only the features selected will be placed in the database or map document. In some cases, of course, passive feature elimination occurs because we are unable to view objects in the field with the instrumentation at hand. Sensitivity to features is also a function of the current state of scientific knowledge. For example, changes in plant and animal species affect what is recorded during census activities; mammal locations that previously could not be pinpointed specifically now can be registered by radiotelemetry; prior to initiation of a national population census, we were unable to collect large-scale information on population changes and socioeconomic variables. Sensitivity to environmental factors and their interactions affects how we conceptualize them, which in turn affects what we select for investigation and later mapping and what is eliminated.

Unlike passive feature elimination, active feature elimination can be used in data collection and in mapping and cartographic database development as well. When we select certain electromagnetic radiation bands for remote sensing,

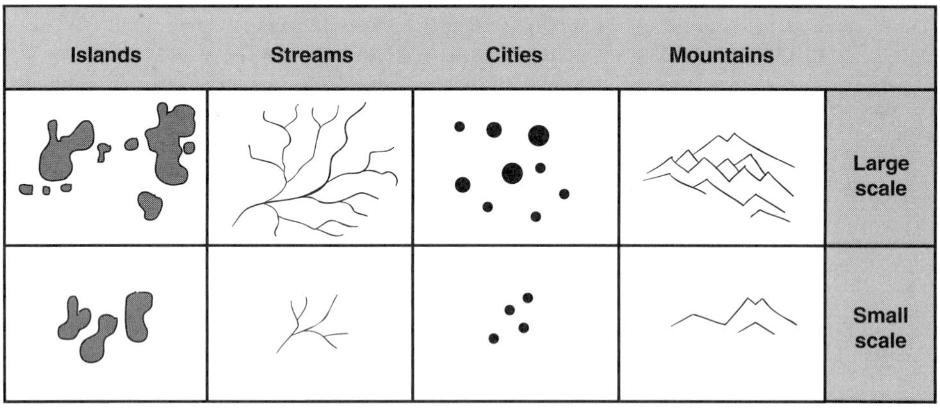

Islands	Streams	Cities	Mountains	
				Large scale
				Small scale

Figure 3.9 Scale change and feature elimination. Comparison points, lines, areas, and surfaces as these features are detailed in large-scale maps and eliminated for small-scale documents. *Source:* P.C. Muehrcke, J.O. Muehrcke *Map Use: Reading, Analysis and Interpretation,* 3rd ed. JP Publications, Madison, WI, © 1992, Figure 3.13, page 81. Used with permission.

we are actively eliminating certain portions from our data set. The sampling schemes we discussed earlier also actively promote selectivity by eliminating large portions of the objects that could be collected. We also perform active feature elimination on the map or digital cartographic database itself, prior to final map production. Here the map acts as the spatial filter, rather than the person collecting the data. Very small towns often do not show up on maps of highly populated states or nations, while towns of the same size may appear on maps of areas low in population density. Similarly, we may eliminate some smaller or less important river tributaries, lakes, or islands during the mapping process because of lack of display space. In all these cases, a set of rules is formulated to determine which are selected and which are not. The rules may be as simple as eliminating a certain proportion of the objects or perhaps every other one; or they may include a set of decision rules (e.g., eliminate towns below a certain population, eliminate the smallest tributaries of stream networks). Whichever set of techniques is applied, the result is a less detailed output. If the product of feature elimination is to be input to a GIS, you create a cartographic database with some data missing.

Another useful method of simplification is called smoothing (Figure 3.10). This process abstracts detailed geometric objects into objects reduced in detail. Much as in caricatures of famous people, important geometric features are retained by representing a given detail as a simplified geometric shape. On maps showing coastal regions, boundaries, sinuous streams, or islands, we may generalize the lines that represent these irregular features so that their existence is recorded, but their spatial detail is minimized to fit on the document. GIS input from these maps will result in less than satisfactory measures of length, shape, area, or other geometric property we may wish to calculate. Because GIS analysis frequently results in cartographic output, however, we may find these two forms of simplification useful when we produce the final results of our analysis.

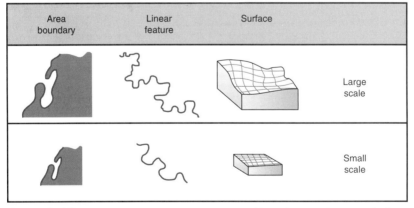

Figure 3.10 Scale change and smoothing. Use of smoothing, including the use of change in dimensionality. Note how features are simplified to only the most representative charcatures.   *Source:* P.C. Muehrcke, J.O. Muehrcke, *Map Use: Reading, Analysis and Interpretation,* 3rd ed. JP Publications, Madison, WI, © 1992, Figure 3.14, page 81. Used with permission.

MAP ABSTRACTION AND CARTOGRAPHIC DATABASES

As we have already seen, map scale restricts the amount of data that can be contained in a single map document. Base maps used for GIS data input have had to be adjusted during compilation to account for these limitations. It is largely on the basis of scale that features must be generalized, spatially displaced, or even abstracted further to allow for map readability (Robinson et al., 1995). As we will see in Chapter 4, when these graphic icons and their associated entities, sometimes called a **cartographic database,** are collected from existing cartographic documents rather than from field or remotely sensed data, they must somehow be transferred to a computerized database.

Cartographic databases move from a higher level of abstraction to a lower level of abstraction, while their counterpart **geographic database** data move from a higher level of abstraction to a lower level (Figure 3.11). That is, when data are obtained from map documents, the symbols representing points, lines, areas, and sometimes surfaces are highly abstract representations of reality. Their size and placement are themselves graphic abstractions that are less precise than the ability of computerized input devices to define them. Therefore, when we move from map document to digital database, we must decide exactly which part of a point, line, or area symbol should be recorded. If the placement of, for example, a point symbol, is physically offset from its actual location to make room for another feature, our GIS input will indicate an accuracy of location that does not exist. In addition, the symbol itself takes up space, and we attempt to locate an exact point for GIS input for this zero-dimensional object; then we must decide whether the center of the symbol is its most accurate location.

Alternatively, if we are creating a geographic database from, say, a survey instrument, we are now faced with the opposite problem. The survey instrument may be accurate to within centimeters or even millimeters of some absolute location. However, the input device is often not capable of reproducing this information with the same level of accuracy.

While each of these issues needs to be addressed individually as we create our GIS database, some of the more difficult problems arise when these two sets of compilation rules interact. For example, a fish and wildlife scientist once attempted to create a database indicating the location of roads relative to a species of bird that inhabits regions in or around the roads. The expedience of creating a road network with available data, in this case from existing 1:24,000

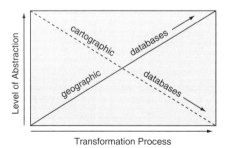

Transformation Process

Figure 3.11 Abstraction for cartographic and geographic databases. Diagram showing how cartographic databases move from higher levels of abstraction of lower levels and geographic databases move from lower abstraction levels to higher.

USGS topographic maps, made the decision to create a cartographic database for this GIS coverage a straightforward one. The birds, in this case burrowing owls, create individual burrows on the ground that do not appear on maps, so a geographic database was created by using a GPS unit with 1 meter accuracy, and then entering the data into the GIS. But when the two coverages were displayed, the locations of burrows that were known to lie well within 10 meters of the roads appeared to be offset by at least 100 meters from the roads. This incompatibility of different compilation rules resulted in failure of the GIS to analyze the interactions between roads and burrowing owls.

The same problem can arise when creating a cartographic database from maps of different scale. As we have already seen, maps at very small scales will require symbols that occupy much larger proportions of space on the ground than will very large-scale maps with the same symbol sizes. Thus the scales of input for a cartographic database should be as nearly identical as possible.

SOME PROBLEMS RELATED TO SPECIFIC THEMATIC MAPS

Not long after you begin to work with GIS, you will encounter a wide array of map types, representing an enormous number of possible themes. Numerous maps are seen frequently because of their ready availability, low cost, and general utility to GIS projects. Because many GIS applications involve natural-resource-related spatial data, probably as a result of the historical roots of GIS, most of these frequently used map types are related to the physical environment. This does not mean that only maps of the physical environment have particular problems; rather, because they are so popular, you need to be aware of the types of problem you will encounter in conjunction with their use. Moreover, many of the problems found with the maps described in the remainder of this chapter can be seen to occur in maps of other types.

Soil Maps

Among the most readily available maps worldwide, and certainly in the United States, are soil maps. In this country, these maps are found in county level soil surveys that are normally obtained free of charge from a local office of the U.S. Department of Agriculture, Soil Conservation Service. The mapping of U.S. soils took on a high degree of importance after the drought years of the 1930s, when whole topsoil layers were blown across hundreds of miles of productive farm- and rangeland, especially in the Great Plains. An effort was begun then to map the agricultural soils throughout the country to provide information both for the individual farmer and for the federal agencies responsible for preserving these farmlands. As a result, soil survey documents are commonplace and often find their way into GIS databases.

Soil maps are created through ground survey and sampling of soil profiles (vertical sections of soil extracted as soil cores). Most often the locations of soil types are recorded as polygons by drawing visually distinct soil patterns directly onto aerial photographs of the study area, generally at a scale of approximately 1:20,000. The content of the soil polygons is determined by testing one or more cores sampled from the polygons. Most soil maps produced in this way are presented directly on copies of the aerial photographs, although some counties have versions available without the aerial photographs as a background. Each soil type is identified by a code that relates to tabular data for that type, a system that makes it easy to develop soil databases.

While these maps have the obvious advantages of availability, large spatial coverage, and easily coded attribute data, they present several severe problems for GIS input and analysis. First, because the maps are drawn directly on unrectified aerial photographs, the polygons are likely to be severely affected by vertical movements of the aircraft bearing the camera, as well as decreased reliability at the margins of the photographs, and distortions of scale caused by topographic changes (Avery, 1973). In short, all the problems associated with using aerial photographs as mapping tools apply to soil maps as well. Of course because the aerial photographs are collected directly from aircraft sensing of the land surface, there has been no projection defined for the maps, making detailed measures difficult to obtain. In addition, because there is no map projection, there is also no formal grid system assigned to these maps. All these conditions make it very difficult to ensure that points on other maps in your GIS will match or be **coregistered** to points on the soil. Thus the validity of any attempt to relate variables on one map to variables on soil maps will largely depend on the degree of accuracy required. In fact, it is difficult to input these maps into a GIS without having a coordinate system and a map projection to begin with. Some GIS software will not even run if a coordinate system and a map projection have not been explicitly defined. To make matters worse, the soils themselves, which are highly variable, do not occur as discrete polygons in nature. Instead, like so many other natural features, soil occurs as an ever changing continuum. The act of placing lines on soils presents a change in dimensionality that also reduces the quality of the soil maps for analysis.

Some of the foregoing problems can be minimized so that soil maps can still contribute to a GIS analysis. One technique involves eliminating the topographic distortion that especially affects areas of high topographic change. This technique today involves a mathematical adjustment for topography and produces a map known as an **orthophotoquad**. This topographically corrected map is more accurate than its uncorrected counterpart, but it still lacks a projection or grid system. For small areas of study you will have to assume that the projection is essentially spherical, so your GIS software will allow it to be entered. For placing your map into a GIS together with other maps to obtain coregistration (a process called **geocoding**), you will have to locate specific features that occur on the ground and can be identified on one or more of the other maps to be included in the GIS. These points can act as **control points** from which you can perform a number of mathematical manipulations that duplicate the use of a **zoom transfer scope** to twist and turn the soil map to match other maps in your database. Neither of these techniques will eliminate all the problems we have discussed, but they will allow the use of soil maps as long as you exercise caution and are aware of the limitations of these maps to represent soils at all.

Zoological Maps

A type of map that has received very little attention among the geographic community despite frequent use among zoologists and wildlife managers is the zoological map, which come in two general types: area distribution and point distribution maps. Decisions concerning when points are used to map animal locations and when areas are used are poorly documented, leaving the GIS analyst to ponder the wide variety of possible sampling techniques. Of course, the fundamental problem with mapping animals is that they move, changing their exact locations from minute to minute. Radiotelemetry is often used to sample point locations of mammals, giving rather accurate locations at prese- lected time intervals. In mapping bird locations, on the other hand, visual sightings at a distance, or even auditory records of bird sounds, often are used to pinpoint positions. Although there are standard techniques for recording these data for scientific research, a problem arises when we attempt to define a point location for an object whose actual location can only be estimated to perhaps tens of meters.

Species range maps, drawn using area symbols to depict the range of certain animals, are sometimes compiled from data collected over decades by many people, and sometimes the data are gathered by one person for a single season, a month, or even a day. Whether using the raw data to produce a geographic database or using the cartographic representation of the data to produce a cartographic database, the same problem remains—how to represent point data as an area inside the GIS. This problem occurs in many other situations when point samples or point locations are being aggregated to produce areas. In such cases relatively simple computer graphic techniques can be applied including the so-called convex hull, a sort of computerized shrink-wrap ap- proach, but these approaches do not necessarily reflect the actual grouping of the organisms into what a biologist would call a range. Alternatively, other techniques, based on mathematics, intuition, and perhaps a little slight of hand, may be more useful for depicting regions based on points (Rapoport, 1982). Whichever you choose, you will need to know something about the behavior of the animal or other moving point object.

Digital Remote Sensing Imagery

Digital remotely sensed data are an ever increasing input to GIS databases, especially where large areas must be analyzed and repeat coverage is necessary due to rapidly changing conditions. Sensors differ widely in the portion of the electromagnetic spectrum used to evaluate earth features. They also provide a wide array of temporal coverages, abilities to produce stereo or pseudostereo images, and modes in which they **quantize** or sample the earth. In addition, they vary in their ability to be electronically manipulated to produce meaningful categories. We have seen how differences in electromagnetic spectra impose a form of active simplification on maps by selecting for a limited range of radiation that can be used. We have also noted how the amount of temporal sampling is related to the remote sensing system used, some allowing greater opportunities than others. Among the more difficult potential problems arising from the use of these devices are those having more to do with their geometry and later data manipulation. We will look at each of these separately.

The most common forms of remotely sensed imagery, though certainly not the only forms, sense the earth through the use of scanning devices that successively evaluate small portions of the earth at a single instant in time. At each instant, the sensor divides or quantizes the earth into bands (as in side-looking airborne radar), or as individual rectangular grids called **pixels** (picture elements) (Figure 3.12). Within the sensor's ability to sense, called its instantaneous field of view (IFOV), pixels can range in size from 20 meters to 1.1 kilometers. Of course airborne sensors can give considerably better spatial resolutions than these. What the sensor records then is a series of average radiometric values for each of the pixels, for each band it is sensing. An image is built up by combining consecutive lines of pixels **(scan lines),** one by one. The pixels and lines are merely collections of radiometric values representing the recorded average for each band. Details of the process and the different capabilities can be found in texts devoted to the subject of remote sensing (Lillesand and Kieffer, 1995).

For the GIS user, the problems stemming from remote sensing devices are twofold. First, the quantization of space into rectangles imposes another level of abstraction or simplification on our earth features. Objects that are much smaller than the pixels cannot be directly identified (a state, called subresolution), yet their presence can have an impact on the amount of radiation a given band produces (called **mixels** or **mixed pixels**). Mixed pixels can often be used to identify collections of features that are substantially different from their surroundings. Where small-scale maps are combined with remotely sensed data, it is safe to assume that subresolution features are not of consequence for the analysis. And, methods have been adopted that enhance the utility of coarse spatial resolution by substituting temporal information from repeated coverages. However, the spatial resolution of the imagery should be considered when inclusion of these data types with other, nonsensor data is being contemplated, especially for analysis of medium- to large-scale maps.

Another major problem associated with the use of satellite remotely sensed data is that the raw data provide little information until they have been analyzed. Analysis procedures fall into two major types: image enhancement and image classification. In image enhancement the raw data are recast or grouped into statistically or visually useful patterns, to improve the visual appearance of features, patterns, edges, or even clusters. For example, edges can be made to appear more prominently by mathematically enhancing their values while downplaying the surrounding values. Or, in some cases, the image can be filtered to remove isolated values that appear too prominently, contributing to a salt-and-pepper appearance not unlike the snow that appears when your television reception is not just right. While these enhancements are quite useful, they are not unlike the classification techniques used in traditional cartography in that they impose another level of abstraction or simplification on the final product. One method of avoiding the need to create nominally classified maps from remotely sensed imagery is to use biophysical data on an interval or ratio scale as direct inputs to the GIS (Jensen, 1983). Such values, obtainable from satellite sensors, include biomass estimates, temperatures, elevations, and

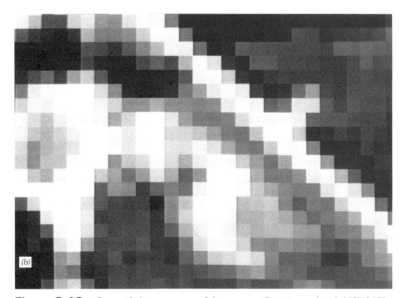

Figure 3.12   Quantizing geographic space. Portion of a *LANDSAT MSS* image showing the quantization of a portion of the earth into rectangular pixels. *(a)* Traditional aerial photograph of earth features. *(b)* The same region as depicted from a *LANDSAT MSS* image showing the quantization into regular pixels.

other, nonclassified data. These inputs are particularly useful for ecological applications of remote sensing, rather than planning uses, most of which still require some form of classified map from the original data.

All classification techniques, as they are employed in remote sensing, have much the same effect in that they are designed to separate out or group pixels into categories to which names can be assigned. The two basic approaches used here are called **supervised** and **unsupervised classification:** in the first case, the computer operator interacts with the program to assign target cells of known category, and in the second, the computer program performs the operations based only on a particular set of algorithms. While decidedly different in their results, both these approaches use a sort of **region growing** method, in which individual pixels are selected and, with other pixels having similar signal responses, are combined into categories.

Although the three approaches just described differ markedly from the way in which people classify aerial photographs, the results are often quite useful. The problem with GIS implementation is that the categories may not relate well to those produced through manual interpretation of aerial photography, historical land cover maps, and other map coverages with which they will need to be compared in the GIS environment. This incompatibility is particularly troublesome if the remotely sensed data are being used as a means of updating the historical maps. It may be difficult as well to compare two different sets of imagery from the same satellite taken at different times under different environmental conditions. Most of these problems have relatively easy solutions as long as the raw data are used for the comparisons, rather than preclassified versions.

Other, minor problems that accompany the use of remotely sensed data are mentioned only in passing. Changing atmospheric conditions—for example, the occurrence of clouds over particularly useful portions of your study area, or differences in the amount of haze in the air—can have an enormous impact on the quality of the imagery and its usefulness in mapping. In addition, because both the satellite and the earth are constantly in motion, some method of using reference points must be established to ensure that the imagery is properly georeferenced to other maps in the GIS database. Finally, you need to consider the sensor's age, since older equipment is subject to sensor drift, which produces slight discrepancies in results that should be the same. All of these potential problems are mentioned again in Chapter 15 when we discuss design.

Still, and this should be carefully noted, satellite remote sensing is a rapidly maturing field of investigation, and the use of satellite imagery should be encouraged, especially for regional analysis. The low price per hectare of land surface covered by these sensors frequently is an advantage far outweighing the problems just examined. In addition, the remote sensing community is developing new and improved methods of adjusting for these difficulties, and many are actively involved in GIS endeavors themselves. This trend can only improve both fields of study.

Vegetation Maps

While today satellite remote sensing is commonly used to produce vegetation maps for large areas, the traditional approach has been to combine aerial photography and ground work. Many potentially useful vegetation maps, especially those produced in previous decades, still exist and have the distinct advantage of providing a baseline for current studies of vegetation. Among these maps are three general categories as they were established in the 1940s and 1950s (Küchler, 1956). Some vegetation maps are based exclusively on species level classifications (floristic vegetation maps), while maps at the other extreme base their classification on vegetative structure (physiognomic vegetation maps) (Küchler, 1949), and others use some combination of these two methods (Küchler, 1955). The wide variety of possible admixtures of floristics and physiognomy has resulted in a curious set of vegetation maps that vary widely, hindering comparison with today's vegetation maps. Even now, despite established conventions (Küchler and Zonneveld, 1988), there is enormous variation in the techniques of sampling and classification methods used for vegetation mapping.

As with most cartographic documents produced under the traditional communication paradigm, vegetation maps reflect an intellectual filter through which the documents were produced, based in large part on their intended purpose (Küchler, 1956). Avoid blindly including vegetation maps as part of a cartographic database, especially if the classification technique is not fully explained. In turn, modern vegetation maps input to the GIS for purposes of comparison with historical documents need to be based on a comparable classification system. In certain ideal circumstances, the field notes of the traditional vegetation mappers will be available, and the GIS user may be able to recreate a geographic database from these materials (DeMers, 1991). This experience should suggest that notebooks be kept for others who may find the details of current mapping projects useful later on. We will come back to this topic when we discuss GIS design in Chapter 15.

Historical Maps

As suggested by the discussion of vegetation maps, historical maps of all kinds are potentially useful for spatiotemporal analysis of portions of the earth (Hodgson and Alexander, 1990; Hunter et al., 1990; Vrana, 1989). This potential should be tempered with the realization that the tools for data collection were different in years past; the purpose of the maps most likely is different; classification systems change; and, as we have seen, the basic knowledge base under which the maps were produced places limits on their reliability.

Ultimately it is the reliability issue that is most often important. If, for example, historical maps are used to determine changes in boundary configuration, the dimensions and placement of the boundary as well as the internal classification of the area bounded should be examined for accuracy. If the classifications themselves are being compared, methods must be developed to ensure cross-classification. And, especially when using historical maps, we need to keep in mind the changes in map abstraction that have a major impact on how these maps, or any maps, might impact on our GIS analysis.

Terms

graphicacy	reductions	analytical paradigm
holistic paradigm	representational fraction (RF)	scale
verbal scale		graphic scale

entities

map projections

conical projections

principal scale

conformal

equivalent projections

Lambert's conformal
  conic projections

Lambert's equal area
  projections

orthographic projections

Cartesian coordinate
  system

false origins

zones

Public Land Survey
  System (PLSS)

class interval selection

geographic database

geocoding

quantize

mixels (mixed pixels)

unsupervised
  classification

communication paradigm

attributes

projection family

azimuthal projections

scale factor (SF)

orthomorphic

equidistant projections

Mercator projections

conformal stereographic
  projections

stereographic projections

gnomonic projections

plane coordinates

easting

false eastings

universal polar
  stereographic (UPS)
  grid

grid changes of
  dimensionality

passive

coregistered

control points

pixels

region growing

map legend

cyclindrical projections

reference globe

angular conformity

equal area projections

standard parallels

transverse projections

Alber's equal area
  projections

azimuthal equidistant
  projections

rectangular coordinates

digitizers

northing

false northings

cartographic database

orthophotoquad

zoom transfer scope

scan lines

supervised classification

Review Questions

1. What is graphicacy? What impact does improved graphicacy have on our ability to function as GIS specialists?

2. What is the communication paradigm? What is its primary purpose? What impact does it have on GIS? Give an example of the use of the communication paradigm for traditional maps.

3. What is the analytical or holistic paradigm? How does it differ from the communication paradigm? Give an example of its use in cartography. What impact does this paradigm have on GIS?

4. What are the basic methods of illustrating scale on a map? Describe them. What are the relative merits of each type of scale as used in GIS?

5. What are some potential problems of putting multiple scales of maps into a GIS database? How could they affect analysis and measurements?

6. What is the purpose of the legend on a map? What impact does a map legend have on the relationships between entities and attributes.

7. What are map projections? What is their purpose? What are the three basic families of map projections?

8. What basic properties of the spherical earth are affected by using map projection? Which types of projection are best for preserving each of these properties?

9. Based on your answer to Question 8, suggest some basic decisions you will need to make in deciding which projections to use for a variety of different GIS analyses.

10. Describe the UTM grid system. What are its advantages and disadvantages for GIS use?

11. What is the state plane coordinate system? What impacts, positive or negative, might the use of this system have on GIS analysis along state boundaries?

12. Describe the U.S. Public Land Survey System. What is a section? A township? What are the problems associated with using PLSS for GIS work?

13. What impact does class interval selection have on GIS input and later analysis?

14. What impact does the physical size of cartographic symbols have on map accuracy? What about the interaction of two or more map symbols?

15. What is the difference between active and passive simplification? What impact does simplification have on the potential use of existing maps to create cartographic databases?

16. What is the difference between a cartographic and a geographic database? What potential problem exists in attempting to produce a GIS database from both these sources?

17. What should you look out for with respect to soil maps, vegetation maps, satellite remote sensing, vegetation maps, and historical maps as they apply to GIS?

18. How can unclassified, remotely sensed data be used as an input to GIS without the need for classification?

References

Avery, T.E., 1973. *Interpretation of Aerial Photographs.* Minneapolis: Burgess Publishing Company.

Claire, C.N., 1968. "State Plane Coordinates by Automatic Data Processing." Coast and Geodetic Survey, Publication 62-4, U.S. Department of Commerce. Washington, DC: U.S. Government Printing Office.

DeMers, M.N., 1991. "Classification and Purpose in Automated Vegetation Maps." *Geographical Review,* 81(3): 267–280.

Hodgson, M.E., and B.E. Alexander, 1990. "Use of Historic Maps in GIS Analysis." *ACSM Technical Papers, ASPRS-ACSM Annual Convention,* 3:109–116.

Hunter, G.J., and I.P. Williamson, 1990. "The Development of a Historical Digital Cadastral Database." *International Journal of Geographical Information Systems,* 4(2):169–179.

Jensen, J.R. 1983. "Biophysical Remote Sensing." *Annals of the Association of American Geographers,* 73:111–132.

Küchler, A.W., 1949. "Physiognomic Classification of Vegetation." *Annals of the Association of American Geographers,* 39:201–210.

Küchler, A.W., 1955. "Comprehensive Method of Mapping Vegetation." *Annals of the Association of American Geographers,* 45:405–415.

Küchler, A.W., 1956. "Classification and Purpose in Vegetation Maps." *Geographical Review,* 46:155–167.

Küchler, A.W., and I.S. Zonneveld, Eds. 1988. *Vegetation Mapping: Handbook of Vegetation Science.* Dordrecht: Kluwer Academic Publishers.

Lillesand, T.M., and R.W.K. Kiefer, 1995. *Remote Sensing.* New York: John Wiley & Sons.

Mitchell, H.C., and L.G. Simmons, 1974. The State Coordinate Systems. U.S. Coast and Geodetic Survey, Special Publication No. 235. Washington, DC: U.S. Government Printing Office. (Originally published in 1945.)

Muehrcke, P.C., and J.O. Muehrcke, 1992. *Map Use: Reading, Analysis, Interpretation.* Madison, WI: J.P. Publications.

Nyerges, T.L., and P. Jankowski, 1989. "A Knowledge Base for Map Projection Selection." *American Cartographer,* 16:29–38.

Pattison, W.D., 1957. "Beginnings of the American Rectangular Land Survey System, 1784–1800." Chicago: University of Chicago, Department of Geography, Research Paper 50.

Rapoport, E.H., 1982. *Areography: Geographical Strategies of Species.* New York: Pergamon Press.

Robinson, A.H., J.L. Morrison, P.C. Muehrcke, A.J. Kimerling, and S.C. Guptill, 1995. *Elements of Cartography,* 6th ed. New York: John Wiley & Sons.

Snyder, J.P., 1988. Map Projections—A Working Manual. U.S. Geological Survey, Professional Paper 1935. Washington, DC: U.S. Government Printing Office.

Töbler, W., 1959. "Automation and Cartography." *Geographical Review,* 49:526–534.

Töbler, W., 1973. "Choropleth Maps Without Class Intervals." *Geographical Analysis,* 5:262–265.

Vrana, R., 1989. "Historical Data as an Explicit Component of Land Information Systems." *International Journal of Geographical Information Systems,* 3(1):33–49.

CHAPTER 4

Cartographic and GIS Data Structures

Thus far we have viewed geography in its traditional forms, either field or exploration oriented, or as cartographically represented and later analyzed through cartographic measurements or separate spatial-analytical and statistical techniques. Together these traditions have contributed to an understanding of how spatially distributed earth features operate. With the advent of the quantitative revolution in science, combined with developments in computer technology, geographers began to experiment with automated methods of inquiry. These methods included both cartographic analysis and spatial analysis of cartographic phenomena. It is difficult, or even impossible, to separate either of these methods of inquiry from the development of GIS. Nor does it serve any useful purpose. Both are part of a larger body of investigative tools for automated geography.

Many commercial GIS packages contain direct links to a variety of statistical packages allowing analysis apart from the map itself. Others use particular arrangements of data (data structures) that allow map-based data to be passed easily to freestanding statistical or analytical software for separate analysis. In both cases the reason is the same—to allow the geographer access to the largest set of techniques possible for answering questions about spatial data. This consolidation of automated techniques is likely to increase as the industry matures and as users become more familiar with the capabilities of automated geography. As GIS professionals, you should welcome these changes and even encourage more mergers to include, for example, geostatistical packages, special modeling software, and the like.

A basic understanding of the structural makeup of the GIS, whether in isolation or in tandem with other software, is essential to enable us to perform our work efficiently. Each system has its own unique structures, methods of representation, and ways of analyzing spatial data. Fortunately, these can be grouped into a relatively small number of basic data structures within which a particular system will fall. Because you will likely move from system to system throughout your career, you should become familiar with all the basic types. You will have plenty of opportunity to focus on the system you use most often as you work with it.

The representation of spatial data is another formalism not unlike the ones we have already examined when moving from actual earth features to a limited set of definable objects called points, lines, areas, and surfaces. Nor is it unlike the cartographic abstractions of these objects and their measurement levels into mappable objects. The difference is merely in how we can represent the data inside a digital computer in such a way that they can be edited, measured, and analyzed, and output in some useful form. In this chapter we will examine some basic computer file structures. Then we will move to database structures that enable large amounts of data to be organized, searched, and analyzed. We will look at the basic concepts involved in the representation of space and its objects by graphic data structures. Then we will develop more comprehensive data models that allow multiple sets of cartographic data and their attributes to be linked to form a complete GIS database system.

As you read this chapter you will be introduced to these structures in increasing complexity. Spend time understanding the major types of cartographic data structure before moving on to the more complex GIS data models. If you have a number of GIS programs available, you should determine which structures each one exemplifies and how the respective programs work. As you examine the programs, ask yourself what the advantages of one type over another might be. Which would be best for saving storage space? Which is more spatially accurate? Which is best for representing points, lines, areas, or surfaces? What are the advantages of each system for geographic questions of the type you will want to ask? The answers to these questions will prove useful to you as an analyst and invaluable if your career path involves system design for potential users of GIS.

LEARNING OBJECTIVES

When you are finished with this chapter you should be able to:

1. Understand the difference between simple graphics (analog or digital) and a map.

2. Know the different types of file structure and indicate the advantages and disadvantages of each for computer search.

3. Be able to identify the differences among network, hierarchical, and relational database structures and know the advantages and disadvantages of each.

4. Be familiar with the terminology of relational database management systems, including such terms as the primary key, tuple, relation, foreign key, relational join, and normal forms.

5. Describe the basic methods of representing graphic entities of a map using raster, vector, and quadtree systems, and understand their advantages and disadvantages.

6. Describe the three basic raster data models used for multiple coverages in a GIS, and understand their advantages or disadvantages.

7. Describe the methods of compact input and storage of raster data and know how they work.

8. Understand the difference between simple and topological vector data models and know their advantages and disadvantages.

9. Describe how the vector data model can be compacted to save computer space.

10. Describe how the TIN model works to store surface data.

11. Understand the basic differences between hybrid and integrated GIS systems, especially with regard to how the data are stored and accessed.

12. Have a basic understanding of how object-oriented GIS systems organize data and analytical techniques.

A QUICK REVIEW OF THE MAP AS AN ABSTRACTION OF SPACE

Before we move into another level of abstraction, one that allows the computer to operate on spatial data, it is best to review how we have moved from the real earth to more abstract views that we can manage intellectually. As we journey through our study areas, whether on foot or using remote sensing tools, we begin the process of abstraction by conceptualizing what we encounter as a group of points, lines, areas, and surfaces. This process, as you remember, is filtered by the questions we want to ask and how we intend to answer them. We make decisions about which objects to take note of and which to ignore. Then we decide on a method of data collection, whether it be a complete census or a sampling procedure. Some objects are assigned names; others are measured at higher levels of measurement (i.e., ordinal, interval, and ratio).

Once we have obtained our data, we proceed to make decisions about representing them in graphic form. We collate and group our data, decide on which projection we will need (if any), what grid system would be best, and so on. In some cases, especially in the absence of GIS, we produce a map directly from the data, only later entering these data into a cartographic database for use inside the GIS. More often today we do enter the data into the GIS directly, creating a geographic database based on direct observations.

The processes just summarized involve first looking over the environment, deciding what a necessary conception of that reality should be, and abstracting it further, either as a map or in computer-compatible form for direct entry into GIS (Figure 4.1). However, the computer forces us to modify how we envision our data. Computers do not think as we know thought, nor do they operate directly on visual or graphic objects as we would draw on a piece of drafting paper. Instead the computer must be addressed by means of a programming language such as FORTRAN, C, or Pascal.

When we develop our concept of space and spatial relationships, we are able to organize our data with a view to making sense of it. The map has for some time been the graphic language we have used to visualize space and its objects. But our graphic language has a distinctly different structure from what is available inside the computer. Take, as a simple example, the process of examining a map to identify the relationships between a lake inside an island covered by trees on its north side and cleared beach on the other. The verbal description

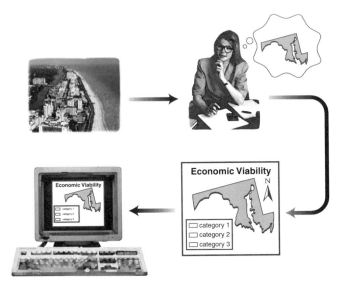

Figure 4.1 Transforming geographic space into a GIS database. Flow diagram showing change in abstraction level from real world to cartographer's conception to cartographic abstraction, and finally to computer abstraction.

immediately engenders a visual image, and the map is easily interpreted with little explicit instruction. We can clearly see that the lake is "inside" the island, that the island is "surrounded" by ocean, that trees occupy the north side of the lake, and that there is a beach on the south side. Again, no explicit instructions are needed for us to evaluate these facts. The computer, however, knows nothing about lakes or islands or forests or beaches or directions. We must create an explicit, rule-based language that allows the computer to use its digital (0 and 1) view of the world to identify the spatial extent of each object, to locate it in a system of coordinates, to separate adjacent objects from one another, and in fact to be able to identify and sort objects by orientation, size, location, and so on. It is not unlike trying to explain in excruciating detail how one manages the task of going shopping. Every detail must be included, from turning a doorknob to operating a vehicle, to navigating roads, locating the store, and deciding what to purchase. Such matters that require analog thought for their accomplishment are very simple to us, but very difficult for the computer.

Fortunately, we as students of GIS are not going to have to begin from scratch to explain all the detail of representing and operating on geographical objects in a spatial context. But, to be good at GIS we should be aware of some of the basic ways others have found to allow the computer to do so. We will begin at an elementary level of computer structures, but not at the machine level. Instead we will begin by examining traditional **computer file structures** that allow for the storing, ordering, and searching of pieces of data. Then we will move to a higher level of organization in the computer called **database structures**— composed of combinations of file structures to allow more complex methods of managing data. Then we will examine the way in which the geographic space can be represented explicitly into a form of **graphic data structure**, and finally we will extend this to include multiple graphic data layers and their databases into what we call GIS.

SOME BASIC COMPUTER FILE STRUCTURES

If you have had a basic computer course or a computer language course, among the first things you learned was simple computer file structures. Files are nothing more than a simple accounting system that allows the machine to keep track of the records of data you give it and retrieve these records in any order you wish. For GIS this is no different. Much of what we do in GIS consists of storing entity and attribute data in a way that permits us to retrieve (for display, for example) any combination of these objects. This requires the computer, using a representational file structure, to be able to store, locate, retrieve, and cross-reference records. In other words, each graphical feature must be stored explicitly, along with its attributes, so that we can select the correct combinations of entities and attributes in a reasonable amount of time. This is identical to having a cross-referenced list of names, addresses, and phone numbers. The list items will, of course, need to be sorted—for example, by alphabetically organizing the names.

Simple Lists

The simplest file structure is called a **simple list**. In our names and addresses example this is much like creating a separate index card for each name in a file like a Rolodex. Rather than organizing the names in any formal order, however, you place the cards in the Rolodex in the order in which they are entered (Figure 4.2). The only advantage to such a file structure is that to add a new

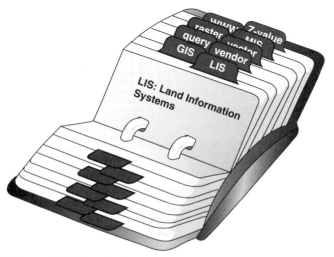

Figure 4.2 Simple list. File structure illustrated as an unordered Rolodex.

record, you simply place it behind all the rest. Clearly, all the cards are there, and an individual name can be located by examining the cards, but the lack of structure makes searching very inefficient. If you have created such a Rolodex you will likely wish you had done it some other way, especially if you have many names to keep track of. Suppose your database contains 200,000 records. If your basic file structure is a simple, unstructured, unordered list structure, you may have to search 200,000 cards to find what you are looking for. If it takes, for example, a second to perform each search, searching will require you to perform as many as $(n + 1)/2$ search operations (Burrough, 1983). This translates into a maximum of $(200,000 + 1)/2$ seconds, or nearly 28 hours to search for one point. Obviously you need to get organized.

Ordered Sequential Files

As you know, most Rolodexes, like telephone directories, are ordered according to the sequence of alphabetic characters (Figure 4.3). This method of ordering allows each record to be compared to each record before or after it to determine whether its alphabetic character is higher or lower in sequence. **Ordered sequential files**, as these are called, can use alphabetic characters, as in our Rolodex example, or numbers, that also occur in recognizable sequences against which individuals can be compared. The normal search strategy here is a sort of divide-and-conquer approach. A search is begun by dividing the file in half and looking first at the item in the middle. If it exactly matches the target combination of numbers or letters, the search is done. If not, the item of interest is compared to each of its neighbors to determine whether the alphanumeric combination is lower or higher. If it is lower, the half containing higher numbers or letters is searched in the same way. If it is higher, the half containing lower numbers or letters is searched by the divide-and-conquer method. In this way,

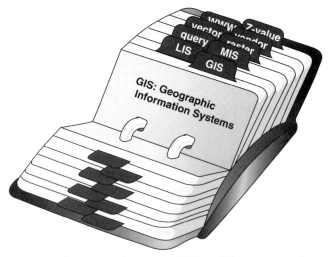

Figure 4.3 Ordered sequential lists. File structure illustrated as an ordered Rolodex. In this case the ordering is produced using alphabetical listings of names.

the program avoids having to search large portions of the file. The number of search operations using this strategy is defined as $\log_2(n + 1)$ operations. In the case of our graphic example we reduce the maximum amount of time to just over 2 hours rather than the nearly 28 hours that would have been needed to search the entire list if each operation took a second to perform. Of course, any computer that takes a full second to perform this operation is not likely to be very useful for GIS. But the proportional savings in time is made quite obvious by this example. The search speed does come at a cost, however, because now every new record must be entered in the correct sequence or your file will quickly resemble the simple list structure, defeating the binary or divide-and-conquer search strategy.

Indexed Files

In both the preceding examples the records were retrieved by examining and comparing a key attribute, say a number sequence or alphabetic sequence. The search strategy was based on the key attributes themselves. In geographic information systems, as in many other situations, the items you want to search are points, lines, and areas, primarily. However, you will seldom find yourself searching for a particular point, line, or area based on its coded numbers. In other words, you will not ask the GIS to display line number 3001 (based on how it was input to the system in the first place). Instead, each point, line, and area entity will often have assigned to it a number of descriptive attributes, just as we have seen before. Typically, a search will consist of finding the entities that match a selected set of attribute criteria. Thus you might ask the GIS to find all study plots in excellent condition for subsequent display or analysis. Or you might want to examine or analyze all study areas in poor condition that that have slopes less than 25%. Because of the possibly large numbers of attributes linked to each entity, a more efficient method of search will be necessary if we are to find specific entities with associated, cross-referenced attributes. Our search method otherwise will rapidly deteriorate into an exhaustive search of all attributes associated with all entities—the same tedious process employed with the simple list file structure (Burrough, 1983). In short, we need an index to our directory (Figure 4.4), much like the Yellow Pages you would use to find a particular type of store.

Indexed files can be developed as **direct files** (Figure 4.5a) or **inverted files** (Figure 4.5b). In direct indexed files the records themselves are used to provide access to other pertinent information. For example, if you search for four-lane highways, the computer will invoke explicit file information, perhaps a code, that tells the exact location of entities bearing the code for four-lane highways. The program search can now be directed to those specific locations or record numbers by creating an index that directly relates the codes for four-lane roads to their locations in the file, and roads that do not meet the four-lane criterion will be ignored.

Further improvements in search speed can be obtained if a formal index is created for a selected attribute to be searched. We could, for example, create an index of road type attributes (unimproved roads, single-lane paved roads, two-lane highways, four-lane highways, etc.) and have this index associated

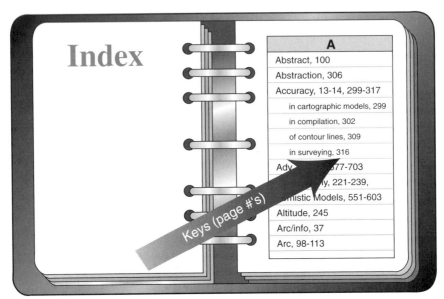

Figure 4.4   Indexed data structure. Files illustrated as an index to a book. The index shows how to find information in the larger file by selecting key features that can be searched for.

with specific entities and their computer locations, analogous to the setup of Figure 4.5c. Because the index is based on possible search criteria, rather than the entities themselves, the information is literally inverted, in that the attributes are the primary search criteria and the entities rely on them for selection. For this reason we call this an indexed inverted file structure.

To create an inverted file structure requires an initial sequential search of the data to order the entities based on the attributes of interest. This search, in turn, has three requirements. First, it requires you to know beforehand the criteria you are likely to be searching. Second, any additions of data will require you to recalculate the index as well as the original data. Finally, if you fail to indicate a particular criterion needed for your search—for example, if you forget to include accident rates as one of your search criteria—you will be required to use sequential search methods to obtain the information you want. This third condition of database design often causes problems in GIS work because sequential searches are both difficult and time-consuming. We will touch more on this subject in Chapter 15 concerning design.

COMPUTER DATABASE STRUCTURES FOR MANAGING DATA

As we have seen, there are three basic types of file structure for the storage, retrieval, and organization of data. However, we seldom store a single file; rather, we compile and work with multiple files. A collection of multiple files is called a **database**. The complexity of working with multiple files in a database requires a more elaborate structure for management, called a database structure or a

Item key	Index Record #	Items
A	1	A(1)
B	$n + 1$	A(2)
C	$(n + 1) + 1$	.
.	.	.
.	.	B(1)
.	.	.

(a)

Quadrat #	Properties # species	Slope (deg)	Aspect (deg)	% Bare	Condition
1	15	8	N-NW	14	Good
2	3	27	N	35	Poor
3	21	5	NE	<5	Excellent
4	11	10	S	20	Fair
5	6	18	SW	15	Poor
6	18	7	NW	10	Good

(b)

Properties	Quadrats
Poor	2, 5
Fair	4
Good	1, 6
Excellent	3

(c)

Figure 4.5 Direct and inverted files. Comparison of *(a)* direct indexed files and *(b)* inverted index files. Note the improvement made by ordering the file to select the index items in the form of an inverted file for the single category of property "condition" *(c)*.

database management system. Although new forms of database structure are being created all the time, there are three basic types with which you should be familiar in the early stage of your GIS education. These are **hierarchical data structures, network systems,** and **relational database structures.** Each will be examined separately.

Hierarchical Data Structures

In many cases there is a relationship among data called a one-to-many or parent–child relationship (Burrough, 1983). The parent-to-child relationship implies that each data element has a direct relationship to a number of symbolic children, and, of course, each "child" is also capable of having an association with his or her "offspring," and so on. As the name implies, the parents and children are directly linked, making access to data both simple and straightforward (Figure 4.6). This type of system is perhaps best exemplified by the hierarchical system of classifying plants and animals found in taxonomic literature.

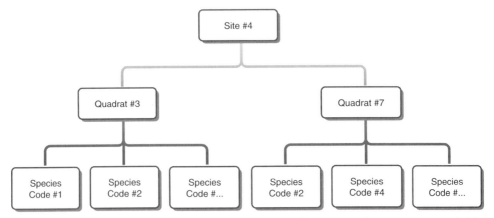

Figure 4.6   Hierarchical database structure. This illustration shows parent-to-child branching based on key attributes. *Source:* Modified from R.G. Healey, "Database Management Systems," Chapter 18 in *Geographical Information Systems, Principles and Applications*, D.J. Maguire, M.F. Goodchild, and D.W. Rhind (eds), Longman Scientific and Technical, Essex, England, © 1991, Figure 18.3, page 256. Used with permission.

Animals, for example, are separated into vertebrate and invertebrate forms—those with a backbone and those without. In turn, the vertebrates have a group called mammals, creatures that nurse their young. Mammals, of course, can be further broken down into other groups. The structure then begins to resemble a family tree, and in fact taxonomists use a nearly identical graphic structure to illustrate the relationships among organisms. A major feature of hierarchical systems exemplified by the taxonomic hierarchy is a direct correlation between one branch and another. The branches are based on formal criteria or key descriptors that act as a decision rule for moving from one branch to another through the structure.

As you might imagine, however, if your criterion or key descriptor information is incomplete, your ability to navigate through the net is severely hampered. In fact the nature of the hierarchical system requires that each relationship be explicitly defined before the structure and its decision rules are developed. A major advantage of such a system is that it is easy to search because the structure is so well defined, and it is relatively easy to expand by adding new branches and formulating new decision rules. If your initial description of the structure is incomplete, however, or if you want to examine the structure based on valid criteria that are not included in the structure, a search becomes impossible. A knowledge of all possible questions that might be asked is absolutely necessary to develop a hierarchical system because questions are used as the basis for developing the decision rules or keys. If the types of search are identified, it is probably safe to assume that the system has a well-established set of linkages based on long-term use under nonautomated conditions, or that the number of questions is relatively limited.

Most of you, for example, have probably used a modern computerized system of bibliographic search. This type of system was developed to simulate the ways in which people looked for books or articles before the use of computers. We might search based on subject, or author, or title, or even a range of catalog numbers placing us in a focused portion of the library holdings. Systems that include the ability to add multiple search criteria or Boolean operations exhibit both the necessities for easy hierarchical development: well-defined criteria and a limited number of search types. Imagine, however, that you know there was a great book on GIS located in the northwest corner of the fifth floor of the university's main library. You don't recall the catalog number, but you would like to find other books located near the one you remember seeing. Of course you would not consider asking the bibliographic system to locate all the books in the northwest corner of the fifth floor of the main library. The technique might have worked in the long run, but as a practical matter, you need to know more about the books in your target area.

Situations like the extremely vague search just described occur more often than not when you are working with information inside a GIS. Among the most difficult things to do is to anticipate all the possible searches you might perform. After all, the GIS database normally contains widely varying types of data as points, lines, or areas, and there are many different thematic maps in the database. One of the most enjoyable features of GIS is that you can try out searches or test out relationships you had not envisioned before you began. Unfortunately, the hierarchical structure is not very good at this because of its rigid key structure.

Burrough (1983) cites a classic example of how a system based on a hierarchical structure fails to comply with a user's request because the information needed is not even included in the system (for additional details, see Beyer, 1984). Not unlike our fifth-floor library search, the director of the Royal Botanical Gardens in Kew wanted to query the institution's botanical database system for all the available plants native to Mexico so that he could examine them before an upcoming field visit. This particular search criterion had not been anticipated, however, and the geographical locations of the plants in the system were not recorded; thus the program was incapable of selecting plants based on location. Clearly the rigid structure of the hierarchical system also would make it difficult to restructure to allow further searches of this nature. Beyond this severe limitation, the hierarchical structure creates large, often cumbersome index files, frequently requiring repeated entries of important keys to perform its searches. This adds substantially to the amount of memory needed to store the data and sometimes contributes to slow access times as well.

Network Systems

As we have seen, searches performed in hierarchical data structures are restricted to the branching network itself. In graphic databases, users frequently need to jump to different portions of the database to acquire entity information based on queries of attributes. Since attribute and entity data may very well be located in different locations, the creation of a hierarchical structure requires that direct links be made between search criteria and the graphic devices used to illustrate the locations in space. While this can be done, the potential numbers of hierarchical branchings and associated keys can get very unwieldy. Such awkwardness is experienced principally because the hierarchical data structure is most useful when there is a one-to-one or a many-to-one relationship among the variables.

Many GIS databases have, in addition to one-to-one and many-to-one relationships, many-to-many relationships, in which a single entity may have many attributes, and each attribute is linked explicitly to many entities. (An area representing a research quadrat for a study site, for example, will have numerous point locations, with multiple plant or animal species associated with each. In addition, each species might be found in more than one quadrat.) To accommodate these relationships, each piece of data can have associated with it an explicit computer structure called a pointer that directs it to all of the other pieces of data to which it relates (Figure 4.7). Rather than being restricted to a branching tree structure, each individual piece of data can be linked directly anywhere in the database, without the existence of a parent–child relationship. Pointers are common features in computer languages such as C and C^{++}, and a basic knowledge of these languages will aid in your understanding of exactly how the devices are used. For our purposes, a graphic visualization should suffice. Figure 4.7 illustrates 2 quadrats (#3 and #7) for study site #4. Notice how the pointers are used to relate individual point locations to their representative species identified for each. Pointers also look back from the species to the locations and finally back to the quadrats in which they are located.

Network systems are generally considered to be an improvement over hierarchical structures for GIS work because they are less rigid and can handle many-to-many relationships. As such, they allow much greater flexibility of search than do the hierarchical structures. Also, unlike the hierarchical structure, they reduce redundancy of data (e.g., coordinate pairs). Their major drawback is that in very complex GIS databases, the number of pointers can get quite large, coming to comprise a substantial portion of storage space. In addition, while linkages between data elements are more flexible, they must still be explicitly defined with the use of pointers. The numerous possible linkages may become an extremely tangled web, often resulting in confusion and missed and incorrect linkages. Novice database and GIS users often are overwhelmed by these conditions, although experienced users can become quite efficient with such systems and often prefer them over other types.

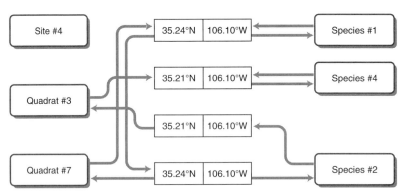

Figure 4.7 Network database structure. This structure allows users to move from data item to data item through a series of pointers. The pointers indicate relationships among data items. *Source:* Modified from R.G. Healey, "Database Management Systems," Chapter 18 in *Geographical Information Systems, Principles and Applications,* D.J. Maguire, M.F. Goodchild, and D.W. Rhind (eds), Longman Scientific and Technical, Essex, England, © 1991, Figure 18.4, page 257. Used with permission.

Relational Database Management Systems

The disadvantages of large numbers of pointers can be avoided by using another database structure: in relational database structures, the data are stored as ordered records or rows of attribute values called **tuples** (pronounced "tooples," to rhyme with "quadruples"). In turn, tuples are grouped with corresponding data rows in a form collectively called **relations** because they retain their respective row positions in each column and are, of course, related to one another (Healey, 1991) (Figure 4.8). Each column then represents the data for a single attribute for the entire dataset. For example, you could have a column of quadrat numbers (a single attribute) organized numerically. In a separate column you would have additional information pertaining to the collector, and yet another showing collection date, and finally a column showing the site number that corresponds to each of the other columns. In this way each item in each column corresponds and can then be related to additional tables as well.

Relational systems are based on a set of mathematical principles called relational algebra (Ullman, 1982) that provides a specific set of rules for the

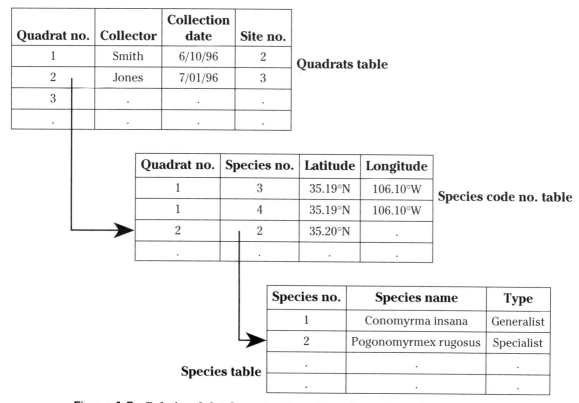

Figure 4.8   Relational database structure. Note the tuples, relations, and primary keys.   *Source:* Modified from R.G. Healey, "Database Management Systems," Chapter 18 in *Geographical Information Systems, Principles and Applications,* D.J. Maguire, M.F. Goodchild, and D.W. Rhind (eds), Longman Scientific and Technical, Essex, England, © 1991, Figure 18.5, page 258. Used with permission.

design and function of these systems. Because relational algebra relies on set theory, each table of relations operates as a set, and as you'll recall, the first rule is that a table can't have any row (tuple) that completely duplicates any other row of data. Because each of these rows must be unique, a single column, or even multiple columns can be used to define the search strategy. Thus, as an example of using a single column to decide your search strategy, you might search for a social security number, telephone number, home address, and so on available in other columns of the same table by selecting a particular name from the first column. This search criterion is called a **primary key** for searching the other columns in the database (Date, 1986). No primary key row can have missing values because missing row values could result in permitting duplicate rows to be stored, thus violating our first rule.

Relational systems are useful because they allow us to collect data in reasonably simple tables, keeping our organizational tasks equally simple. When we need to, we can match data from one table to corresponding (same-row) data in another table by the use of a linking mechanism called a **relational join**. Because of the predominance of the use of relational systems in GIS, and because of the rather large databases produced for GIS, this process is a common one and you should pay close attention to it. Any number of tables can be "related." The process is one of matching the primary key (column) in one table to another column in a second table. The column in the second table to which the primary key is linked is called a **foreign key**. Again, the related row values are assumed to be in the same positions, to ensure that they correspond. This link means that all the columns in the second table are now related to the columns in the first table. In this way, each table can be kept simple, making accounting easier. You can link a third table by finding a key column in the second table that acts as the primary key to a corresponding key column (now called the foreign key) in the third table. The process can continue, connecting each of the simple tables to allow for quite complex searches, while maintaining a very simple, well-defined, and easily developed set of tables. This approach eliminates the confusion found in database development using network systems.

To allow us to perform relational joins, each table must have at least one column in common with each other table we are trying to relate to. This redundancy is what allows the relational joins in the first place. Whenever possible, however, the amount of redundancy should be reduced. A set of rules called **normal forms** has been established (Codd, 1970), to indicate the forms your tables should take. We will discuss three basic normal forms; there have been some additions, but they are really more refinements than normal forms (Fagin, 1979).

The **first normal form** states that the table must contain columns and rows, and because the columns are going to be used as search keys, there should be only a single value in each row location. Imagine how difficult it would be to search for information by name if the name column had multiple values in each row location.

The **second normal form** requires that every column that is not a primary key be totally dependent on the primary key. This simplifies the tables and reduces redundancy by imposing the restriction that each column of data be findable only through its own primary key. If you wish to find a given column using other relationships, you can use the relational join rather than placing the column over and over again in separate tables to make sure it can be found.

The **third normal form,** which is related to the second normal form, states that columns that are not primary keys must "depend" on the primary key, whereas the primary key does not depend on any nonprimary key. In other words, you must use the primary key to find the other columns, but you don't need the other columns to search for values in the primary key column. Again, the idea is to reduce redundancy—to ensure that the smallest number of columns are produced.

The rules of the normal forms were summed up by Kent (1983), who indicated that each value of a table represents something important "about the primary key, the whole primary key, and nothing but the primary key." For the most part, these rules are highly useful and should be rigorously enforced. Having said that, of course, it is necessary to admit that there will always be circumstances when rigid enforcement will be impossible or will hamper system performance (Healey, 1991).

GRAPHIC REPRESENTATION OF ENTITIES AND ATTRIBUTES

Thus far we have concentrated on data structures that have little to do with the graphic representation of cartographic or geographic objects as we have envisioned and abstracted them. While all three of the systems mentioned could be used to manage graphics, they tell us little about how the graphics themselves will be represented in the GIS. We know that the human mind is capable of producing a graphic abstraction of space and objects in space. This representation is actually quite sophisticated, as you will see when we attempt to make the jump to computer handling of graphic devices. A primary difficulty appears because our graphic devices contain an implied set of relationships about the elements contained on the paper. Lines are connected to other lines and together are linked to create areas or polygons. The lines are related to each other in space through angles and distances. Some are connected while others are not. Some polygons have neighbors while others are isolated. The list of possible relationships that can be contained on a graphic diagram is virtually endless. We need to be able to find a way to represent each object and each relationship between objects as an explicit set of rules, to permit the computer to "recognize" that all these points, lines, and areas represent something on the earth, that they exist in an explicit place in space, and that those explicit locations are also related to other objects within space in an absolute as well as a relative sense. We may even want to explain to the computer that a polygon has an immediate neighbor to its left, and that neighbor may share points and lines. In other words, we need to create a language of spatial relationships.

There are two fundamental methods of indicating geographical space. The first method serves to **quantize** or divide space as a series of packets or units, each of which represents a limited, but defined, amount of the earth's surface. This **raster** method can define these units in any reasonable geometric shape, as long as the shapes can be interconnected to create a planar surface representing all the space in a single study area. Although a wide variety of raster shapes are possible—for example, triangles or hexagons—it is generally simpler to use a series of rectangles or, more often, squares called **grid cells**. Grid cells or other raster forms generally are uniform in size, but this is not absolutely necessary, as we will see when we briefly review a less often used system called **quadtrees**. For the sake of simplicity, we will assume that all grid cells are the same size, and therefore each occupies the same amount of geographic space as any other. We will further restrict our discussion to square grid cells, again for simplicity.

Raster data structures do not provide precise locational information because geographic space is now divided into discrete grids, much as we divide a checkerboard into uniform squares. Instead of representing points with their absolute locations, they are represented as a single grid cell (Figure 4.9). The assumption is that somewhere inside that grid cell, a point object can be found. This is an additional form of change of dimensionality abstraction that requires us to illustrate a zero-dimensional object with a two-dimensional data structure. Likewise, lines, or one-dimensional objects, are represented as a series of connected grid cells. Again, we are changing our dimensionality from one-dimensional objects to two-dimensional data structures. Each point along the line is represented by a grid cell, meaning that any point along the line must occur somewhere within one of the displayed grid cells. It is easy to see that this form of data structure produces a rather stepped appearance when it is used to represent very irregular lines (Figure 4.9). This stepped appearance is also obvious when we represent areas with grid cells (Figure 4.9). All points inside the area that is bounded by a close set of lines must occur within one of the grid cells to be represented as part of the same area. The more irregular the area, the more stepped the appearance.

In grid-based or raster systems there are two general ways of including attribute data for each entity (object). The simplest is to assign a single number,

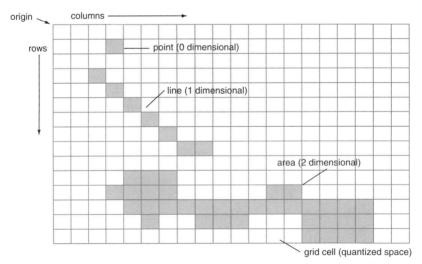

Figure 4.9   Basic raster graphic data representation. The structure illustrates points, lines, and areas as quantized units of geographical area. Grid structures do not allow precise locational information.

representing an attribute (e.g., a class of land cover), for each grid cell location. By positioning these numbers, we ultimately are allowing the position of the attribute value to act as the default location for the entity. For example, if we assign a code number of 10 to represent water, then list this as the first number in the X or column direction and the first in the Y or row direction, by default the upper left grid cell is the location of a portion of the earth representing water. In this way, each grid cell can hold only a single attribute value for a given map. An alternative approach, actually an extension of the one just discussed, is to link the grid cells to a database management system, with the result that more than one attribute can be represented by a single grid cell. This approach is becoming more prevalent because it reduces the amount of data that must be stored and because it can easily be linked to other data structures that also rely on database management systems to store, search, and manipulate data.

Although absolute location is not explicitly part of the raster data structure, it is implied by the relative locations of the grid cells. Thus, a line is represented by grids at particular locations relative to one another, and areas are represented by the grid cells that form as a result of being adjacent to one another or connecting at points of one another. As you may have surmised, the larger the grid cell, the more land area is contained within it—a concept called **resolution**. The more land area contained—that is, the coarser the resolution of the grid—the less we know about the absolute position of points, lines, and areas represented by this structure.

Raster structures, especially square grid cells, are pieced together to represent an entire area. To do this we must assume that the area is more or less flat. In other words, the normal coordinate system we are using is a Cartesian coordinate system (see Chapter 3), and the grid cells themselves approximate the locations in space. As we have seen, Cartesian coordinate systems employ a map projection to permit an approximation of the three-dimensional shape of a portion of the earth. The grid system representation may have embedded in it a coordinate system that better approximates absolute location. For example, the pixels from satellite imagery (see Chapter 3) have an associated projection, and a more exact grid can be placed on them for reference. In general, however, exact measurements based on any raster structure are difficult. When exact measures are needed, therefore, raster structures are less often used than other types.

Raster data structures may seem to be rather undesirable because of the lack of absolute locational information. However, quite the contrary is true. Raster data structures have numerous advantages over other structures. Notably, they are relatively easy to conceptualize as a method of representing space. Some of you may have seen the character printouts of cartoon characters that were so prevalent on computer-produced calendars in the 1960s and 1970s. If not, you are no doubt familiar with television, a technology that is quite analogous to the raster representation of earth features in a GIS. Few of us have difficulties identifying actors or other images on the TV screen, even though everything is represented by a series of dots or pixels. In fact, the relationship between the pixel used in much remote sensing and the grid cell used in GIS allows data from satellites to be readily incorporated into raster-based GIS without any changes. This is another advantage of the raster data structures over the alternatives. Another nice feature of grid-based systems, as you will see in more detail in Chapter 12, is that many functions, especially those involving surfaces and overlay operations, are simple to perform with this type of data structure.

Among the major disadvantages of the raster data structure are the already mentioned problem of reduced spatial accuracy, which decreases the reliability of area and distance measures, and the need for large storage capacity, associated with having to record every grid cell as a numerical value. The latter problem is less serious than it once was because of vast improvements in computer hardware storage capabilities. In addition, as you will see later in this chapter, there are methods of reducing the storage needed by compacting groups of grid cells into more integrated forms. Although storage is no longer a major limitation to the use of raster structures, however, even the fastest computers can be slowed to a crawl if highly complex calculations are performed on very large raster databases.

The second method of indicating geographical space, called **vector**, allows us to give specific spatial locations explicitly. Here it is assumed that geographic space is continuous, rather than being quantized as smaller discrete grids. This perspective is acquired by associating points as a single set of coordinates (X and Y) in coordinate space, lines as connected sequences of coordinate pairs, and areas as sequences of interconnected lines whose first and last coordinate points are the same (Figure 4.10). Anything that has a single (X, Y) coordinate pair not physically connected to any other coordinate pair is a point (zero-dimensional) entity.

The vector data structure is much more representative of dimensionality as it would appear on a map. To give it the utility of a map, however, we combine the entity data with associated attribute data kept in a separate file, perhaps in a database management system, then link them together. This extends the structure beyond a simple graphic caricature of the objects, making it more maplike, and more representative of the earth surface we are modeling. Whereas in the raster structure we explicitly stored the attribute and implied its location based on its position, in vector representation we take a quite different approach, explicitly storing the entities without their attributes, and relying on the link to the separate attribute database.

In vector data structures, a line consists of two or more coordinate pairs, again storing the attributes for that line in a separate file. For straight lines, two coordinate pairs are enough to show location and orientation in space. More complex lines will require a number of line segments, each beginning and ending with a coordinate pair. For complex lines, the number of line segments must be increased to accommodate the many changes in angles. The shorter the line segments, the more exactly they will represent the complex line. Thus we see that although vector data structures are more representative of the locations

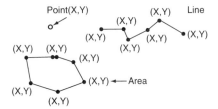

Figure 4.10   Basic vector graphic data representation. The structure shows points as individual coordinate pairs, lines as groups of two coordinate pairs, and areas as connected lines having identical beginning and ending coordinates.

of objects in space, they are not exact. They are still an abstraction of geographic space.

While some lines act alone and contain specific attribute information that describes their character, other, more complex collections of lines, called **networks,** add a dimension of attribute characters. Thus a road network not only contains information about the type of road or similar variables, it will indicate also, for example, that travel is possible only in a particular direction. This information must be extended to each connecting line segment to advise the user that movement can continue along each segment until the attributes change, perhaps when a one-way street becomes a two-way street. Other codes linking these line segments might include information about the nodes that connect them. For example, one node might indicate the existence of a stop sign, a traffic signal, or a sign prohibiting U-turns. All these attributes must be connected throughout the network so that the computer knows the inherent real-world relationships that are being modeled within the network. Such explicit information about connectivity and relative spatial relationships is called **topology**, a topic we will return to when we look at the vector data models we can produce from the basic vector data structure.

Like line entities, area entities can be produced in the vector data structure. By connecting pairs of coordinates into lines and organizing the lines into a looping form, where the first coordinate pair on the first line segment is the same as the last coordinate pair on the last line segment, we create an area or **polygon**. As with point and line entities, the polygon will also have associated with it a separate file that contains data about the attributes or characteristics of the polygon. Again, this convention improves the simple graphic illustration of area entities, making it possible for them to better represent the abstraction of area patterns we observe on the earth's surface.

Generally speaking, cartographers tend to prefer vector data structures because of their similarity to the graphic structures most commonly associated with analog maps. With some limitations, output from vector data structures resembles quite closely the hand-drawn map. But simple map output is not the major focus of GIS; rather, it is the ability to measure and analyze cartographically organized data. As we have already seen, even when these graphic data structures are physically viewed, there is a need to find a way to combine entities with their descriptive attributes. We have alluded to the use of external files or database systems in this connection. The data structures must be developed to allow such links either directly or indirectly. In addition, there are many other facets of graphic structure as an analyzable maplike system. We need to move from simple data structures to what are often termed data models, which will act more like the map in terms of capabilities to be analyzed. Some generic types will be illustrated for both raster and vector data structures.

GIS DATA MODELS FOR MULTIPLE COVERAGES

While simple raster and vector data structures provide us with a means of depicting spatial phenomena in a single map environment, there is still a need to develop more complex versions, called data models, to accommodate the necessary interactions of objects in the database, to link entities and attributes, and to allow multiple map coverages to be analyzed in combination. We will look at raster data models first and then proceed to vector data models to illustrate some basic methods that can be used to allow us to perform the multiple analyses we would have performed without automation. Then we will go one step further to illustrate some of the ways of combining these data models into systems—in this case, geographic information systems.

Raster Models

As indicated in our initial discussion of raster data structures, the simplest approach is to use grid cells to represent quantized portions of the earth. Each grid cell is represented by one and only one attribute value, again in its simplest form. To create a thematic map of grid cells, we collect data about a particular topic (theme) in the form of a two-dimensional array of grid cells, where each grid cell represents an attribute of the individual theme. This two-dimensional array (essentially a 2×2 matrix, as used in matrix algebra) is called a **coverage.** We can use coverages to separate individual types of subject matter (land use, vegetative cover, soil type, surface geology, hydrology, etc.) so that their respective attributes are easily identified. In addition, this approach forces us to focus our attention on the objects, patterns, and interrelationships of each theme without unnecessary confusion. Because we will most often be concerned about the relationships of one theme, such as soil type, with another theme, say vegetation, we create a separate coverage for each additional theme. We can then stack these maps into a form of three-dimensional structure, where the combinations of all the themes might adequately model all the important attributes for an entire area, based on what we have selected as our geographic filter or study framework. As such, if we are interested only in physical phenomena, each of the important components of the physical geography will be represented separately, but together they will give us a more complete, three-dimensional view of the study area.

There are a number of ways of forcing a computer to store and reference the individual grid cell values, their attributes, coverage names, and legends. Among the first attempts at conceptualizing this could be called the **GRID/ LUNR/MAGI** approach (Burrough, 1983) (Figure 4.11a) because each of these early raster GIS systems used this method. In this model each grid cell is referenced or addressed individually and is associated with identically positioned grid cells in all other coverages, rather like a vertical column of grid cells, each dealing with a separate theme. Comparisons between coverages are therefore performed on a single column at a time. For example, to compare soil attributes in one coverage with vegetation attributes in a second coverage, each X and Y location must be examined individually. So a soil grid cell at location $X10$–$Y10$ will be compared to its vegetation counterpart at location $X10$–$Y10$. You might be able to envision this by imagining a geological core in which each rock type is lying directly on top of the next, and you know that to get a picture of the entire study area, it will be necessary to put a large number of cores together. The advantage, of course, is that computational comparison of multiple themes or coverages for each grid cell location is relatively easy. This is a reasonable approach and has proven successful, but it limits the

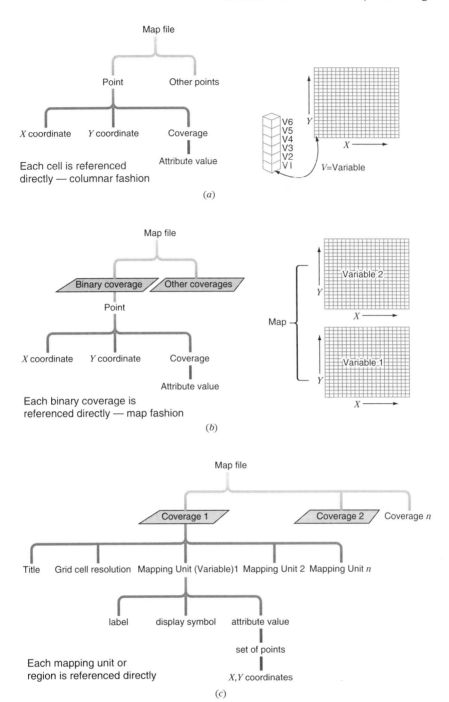

Figure 4.11   Three raster data models for managing multiple sets of grid coverages. *(a)* The GRID/LUNR/MAGI model. *(b)* The IMGRID GIS data model. *(c)* The Map Analysis Package (MAP) GIS data model.

efficient examination of relationships of themes to one-to-one relationships within the spatial framework. In other words, it is awkward to compare groups in one coverage to groups in another coverage because each grid cell location must be addressed individually. Moreover, the approach is counterintuitive in that the representation is vertical, rather than horizontal, which would more closely resemble our notion of maps.

Think carefully about how you might want to alter this approach. Perhaps you could think of a checkerboard, with its red and black squares. If each of these squares is taken to represent a simple map of land cover (e.g., red = water and black = land), we have produced a simple coverage. But how are the attributes of our land cover coverage physically connected? We can pick up the entire checkerboard because it is a physically connected structure. Likewise, when we pick up a thematic map, it too represents all the different changes in the theme as a single, connected object. The similarity between the checkerboard as a single unit of play for a game and the map as a single unit of storage for spatial information is a natural one.

In fact, with a slight modification of the checkerboard analogy we can examine the second basic raster data model, which we will call the **IMGRID** data model (Figure 4.11*b*) because of its use in that early GIS system (Burrough, 1983). Here, we will assume that the red squares on our checkerboard are water and the black squares are simply "not water." We have simplified the theme of our checkerboard map to contain a single attribute, rather than just a theme. In this way we do not need to store a wide range of values for each coverage. Instead we can use the number 1 (red squares) to represent water and 0 (black squares) to indicate the absence of water. How would we be able to represent a thematic map of land use that contained, say, four categories: recreation, agriculture, industry, and residences? Each of these four attributes would have to be separated out as an individual layer. One layer would stand for agriculture only, with 1s and 0s representing the presence or absence of this activity for each grid cell. Recreation, industry, and residences would be represented the same way, with each variable referenced directly, rather than referencing the grid cell as we did in the GRID/LUNR/MAGI data model. Finally, the coverages would be combined vertically, or in column fashion, to produce a single theme or coverage, much as red, yellow, green, and blue printing plates are combined to create a single color image.

The major advantages of the IMGRID system are twofold. First, we have a contiguous object that more closely resembles how we think about a map. That is, our primary storage object is a two-dimensional array of numbers, rather than a column of numbers, for different themes. Second, we reduce the numbers that must be contained in each coverage to 0s and 1s. This will certainly simplify our computations and will eliminate the need for complex map legends, as well. In fact, since each variable is uniquely identified, we will not be limited to assigning a single attribute value to a single grid cell, and this is a third advantage. For example, in a given grid cell we may have some agriculture and some recreation. Because each of these attributes of the land use theme is separated out, we can show that both land uses occur somewhere within the space contained in that grid cell. Of course we may encounter difficulties when creating our final thematic coverage if multiple values occur in individual cells. To avoid such problems, we must be able to ensure that each grid cell has only a single value for each variable.

The IMGRID model is seen to be more intuitive from a map abstraction viewpoint, and it forces us to be very specific about the attributes to be contained in each coverage. Moreover, it offers the advantage of using the coverage as the direct object of reference for the computer. Its limitations stem primarily from the problem of data explosion. Imagine for a moment that you have a database that will be composed of 50 themes (not an unusually large number for GIS work). Each theme must be separated out into binary (0 and 1) coverages based on individual attributes within each theme.

Suppose that there is an average of 10 categories for each theme. To represent this rather modest database, you will need a total of 10 × 50 or 500 coverages. While available storage devices can certainly manage such volumes, you need to look for a more efficient way to represent your database, one that doesn't give you so many maps to manage and keep track of. Examining this approach further, imagine how many values must be modified and recoded to create a new theme. For example, to combine 10 binary coverages to create a new thematic coverage with 10 categories, you would have to separate the thematic coverage into each of 10 new binary coverages. Thus for a simple operation you had to combine 10 grid cell values, and to create additional thematic coverage it's been necessary to produce 10 new values of 0 and 1 for each variable. This is a rather tedious approach.

Our third raster model, which we will call the **MAP** model after the system developed by Dana Tomlin (Burrough, 1983) for his doctoral dissertation, formally integrates the advantages of the two raster data structure methods we've examined. In this data model (Figure 4.11c), each thematic coverage is recorded and accessed separately by map name or title. This is accomplished by recording each variable, or mapping unit, of the coverage's theme as a separate number code or label, which can be accessed individually when the coverage is retrieved. The label corresponds to a portion of the legend and has its own symbol assigned to it. In this way, it is easy to perform operations on individual grid cells and groups of similar grid cells, and the resulting changes in value require rewriting only a single number per mapping unit, thus simplifying the computations. The overall major improvement is that the MAP method allows ready manipulation of the data in a many-to-one relationship of the attribute values and the sets of grid cells.

The MAP data model is among the most copied raster GIS models in the marketplace. It is found in many forms, from its original, mainframe version to Macintosh and PC versions and modern UNIX-based workstation versions. Its flexibility and ease of use have made it very easy to learn as a teaching version of GIS; it can perform as a major module in commercial GIS packages as well, and even as a complete and operational raster GIS package.

While raster GIS systems have traditionally been developed to allow single attributes to be stored individually for each grid cell, some have evolved to include direct links to existing database management systems. This approach extends the utility of the raster GIS by minimizing the number of coverages and substituting multiple variables for each grid cell in each coverage. Such extensions to the raster data model have also allowed direct linkage to existing GIS systems that use a vector graphic data structure. Because such integrated raster/vector systems include modules that convert back and forth from raster to vector, the user is able to operate with all the advantages of both data structures. The conversion process is often quite transparent, allowing the user to perform the analyses needed without concern for the original data structure. This feature is particularly important because it is strengthening the relationship between traditional digital image processing software, used to manipulate grid-cell-based, remotely sensed data and GIS software. Many software systems already have both sets of capabilities, and still more are likely in the future. Together with the linkage with existing statistical packages, we are rapidly approaching systems that operate with a superset of spatial analytical techniques, resulting in a maturing of automated geography.

Compact Storing of Raster Data

Before we leave our discussion of raster data models, we shall look at four common methods of storing raster data that result in substantial savings in disk space. These require a data model similar to the MAP data model that allows groups of grid cells to be accessed directly, because the compactions generally operate by reducing the information content of these groups of cells to the absolute minimum needed to represent them as a unit. Compact methods for storing raster data certainly operate under the storage and editing subsystem of a GIS, but they can also be applied directly during the input phase of the GIS operation. We will revisit these methods in the next chapter, when we discuss data input. The approaches discussed in this section are illustrated in Figure 4.12.

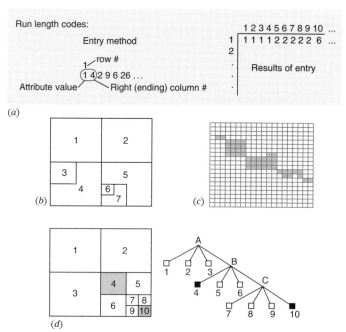

Figure 4.12 Compact data models. Methods of compacting raster data to preserve storage: *(a)* run-length codes, *(b)* raster chain codes, *(c)* block codes, and *(d)* the rather unique structure called quadtrees.

The first method of compacting raster data is a process called **run-length codes**. In the old-fashioned method, raster data were input into a GIS by preparing a clear acetate grid, referencing it, and overlaying it on the map to be encoded. Each grid cell has a numerical value corresponding to a category of data on the map that must be input (generally typed) into the computer. For a map of perhaps 200 × 200 grid cells, someone would have to type 40,000 numbers into the computer. If your instructor hears you giggling at this point, don't be surprised if you find yourself doing it, as an exercise in historical GIS, or perhaps a lesson in humility In fact, you might try it anyway, if you have access to a raster-based GIS. As you begin typing, you will quickly see patterns emerging from the data that present opportunities for reducing your workload. Specifically, there are long strings of the same number in each row. Think how much time you could save if for a given row, you could just tell the computer that starting at column 8 all the numbers are 1s, representing some map variable, until you get to column 56, then at column 57 the numbers are 2s until the end of the row. Indeed, you could also save a great deal of space by simply giving starting and ending points for each string and the value that should be stored for that string. That's run-length codes.

Of course, this method limits you to operation on a row-by-row basis. What if you could tell the computer to begin at a single grid cell with value 1, then go in a particular direction, say vertically, 27 grid cells, and then change to a different grid cell value. This would allow you to code strings in any direction. But the principle can be extended even further. Suppose you see large groups of grid cells that represent an area. If you started at one corner, giving its starting position and grid cell value, then moved in cardinal directions along the area, storing a number representing the direction and another indicating the number of grid cells moved, you could then store whole areas using very few numbers. In this way you would save even more space, and, of course, typing time. This method relies on what are called **raster chain codes,** which literally run a chain of grid cells around the border of each area. In other words, you assign origins, based on *X* and *Y* position, a grid cell value for the entire area, and then directional vectors, showing where to move next, where to turn, and how far to go. Usually the vectors include nothing more than the number of grid cells and the vector direction based on a simple coding scheme, where 0,1,2, and 3 could indicate north, south, east, and west, respectively.

There are two remaining approaches to reducing the storage necessary for grid-based systems, both relying on a square collection of grid cells as the primary unit of storage. The first, called **block codes**, is a modification of run-length codes. Instead of giving starting and ending points, plus a grid cell code, we select square groups of cells and assign a starting point, say the center or a corner, pick a grid cell value, and tell the computer how wide the square of grid cells is, based on the number of cells. As you can see, this is really a two-dimensional run-length code. Each square group of grid cells, including individual grid cells, can be stored in this way with a minimum amount of numbers. Of course, if your coverage has very few large square groups of cells, this method is not a major improvement for storage. But then, run-length codes also become somewhat cumbersome if there are few long runs or strings of the same value. Most thematic maps, however, have fairly large numbers of these groups, and black code methods are very effective methods of reducing storage.

Quadtrees, our final method of compact storage, is a somewhat more difficult approach, and your instructor may elect not to cover it. Still, at least one commercial system called SPANS, from Tydac, and one experimental system called Quilt (Shaffer, Samet, and Nelson, 1987) are based on this scheme. Like block codes, quadtrees operates on square groups of cells; in this case the entire map is successively divided into uniform square groups of grid cells with the same attribute value. Starting with the entire map as entry point, the map is then divided into four quadrants (NW, NE, SW, and SE). If any of these quadrants is homogeneous (i.e., contains grid cells with the same value), that quadrant is stored and no further subdivision is necessary. Each remaining quadrant is further divided into four quadrants, again NW, NE, SW, and SE. Again, each quadrant is examined for homogeneity. All homogeneous quadrants are again stored, and each of the remaining quadrants is further divided and tested in the same way until the entire map is stored as square groups of cells, each with the same attribute value. In the quadtree structure, the smallest unit of representation is a single grid cell (Burrough, 1983).

Systems based on quadtrees are called variable resolution systems because they can operate at any level of quadtree subdivision. If you do not need high resolution for your computations—in other words, if detail is not essential—you can use a rather coarse level of resolution (or quadtree subdivision) in your analysis. Thus users can decide how fine the resolution needs to be for various manipulations. In addition, because of the compactness of storage from this method, very large databases, perhaps continental or even global scales, can be stored in a single system.

The major difficulty with the quadtree structure is in the method by which it separates the grid cells into regions. In block codes, the decision was based entirely on the existence of square groups of homogeneous grid cells, regardless of where they were located on the map. With quadtrees, the subdivision is preset to the four quadrants (NW, NE, SW, SE), resulting in some otherwise homogeneous regions lying in two or more different quadrants. This results in computational difficulties for analysis of shape and pattern that must be overcome through rather complex computational methods that are beyond the scope of this book. GIS software using the quadtree data model operates under workstation and PC platforms and uses multiple operating systems. Such programs are in use worldwide and offer some interesting opportunities, especially for those needing to use very large databases.

Vector Models

As we have already seen, vector data structures allow the representation of geographic space in a way that is somewhat more intuitive and certainly more reminiscent of the familiar analog map. You also remember that they represent the spatial location of items explicitly, most often storing the attributes in another file for later access. Since the relationships between individual entities are implicit rather than explicit, the space intervening between graphic entities does not have to be stored. There are several ways in which vector data structures can be put together into a vector data model, enabling us to examine the relationships between variables in a single coverage or among variables in different coverages. We will look at a range of these using three basic types as examples: spaghetti models, topological models, and vector chain codes. Although there are others, and many variations on each type, these should suffice to give you an overview of what is available for vector GIS systems.

The simplest vector data structure, called the **spaghetti model** (Dangermond, 1982) (Figure 4.13), is essentially a one-for-one translation of the graphic image we would normally encounter on a map. It is probably the one most of us envision as the most natural or most logical, primarily because the map is maintained as the conceptual model. Although the name may seem odd, it is actually quite representational. If you imagine covering each graphic object on our analog map with a piece of spaghetti (al dente), you will have a pretty good idea of how the model works. Each piece of spaghetti acts as a single entity— very short ones for points, longer ones for straight-line segments, and collections of line segments that come together at the beginnings and endings of surrounding areas. Each entity is a single, logical record in the computer, coded as variable length strings of (*X, Y*) coordinate pairs.

If we imagine a large collection of pieces of spaghetti, each straight piece has a beginning and an ending set of coordinates. Any polygons that lie adjacent to each other must have separate pieces of spaghetti for adjacent sides. That is, no two adjacent polygons share the same string of spaghetti. Instead, each side of each polygon is uniquely defined by its own set of lines and coordinate pairs. Of course, adjacent sides of polygons, even though they are recorded separately in the computer, should have the same coordinates.

Because the data model is conceptualized as a one-for-one translation of the analog map, the relationships (topology) between and among objects—for example, the locations of adjacent polygons—are implied, rather than being explicitly coded in the computer. In addition, all relationships among all objects must be calculated independently. A result of this lack of explicit topology is enormous computational overhead, making measurements and analysis difficult. Because it so closely resembles the analog map, however, the spaghetti

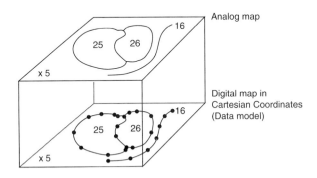

Data Structure

Feature	Number	Location
Point	5	x,y (single pair)
Line	16	(string of x,y coordinate pairs)
Polygon	25	(closed loop of x,y coordinate pairs where first and last pair are the same)
	26	(closed loop sharing coordinates with adjacent polygons to form a data structure)

Figure 4.13 Spaghetti vector data model. There is no explicit topological information, but the model is a direct translation of the graphic image. *Source:* Figure derived from Environmental Systems Research Institute, Inc. (ESRI) drawings and tabular data.

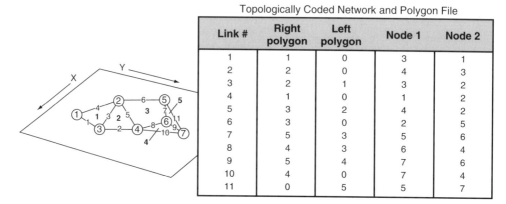

Topologically Coded Network and Polygon File

Link #	Right polygon	Left polygon	Node 1	Node 2
1	1	0	3	1
2	2	0	4	3
3	2	1	3	2
4	1	0	1	2
5	3	2	4	2
6	3	0	2	5
7	5	3	5	6
8	4	3	6	4
9	5	4	7	6
10	4	0	7	4
11	0	5	5	7

X,Y Coordinate Node File

Link #	X Coordinate	Y Coordinate
1	19	6
2	15	15
3	27	13
4	24	19
5	6	24
6	20	28
7	22	36

Figure 4.14 Topological vector data model. Note the inclusion of explicit information concerning connected points, lines, and polygons. *Source:* Figure derived from Environmental Systems Research Institute, Inc. (ESRI) drawings and tabular data.

model is relatively efficient as a method of cartographic display and is still used quite often in computer assisted cartography when analysis is not the primary objective. The representation is quite similar, as well, to that found in many plotting devices, making the translation of the spaghetti model to the plotter language easy and efficient. Plotting of spaghetti data models is usually quite fast compared to some others.

Whereas absence of topology prevents the spaghetti structure from offering ease of analysis, the next vector data model includes it. As you might have guessed, we call models of this type **topological models** (Dangermond, 1982) (Figure 4.14). To allow advanced analytical techniques to be performed easily, we want to provide the computer with as much explicit spatial information as possible. Just as a numeric coprocessor in a personal computer incorporates a great many precalculated mathematical operations, the topological data model incorporates solutions to some of the more often used operations used in advanced GIS analytical techniques. This is done by explicitly recording adjacency information into the data structure, to eliminate the need to determine it for multiple operations. Each line segment, the basic logical entity in topological data structures, begins and ends when it either contacts or intersects another line, or when there is a change in direction of the line. Each line then has two sets of numbers: a pair of coordinates and an associated node number. The **node** is more than just a point; it is the intersection of two or more lines, and its number is used to refer to any line to which it is connected. In addition, each line segment, called a link, has its own identification number that is used as a pointer to indicate which set of nodes represent its beginning and ending. Polygons, composed of these links, also have identification codes that relate back to the link numbers. Each link in the polygon now is capable of looking left and right at the polygon numbers to see which two polygons are adjacent to each other along its length. In fact, the "left and right polygons" are also stored explicitly, so that even this tedious step is eliminated. This design feature allows the computer to know the actual relationships among all its graphic parts. In other words, we have a vector data model that more closely approximates how we as map readers identify the spatial relationships contained in an analog map document.

A number of topological data models have been developed and are in common use. All are slightly different, and we shall look at some of the more common ones to see how the implementations of the data model might be performed. Perhaps the best-known topological data model is the **GBF/DIME** (geographic base file/dual independent map encoding) model created by the U.S. Bureau of the Census to automate the storage of street map data for the decennial census (U.S. Department of Commerce, Bureau of the Census, 1969) (Figure 4.15a). In this case the straight-line segments represent streets, rivers, railroad lines, and so on (Peuquet, 1984). In this topological data structure, each segment ends when it either changes direction or intersects another line, and the nodes are identified with codes. In addition to the basic topological model, the GBF/DIME model assigns a directional code in the form of a From node and a To node (i.e., low value node to higher value node in a sequence). This approach makes it easy to check for missing nodes (see Chapter 6) during the editing process. If, for example, you want to see whether a polygon is missing any links, simply match the To node of one line to the From node of the preceding link. If the nodes do not completely surround an area, a node is missing (Peuquet, 1984).

As an additional useful feature of the GBF/DIME system, both the street addresses and UTM coordinates for each link are explicitly defined, permitting street addresses to be accessed by geographic coordinates. However, this data model suffers the same basic problem that dogs the generic topological model and of course the spaghetti model as well. Since there is no particular order in which the line segments occur in the system, to search for a particular line segment, the program must perform a tedious sequential search of the entire database. As you remember, this is the slowest possible way to search for records in a computer. The GBF/DIME system, moreover, is based on the idea of graph theory: it doesn't matter whether the line connecting any two points is curved or straight. Thus a side of a polygon serving to indicate a curved river boundary would not be stored as a curved line, but rather as a straight line between two points, and the resulting model would lack the geographic specificity we expect of an analog map.

Some of the problems of the GBF/DIME system have been eliminated with the development of another system, or the topologically integrated geographic encoding and referencing system, or **TIGER**) (Marx, 1986), designed for use with the 1990 U.S. census (Figure 4.15b). In this system, points, lines, and areas can be explicitly addressed, and therefore census blocks can be retrieved directly by block number rather than by relying on the adjacency information contained in the links. In addition, since the model does not rely on graph theory, real-world features such as meandering streams and irregular coastlines are given a graphic portrayal more representative of their true geographic

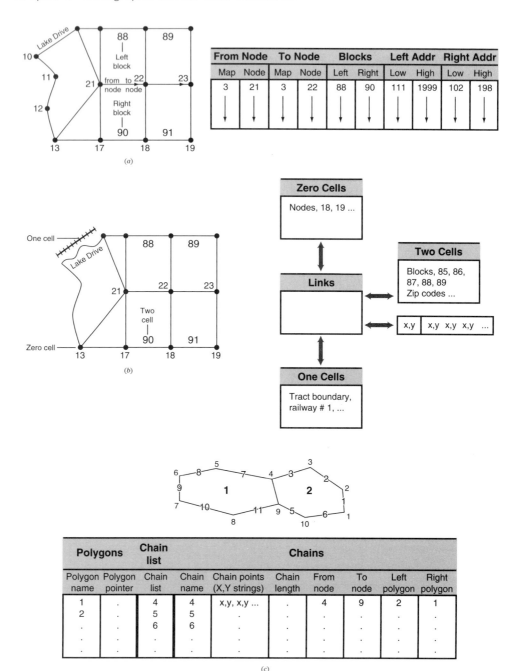

Figure 4.15 DIME files and TIGER files. Example of topological vector data models: *(a)* the GBF/DIME model, *(b)* the TIGER model, and *(c)* the POLYVRT model. *Source:* Parts *(a)* and *(b)* modified from K.C. Clark, *Analytical and Computer Cartography,* © 1990, Prentitice Hall, Inc., Englewood Cliffs, NJ. Used with permission.

shape (Clarke, 1990). Thus TIGER files are more generally useful for non-census-related research.

Another data model, developed by Peucker and Chrisman (1975) and later implemented at the Harvard Laboratory for Computer Graphics (Peuquet, 1984), was called the **POLYVRT** (POLYgon conVeRTer) model (Figure 4.15c). As with the TIGER system, it eliminates storage and search inefficiencies of the basic topological model by separately storing each type of entity (points, lines, polygons). These separate objects are then linked in a hierarchical data structure with points relating to lines, which in turn are related to polygons, all through the use of pointers. Each collection of line segments, collectively called **chains** in this model, begins and ends with specific nodes (intersections between two chains). And as in the GBF/DIME system, each chain contains explicit directional information in the form of To–From nodes as well as left–right polygons (Figure 4.15*c*).

Like TIGER, POLYVRT has the advantage of allowing selective retrieval of specific entity types: you can select points or lines or polygons at will by identifying them based on their codes (which of course are connected to records of their attributes). An additional advantage of POLYVRT is that because lines bounding polygons are explicitly recorded as chains of individual line segments, the individual line segments do not have to be accessed to find the beginning and ending of a particular polygon. Instead, the chains can be accessed directly, saving time for searches (Peuquet, 1984). Because in POLYVRT chain lists bounding polygons are explicitly stored and linked through pointers to each polygon code, the size of the database is largely controlled by the number of polygons, rather than by the complexity of the polygon shapes. This makes storage and retrieval operations more efficient, especially when highly complex polygonal shapes found in many natural features are encountered (Peuquet, 1984). The major drawback of POLYVRT is that it is difficult to detect an incorrect pointer for a given polygon until the polygon has actually been retrieved, and even then you must know what the polygon is meant to represent.

Compacting Vector Data Models

When we looked at raster data models we discovered that data can be compacted to reduce storage space in a number of ways. While vector data models are generally more efficient at storing large amounts of geographic space, it is still necessary to consider reductions. In fact, a simple codification process developed more than a century ago by Sir Francis Galton (1884) is relatively similar to the compaction technique we are about to see. It might be useful to travel back in time to accompany the English scientist as he tried to develop a shorthand scheme for recording directions during geographical excursions. The form Galton devised is simplicity itself. He simply applied eight numerical values: one for each of the cardinal compass directions and one for each of the intermediate directions, northeast, southeast, southwest, and northwest (Figure 4.16*a*).

A surprisingly similar coding scheme, developed in our time, is known as **Freeman–Hoffman chain codes** (Freeman, 1974) (Figure 4.16*b*). Eight unique directional vectors are assigned the numbers 0 through 7. As Galton had done for ground navigation on his journeys, the Freeman–Hoffman method assigned these vectors in the same four cardinal directions and their diagonals. By assigning a length value to each vector, individual line entities can be given a shorthand to show where they begin, how long they are, in which direction they are drawn, and where the vector changes direction. There are many variations on this theme, including increasing the codes to 16 (Figure 4.16*c*) or even

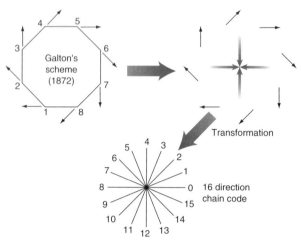

Figure 4.16   Chain codes. Comparison of compact models for direction and path finding developed by Sir Francis Galton and refined in the Freeman–Hoffman chain codes. Note the strong similarity between the older and more recent models.

32 values rather than 8, to enhance accuracy. But the result is the same—reduced storage for vector databases.

While the chain-code models produce significant improvements in storage, they are essentially compact spaghetti models and contain no explicit topological information. This limits their usefulness to storage, retrieval, and output functions because of the analytical limitations of nontopological data structures. In addition, the way the lines and polygons are encoded as vectors, performing coordinate transformations, especially rotation, leads to heavy cost computing overhead. Chain-code models are good for distance and shape calculations because much of this information is already part of the directional vectors themselves. In addition, because the approach is so similar to the way vector plotters operate (see Chapter 14), the models are efficient for producing rapid plotter output.

A Vector Model to Represent Surfaces

We have largely ignored surfaces thus far, even though they are a fundamental feature we will want to model with a GIS. They differ greatly, however, in their manner of representation, especially in vector. In raster, the geographic space is assumed to be discrete in that each grid cell occupies a specific area. Within that discretized or quantized space, a grid cell can have encoded as an attribute the absolute elevational value that is most representative of the grid cell. This might be the highest or lowest value, or even an average elevational value for the grid cell. As such, the existing raster data structures are quite capable of handling surface data.

In vector, however, the picture is quite different. As you remember, much of the space between the graphical entities is implied rather than explicitly de-

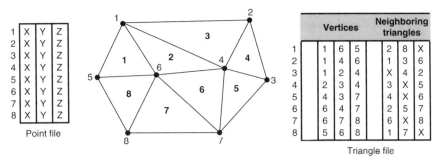

Figure 4.17   TIN model. In a vector system surfaces are represented by connecting points of known elevation into triangulated flat surfaces. The model is called a triangulated irregular network (TIN) model.

fined. To define this space explicitly as a surface, we must quantize the surface in a way that retains major changes in surface information and implies areas of identical elevational data. For a simple way to envision this, consider how mineralogists or crystallographers describe minerals. Each mineral is said to have a series of smooth faces connected by points and lines that show major changes in its structure. In a similar way, we can envision a topographic surface to be a crude mineral, with flat surfaces, edges, and vertices (Figure 4.17). Thus we can model a surface by creating a series of either regularly or irregularly placed points that act as vertices. Each point has an explicit topographic value. We connect any three points to represent an area of uniform topography, thereby creating what essentially amounts to a crystallographic model of our surface.

This model, called a **triangulated irregular network (TIN),** allows us to record topographic data as points in a regular or irregular grid. Then when a three-dimensional view is necessary, the grid can be converted to the TIN or crystalline shape. In addition, the point data can be used for performing the typical processes of representing the surface as a series of contour lines, with all the interpolation procedures needed for surface analysis. This particularly elegant means of representing surfaces was, in fact, used as the primary data structure for earlier systems that relied heavily on surface data (DeMers and Fisher, 1991). We will return to the TIN model in Chapter 10.

Hybrid and Integrated Systems

We have increased in complexity from file structures to database management systems to spatial data models. We now need to move one step further to complete systems. Most raster systems are so simple that the data model itself provides a relatively complete description. In vector systems, however, there are two primary approaches to integrating the graphics elements of the data model to a database management system. It is useful to examine these two models, not only because of the basic differences in approach, but because of the overall prevalence of vector GIS systems in the marketplace. The two major forms of vector GIS are **integrated systems** and **hybrid systems;** both accommodate the linkage between spatial entities and their attributes.

Figure 4.18 Hybrid vector GIS system. Pointers connect the entity files to the database management system that contains the attribute information.

The hybrid GIS data model is an acknowledgment that while graphic data structures and models are efficient at handling entity data, they lack the computational sophistication to manage the attribute data with the same efficiency (Aronson, 1985; Morehouse, 1985). Alternatively, database management systems are well recognized for their ability to manage attribute types of data but are not well adapted to graphics entities. It seems only logical that these two technologies, if linked through software, would provide the best of both modes. To implement this approach, the coordinate and topological data required for graphics are stored as a separate set of files (Figure 4.18). The attribute tables, carrying all the necessary attribute data for each graphic entity, are also stored separately within existing commercial database management system software. Linkage is performed by storing identification codes (i.e., polygon identification codes) as a column of data in the attributes database. In this way, the column is directly associated with the attribute codes contained in the table. Because the hybrid model allows raster and vector data types to be operated on in the same system, this approach is sometimes used to link grid-based modules to the database management system within the overall GIS structure. Moreover, since multiple attributes can be stored in the database management system, analytical capabilities are increased and storage space is saved. Hybrid systems include the CAD-based INTERGRAPH IGDS/DMRS and MICROSTATION-32, the vector/topological-based ARC/INFO, GEOVISION, and INTERGRAPH MICROSTATION GIS, and at least one quadtree-based system: SPANS. An examination of these systems is beyond the scope of this text. Instead the reader is directed to GIS reference works, for example, the chapter on database management systems by Healey (1991), which also appears in the two-volume work edited by Maquire, Goodchild, and Rhind (1991).

The second major type of GIS model is the integrated data model, another spatial database management system approach. In this case the GIS serves as the query processor, but it is more closely integrated with the database management system than in the hybrid system (Guptill, 1987; Morehouse, 1989). Usually based on vector/topological data models, the integrated system stores map coordinate data (entities) as relational tables, together with separate tables containing topological data as separate tables within the same database (Figure 4.19). Attributes are also stored in the same database and can be placed in the same tables as the graphic entities. Or, as noted earlier, they can be stored as separate tables and accessed through the use of relational joins (Healey, 1991).

There are two ways of storing coordinate information as relational tables. The first records individual (X, Y) coordinate pairs, representing points as well

Point ID	x	y
1	25.4	27.5
2	24.8	45.8
3	27.8	50.4
4	28.9	43.7
.	.	.
.	.	.
.	.	.

Point table

Polygon ID	Line ID
34	12
34	14
34	17
34	21
36	78
36	84
.	.

Polygon table

Polygon ID	Attribute
34	104
36	624
37	108
38	107
30	53
41	91
.	.

Attribute table

Relational join

Figure 4.19 Integrated GIS system. Note how a single database can be configured to contain separate files for entities and attributes.

as line and polygon terminators and vertices as individual atomic elements or rows in the database. This approach conforms to Codd's normal forms, but it makes search quite difficult because each element must be recomposed from its atomic format to create whole polygons or groups of polygons. Most GIS applications access large groups of entity elements for display purposes, a function used more often than users might think as they review the results of intermediate analytical steps. To avoid this approach, the integrated model could code whole strings or collections of coordinate information in the tables. As such, a single polygon could be described with its ID code in one column and a list of lines, addressed by code, in another. Then the lines, which would be identified by code in a separate column of a line table, would describe the polygon's locations with a number of coordinate pairs. This approach reduces overhead for retrieval and display purposes, but it violates the first normal form. From the user's perspective this is generally not a serious problem, and grouping these nonatomic strings of data as one-dimensional arrays provides the advantage of enhanced system performance (Dimmick, 1985), while more closely following the rules of first normal form (Sinha and Waugh, 1988).

The choice of hybrid versus integrated system for the vast majority of users is less technical than pragmatic. Each has some advantages over the other, but especially as we move toward higher powered workstations, networking, and distributed computing, both can supply a wide spectrum of analytical power. For most of us, at least as novices, the choice of system will be made for us. Those in the enviable position of choosing their system want to decide on which one best fits present equipment and future networking needs. Both system types will continue to improve, and it will be necessary to ask the vendors for detailed specifications and even performance evaluation tests on given hardware configurations.

In addition to the typical high-level models already discussed, a third, called **object-oriented database management systems,** is emerging as an alternative approach. This model is an extension of the integrated GIS model that incorporates a spatial query language (Healy, 1991) and reflects recognition of the importance of being able to access both the cartographic database and the operations to be performed. Conceptually, such a GIS system is nearly identical to the object-oriented approach to computer programming (Aronson, 1987).

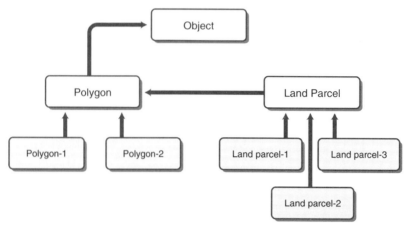

Figure 4.20   Object oriented GIS system. Example of a hierarchy of object classes as they might be configured by an object-oriented GIS. *Source:* Modified from R.G. Healey, "Database Management Systems," Chapter 18 in *Geographic Information Systems, Principles and Applications*, D.J. Maguire, M.F. Goodchild, D.W. Rhind (eds.), Longman Scientific and Technical, Essex, England, © 1991. Used with permission.

There is little agreement about the exact meaning of "object oriented," but we know that an object is an entity that has a condition or state that is represented by both local variables and a set of operations that will act on the object. Because it belongs to a set of objects and operations, each individual object can be thought of as a member of that set (i.e., a set defined by local variables and operations at the same time). Each of these classes of objects belongs to a larger group called a superclass and inherits properties from that superclass— in the same manner that humans (as a class of objects) inherit characteristics of a larger set called mammals. For GIS the concept might be exemplified by an object class called *polygon* that gives to each polygon in the database its many properties (e.g., local variables like number of polygons; lists of nodes, arcs, and areas; operational procedures for calculating centroids, drawing polygons, overlaying polygons, etc.) (Figure 4.20).

In addition, in the GIS context the object class called *polygon* acts as a superclass for another set of objects called *land parcel*. Thus land parcel objects also inherit the variables and operational instructions for the superclass polygon, as well as having many of their own characteristics (e.g., categories of land parcels, values, owners and operational procedures allowing transfer of ownership, rezoning operations). This explicit linking of variables and operations, together with the property of inheritance, is meant to more closely resemble the way actual geographical queries are performed. It also provides a method by which changes in one set of objects can be immediately reflected by changes in related objects.

An object-oriented GIS is the INTERGRAPH TIGRIS system (Herring, 1987), which is based on object-oriented programming rather than the newly developing object-oriented database management system. This is clearly an emerging technology for GIS, but while object-oriented approaches provide some potentially powerful tools for GIS modeling, they are not widely available for consumers. The lack of availability should not dissuade those who wish to experiment with their use, especially if the organization budget allows for multiple system capability.

Terms

computer file structures

simple list

direct files

database management system

relational database structures

primary key

normal forms

third normal form

grid cells

vector

polygon

IMGRID

raster chain codes

topological models

TIGER

Freeman–Hoffman chain codes

integrated systems

database structures

ordered sequential files

inverted files

hierarchical data structures

tuples

relational join

first normal form

quantize

quadtrees

networks

coverage

MAP

block codes

node

chains

triangulated irregular network (TIN)

hybrid systems

graphic data structure

indexed files

database

network systems

relations

foreign key

second normal form

raster

resolution

topology

GRID/LUNR/MAGI

run-length codes

spaghetti model

GBF/DIME

POLYVRT

object-oriented database management systems

Review Questions

1. Explain the fundamental difference between a simple set of graphics and a map in terms of how each represents our environment. What is so difficult about transferring a map to a computer?

2. Why, since we're mostly going to use software, not program it, do we need to know about basic computer file structures, database structures, and graphic data structures?

3. What is the difference between simple list file structures and ordered sequential file structures? Which is more efficient at adding records? Which is more useful for sorting and retrieving records? Give an example of how an ordered sequential file structure works.

4. What are indexed files? How do they differ from ordered sequential files? What are the advantages of indexed files? What is the difference between direct and inverted indexed file structure? Which is more efficient at handling data retrieval?

5. What is a hierarchical database structure? How does it work? Give an example. What are its limitations, especially for GIS?

6. What are network database structures? How do they keep track of records without a hierarchical structure? What are their advantages or disadvantages over hierarchical systems?

7. What is a relational database management system? How does it work. What advantages or disadvantages might it have over database management systems of other types?

8. What is a primary key? A tuple? A relation? A foreign key? A relational join?

9. What are normal forms? List the first three normal forms and describe the restrictions they place on the database management system.

10. Describe the process of quantizing space into equal-sized rasters called grid cells. What impact does grid cell size have on the locational accuracy? How would you store points, lines, and polygons using a raster system?

11. What are some possible advantages and disadvantages of using a raster GIS as opposed to vector?

12. Describe the vector data structure. How does it differ from raster in its ability to locate objects in space? How does it deal with the space between objects and other spatial relationships, as opposed to raster?

13. Describe the three raster GIS models. Explain the advantages and disadvantages of each.

14. Describe the methods of compact storage for raster models. Why are they needed, anyway? What is the major problem with quadtree representation of earth features as squares, compared to the block codes method?

15. Describe the spaghetti vector data model. What are its advantages and disadvantages?

16. Describe the general topological vector data model. How does it differ from the spaghetti model? How is this difference achieved? What are some examples of topological models? Describe their differences. What advantages or disadvantages does each have?

17. What is a common method of compacting vector data models? How does it work? What nineteenth-century geographer developed a similar scheme?

18. Describe the TIN model. How does it quantize space differently from raster models? Why is it necessary to develop such a model for vector GIS?

19. What is the major difference between hybrid and integrated GIS systems? Diagram data storage and access in each system.

20. Explain, in general terms, what an object-oriented GIS is and indicate its potential advantages over other systems.

References

Aronson, P., 1985. "Applying Software Engineering to a General Purpose Geographic Information System," In *Proceedings of AUTOCARTO 7*. Falls Church, VA: ASPRS, pp. 23–31.

Aronson, P., 1987. "Attribute Handling for Geographic Information Systems." In *Proceedings of AUTOCARTO 8.* Falls Church, VA: ASPRS, pp. 346–355.

Beyer, R.I., 1984. "A Database for a Botanical Plant Collection," In *Computer-aided Landscape Design: Principles and Practice,* B.M. Evans, Ed., Scotland: Landscape Institute, pp. 134–141.

Burrough, P.A., 1983. *Geographical Information Systems for Natural Resources Assessment.* New York: Oxford University Press.

Clarke, K.C.,1990. *Analytical and Computer Cartography.* Englewood Cliffs, NJ: Prentice-Hall.

Codd, E.F., 1970. "A Relational Model of Data for Large Shared Data Banks." *Communications of the Association for Computing Machinery,* 13(6):377–387.

Dangermond, J., 1982. "A Classification of Software Components Commonly Used in Geographic Information Systems." In *Proceedings of the U.S.–Australia Workshop on the Design and Implementation of Computer-Based Geographic Information Systems,* Honolulu, HI, pp. 70–91.

Date, C.J., 1985. *An Introduction to Database Systems,* Vol. II. Reading, MA: Addison-Wesley.

DeMers, M.N., and P.F. Fisher, 1991. "Comparative Evolution of Statewide Geographic Information Systems in Ohio." *International Journal of Geographical Information Systems,* 5(4):469–485.

Dimmick, S., 1985. *Pro-FORTRAN User Guide.* Menlo Park, CA: Oracle Corporation.

Fagin, R.,1979. "Normal Forms and Relational Database Systems." In *Proceedings of the ACM SIG-MOD International Conference on Management of Data,* pp. 153–160.

Freeman, H., 1974. "Computer Processing of Line-Drawing Images." *Computing Surveys,* 6:57–97.

Galton, F.,1884. *Art of Travel; or, Shifts and Contrivances Available in Wild Countries.* New York: David & Charles Reprints.

Guptill, C.,1987. "Desirable Characteristics of a Spatial Database Management System." In *Proceedings of AUTOCARTO 8.* Falls Church, VA: ASPRS, pp. 278–281.

Healey, R.G., 1991. "Database Management Systems." In *Geographical Information Systems: Principles and Application,* D.J. Maguire, M.F. Goodchild, and D.W. Rhind, Eds. Essex: Longman Scientific & Technical.

Herring, J.R., 1987. "TIGRIS: Topologically Integrated Geographic Information System." In *Proceedings of AUTOCARTO 8.* Falls Church, VA: ASPRS, pp. 282–291.

Kent, W. (1983), A simple guide to five normal forms in relational database theory. Communications of the Association for Computing Machinery 26(2): 120–125.

Maquire, D.J., M.F. Goodchild, and D.W. Rhind, Eds., 1991. *Geographical Information Systems: Principles and Applications.* Essex: Longman Scientific & Technical.

Marx, R.W., 1986. "The TIGER System: Automating the Geographic Structure of the United States Census." *Government Publications Review,* 13:181–201.

Morehouse, S., 1985. "ARC/INFO: A Geo-relational Model for Spatial Information." *Proceedings of AUTOCARTO 8.* Falls Church, VA: ASPRS, pp. 388–397.

Morehouse, S., 1989. "The Architecture of ARC/INFO." In *Proceedings of AUTOCARTO 9.* Falls Church, VA: ASPRS, pp. 266–277.

Peucker, T., and N. Chrisman,1975. "Cartographic Data Structures." *American Cartographer,* 2:55–69.

Peuquet, D.J., 1984. "A Conceptual Framework and Comparison of Spatial Data Models." *Cartographica,* 21:66–113.

Shaffer, C.A., H. Samet, and R.C. Nelson, 1987. "QUILT: A Geographic Information System Based on Quadtrees." College Park: University of Maryland, Center for Automation Research, CAR-TR-307, CS-TR-1885, DAAL02-87-K-0019.

Sinha, A.K., and T.C. Waugh, 1988. "Aspects of the Implementation of the GEOVIEW Design." *International Journal of Geographical Information Systems,* 2:91–100.

Ullman, J.D., 1982. *Principles of Database Systems.* Rockville, MD: Computer Science Press.

U.S. Department of Commerce, Bureau of the Census, 1969. The DIME Geocoding System. In Report No. 4, Census Use Study. Washington, DC: Government Printing Office.

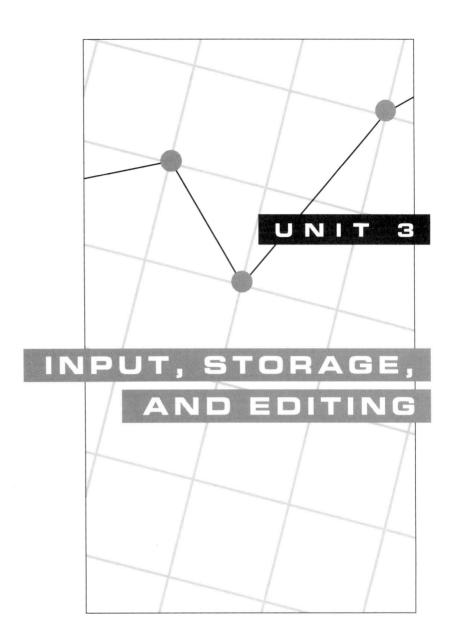

UNIT 3

INPUT, STORAGE, AND EDITING

CHAPTER 5

GIS Data Input

The preparations for our journey are nearing completion. We have examined our tool kit and developed a conceptual framework in which to examine our world; now we need to begin putting our tools together in a coherent, usable package. We will do this by preparing our maps, much as we would if we were setting off on an actual exploration on the ground. In this case, of course, we will visit a digital world where we can examine the earth's features and their relationships efficiently. To prepare for a field excursion we must decide which maps we need, determine the appropriate scale, collect the documents themselves, and mark important places and routes that will guide us in our travels. For our digital journey we must begin to collect our maps and all the important features, format them for input to our digital universe, and input them into a coherent package, or GIS database. Then we will be able to explore their digital features and relationships.

As with traveling, the anticipation of the journey often overshadows the importance of the preparation. We want to begin exploring our digital world, asking questions, observing the myriad patterns that can occur. But, if we do not prepare properly we may very well find ourselves lost, wandering aimlessly through unmarked trails and along dangerous precipices. Preparation is not exciting. It is tedious and often difficult and time-consuming. In GIS, just building the database often occupies as much as 75% of our time. In commercial applications this means that as much as 75% of the costs of operating a GIS system is also contained in this operation, together with editing, which we will cover in our next chapter.

Try to remember that good preparation makes for pleasant and successful results, whether for field excursions or digital ones. Take the time to learn the techniques you will need, and train yourself to spot potential problems that detract from a well prepared GIS database. While input is a slow and often painful process, care taken to do it right the first time will result in large savings in editing time later. In fact, finding and correcting a single error can sometimes consume much more effort than would have been needed to input the data correctly in the first place. And, of course, the less time you spend editing, the sooner you can begin exploring.

There are many ways to input data. Some may seem primitive, such as placing a Mylar grid overlay on a map. Others are somewhat less primitive because you use digital input devices called digitizers. Still others are very advanced and include highly sophisticated and expensive automated scanners. Your instructor may have you go through the process of preparing a map and then inputting it into the computer. The devices used will vary based on what is available. If you do not have an exercise on digitizing in your course, you will most likely have access to people who are going through the process as part of a research project. If so, take some time to watch the process. You might time the digitization of a portion of a map or even a whole map, to get an idea of how long the process would take for an entire database with many map coverages. This will at least give you an appreciation of how the job is done, and how much of the GIS practitioner's time is spent using this GIS subsystem. It will also allow you to appreciate how much time can be saved by purchasing digitized data if they are available.

LEARNING OBJECTIVES

When you are finished with this chapter you should be able to:

1. Know the four primary functions of GIS data input.

2. Know how to decide on the input devices you will need.

3. Be aware of the potential problems of converting to and from different data structures.

4. Understand the transformations that take place during data input.

5. Know the procedures for translation, rotation, and scale change necessary for modifications in GIS data coverages.

6. Understand the procedures of map preparation and the importance of those map preparation procedures in map input.

7. Have a feel for what data and how much data need to be input and why they need to be input to a GIS.

8. Understand the relationship between input scale and projection on GIS error.

9. Be familiar with the basic procedures for digitizing and the importance of attribute data to the overall quality of the GIS database.

10. Be familiar with the four basic methods of raster data input, with their advantages and disadvantages, and with the data types that should be input with the respective methods.

11. Understand some of the technical problems of using aerial photography for GIS input.

12. Be aware of the potential for and the problems of external databases for GIS input.

THE INPUT SUBSYSTEM

Before we can use any of the data structures, models, and systems we examined in Chapter 4, we must convert our abstracted reality into a format usable by the computer. The methods by which this is done will depend to some degree on the equipment available and on the particular system at hand. No matter what system we have or how we intend to get the spatial data into it, our GIS input subsystem will share some attributes with all others. First, of course, it is designed to transfer the data (both entities and attributes) into the computer. Second, it must allow us to account for at least one of the two fundamental ways in which we can view geographic entities, that is, as grid cells or as vector objects. Third, it must have available a form of Cartesian grid, whether projected from the reference globe or assumed to be strictly a Cartesian representation of space. Finally, it must have the capability link to the storage and editing system, to ensure that what we input is saved and retrievable and that mistakes can be eliminated and changes made when necessary.

Input Devices

Many different types of device have been and are being used for general input of any data into a computer. Most, if not all, are used to a lesser or greater degree today in modern GIS input. Perhaps the first approach to cartographic input to a mapping or GIS package was a tedious, error-prone technique for producing a square grid on a clear plastic material, such as Mylar, then transferring (typing) the data, grid cell by grid cell, into the computer. In most cases, a numerical value was assigned for each grid cell, which in turn was input by typing the numbers into the computer one at a time. This called for a systematic method of determining where, within the grid cell, the object being encoded was located. For example, you might choose the central point of each grid cell as the spatial locator identifying where, in space, the data were to be placed. Or, you might choose any of the four corners (assuming a square grid cell raster). While it is quite obvious that knowing the whereabouts of the spatial reference point for each grid cell is vital to vector-based systems that operate on an explicit locational framework, it is also important to decide this for raster data, which will be represented inside the computer as grid cells, one for each item. Imagine, for example, trying to measure the distance from one place to another, based on the number of grid cells: you would have to know whether you are measuring from the sides of the grid cell or from the centers. After all, remember that a grid cell by definition occupies space. And of course, the more space represented by a single grid (i.e., the coarser the resolution), the more important this factor becomes.

While most of us will not be forced to input data using an acetate grid overlay, some very small operations still do it, as do some universities, to demonstrate the digitizing process at a grassroots level. Generally, however, you will have access to a more sophisticated technology. To input spatial data manually, the use of a digitizer is standard. The digitizer is a somewhat more advanced and much more accurate version of an input device used in nearly all modern personal computers—the mouse, which the user moves freely along pretty much any surface. Inside the mouse are sensors that respond to the motion of a rubber ball encased in the frame of the device. All the locational information is contained within the mouse itself. To enhance the accuracy of such a device, the degree of movement for a digitizer is recorded through the use of an electronically active grid within the digitizing tablet. A mouselike device, often called a puck, is connected to the tablet and is moved along the tablet to different locations on a map that is attached to the tablet. Digitizing pucks contain a crosshair device, encased in glass or clear plastic, that allows the operator to place the puck exactly over individual map elements. In addition, the puck has buttons (the number depends on the sophistication of the digitizer system) that indicate the beginnings and endings of lines or polygons, or explicitly define left and right polygons, and so on. How the digitizer buttons are used is largely determined by the GIS software requirements.

Digitizing tablets come with rigid or flexible tablets and can range from small page-sized formats to very large formats capable of accepting good-sized maps with room to spare (Figure 5.1). Some of the larger format digitizers also have adjustable stands that can vary the elevation of the tablet from the ground and change the angle of the tablet to make digitizing easier for the operator. Sizes and formats are determined in part by the general size of documents to be input to the GIS and in part by budgetary constraints. Generally the smaller the digitizer, the lower the cost. In addition, there is a direct relationship between cost and digitizer accuracy. The lower priced digitizers frequently give adequate results, and as technology improves, the prices of these devices will continue to drop. Modern digitizers can provide resolutions of 0.001 inch, with an accuracy that approaches 0.003 inches for an area of 42 inches by 60 inches (Cameron, 1982).

Factors that will prove useful in selecting a digitizer include **stability, repeatability, linearity, resolution,** and **skew.** Stability deals with the tendency of the exact reading of the digitizer to change as the machine warms up. For

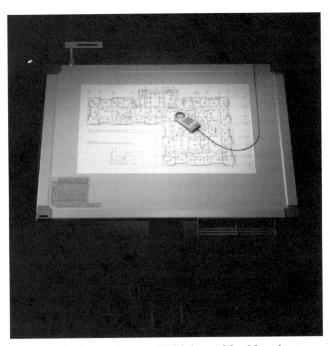

Figure 5.1   Large-format digitizing table. Most large tablets allow the user to raise and lower the table and adjust the angle.

the first-time digitizer operator, it can be most disconcerting to watch the values change while the puck sits in one place. The simplest solution is to allow the digitizer table to come up to operating temperature before using it. If the drift continues when the tablet is warm, the tablet may need to be repaired or replaced.

Repeatability is a synonym for precision. If you place the puck in the same exact location twice, how close will the first and second readouts be? Good digitizers should be repeatable to about 0.001 inch (Cameron, 1982). Linearity is a measure of the ability of the digitizer to be within a specified distance (tolerance) of the correct value as the puck is moved over large distances. A linearity of 0.003 inch measured over 60 inches is common with today's modern equipment. Resolution is the ability of the digitizer to record increments of space. In other words, the smaller the units of measure it can handle, the better its resolution. This is not unlike the resolution of a camera or a remote sensing system. Resolutions of 0.001 inch are quite good but may prove to be unnecessary for much GIS work. Finally, skew is a measure of the squareness of the results on a tablet: Do coordinates located at the four corners of your digitizer produce a true rectangle, as intended? Some portions of the digitizer may begin to wear out, especially toward the edges, reducing the ability to digitize on the entire tablet, perhaps even compromising the quality of the input.

Scanners are more advanced input devices that perform much of the work of the entity input process with little help or intervention from the user. Scanners come in two types, line-following scanners and drum scanners (Figure 5.2). Line-following scanners, as the name implies, are placed on a line and move on small wheels, tracked by a laser or other guiding mechanism. At different intervals (either time intervals or distance intervals), these devices send a signal to the computer that records the digitizer coordinates at each of the sample locations. Line-following scanners require more technician intervention because they must be manually placed on each new line to keep the scanning process going. They also suffer from two shortcomings. First, complete automation requires the device to use a simple time or distance measure within which to record or sample the lines followed. As you remember from Chapter 2, however, this is not always the best way to abstract a line. For very complex lines, more points should be sampled than is necessary for relatively straight lines. Additional operator intervention is required to ensure that sufficient points are sampled. The second problem is a more difficult one for the line scanner to deal with. Say you are using a line scanner on a topographic map to digitize contour lines that converge along a cliff (a conventional method of depicting cliff faces with contour lines). When the lines diverge on the opposite

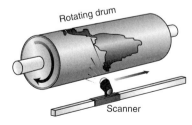

Figure 5.2 Principles of map scanning. Diagram of a drum scanner showing the rotation of the drum and the movement of the scanning device along the drum.

side of this feature, however, it is quite common for the scanner to get confused about where it should go next. This problem can get even worse if the topography under an overhang is represented by dashed lines, which the scanner may not be able to find because of the gaps in the lines or because the color is lighter and has less contrast than the original contour line.

The other form of scanner, the drum scanner, uses a more rasterlike approach that is really closer to vector mode. The map document is placed on a drum that rotates at the same time that a sensor device moves along the frame at right angles to the direction of rotation. In this way, the entire document is scanned one line at a time. Each location on the map is recorded, even if there are no cartographic objects present. The result is a detailed raster image of the entire document. Drum scanners can give monochromatic or color output. For color, each of the three primary colors must be scanned individually. Both monochromatic and color output must be converted to vector format if that is what your GIS requires; both produce very large data files.

Scanning is generally very expensive compared to manual digitizing. The devices themselves cost more than a small consulting firm's first-year operating budget. In addition, editing can take nearly as long as manual digitizing would have taken, especially if the document was very complex to begin with. I have seen scanning operators spend days trying to find missing lines, or connecting lines that were supposed to be continuous. As the technology improves, the amount of editing necessary will undoubtedly decrease. In the meantime, don't believe that scanners are going to remove the human from the input process. Certainly, even with the use of drum scanners, the editing process will require human intervention. In short, at least for the very near future, drum scanners should be thought of as time-saving devices only when maps are clear, show good contrast, and contain a relatively simple amount of content. Drum scanners are most often used in large-scale, production-oriented firms specializing in line graphics. You can expect to be doing manual digitizing for many maps for which the drum and line scanners are unsuitable or for which the cost is prohibitive.

Raster, Vector, or Both

No matter which approach is chosen for GIS data input, it is necessary to determine at the outset whether you will be using a raster or vector GIS, as well as whether your GIS is capable of converting back and forth within these data types. Some programs, especially those that evolved primarily to handle remotely sensed data, operate predominantly on the grid data structure, while others operate primarily on vector data. Most of the commercial programs allow interconversion. In addition, data types among GIS software are becoming more alike in general format as the clear separation between the two is becoming less profound.

The type of input device will also have an impact on how you will operate on the data and on whether format changes will be needed. As we have seen, line-following scanners tend to produce vector output, while drum scanners will produce raster output. Hand digitizers are largely controlled by the software system that drives them, providing output in the appropriate format.

While conversion between raster and vector is fairly common, there are some things you should remember. Most often, when you convert from vector to raster, the results are visually satisfactory, but the techniques for the conversion can produce results that are not satisfactory for the attributes each grid cell represents. This is particularly true along the edges of areas, where the user seldom knows the decision rules concerning how the space is to be quantized. (See the later discussion on raster input.) Alternatively, by converting from raster to vector, you may preserve the vast majority of the attribute data, but the visual results will often reflect the blocky, steplike format of the grid cells from which conversion proceeded. Some algorithms are available for smoothing this blocky appearance with the use of a mathematically based graphics technique called **splines.** Without going into the gory details, this is simply a graphic approach that smooths out jagged lines and sharp edges. If you have access to a CAD system, you can sample this effect by using the spline feature to enter a stickman on the screen. The result will look much like the Pillsbury Dough Boy, with fat, rounded surfaces.

Before entering data with your system, especially if you need to convert back and forth, try to find out how the conversion takes place by talking to the vendor or consulting the documentation. You may want to develop a small database against which to compare test results. This check may not allow you to determine the actual algorithms used but will indicate the nature of the results. You can then determine which system is the most appropriate for your needs.

Reference Frameworks and Transformations

Digitizers are designed to input existing maps, although they can also be used to input data from aerial photography and other analog products of remote sensing equipment. We will save our discussion of the role of remote sensing in data input and concentrate instead on the typical cartographic document from which we will produce a cartographic database. As we have seen, maps are representations of a three-dimensional reference globe projected onto a flat surface. That is, our geographical data have already been transformed from their original form as part of a spherical planet to a two-dimensional surface, with all the accompanying shape, area, distance, and angular deformations. When we digitize this map, we reduce this sophisticated projection to a set of Cartesian (in this case digitizer) coordinates. Before we do this we will normally provide the GIS software with information about the type of projection used and even specific information about the grid system and the zone or zones of origin within which it was produced. This preparation will allow us to transform the map to its original projection after it is input. In fact, the GIS will produce a number of transformations of this kind as we project from the Cartesian coordinates on the digitizer to the two-dimensional map projection coordinates, and from there through a process called the **inverse map projection** to three-dimensional latitude and longitude coordinates. From there we will eventually need to reverse the process to produce Cartesian coordinates for the output device (Figure 5.3).

To perform these various projections and manipulations, the GIS software,

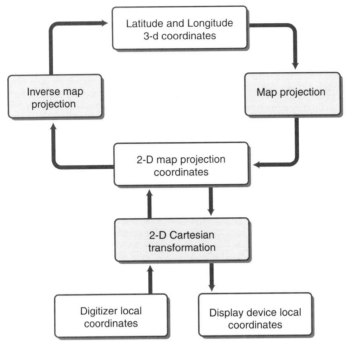

Figure 5.3   GIS coordinate transformations. Transformation steps from Cartesian (digitizer) coordinates to two-dimensional projected map coordinates, through an inverse map projection to longitude and latitude coordinates. To produce map output, the process is reversed through the projection process and finally to the necessary Cartesian (display device) coordinates.   *Source:* Duane F. Marble, Department of Geography, The Ohio State University, Columbus, OH.

in large part within the input subsystem, will need to perform a number of graphic manipulations. The three primary processes that must occur, often simultaneously, are **translation, rotation,** and **scale change.** Translation is simply the movement of parts or all of a graphic object to a different location on the Cartesian surface. This is done by adding or subtracting the coordinate values necessary for the X and Y coordinates of the object (Figure 5.4*a*). In other words, the new X coordinate X' for each graphic object will be equal to the original X coordinate plus some value T_X, and the new Y coordinate Y' for each graphic object will be equal to the original Y coordinate plus some value T_Y.

$$X' = X + T_X \quad \text{and} \quad Y' = Y + T_Y$$

where T_X and T_Y can be either positive or negative.

Scale change is also relatively useful because of the need to compare differently scaled maps and to output in different scales as well (Figure 5.4*b*). This is done by multiplying the overall X coordinate extent (e.g., the two sets of X coordinates for a line segment) is multiplied by a scale factor s_X, and each set of Y coordinates by a Y scale factor s_Y.

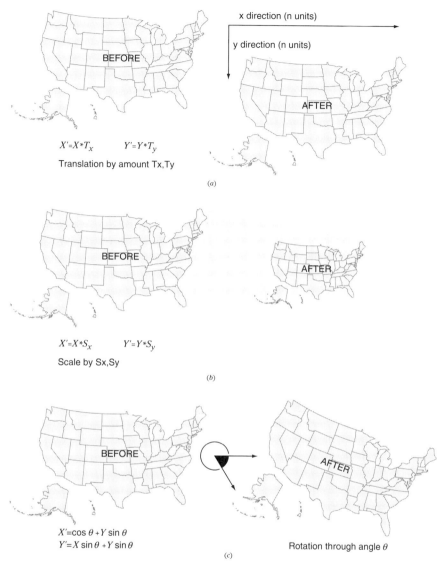

$X'=X*T_x$ $Y'=Y*T_y$

Translation by amount Tx,Ty

(a)

$X'=X*S_x$ $Y'=Y*S_y$

Scale by Sx,Sy

(b)

$X'=\cos\theta+Y\sin\theta$
$Y'=X\sin\theta+Y\sin\theta$

Rotation through angle θ

(c)

Figure 5.4 Translation, scale change, and rotation. The three basic graphic transformations necessary to make the projection transformations: *(a)* translation, or object movement within the coordinate space, *(b)* scale change, to adjust the object size, and *(c)* rotation, to reorient the object within the coordinate space.

$$X' = X\,s_X \quad \text{and} \quad Y' = Y\,s_Y$$

where S_X and S_Y represent the amount or percentage of scale change.

Rotation is used frequently during the process of projection and inverse projection. It is accomplished by using basic trigonometry (Figure 5.4c). For X coordinate locations, the new X location X' is found by multiplying it by the cosine of the new angle (θ), then adding that value to the original Y coordinates multiplied by the sine of theta (sin θ). The new Y coordinate locations Y' are found by multiplying the negative of the original X value by the sine of the angle and again adding that to the product of Y coordinate and $\sin \theta$:

$$X' = X \cos \theta + Y \sin \theta \quad \text{and} \quad Y' = -X \sin \theta + Y \sin \theta$$

where θ is the angular displacement wanted.

All necessary transformations can be performed with three basic types of graphic manipulation.

Map Preparation and the Digitizing Process

As we just saw, we cannot begin digitizing until we have provided the GIS software with pertinent information regarding the projection, the grid, and so on. This is a part of the map preparation process so often neglected, yet so important to developing a useful database. Many software packages will require you to provide this information before you can begin, although some will allow you to input the information later. In either case, you should prepare the information beforehand and keep it handy so that you will always know what it is and where to find it.

It is also a good idea, before you place your map on the digitizing tablet to begin the input process, to prepare the map by making appropriate marks directly on the map document or on a firmly attached clear plastic covering, to identify the exact locations you will be digitizing. Remember, there will be many curved lines, and you will have to reduce the curves to short, straight-line segments. While many prefer to digitize freehand, eliminating this useful step, if you know all the points that will be digitized (which are the beginnings and which are the endings of polygons, which are nodes, etc.), you will not need to perform this tedious procedure while you are digitizing.

Because digitizing is monotonous work, you probably will digitize the map in portions rather than all at once. This is all the more reason to prepare the map ahead of time by outlining directly on the map portions you expect to be digitizing at each session. However, because you will probably be digitizing in multiple sessions, and sometimes you will have to remove the map to allow others to use the equipment, you will need to tell the computer software where your map area is and what its coordinates are. These first points, called **registration points** or **tick marks,** will be entered in digitizer inches as well as in map coordinates. They also should be marked on the map as part of your map preparation, so you, as well as the computer, know your starting points. The registration points provide an outside frame for the document and should be outside any graphic object you will digitize, including a neat line if you will include that in your database. Usually three points, located on the corners of a rectangle, are needed to define the map area for the software. Some software can get by with only two points if they are located on a diagonal. In this case, the software assumes that the outside boundary is a rectangle and infers the other two corners. No matter which technique you are required to use for your software, the accurate locations of these registration or tick marks is absolutely essential to ensure good quality. Special care should be taken to locate your tick marks precisely. It is a good idea to double-check these because if registration marks are misplaced, virtually all the remaining digitizing will be erroneous.

Other map preparations include a clear definition of the order in which you intend to digitize; a systematic approach to identify the portions of the map to be digitized at each session has been mentioned. It is also a good idea to develop a method of identifying which areas (sections, lines, points, etc.) have already been digitized. Taking periodic breaks to mark the map with a felt-tip pen helps you to keep track of your progress as well as giving you a rest. Your software may require you to identify nodes, left and right polygons, and so on, depending on its sophistication and graphic data model it uses. These should be placed directly on the map document as well, to spare you from having to stop frequently to retrieve the information.

Most digitizing software provides editing capabilities to help you identify your mistakes. In fact, some software allows you to use the editing system to do the digitizing, thus allowing you to edit as you digitize. We will discuss the types of error that can occur and how to correct them in the next chapter. In the meanwhile, most digitizing software includes a feature that allows for the occasional shaky hand, namely, a sort of fudge factor, sometimes called a **fuzzy tolerance** (it can have other names). The inclusion of this feature reflects the assumption that you will not be able to place the crosshairs of the digitizing puck over exactly the same location twice, as required to digitize perfectly the beginning and ending of a polygon. This is because humans generally do not have the manual dexterity to execute such precise small-scale motions, and, of course, the digitizer itself has limits. The fuzzy tolerance may have to be set before digitizing, or it may be set afterward during the editing process. In any event, you should be sure to set the tolerance to a level that is reasonable for the detail of the map as well as for the shakiness of your hand. Small fuzzy tolerances are less forgiving of digitizing errors and may result in gaps between points that were meant to be connected. Alternatively, a fuzzy tolerance that is too large will result in the merger of points and lines that are real because the software assumes the gaps were simply mistakes. We will talk more about this when we discuss the storage and editing subsystem in the next chapter. Chapter 6 offers other digitizing hints regarding techniques to improve the likelihood of producing a good, clean product. In addition, you might want to consult an article by Marble et al. (1990) that discusses the entire digitizing system, especially with regard to organizations and commercial operations. This provides an excellent overview of the process.

Final map preparations deal primarily with the tendency of the source material to shrink and swell with changes in temperature and humidity. A stable material, such as plastic, is preferred to paper for digitizing. Although plastic products also will shrink and swell with changes in temperature, they react far less than paper. In addition, plastic products are far less affected by changes in humidity. While this matter of digitizing media may seem trivial, look around your room for especially large paper posters attached with thumbtacks or staples along the borders. If you place your hand in the center of these posters, you will find as much as an inch or two of play in the paper. The whole poster may be sagging pitifully from the thumbtacks. This is most likely not because the poster was stretched tightly when it was hung, but because it has expanded as a result of changes in temperature and humidity and is being pulled downward from the force of gravity.

There are several ways to limit the amount of digitizing error based on media fragility. First, the room should be air-conditioned to maintain a standard cool temperature and low humidity level. The material you are about to digitize should be allowed to stay in the room for several hours, unrolled (it is a good idea to avoid using folded maps because creases reduce accuracy of such documents severely). Once acclimated to the room conditions, the map can be taped to the digitizer tablet using a removable tape material such as drafting tape. Do not use masking, clear plastic, or other very sticky tapes that will tear the document or even remove information when the document is moved. In addition, very sticky tape may make it difficult to remove the document, stressing it unnecessarily and possibly resulting in stretched media. When you tape the map to the digitizer, place it several inches from the edges to avoid any possible skew (horizontal distortion), which is most likely to occur there. The map should also be placed in a manner that does not cause you to stretch to locate the puck on the map objects, since this can put stress on the document and can also limit your freedom of movement, adding more error to your database. When you digitize a map in multiple sessions, be sure to keep the document in the climate-controlled room to avoid expansion and contraction.

WHAT TO INPUT

Now that we have some basic guidelines on how to digitize, especially how to avoid errors during digitizing, we can begin to select the appropriate data for input. Most texts, and even most software manuals, give little guidance in this regard. It is much like beginning a journey by hearing a guide warn you to be sure all your equipment is packed carefully, without giving you a clue as to what the equipment is. Each journey into the digital world is unique, however; each environment requires different coverages, and each purpose for which we journey into the GIS demands a separate set of criteria. We will attempt to sort this out to provide a simple set of guidelines, applicable under any circumstance.

A major factor guiding what cartographers put on the map and how it is produced is the target audience—the user. The same can be said for producing cartographic and geographic databases for map analysis within a GIS. Historically, a common practice in many GIS systems, including those designed for whole states, has been to input everything (DeMers and Fisher, 1991; Fisher and DeMers, 1990). As we will examine in Chapter 15, this very often destines a system to failure. Thus rule one is to determine why you are building the GIS database in the first place. This will at least limit the input to coverages that are likely to be used. While a really nice map of quaternary geology may seem natural as an input coverage, first because it exists, and second because of its quality, it is unlikely to be at all useful for a study of atmospheric pollution caused by factory smokestacks. What you should observe from this is that the coverages should be directly related to the modeling and analysis you plan to perform, and the results you need. Anticipation of future needs may seem to be a good reason to input your quaternary geology map, but you would be better advised to keep such materials in a separate file, or store them for later input to the GIS, should the need arise.

The need to anticipate what coverages will be required in the future does present a problem, particularly if either you or your client has only a larval view of what is to be done. A serendipitous approach to GIS may prove to be fun, but a GIS begun under uncertain guidelines is unlikely to produce reasonable results without substantial reworking, corrections, enhancements, and work-arounds. Such an approach becomes costly. Perhaps databases should be developed without a very clear understanding of the intended product (sometimes called **spatial information product**), only when the primary objective is to determine possible relationships among coverage variables as part of initial scientific hypothesis formulation. This is not the way to proceed in the commercial environment. So, rule two, which relates to the first, is that you must define your goals as specifically as possible before selecting the coverages.

Even under very specific goals, with known spatial information products, there will be multiple ways of obtaining the available data in some cases. For example, coordinate locations and elevation variables can now be obtained with the use of GPS units. But they may also already be available on maps in a reasonably accurate format. Or data on current land use may be obtainable from ground surveys, aerial photographs, satellite output, airborne scanners, or any number of other sources. There is no easy answer to which should be used. But while there is no recipe for success, there is a recipe for failure. Which leads us to rule three: avoid the use of exotic sources of data when conventional sources are available, especially if the latter provide a similar level of accuracy. You will need to define "exotic" for your own particular project. In general, I like to use a practical definition, applying the term to any data source I am not familiar with. If you or another team member is familiar with a given set of data and can comfortably use it correctly, and if it increases the utility or accuracy of your database, it should be applied. If all your multiple data sources for a particular theme or coverage are in traditional form, then invoke rule four: use the best, most accurate data necessary for your task.

You should keep in mind that "accuracy" in this situation refers to the necessary accuracy, not an absolute accuracy. If you don't need 1-centimeter contours for your topography, for example, use topographic data that most closely matches your level of observation. While having an extremely detailed map of any coverage may seem advantageous, it is costly to input; it also slows analysis and may even make analysis more difficult. An example of the use of 30-meter TM data from the *LANDSAT* satellite as compared to 80-meter MSS data from the same source might prove instructive. Suppose your purpose is to identify large fields of grain. Since the enhanced spatial resolution of 30-meter data has been known to produce many, difficult-to-separate categories over an area that is essentially all grain fields, the better resolution would confuse your situation rather than simplifying it. And, of course, the computing and human resources needed to provide clarification would increase the overall cost of the system. Thus we have rule five: remember the law of diminishing returns when deciding on data accuracy levels.

Another question concerning what to input deals somewhat with the last topic about the data source. Most thematic maps (e.g., USGS topographic maps) contain ancillary data on roads and other anthropogenic features that may very well be useful as input to the GIS. Whenever possible, and when the quality of the data dictates it, you should input these data as separate coverages from the same map sheet. That's rule six. This rule does not restrict the use of other sources when they are of superior quality or accuracy, but it has two advantages. First, because the data are on a single map, you don't need to go to multiple map sheets, and then repeat all the preliminary steps in map preparation. Second, because the data are on the same map sheet, they are already georeferenced, reducing the need to perform this sometimes difficult task later.

The final general rule, rule seven, is that each coverage should be as specific as possible. A coverage should be as thematically specific as possible, without relying on an IMGRID-like binary system. The more specific a coverage, the easier it is to search if you need to know something pertaining to data contained in a single coverage. In addition, when you perform operations like overlay (see Chapter 13), it is easier to keep track of the process if you are completely familiar with the data. Overlay operations are simplified for the computer program itself if there are no extraneous data in a given coverage.

We can summarize the seven rules in a very few simple statements. First, define your purpose. Then be sure your maps address the purpose. Use the most accurate maps needed for the purpose—not too accurate for your needs and not too inaccurate to do the job. Keep your coverages simple, and use the same map to obtain these simple coverages whenever reasonable and prudent, to avoid the need to georeference them. Above all, think about your project before you begin inputting data. Data input takes time and costs money.

HOW MUCH TO INPUT

Related to the types of data you will input is the question of how much. Again, to use our field trip analogy, as you prepare for your journey you will need to know how much food to pack, not just what kind. Too much food and you will find yourself carrying unnecessary weight for the whole trip. Too little and your journey will end prematurely because you will have to return to a source of food. In like manner, too much data input and your GIS must bear the added weight of the data throughout the project life cycle; too little data and you may find yourself unable to answer questions about matters you had planned to cover.

As with preparing packs for a trip, the input of data into the computer is a sampling process. In vector GIS each line you input will likely have some curves. To produce a reasonable facsimile of the line using straight-line segments, you will decide thousands of times where to place the digitizer puck and where to record data. This process is very similar to that of line generalization, encountered in our earlier discussion of cartography. A simple rule of thumb is to take more samples (i.e., record more points) for very complex objects than for simple lines (Figure 5.5). The locations of a straight line can be recorded accurately with only two points: one at the beginning and one at the end. I have seen straight-line segments outlining the external boundaries of a standard quadrangle map with as many as 2000 digitized points. Not only does this clutter the computer with tons of unnecessary data points, and slow computation later on, but it is highly unlikely that the straight line will appear straight when output.

Line and polygon complexity can be compared to information. The more the line changes direction, the more information it conveys (this can also be true for surfaces, but we will discuss that later). And the more densely packed the points, lines, and areas, the more information content the map has. The higher

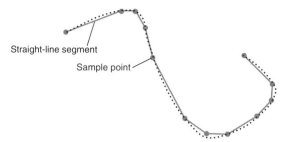

Figure 5.5 Sampling a complex line. An example of sampling for digitizing a line. Points to be digitized are selected based on change of direction of the line. Each direction change indicates an additional piece of information the map contains.

the information content, the higher the sampling rate needs to be. Shannon (1948), who used this science, known as **information theory,** to examine the sampling of voice information for telephone communication to ensure that a complete voice signal was received, discovered a useful rate of sampling that can easily be applied to the digitizing process. In general, in voice transmission technology, the smallest object that is to be included in a system should be recorded at least twice. So, for each change of line direction (i.e., for each piece of information), you should record at least two data points. This is all the more reason for careful map preparation. You should also keep in mind that because for each entity input to the GIS, there is going to be a need for attribute data as well, a direct relationship exists between the complexity of the map, or the volume of data in the map, and the spatial data handling problems (Calkins, 1975).

The idea of information content also can be applied to raster data. Once again, the general rule is this: the smaller the object to be identified in your database, based on your modeling needs, the smaller the grid cells need to be (DeMers, 1992). This principle often determines the selection of grid cell size (resolution) for the entire database. Of course information theory can also be applied to the input of raster data, but keep in mind that grid cells are two-dimensional rather than one-dimensional. The information content is a change in attribute type as we move across the map. Let us say that you want to use grid cells to represent farms displayed on a map. If the smallest farm is, say, 40 hectares, you would need to sample this area at least four times to ensure that it has been captured in your GIS. Stated differently, it means that the grid cells will need to be 10 hectares or smaller to allow the capture of the 40-hectare object in your GIS database. This also assumes, by the way, that four grid cells will surround the field itself. Suppose the field lies along a long, skinny riparian area. Although it is 40 hectares in total area, it is also spread out as a linear object, reducing the chances that the entire field will be input into your GIS. This aspect of the process is somewhat dictated by the method by which you input the grid cells. And we will cover that in a later section. The general rule of thumb, however, remains the same: sample more for more information content.

Whether in raster or vector, sampling is dependent on the amount of area covered by the map and the use for which the data are input. Small-scale maps, those covering large amounts of space, contain a much more abstract view of the land surface being mapped. In addition, the lines and symbols on the map take up physical space, as we discussed in Chapter 3. The amount of error contained in a symbol is dependent on the scale of the map on which it is placed. Lines on small-scale maps take up more land space than same-sized lines on large-scale maps. This physical condition, called **scale-dependent error,** is an indication that the amount of error is directly related to the scale of the map and needs to be considered during the map preparation phase prior to digitizing.

METHODS OF VECTOR INPUT

As indicated earlier, there are numerous instruments available to input vector data into a GIS. We will restrict our discussion to manual digitizing because of the frequency with which it occurs. After preparing your map and placing it on the digitizing tablet, you will need to use the digitizer puck to locate and record the registration marks. Some software requires these marks to be recorded in a specific sequence, while other packages do not. The software documentation and/or the software itself will indicate the requirements. In addition, the software package will indicate which numbered keys you need to enter for specific object types. Some numbered keys will be used to indicate the location of a point entity, others for beginning and ending of line segments, still others for the closure of a polygon. Many digitizing errors, especially novices, are due to pushing the wrong-numbered button.

The exact method of digitizing is also related to the data structure on which the software operates (Chrisman, 1987). Some (e.g., the POLYVRT structure) require you to indicate the locations of nodes, while others do not. Some want you to encode explicit topological data while you digitize, others will use software to build in the topology after the database has been produced. The rules are different for each package, and you will have to look over the appropriate documentation to determine these strategies ahead of time. This work can be considered to be part of the preparation for digitizing rather than the digitizing. Remember to keep the rules uppermost in your mind as you digitize to ensure that your lines don't become nodes and your nodes don't become just points. Some will find it useful to tape a crib sheet to the corner of the digitizing tablet until they become more familiar with the process.

In vector GIS the attribute data are most often typed in using the computer keyboard. While simply typing data in is extremely simple, the task requires the same care used for the inputting of the entities. This is true for two reasons. First, it is very easy to make typing errors (why else do word processors have spell checkers). And second, and perhaps the most potentially problematic, the attributes must be attached to the entities that are to be associated with each other. As we will see in the next chapter, these are some of the more difficult errors to detect because they cannot always be picked up visually and often do not appear until an analysis is under way. A good practice is to check attributes while they are being input, perhaps by taking frequent short breaks to look them over. Time spent doing this on input will save a great deal of time during editing.

METHODS OF RASTER INPUT

Raster data input and its vector counterpart involve different strategies. In the first place, as we have seen in our discussion of how much to input, we must decide how much area should be occupied by each grid cell. This decision must be made prior to digitizing or Mylar overlay input, to tell the digitizer how big the grid cells will be, or to let the person know how big to make the Mylar squares. In addition, we must decide whether it is appropriate to use a method of encoding that shortens the process, such as run-length encoding or block encoding. While compacting methods are good at reducing data volume, their use in encoding may be even more important because of the reduction of input time. Some raster GIS systems that either do not support digitizer input or support both keyboard and digitizer input methods have commands that allow for data to be input by typing attributes set up as strings or blocks. You can refer to your software documentation to determine what these commands are and how they are used. Once you have selected a method, you will need to decide how each grid cell will represent the different themes that will occur. Beyond the grid cell resolution, this may be the most important decision you will have to make. Let's examine it in detail.

As we've seen, raster input is sometimes still done with the use of a Mylar grid overlay, with the attributes typed in one at a time. The wide availability of digitizers is rapidly replacing this difficult approach to raster input, but its use clearly illustrates the different methods by which digitizing software can be used to input grid cells. Normally the digitizer records vectors that are then converted to raster format. The entities and attributes are generally entered simultaneously. This approach, requiring vector-to-raster conversion, can sometimes be iffy depending on the sophistication of the digitizing software used. The problem arises primarily in the raster-to-vector conversion process, which often is not well documented by the software vendor. Most often, difficulties extend from digitizing adjacent areas using vector lines that are then converted to two separable polygons. In such cases the software must decide which polygon will contain the grid cell through which the line runs. (Remember that in many grid-based systems a grid cell can contain only a single value.) The decision is sometimes based on the "last come, last coded" rule. That is, when the same line is digitized first for one polygon, and then again for the second polygon, the second polygon will be assigned to that grid cell. In a more computer-oriented approach, the assignment will depend on which value occurs first inside the computer. While there are convenient methods of making such determinations computationally, they often produce unacceptable errors along edges, especially if the grid cell resolution is coarse. For very large raster databases this level of inaccuracy may not present a substantial percentage of error compared to the entire database, but measurements of shape and area will be compromised. A better method, at least from the standpoint of the utility of the database, would be to allow the system to select from one of four systematic input methodologies outlined by Berry and Tomlin (1984).

In the first of these methods, the **presence/absence method,** for each grid cell on each coverage a decision is made based on whether the selected entity exists within the given grid cell, hence, the name "presence/absence" (Figure 5.6a). A major advantage of this method is that decisions are easy. No measurements are necessary. A simple Boolean operator either is there or isn't.

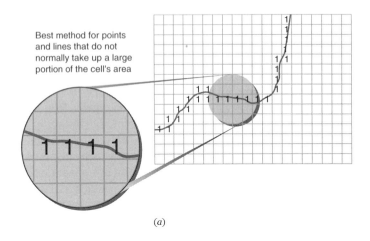

Best method for points and lines that do not normally take up a large portion of the cell's area

(a)

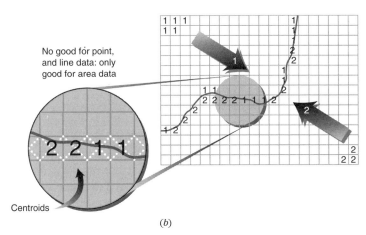

No good for point, and line data: only good for area data

Centroids

(b)

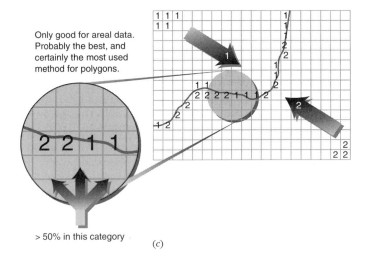

Only good for areal data. Probably the best, and certainly the most used method for polygons.

> 50% in this category

(c)

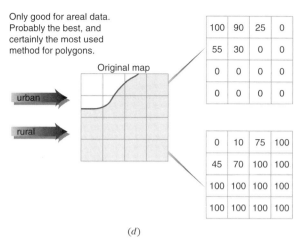

(d)

Figure 5.6 Methods of raster data input. The four basic methods of raster data input as defined by Berry and Tomlin (1984): *(a)* presence/absence, *(b)* centroid-of-cell method, *(c)* dominant type method, and *(d)* percent occurrence method.

Presence/absence is the best method, in fact the only useful method, for coding points and lines for grid systems, because these entities do not normally take up a large portion of a cell's area. Thus if a road crosses through a grid cell, its presence is recorded with an attribute code (a number); if it does not, it is ignored.

The second method of raster data input is called the **centroid-of-cell method** (Figure 5.6*b*). Here the presence of an entity is recorded only if a portion of it occurs directly at the central point of each grid cell. Clearly this requires substantial calculations, since each central point will need to be calculated for each grid cell, and then the object will have to be compared with the location of that point. Of course, if you are using a clear Mylar grid overlay, this becomes a visual, rather than a computational approach, but still you must evaluate each grid cell centroid against each entity. As you might imagine, the chances of point entities and line entities passing directly through the center of a given grid cell are minute compared to the chances they will occur anywhere else. Therefore, the use of the centroid-of-cell method should be restricted to use for polygonal entities.

A more common method of coding area data is also generally considered to be the best method. This so-called **dominant type method** (Figure 5.6*c*) encodes the presence of an entity if it occupies more than 50% of the grid cell. Under most circumstances the decision is straightforward and the coding is reasonably representative of what is there. It seems logical that if you are restricted to a single category for each grid cell, the one that occupies the most space should be coded. Computationally, this requires the computer to determine the maximum amount of each polygon for each grid cell—again a computationally expensive strategy. Performed manually with a Mylar grid, the method is reasonably easy, since each grid cell can be given a quick visual inspection. In two cases, however, this approach may not be satisfactory. First, there may be highly irregular or elongate polygon shapes, and features that start out long and sinuous, like a stream, covering little of the grid cell, then become quite bulbous within the same cell. It is difficult to decide such cases by visual inspection, and computational approaches making use of a digitizer are likely to be preferable. In the second problem case, three or more polygon types converge in an irregular pattern within a single grid cell. With three or more polygons in a grid cell, the chances are slim that any one will occupy more than half of the area under the cell, and so you cannot use the 50% value as your cutoff point for input. Instead, the amount of area for each must be visually or computationally determined, and the largest one encoded. This situation is likely to arise when you are encoding a map that is at a small scale but retains a high amount of areal detail—for example, a detailed land use or land cover map produced as a hard copy map from satellite data at a scale of perhaps 1:1,000,000. Difficulties may arise, as well, when you are trying to encode a map of physical features such as soils, or vegetation. Natural features tend to be quite irregular and their polygons much more sinuous and intermingling. While these examples do not mean that all detailed or complex maps are to be avoided as raster data input, they may suggest the use of centroid of cell as an alternative to simplifying the input process. This is especially true if the complexity occurs uniformly throughout the map.

A final method of raster data input, called **percent occurrence,** is also used exclusively for polygonal data (Figure 5.6*d*). The idea is to give more detail, not by coding just the existence of each attribute but, rather, by separating each attribute out as a separate coverage, then recording the percentage of the area of each grid cell it occupies. For example, a map of land use divided into urban and rural categories would be separated into two more specific coverages, one urban and one rural. The percentages urban and rural would be recorded for each grid cell, with 0s entered to indicate nonappearance of a category. If the percentages are accurately calculated, the urban and rural maps should be perfect complements.

The percent occurrence method offers the advantage of more detailed data about each attribute. However, the disadvantages are numerous. First, of course, both for automated and visual approaches, the amount of decision making is greatly increased. Second, we need to cross-check both complementary coverages to ensure that the percentages for each grid cell add up to 100%. Finally, as in the case of the dominant type method, we have the problem of multiple categories in a single grid cell. In this case, however, this problem may be even more severe because the more categories we have, the more coverages we produce: a map of land use with 15 attributes explodes into 15 coverages, each one representing a single attribute. This problem of data explosion is the same one we encountered in the GRID/LUNR/MAGI raster data model, which records each attribute separately, but only as 0s and 1s. Before the percent occurrences method is selected, the advantages of the additional data detail should be weighed carefully against the input problems.

REMOTE SENSING AS A SPECIAL CASE OF RASTER DATA INPUT

As we saw in Chapter 2, a common raster input to a GIS consists of digital remotely sensed data, which can be quite useful. These data are not dominant, however, with respect to the many other sources of input, such as traditional cartographic products, digital elevation data, digital line graphs from topographic maps, digital land use data, and digital soils data. In addition, the raster appearance of remotely sensed data may give the impression that software designed to manipulate these raster data is by definition a GIS. Although many of the algorithms are shared by image processing and GIS, and both contribute to the field of automated geography, a GIS should not be viewed as one stage in the manipulation or analysis of electromagnetic data obtained from remote sensing devices. This view is very limiting, and it ignores the ability of GIS to operate independently of remotely sensed data, as well as the unique analytical capabilities that give it the power to, for example, analyze networks for transportation studies. Instead, GIS and image processing software should be thought of as complementary technologies, with image processing software operating primarily (although not exclusively) on electromagnetic data and GIS as a more integrative tool, using a wide array of data types and data sources.

The utility of digital and other forms of remote sensing as data sources for GIS is unquestioned, especially for tasks such as rapid data updating and building temporal databases for large areas. Most satellite remote sensing data, as we saw in Chapter 3, are based on a raster format, with each grid cell (pixel) recording electromagnetic radiation as a number of radiometric values. The number is dependent on the type of system used. For example, *LANDSAT* data have a radiometric resolution of 256 gray scale levels, while AVHRR data derived from the *GOES* weather satellite have 1024 radiometric levels. In either case, input to a raster GIS is easily accomplished because of the similarity of data structure. Even so, the raster data structure of satellite remotely sensed data should not drive the choice of a raster GIS data model over a vector model. Rather, the choice should be based on the use for which the database is being built. In addition, when any kind of remotely sensed data are being considered as input to the GIS, the data must be evaluated for cost, utility, and accuracy compared to data from alternate sources. Remember input rule three: avoid exotic data whenever possible. Of course a familiarity with remotely sensed data and its spatial, spectral, and radiometric characteristics may make the choice of remotely sensed data advantageous. Let's briefly examine the sources of remotely sensed data inputs and review some of the characteristics of the data discussed in Chapter 2.

Today, Aerial photography is not considered to be an exotic source of remotely sensed data. In fact, aerial photography has long been a primary source of base map data for many common products. The USGS topographic maps are largely compiled and revised from viewing stereo pairs of aerial photographs, for example, and we have mentioned the use of aerial photography as a base document for soil maps. Because many maps come from aerial photography, it is wise to find out whether maps derived from such photographs already exist before choosing to use this mode as direct input. Of course the choice may also reflect the data that were mapped from the aerial photography as well as the classification system used. The two primary stumbling blocks to the use of aerial photography as direct input to the GIS are the relationship between the classification needed and an ability to obtain those classes from the photography and, of course, the problems of rectification and lack of reference grid.

You should consult a text on general remote sensing or a specific text on aerial photography before beginning the GIS process. Prior to input you need to determine how much distortion your project can tolerate and what specific categories of data you will need to perform your analysis. When these categories are known, they can easily be matched to the scale and spectral sensitivity of your photography.

A special type of aerial photograph deserves mention because the images do not contain the scale, relief, and tilt distortions normally characteristic of aerial photographs. These products, called orthophotographs, or orthophotoquads if they are based on the areas occupied by topographic quadrangle maps, are photographic images of the earth that resemble maps in that they have a single scale. Orthophotographs are subjected to a process called **differential rectification,** which involves point-by-point correction of the scale and relief displacements normally caused by differences in elevation between the aircraft and the topography over which it flies. While a detailed description is beyond the scope of this book, it is important to note the availability of these products (both in analog and now in digital forms) as sources of input. The reader is referred to Lillesand and Kiefer (1995) for more information on rectification. If digital orthophotographs are not available, the analog versions can serve as an excellent source of manual data input to the GIS.

In our discussion of digital remote sensing in Chapter 2 we noted that there are generally two major products derived for input to the GIS: digitally enhanced imagery (designed to highlight certain features for analysis, such as edges) and classified images (obtained through complex computer manipulations designed to replace the visual analyst as a classifier of features). As an input to GIS, these classified images will most likely be used to update and/or compare their classification to classified data already inside the GIS. Even when two sets of classified data are obtained through digital remotely sensed image sources, the comparison is difficult. In fact, remote sensing scientists often prefer that these comparisons be made on the raw, unclassified images, because of the potential for confusion in classification (Haddad, 1992). Indeed, direct comparisons of classified digital data to maps classified from aerial photography or historical data are very difficult. Let's look at some of the technical difficulties associated with remotely sensed data input to the GIS as defined by Marble (1981) and Marble and Peuquet (1983).

Satellite data require preprocessing to remove geometric and radiometric flaws resulting from the interaction of two moving bodies (the satellite and the earth), sensor drift as the satellite systems age, and differences in atmospheric conditions. Techniques for correcting radiometric difficulties are readily available with most digital image processing software, and the necessary equations are quite easily obtained. For GIS input, the major problem related to preprocessing is a need to obtain geometrically correct ground positions for the imagery. This geometric correction requires a number of **ground control points (GCPs)** within the image, to place it in a correct spherical coordinate space on the surface of the earth. There should be a reasonable number (the more the better) of GCPs, and they should be evenly spaced. In some areas it is reasonably easy to use devices such as GPS units to obtain very precise locations, but the ground points must also be observable on the imagery.

Obtaining adequate GCPs may be quite difficult, especially in areas like the tropics where the forest canopy or overstory is so dense that the GPS field unit has no direct line of sight to the satellite. In addition, even if line of sight exists, it is often difficult in such areas to find features that are distinguishable from the imagery. Care should be taken when using imagery lacking in GCPs because their absence degrades the coordinate accuracy severely, especially at the margins of classified areas. The issue of GCP accuracy should indicate the importance of coordinate accuracy to the overall functioning of a GIS. A proper geodetic framework will always improve the utility of a GIS to perform its measurement and other analytical functions.

We have already touched on the second major technical problem with using digital remotely sensed data, that of classification. If your project does not use direct biophysical inputs to the GIS, you most likely will have to convert from interval or ratio data to a nominal classification scheme. Remember that the classification process used in image processing software is often based on the ease with which it can be obtained. That is, the supervised and unsupervised classification methods are designed to improve the ability to obtain an adequate classification based on what is in the image; the utility of the image in an integrated GIS database is not always addressed. Supervised classification, requiring human intervention, is generally an improvement over unsupervised classification, since the process can be more easily manipulated to conform to user needs, rather than being based strictly on the statistical characteristics of the data. Nevertheless, the question of correspondence to classifications for comparable coverages remains a major consideration.

The classification of satellite imagery also implies that the results are accurate, not just compatible with existing coverages. Lillesand and Kiefer (1995) have shown that the ability of image processing software to produce classifications far exceeds our ability to assess the accuracy of the classification. This is especially true of comparisons of multiple-date imagery, where the accuracy of each data set must also be compared to the accuracy of the amount of change between the two dates. It is often best to use ancillary data to produce the classes from digital imagery for input to a GIS. The insertion of topographic data, preclassified data, rule sets, and other techniques will generally result in a much improved classification, and one that more easily conforms to the corresponding GIS coverages.

The final major set of technical problems associated with using digital remotely sensed data as an input source for GIS could more aptly be called institutional problems in that they hamper the process rather than produce error. Lauer et al. (1991) indicate six basic institutional issues that negatively influence the use of remotely sensed data. These institutional issues have been seen as more problematic than the technical problems in the integration of this data source within a GIS. We will look at the most important of these.

The first institutional problem is the general lack of availability of remotely sensed data. Although there are several major sources, the acquisition of remotely sensed data often requires the user to be familiar with the process of obtaining the data in the first place. Once the procedures are understood, the problem is one of obtaining imagery for a particular date within a given study area at a moment of minimum cloud cover. Mechanisms exist to have satellites view particular regions, at a cost to the user, and with a specified minimum cloud cover. In areas that experience persistent and substantial cloud cover, it is often necessary to piece together multiple sets of imagery, taken at different times, to obtain a cloud-free image. This process presents additional technical difficulties because of the differences in atmospheric conditions, and even changes on the ground resulting from the flowering or dying off of vegetation.

In other cases, two or more contiguous satellite images may have to be pieced together to cover a large study area completely. If, however, the reflectances obtained from these two images are substantially dissimilar, there will be an obvious line between them, and the classification process will be degraded along that margin. Finally, the unavailability in historical archives of imagery covering earlier dates may result in temporal gaps affecting the performance of spatiotemporal analysis. A primary reason for many of these institutional problems is that the data and in most cases the satellite systems themselves were originally designed as experimental rather than operational systems.

With the operationalization of the U.S. land satellite program, a different institutional problem arose. Rather than the government bearing the cost of satellite operation, as well as the cost of data dissemination, the customer now absorbs these costs through increased charges for imagery. This limits the user community to organizations whose budget can withstand $5000–$6000 price tags for *LANDSAT* thematic mapper data, for example. A result is that fewer GIS operations are likely to consider obtaining and using large quantities of these data in their day-to-day operations.

A third institutional problem with remotely sensed data input also involves money. Until recently, the hardware and software costs just to manipulate the data were too high for many would-be users. The ready availability of lower priced image processing software running on standard computing hardware has improved this situation considerably. However, the costs of manipulating remotely sensed imagery are also linked to the availability of technically trained, remotely sensed data, especially those capable of linking the raw data to other GIS coverages. Every attempt should be made to integrate the classified data with other available GIS coverages, rather than placing the burden on the GIS user. This further illustrates the need for students of GIS to become familiar with the techniques and data available to the remote sensing community.

The foregoing considerations lead directly to education and training, a problem for both GIS and remote sensing. Today technical institutions and image processing and GIS vendor companies offer programs that provide extensive, hands-on experience with the use of a particular remote sensing or GIS system. While those who complete the programs may develop an in-depth knowledge of how particular systems operate and how to perform each individual function, it is short-sighted to reduce the complexity of the discipline to one of system operations. People trained and experienced on one system are capable of producing reasonable results from complex models as long as the conceptual nature of the model is explained in enough detail. However, technicians will seldom have the conceptual knowledge necessary to formulate the solution to a problem if they don't also know how the model should work and what the GIS or image processing system is capable of doing. As you might guess, communications between, for example, an environmental scientist, who knows nothing about system requirements and a technician who knows nothing about the environment lead to inadequate, and often quite incorrect, results from analysis. While many job advertisements today list knowledge of specific systems, a person who knows the concepts and can find a way to implement them is generally going to prove more valuable. As a result, arguments over technical versus conceptual training are likely to continue in the remote sensing and GIS communities, as evidenced by another of the six institutional problems—that of professional certification. Much of this discussion entails turf conflict over who should be certified and who should not. It is of little concern to us here.

A final institutional problem of remotely sensed data input features organizational infrastructure. Neither GIS nor remote sensing has a clearly defined, well-organized, adequately funded sponsor within the federal establishment. This was clearly evident when, during the 1980s, there was an effort in the U.S. government to discontinue the *LANDSAT* satellite program. Without a flood of letters from individual users and scientists, the future of the program would likely have been very short. Thus there is a need for a champion for remote sensing and GIS to allow mutual technical improvements that will enhance the utility of both technologies.

EXTERNAL DATABASES

An efficient method of building a GIS database is to limit the amount of time and costs necessary to develop databases in the first place. Fortunately, an increasing supply of digital databases is becoming available. Digital elevation models, digital orthophotoquads, and digital line graphs are available from the USGS as well as third-party vendors (See Appendix A). The U.S. Bureau of the Census has TIGER and DIME files, as does its Canadian counterpart (see Appendix A). The U.S. Department of Agriculture makes soil maps available in digital form. There are, of course, many more examples. In fact, members of an increasingly active and visible group of digital data entrepreneurs promise to provide much needed data to GIS analysts. These organizations, whose ads appear in such commercial GIS magazines as *GIS World* and *GeoInfo Systems,* are filling a significant gap in the GIS infrastructure.

But the availability of databases also introduces other problems—including some you will also encounter as you input data to your GIS. We will examine these technical problems with a view toward how we, as potential database providers, might avoid them. The first is the physical format of the media. Countless hours can be spent trying to obtain digital data in the proper format. As we will see in the next chapter, there are many formats from 9-inch tape formats to 8 mm tape formats to CD-ROM technology, and the like. You must be able to get the data in a format compatible with your retrieval equipment. While this may seem obvious, but unless you are quite specific about your needs, a vendor may provide data in a standard (unusable) format. Even if the media are the same, there are numerous formats in which data can be provided. This need not lead us to a detailed discussion of data formats and data exchange standards (Moellering, 1992), but it should suggest that you be aware of the format your current system requires or produces. You will learn the technical details as you gain experience with your system.

A more insidious problem with external databases deals with the quality of the data. While some third-party data vendors may provide easier access for data than could be obtained from government bodies, you need to be aware that the service may not supply data in the original format. Some data, no matter the source, will be filled with easily viewable errors, some systematic and correctable, some not. You need to be aware of the quality control procedures used by each vendor. In addition, you need to know what your options are concerning return of poor quality data. Ask where the data were obtained. Were they created in-house by qualified professionals, or were they obtained from digital sweatshops, frequently operated by severely underpaid and untrained individuals? All these questions are vital to the utility of the data. Unfortunately, current standards and practices are generally quite low. Some vendors are unwilling to define their quality control procedures; others are unable to do so because such procedures are lacking. Even descriptions of the data themselves are often inadequate or wrong. As in any purchase, you should demand a complete accounting of what you are receiving. Details should include the specific format provided, the quality control procedures under which it was developed, the quality you can expect, the return policies of the vendor, and any other pertinent information that will help guarantee successful integration of the data into your GIS. Vendors that fail to comply with these requests should be considered suspect.

A major problem that is often encountered with the use of an external database is one you should also take to heart as you prepare your own databases. Databases require information about their own content; such **metadata** or **data dictionary** material amounts to information about information. There are two general forms, active and passive. Passive data dictionaries or metadata might include scale, resolution, the names of the data fields in the database, the codes used, and what they mean. Imagine the GIS applications person obtaining from a vendor a database that includes a category called "wetlands." The definition may seem self-evident to you, but you need to know more about the vendor's criteria for establishing this category. Remember, one person's wetland may be another person's watered lawn. The metadata should provide enough detail to ensure that any analysis based on it will be valid. This, of course, should remind you to keep a clear and concise record of your operations in a form that will enable someone unfamiliar with your original database input procedures to recreate them.

Active data dictionaries operate on the GIS database by performing checks for correctly coded inquiries. For example, if your vector GIS database management system is set up to allow only a four-digit code for a particular entity, the active data dictionary could check each inquiry to determine whether this four-digit limitation is uniformly met. Such checks are quite useful for allowing proper functioning of the system and for preventing erroneous results from incorrect input requests.

Beyond their technical problems, external databases are accompanied by certain fundamental legal and institutional problems. We will look at these in more detail later on, but they need to be mentioned for completeness here. A major institutional problem is that external databases are often hard to find, especially if they were produced in government agencies that may be tasked with dissemination, but not promotion and advertising. There are presently no major efforts to consolidate GIS database catalogs to facilitate searching, although more electronic newsnets and home pages are being established all the time. This is being done on a piecemeal basis, however, and ignorance of available databases often results in costly data redundancy. The cost of the data is also an institutional problem that may limit access. It is not the overall cost per unit that is prohibitive, but the frequent practice of providing data in large units that cover far more than a user needs.

Among the thornier issues facing the GIS user today is the fairness of having to pay for data produced through public funds. Tied to this, of course, are the twofold problems of data access and data security (Dando, 1991; Davies, 1982; Rhind, 1992). While many believe that public data should be readily accessible by the general public, the problem of sensitive data, such as the specific locations of endangered species or military storage facilities, makes the issue much less simple. Even when data can be obtained through the U.S. Freedom of Information Act, the time required to complete the legal documentation may exceed the life cycle of the GIS project for which the data are needed. These problems will not be solved easily, but you will encounter them as you continue on your journey in automated exploration.

Terms

stability	repeatability	linearity
resolution	skew	splines
translation	rotation	inverse map projection
registration points	tick marks	scale change
spatial information product	information theory	fuzzy tolerance
presence/absence method	centroid-of-cell method	scale-dependent error
percent occurrence	differential rectification	dominant type method
data dictionary	ground control points (GCPs)	metadata

Review Questions

1. What four characteristics are shared by every method of GIS data input?

2. What are the five factors, beyond cost, that should be examined in deciding on a digitizer? Define each.

3. What are the fundamental differences between line-following scanners and drum scanners? What potential problems relating to the map document itself act as sources of error for line-following scanners?

4. Converting between raster and vector data structures can produce results that degrade the quality of the initial input. What primary problem might occur during conversion from vector to raster? Where is this most likely to occur? What problem often arises during conversion from raster to vector?

5. Describe the transformation processes involved in moving from the digitized map to three-dimensional coordinates and finally to an output map. What is an inverse map projection? How does it relate to the map projection process?

6. Illustrate the processes of translation, rotation, and scale change? Why are these important to the input subsystem of a GIS?

7. Why is map preparation important to GIS data input? What are registration points or tick marks used for? Why are they needed?

8. Why should you mark your map before digitizing? What kinds of information should you include on your prepared map document? Why do you

need to provide information about projection and grid system when the digitizer is set up for planar or Cartesian coordinates?

9. What is fuzzy tolerance? Why is it important? What difficulties can arise if it is set too low? Too high?

10. What is the potential impact of the cartographic medium on the digitizing process? How does this relate to temperature and humidity? What can be done to reduce errors due to media distortion?

11. How do you decide what to input to the GIS? What is a spatial information product and how does it relate to GIS input?

12. List and explain the seven rules determining what should be input to the GIS.

13. What is a good rule of thumb to help you determine how much to input? What do we mean when we say that digitizing is a sampling procedure?

14. What does data input have to do with information theory? What general rule does information theory suggest for data input? Suggest situations in which this general rule is less useful than one might suspect.

15. What is scale-dependent error? How does it relate to the input subsystem of a GIS?

16. Why do you need to read the software manuals in regard to vector data input? Isn't this just a simple matter of pointing to a point and pushing a button? What are the numbered buttons on the digitizer puck used for? What can you do while you are digitizing to reduce editing time later on?

17. What is so critical about attribute data input in the vector domain? What is the primary problem that can arise if it is not done carefully? Why is this situation so hard to detect later?

18. What are the four basic methods of raster data input? How do they differ? What are the advantages and disadvantages of each? Which is (are) best for point and line data? Which results in data explosion?

19. What are the technical problems involved in using aerial photography for GIS input? How about digital satellite data? What are ground control points, and why are they important in using digital satellite data? What institutional problems are involved in using remotely sensed data as GIS input?

20. What positive impact are external databases likely to have on the growth of the GIS industry? What are some of the major technical and institutional problems with using external databases? Why is the data dictionary or metadata so important? What is the difference between active and passive data dictionaries?

References

Berry, J.K., and C. Dana Tomlin, 1984. Geographic Information Analysis Workshop. New Haven, CT: Yale School of Forestry.

Calkins, H., 1975. "Creating Large Digital Files from Mapped Data." In *Proceedings of the UNESCO Conference on Computer Mapping of Natural Resources,* Mexico City.

Cameron, E.A., 1982. "Manual Digitizing Systems." Paper presented to the ACSM/ASP National Meeting.

Chrisman, N.R., 1987. "Efficient Digitizing Through the Combination of Appropriate Hardware and Software for Error Detection and Editing." *International Journal of Geographical Information Systems,* 1:265–277.

Dando, L.P. 1991. "Open Records Law, GIS, and Copyright Protection: Life After Feist." In *URISA Proceedings,* pp. 1–17.

Davies, J. 1982. "Copyright and the Electronic Map." *Cartographic Journal,* 19:135–136.

DeMers, M.N., 1992. "Resolution Tolerance in an Automated Forest Land Evaluation Model." *Computers, Environment and Urban Systems,* 16:389–401.

DeMers, M.N., and P.F. Fisher, 1991. "Comparative Evolution of Statewide Geographic Information Systems in Ohio." *International Journal of Geographical Information Systems,* 5(4):469–485.

Fisher, P.F., and M.N. DeMers, 1990. "The Institutional Context of GIS: A Model for Development." In *Proceedings of AUTOCARTO 9.* Falls Church, VA: ASPRS, pp. 775–780.

Haddad, K.D., 1992. "CoastWatch Change Analysis Program (C-CAP) Remote Sensing and GIS Protocols." In *Global Change and Education,* Vol. 1, ASPRS/ACSM 92 Technical Papers, Bethesda, MD, pp. 58–69.

Lauer, D.T., J.E. Estes, J. R. Jensen, and D. D. Greenlee, 1991. "Institutional Issues Affecting the Integration and Use of Remotely Sensed Data and Geographic Information Systems." *Photogrammetric Engineering and Remote Sensing,* 57(6):647–654.

Lillesand, T.M. and R.W. Kiefer, 1995. *Remote Sensing and Image Interpretation.* New York: John Wiley & Sons.

Marble, D., J. P. Lauzon, and M. McGranaghan, 1990. "Development of a Conceptual Model of the Manual Digitizing Process." In *Introductory Readings in Geographic Information Systems,* D. J. Peuquet and D. F. Marble, Eds. London: Taylor & Francis, pp. 341–352.

Marble, D., 1981. "Some Problems in the Integration of Remote Sensing and Geographic Information Systems." In *Proceedings of the LANDSAT '81 Conference,* Canberra, Australia.

Marble, D., and D. Peuquet, 1983. "Geographic Information Systems and Remote Sensing." In *The Manual of Remote Sensing,* 2nd ed. Falls Church, VA: American Society of Photogrammetry, Chapter 22.

Moellering, H, Ed., 1992. *Spatial Data Transfer Standards: Current International Status.* London: Elsevier Applied Science.

Rhind, D., 1992. "Data Access, Charging and Copyright and Their Implications for GIS." *International Journal of Geographical Information Systems,* 6(1):13–30.

Shannon, C.E., 1948. "The Mathematical Theory of Communication." Bell System Technical Journal, 27:379–423, 623–656.

CHAPTER 6

Data Storage
and Editing

We have completed our preparations for the exploration of our digital world. Before we begin, however, it is best to examine these arrangements. Have we forgotten anything? Are any of our preparations in error? Do we have access to the data we input? Any analyses we perform must be based on good data, correctly organized and in the proper format. In our digital environment the preparations are extensive, consuming a large portion of what we do. Each point, line, and polygon must be correctly entered; otherwise, we'll find ourselves traveling paths with no ends, searching for polygons that should be there but aren't. We need to be sure that the correct attributes are attached to each entity, to spare ourselves from searching for relationships that cannot exist. For each coverage we produce, there is the potential for error. And our databases will most often include many coverages, thus increasing the risk of problems along the way.

The storage and editing subsystem of our GIS provides a variety of tools for storing and maintaining the digital representation of our study area. It also provides tools for examining each coverage for mistakes that may have crept into our preparations. Before we can successfully use these tools, we need to know what these possible mistakes are and how they can be discovered and corrected. If we have been careful in our input, we should encounter relatively few errors. As we've seen, however, even the selection of an improper fuzzy tolerance level can produce errors. Many of these errors will not appear until the GIS has completed its task of organizing complete coverages. In raster, for example, we may need to display each coverage to isolate illogical or out-of-place grid cells as we compare them to the input documents. In vector systems, we may have to build in topology after the initial data input, to help us to pinpoint any polygons that don't completely connect, lines that end in the wrong place, or points that occur where they should not. In the case of entity–attribute agreement, we may need to output sample portions of our map for comparison against the original input material.

As you can see, there are many aspects of error detection and correction. While you read through this chapter you will encounter a set of terminology that will be useful not only for building your geographical language filter, but for identifying the appropriate techniques with which to find errors and make corrections. When you encounter each type of error, try to relate it back to Chapter 5 on input. Ask yourself how such errors could be avoided in the first place through proper planning and pre-input preparations. Compile a list of special techniques that will facilitate easier error detection and subsequent identification for each type of error you encounter. These will sometimes be standard approaches, or they may be tailored to your own needs, the types of data you most often input, and the mistakes you tend to make systematically. This is not unlike creating a set of macros for your word processor to handle typos you make frequently. Perhaps, for example, you consistently tend to digitize your lines just short of where they should intersect with other entities, when you are working in vector. Or you may make errors when you input attribute codes because you type more slowly than you read. It is a good idea to practice your input methods on sample databases and keep track of the kinds of error you make. This simple precaution can eliminate many errors before they occur because your awareness of your tendencies will make you more careful when you digitize.

Some of you may have had the opportunity to work with GIS systems prior to your formal coursework. This experience often results in an almost unendurable impatience as you anticipate "doing" GIS analysis rather than wasting time with this editing stuff. But remember that while you may have seen some of these errors, and even had the opportunity to correct them, many GIS professionals struggle for years as they continue to make them over and over again. In the often hectic world of commercial GIS applications, there is seldom time to systematically examine your work habits as you rush to finish a project that is due yesterday. Take the time now, when the extra time spent does not cost you or a client money. At the same time, share your experiences with your classmates and your instructor. This will help you to recognize your own consistent patterns of error production and to devise methods of correction. It will also add immensely to the level of understanding for the entire class. There may well come a time when you will have to help train a new employee in database production. This experience will be invaluable to you and to your GIS operation.

LEARNING OBJECTIVES

When you are finished with this chapter you should be able to:

1. Know what tiling is and what its purpose is.

2. Understand the three basic types of error that can occur in GIS, how these errors are edited, and the importance of editing in GIS databases.

3. List six areas of entity error with specific examples and suggest how they might occur, how they are edited, and how they can be avoided.

4. Describe the types of attribute error in both vector and raster, suggesting how they might occur and how they are detected and corrected.

5. Describe the process of converting projections using a vector GIS.

6. Understand the idea of edge matching and why it is needed.

7. Describe the process of conflation, explain why it is needed, and discuss conceptually how it is executed.

8. Describe the process and the purpose of templating.

STORAGE OF GIS DATABASES

An analysis of the precise computer methods for storing GIS databases is well beyond the scope of this text, as are the ever increasing types of hardware technology used to record the data. The methods themselves are also highly dependent on the data model used in your system (refer to Chapter 4). Still, the storage portion of the storage and editing subsystem is worth mentioning, at least as it relates to the necessity of editing and updating databases.

In raster systems the attribute values for the grid cells are the primary data stored in the computer, usually on a hard drive, whether you are using a UNIX-based workstation or a personal computer. The locations of each grid cell are catalogued by their positions relative to the order in which they are placed in columns and rows. In other words, their positions are relative to the locations of the other grid cells. For this reason, editing is primarily concerned with the correct relative positions of each grid cell. Some raster systems, as we saw in Chapter 4, use compact methods of storage such as run-length codes, block codes, raster chain codes, and quadtrees. To effectively examine the relative positions of individual grid cells, most often you must be able to retrieve the data from storage for display in a manner that allows each individual grid cell to be identified separately by column and row position as well as by attribute code.

If your raster system allows a linkage to a database management system, the matter becomes somewhat more complicated in that each grid cell has attached to it a number of different attribute codes. Depending on how this is done with your particular GIS, you may have to display and analyze each set of coverage attributes as a separate map. Others may provide you with the ability to list the attribute codes for each grid cell as you examine it. You need to become aware of your system's editing capabilities and approaches.

In vector, the entities and attributes are either stored as individual tables within a single database or as separate databases, linked by a series of pointers. The separation of entities and attributes requires you to look at the editing procedures applied to entities, attributes, and databases. You can retrieve the graphic entities separately and display them to identify missing objects, incomplete links, and polygons. By retrieving the attribute tables, you will be able to examine them apart from their linked entities to determine whether you have typos, incorrect code sequences, or even the wrong attributes in the wrong columns of your table. Finally, you will be able to retrieve part or all of your database (i.e., parts of the graphics and/or parts of the database), to examine both the entities and the attributes for agreement. You will most often be able to isolate individual entities and display, on the same screen, the attribute values you desire.

Many vector GIS systems also allow you to separately store portions of your database as large, predefined subsections for archival purposes. This process called **tiling** is most often used to reduce the volume of data needed for the analysis of extremely large databases. Say, for example, that you are creating a detailed database for an entire county. You may wish to divide the database into smaller tiles based perhaps on the coordinates of the individual maps (such as topographic sheets) that you used for input. While tiling does not require that you use such a formal framework, many find it useful for the sake of keeping better track of the tiles that exist. In addition, some analyses may require only that you select a particular portion of the database to work on, upon entering the GIS. Tiling the portions individually means that you retrieve just the portion of the overall database with which you are going to work, thus reducing your computational overhead and increasing system response. Another important purpose of tiling is to allow a system administrator to have final control over the editing and updating process by permitting only certain sections of the database to be operated on when needed. Even when small portions are released for editing and updating, the system maintains an original copy of the preedited database until the system administrator is satisfied that the updates and edits are correct. Thus by limiting access to those who are qualified to make changes, corruption of an entire database can be prevented.

Tiling not only provides needed data security and reduces the amount of data; it also acts as a screening mechanism for the entire editing process. Most often the database is completely cleaned and edited prior to tiling and archiving, and as a rule access is gained for updating and analysis. This is not always the case, however, and you will need to select the appropriate tiles to do your editing. In some cases you may need to perform the process called **edge matching**—operating on more than one tile at a time to ensure that there is a correct match between the two tiles for entities that extend across the tile boundaries.

In general, today's GIS software, whether raster, vector, or quadtree, provides a visual display mechanism that will enhance your ability to visualize the errors. The exact methods will depend on the data model you are using and the sophistication of your system. Since most systems allow interactive editing within this visualization subsystem, it should be possible to correct errors as you detect each one. This is a far cry from the old days of computer assisted cartography, in which a hard-copy output of each entity's coordinates had to be produced and compared against the written coordinates typed in originally. Still, despite the sophistication of modern GIS software and its ability to find some obvious errors, the process is not completely automatic. You will need to interact closely with the software to both detect and edit the coverages. This is all the more reason for ensuring that your map preparations prior to input are complete. The prepared documents will often be used as a form of truth set against which to evaluate the digital database.

THE IMPORTANCE OF EDITING THE GIS DATABASE

While some errors might occur as a result of computational miscalculations and rounding error in the GIS software, and this does happen from time to time, most database errors result from improper input. Even with the most meticulous map preparation procedures, the finest equipment, and the best trained input technicians, mistakes will happen. Causes include simply pushing the wrong button on the ditigizer puck, a hand made shaky by fatigue, typing errors during attributes input, and even registration difficulties. Actually, the potential sources for error are numerous and include problems with the input documents themselves (Laurini, 1994). But the most annoying aspect of input errors is not their source; rather, since such mistakes are generally very small and extremely difficult to find even with the best GIS software, correcting them is time-consuming and costly. It is not unlikely that more time will be spent correcting even a small number of errors than was used to prepare your maps and input them to the system. Your instructor may provide an exercise in editing to acquaint you with the frustration of trying for an hour to find a single entity error, which can then be corrected. Such a lesson is a good reminder to prepare thoroughly before you input, because editing is even more tedious than the input step.

We have discussed three general types of error. I'll repeat them now, so they will be prominent in your mind. The first type of error you will encounter deals primarily with vector systems and is called **entity error** (positional error). Entity error can come in three different forms: missing entities, incorrectly placed entities, and disordered entities. We will discuss these in more detail later on. The second type of error is **attribute error**. Attribute error occurs in both vector and raster systems, often with the same frequency. Most often attributes are typed in, and the sheer volume of typing required for large databases often constitutes a major source of error. In vector systems attribute errors include using the wrong code for an attribute, as well as misspellings, which make an attribute impossible to retrieve if a query uses the correct spelling. In raster, the input most often consists of attributes, so the result of typing a wrong code number or placing it in the wrong grid location is a map that displays these incorrectly coded grid cells in the wrong place. Such incorrectly placed attribute data comprise the third kind of error, **entity–attribute agreement error** (logical consistency), which also occurs in vector systems when correctly typed codes are attached to the wrong entities.

Of the three basic types of error found in GIS databases, the latter two, both relating in some degree to attributes, are the most difficult to find. Mistyped attributes placed in correct locations (e.g., in a particular place in an external database) might be found if an active data dictionary is part of the system. This feature generally is helpful if you have violated a rule already established for the data dictionary, such as putting in numbers in locations requiring letters, or putting 80% of a five-digit code in a four-digit slot. However, misspelled attributes may not be found until you actually perform an analysis. Entity–attribute agreement error is often even more difficult to find than misspellings or incorrect codes. In raster, the only way you can observe problems of this type is to display the map to identify misplaced grid cells. With vector, you will most often be able to point to entities and display their attributes on your monitor. However, the GIS is not likely to be able to tell you that you have the wrong attribute attached to a particular entity. Instead, you will need to have a copy of your input map beside you as you display or highlight each entity.

As you might guess, if you created a very complex database, you may have to spend months evaluating each of the thousands of entities and making comparisons to your input document. It is far better to check these errors in small groups, as you input the database. For one thing, you will be more familiar with the data when you are inputting them than you will be if you go back later. In addition, the input document is already there in front of you. For this reason, some software vendors allow you to use the editing portion of the GIS as a method of input, rather than the input subsystem. Some vendors take an alternate approach and build the editing capabilities into the input subsystem. In either case, you can examine your map for entity as well as attribute errors and determine agreement problems as they occur. While these steps will slow the input process, once again remember that it is much better to do it right the first time than to spend hours correcting mistakes after an entire coverage has been encoded.

Although you've heard this before, it won't hurt to hear it again. Error-prone data will contribute to error-prone analysis. And while simple errors may seem quite innocuous, even the simplest can produce results that are grossly incorrect. As a simple example, imagine a database containing over 8000 polygons, some of which illustrate the locations of highly toxic materials; a single polygon (let's say it's polygon number 2003) bears an incorrect attribute code indicating that the location contains no toxic material. In your analysis you are searching for the polygons that correspond to the highest rates of cancer mortality in an area. As it turns out, the highest rate of cancer mortality on one coverage corresponds to polygon 2003 in your toxic substance coverage. Thus although common sense would lead you to suspect a direct spatial correspondence between the highest cancer mortality and very high toxic substances, your analysis fails to demonstrate this. In the short term, your GIS analysis has produced incorrect results. In the long term, there is a likelihood that decisions concerning the cleanup of toxic substances for health reasons will be incorrect. Such incorrect decisions are among the topics covered in an ongoing discussion about the legal liability of GIS databases as tools for decision making (Epstein, 1989; Seipel, 1989). And while this hypothetical example may seem extreme, it should point up constant possibility that minor errors in data can produce major errors from analysis. Fear of litigation may be as good a reason as any to spend the time necessary to ensure database integrity and accuracy.

DETECTING AND EDITING ERRORS OF DIFFERENT TYPES

As we have seen, a GIS database is subject to errors involving entity, attribute, and entity–attribute agreement. While these three are distinctly different, in the following discussion we will look at entity errors first, considering attribute errors and entity–attribute agreement errors together in a second group. Most often attribute errors are detected and identified because of a failure of the entities and attributes to agree. While this is not always the case, the detection of pure attribute errors is most commonly performed by producing tabular listings of the attributes or their tables. Although this is certainly part of the process of error correction, a complete description is unnecessary. Examples for each type of GIS would take up a large portion of the book, so we will see how errors are detected for a common system or two, and you can modify your precise procedures based on the GIS you use.

Entity Errors: Vector

Upon completion of the digitization, GIS systems require you to perform an operation that builds topology, unless this was part of the digitizing procedure itself. In either case, the topology, by providing explicit information about the relationships of the entities in your database, should permit you to identify the types of entity error in your coverages. Some of these errors will be pointed out through text-based error flags telling you that you have a problem. Others must be interpreted, by looking at database statistics concerning the numbers and types of entities, or by inspecting the graphics displayed on the screen for error the GIS is not designed to detect. You will be looking for six general types of errors represented by the negative case of the following statements (Environmental Systems Research Institute, 1992).

1. All entities that should have been entered are present.

2. No extra entities have been digitized.

3. The entities are in the right place and are of the correct shape and size.

4. All entities that are supposed to be connected to each other are.

5. All polygons have only a single label point to identify them.

6. All entities are within the outside boundary identified with registration marks.

A major commercial GIS should be able to provide these general topological relationships, and you can use them to identify errors. A useful procedure for comparing the entities you digitized and the original map document is to produce a monitor display or even a hard-copy plot of the preedited database. The latter will allow you to physically overlay the two maps using a standard light table. In addition, many GIS systems provide a number of symbols that indicate some errors. To save time, you should become familiar with these before you begin to edit. Let's now go through and identify specific types of error we can find related to the six general types.

As you remember from our discussion of vector data models, nodes are special points that indicate a link between lines composed of individual line segments. In such vector data models as POLYVRT and DIME, for example, the nodes are often described as to nodes and from nodes, indicating the overall extent of the line feature. Nodes are not just points between line segments that show directional changes in the line; in addition, they have specific topological meaning. Nodes may be used to identify the existence of an intersection between two streets, or a connection between a stream and a lake, but they should not occur at every line segment along a line or a polygon. Thus the first type of error that can be detected entails false nodes called **pseudo nodes**, which occur where a line connects with itself (an **island pseudo node**: Figure 6.1a) or where two lines intersect along a parallel path rather than crossing. Your GIS should be able to flag the existence of pseudo nodes by means of an easily identifiable graphic symbol. When you build your first GIS database you may be perplexed by an abundance of pseudo nodes showing up on your coverage. Before you panic you should be aware that some pseudo nodes are not errors

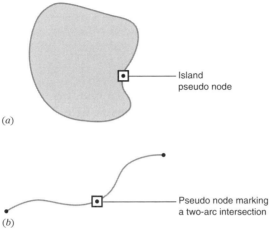

(a)

(b)

Figure 6.1 Pseudo nodes. Two types of pseudo node occur as errors, island pseudo nodes that indicate the presence of an island polygon inside another larger polygon when this is not the case *(a)*, and pseudo nodes that occur between two adjacent line segments of a single line *(b)*. Some pseudo nodes are permitted when there really is an island polygon (spatial pseudo nodes) and those that indicate attribute changes along a line, such as a speed limit change (attribute pseudo nodes). *Source:* Figure derived from Environmental Systems Research Institute, Inc., (ESRI) drawings.

but are merely flags indicating the presence of potential problems. A pseudo node connecting a line with itself may simply be the beginning and ending of an island polygon (sometimes called a **spatial pseudo node**), in which case its flag can be ignored. Or, for two line segments with an intervening node (also known as an **attribute pseudo node**), the node may indicate something as simple as a speed limit change (Figure 6.1*b*).

Spatial pseudo nodes that are not the result of your efforts to produce an island polygon (i.e., a polygon within a polygon) are most often due to a misplaced data point or to pushing the wrong button on the digitizing puck. In other words, either you were trying to create a nonclosing structure but placed the puck in the wrong place, or you were trying to create a polygon that connected to other polygons (i.e., had neighbors that share line segments) but pushed the wrong button. As a means of avoiding improper spatial nodes, you can number your points when preparing your map or use a special code or symbol to indicate where your nodes are going to be. A good practice is to use a numeric code that is identical to the numbers on your digitizing puck that correspond to nodes. This procedure will also reduce the likelihood of producing an erroneous pseudo node for line entities.

If your software indicates that your coverage has one or more pseudo nodes, the prepared map document can be used to help you correct any mistakes. First, you need to determine whether the pseudo nodes are in fact errors. Correct pseudo nodes (i.e., those present for a particular purpose) can be ignored. Incorrect nodes can be corrected by either selecting them individually

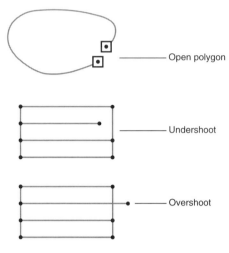

Open polygon

Undershoot

Overshoot

Figure 6.2 Node errors. Illegal dangling nodes come in three basic types: those that are a result of a failed polygon closure, those classified as undershoots because a node falls short of an object to which it is meant to be connected, and overshoots, where a node lies beyond an object to which it is supposed to be connected. *Source:* Figure derived from Environmental Systems Research Institute, Inc. (ESRI) drawings.

and deleting when necessary, or by adding nodes where needed to convert to a polygon an island that is attached to other polygons. The process is generally easy to perform with commercial software.

Another common node error, called the **dangling node**, can be defined as a single node connected to a single line entity (Figure 6.2). Recall that in some GIS systems you most often need to have a from node and a to node, rather than just a single node. Dangling nodes, sometimes just called dangles, can result from three possible mistakes: failure to close a polygon, failure to connect the node to the object it was supposed to be connected to (called an **undershoot**), or going beyond the entity you were supposed to connect to (called an **overshoot**). In some cases the problem is a result of incorrect placement of the digitizing puck, while in others the fault lies in a fuzzy tolerance distance not set large enough to account for the level of digitizing accuracy you are normally capable of producing. Setting the fuzzy tolerance properly is, of course, one way of avoiding this problem, and map preparation is another. It is generally easier to find overshoots than undershoots. If you have a tendency to produce dangling nodes, a good practice is to overshoot rather than undershoot the line you are trying to connect. While accurate digitizing would be even better, this method has proven useful for some who produce this error frequently.

Dangling node errors generally are identified by a graphic symbol different from the one used for pseudo nodes. In addition, if your dangles indicate an open polygon, the GIS will alert you by telling you the number of complete polygons in the database: if it differs from the count you had prepared prior to digitizing, you know you need to look for these dangles as incomplete polygons. Corrections are again quite simple. For undershoots, the node is identified and is moved or "snapped" to the object to which it should have been connected. Overshoot errors are corrected by identifying the intended line intersection point and "clipping" the line so that it connects where it is supposed to. In the case of an open polygon, you merely move one of the nodes to connect with the other. Most often the GIS will eliminate one of the nodes when this has been done.

As with pseudo nodes, some dangles are intentionally input to the GIS for a particular purpose. Most often these nodes serve as indicators of something

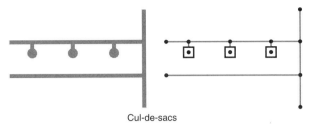

Cul-de-sacs

Figure 6.3 Acceptable dangling nodes. Legal dangling nodes created to indicate the existence of cul-de-sacs along a residential street. *Source:* Figure derived from Environmental Systems Research Institute, Inc. (ESRI) drawings.

important at the end of a line or arc. For example, you might use nodes to indicate the locations of residential cul-de-sacs (Figure 6.3). In even more unusual circumstances, a line used to indicate the location of a multistory building may contain numerous nodes, each one indicating the location of a separate floor. While this abstraction deviates from the normal cartographic form, it illustrates the potential for dangling nodes to be legitimate objects within the database, rather than existing as errors.

While you are digitizing polygons, you will need to indicate a point inside each polygon that will act as a locator for a label on which you will display text information about the polygon. You need only one label point per polygon, but you do need it. Two types of errors can occur relating to label points in polygons: **missing labels** and **too many labels** (Figure 6.4). Both these errors are most often caused by failure to keep track of the digitizing process. While good map preparation will reduce the occurrence of label error, most often the problem is caused by confusion, disruption in the digitizing process, or fatigue. Fortunately such errors are very easy to find. They are, once again, indicated by a graphic device distinguishing them from other error types. Editing is simply a matter of adding label points where necessary and deleting them where superfluous.

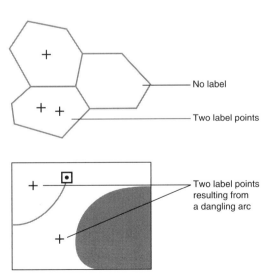

No label

Two label points

Two label points resulting from a dangling arc

Figure 6.4 Label errors. Inside each polygon there must be a single point, somewhere, to which attributes can be attached. Errors result when polygons have no labels or more than one label.

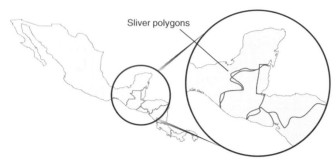

Figure 6.5   Silver polygons. These result from poor digitizing along common boundaries where the line must be
digitized more than once. Highly irregular national boundaries, as in Central America, are particularly vulnerable to
such digitizing fuzziness.

Another type of digitizing error most commonly occurs when the software
uses a vector data model that treats each polygon as a separate entity. In such
cases you are required to digitize the adjacent lines between polygons more
than once. Failure to place the digitizing puck at exactly the correct location
for each point along that line will often result in a series of tiny graphic polygons
called **sliver polygons** (Figure 6.5), an effect I like to call the Frankenstein effect
because it resembles a really bad stitching job. Sliver polygons can also occur
as a result of overlay operations (see Chapter 12) or when each of two adjacent
maps is input through a separate projection (see later section in this chapter).
We will limit our discussion to the sliver polygons produced through the input
process.

Of course the easiest way to avoid sliver polygons on input is to use a GIS
that does not require digitizing the same line twice, and in fact this requirement
is becoming less common. At times, however, you will accidentally digitize the
same line twice. The result is the same: sliver polygons. The method of finding
sliver polygons depends somewhat on whether you actually completed the
adjacent line with nodes that are effectively placed on top of each other. If you
digitized the same line twice by accident, you may also have a dangling node
hanging around because an unneeded line has been created. In this case the
line can be removed, eliminating the problem.

Finding slivers in the absence of a dangling node is more difficult. One way
is to compare the number of polygons produced in your digital coverage with
that of the original input map. It is often very difficult to locate slivers, however,
even though you know they are there, somewhere. Most often you have to move
through your image, searching for suspect polygon boundaries, then zoom in
to see the slivers. In some cases you may have to zoom in several times to see
them. Frustratingly, you often don't know until you see the slivers whether you
are zooming in on a single line or on sliver polygons.

Sometimes when you have a series of very tiny polygons but no dangling
node, it is possible to adjust (coarsen) the fuzzy tolerance you used during data
input. If this measure is possible within the editing subsystem, the software will
remove the slivers automatically.

A separate problem related to polygons is the production of **weird polygons**,

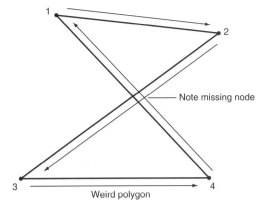

Figure 6.6 Weird polygons. Example of the creation of a weird polygon by digitizing points out of sequence. While graphically we seem to have two polygons, the point at which the lines cross in the middle does not have a node.

which are defined as polygons with missing nodes (Figure 6.6). In this case the polygon is a graphic artifact that appears to be a true polygon but is missing one or more nodes. Generally this occurs when two or more lines cross over, producing the semblance of a polygon. The most frequent cause of this error is a point digitized in the wrong place or in the wrong order. For example, let's say you have a rectangular polygon that requires only four points to define it (Figure 6.6). You would want to start at the upper left-hand node, move to the upper right, then lower right, and lower left, and end at the upper left where you began. However, perhaps because you did not number these points during your map preparation procedures, you instead go from upper left to upper right, and then mistakenly go to the lower left and then the lower right (in a sort of figure-eight arrangement), finishing at the upper left where you had begun. While you digitized all the points correctly, your polygon looks more like an hourglass than a rectangle. In effect it appears like two triangular polygons connected at a central point. However, the central point has no node.

As you can surmise from the example of Figure 6.6, a simple way of avoiding this problem is to number the input points. Even if you don't, however, you can avoid the problem by establishing a set pattern for digitizing polygons. For example, you may decide always to digitize polygons in a clockwise fashion. Such a systematic procedure for digitizing will most often keep you from omitting essential nodes. As an aside, many users also take this systematic approach in deciding which portions of a map to digitize when multiple sessions will be required to complete the procedure. It is a good habit to get into.

Detecting weird polygons is difficult but is not impossible. One straightforward method is to highlight the nodes and display them as part of the polygon coverage. In this way, the areas that should have nodes but don't will stand out from those that are properly digitized. Editing the error involves moving the lines to the correct locations, thereby placing the nodes in the correct sequence. Sometimes it is easier to simply remove the offending lines and use the editing subsystem to redigitize the points in the correct sequence.

The errors we have seen thus far are among the easiest vector mistakes to find; as a rule, you can make the necessary adjustments without plotting out your map to identify the errors. A more annoying group of entity problems is identified by numbers 1,2, 3, and 6 in the preceding list. Problems of missing objects, additional objects, and displaced or misshapen objects are easiest to

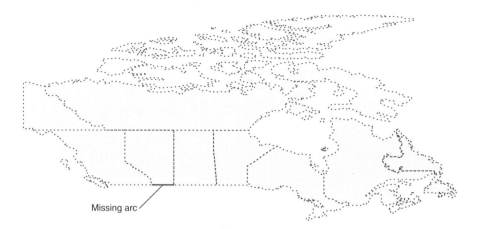

Figure 6.7 Missing arc. Illustration of the use of source map comparison to identify common graphical errors. Notice the missing arc in the digitized version of this map of Canada.

detect by plotting the digitized coverage at the same scale used for input (Figure 6.7). If you overlay the original document and the plot on a light table, you can identify problem areas. Nearly all these errors are due to lack of preparation, although the problems of disruptions and fatigue will always play a part in their occurrence. Correcting them requires you to mark the problem areas on the map document, preferably by noting the exact nature of the problem and how you will correct it. If you are missing an object, mark it by indicating, in order, the points, lines, and polygons that need to be digitized, including any other information concerning locations of nodes and other topological information that may be needed. Additional objects should be marked for removal. For objects that extend beyond the map extent borders defined by your registration marks, the points need to be removed and redigitized. Objects that are misshapen or incorrectly placed but still within the map extent can most often be individually selected and moved, without being redigitized.

In all these cases, consult your GIS software manual to determine the exact commands needed to perform the operations. Because you are now aware of the typical graphical problems that can occur, you can easily find the appropriate instructions in your documentation. But, before we leave the vector entity error procedures, there are a couple of caveats. First, remember that you have made changes to your entities, and the computer needs to understand that these changes may modify the spatial relationships you originally input. Most GIS software will require you to enter one or more commands to indicate that the changes have been made. The software will most likely also have to be instructed to rebuild the topological structures based on the new entities. Finally, you must save and store your new map. Forgetting to save changes to a coverage representing hours of editing is a mistake you will not want to repeat.

Attribute Errors: Raster and Vector

As we have seen, attribute errors, including attribute–entity agreement, are among the most difficult to detect. This is primarily because the GIS does not know which attributes are correct and which are not. Because vector entity and raster grid attributes (and their assumed entity locations) differ significantly from application to application, and because there is no attribute equivalent of topology, there are no rules against which the GIS can check your accuracy. By this I mean that there are no explicit rules to indicate that a particular attribute occurs with a particular, stable pattern with respect to its neighbors. If there were, much of what we do in analytical operations of the GIS would be redundant. In fact, it is the search for these consistent patterns that often drives the analysis in the first place. Perhaps, with a few decades of research, we may be able to define some of these, but in the meanwhile we are forced to compare the attributes in our digital database to the original map document to determine most of the attribute errors that might occur.

Missing attributes are perhaps the only attribute errors that are detectable without direct comparison to the input document. In raster these tend to occur as maps that are missing whole rows or columns or portions of rows or columns. These are detectable because familiarity with the original map shape will alert us to the absence of rows or columns of grid cells that substantially alters the overall shape (Figure 6.8). Missing rows and columns in raster are most often caused by missing one's place while typing the grid cells values; seldom will this problem occur when a digitizer is used. If you are typing in grid cell values and miss several of the attribute values, quite often an unusual value will appear in your map, one that has little correspondence to the rest of the data. This data value is probably a row indicator value that tells the GIS which row number you are on as you input the grid cell attribute values. From that point on, the grid cell values will be in the wrong place. A simple way to avoid this type of error is to display the text equivalent of the attributes prior to completing the input. This is very easy if you are using a word processor for creating the database, and you will be able to make the necessary corrections prior to database development. Alternatively, if you are using the software to input the

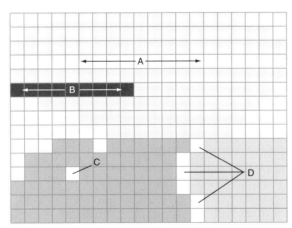

Figure 6.8 Raster attribute errors. Common raster attribute errors that are identified on the basis of how they distort the coverage: A, missing row; B, incorrect or misplaced attributes (appearing as one or more rows of vastly different values); C, incorrect attributes occurring singly; D, attributes errors occurring along area margins (most often caused by digitizing problems).

values by typing them in, the GIS most likely will indicate the row numbers as you enter the data row by row. If you are typing in the codes without the assistance of the software, the potential for missing values is increased. By having someone read the data to you as you enter the values from the keyboard, however, you can focus on your typing, minimizing this and other types of attribute error as well.

In raster, if you miss a complete row of data, or more than one row of data, your map will appear shorter than it should. Visually this discrepancy is more difficult to spot, but the software most likely will be able to tell you how many rows, columns, and even total grid cells the database contains. This response is easily checked against a known value for the complete database. The specific locations of such missing rows are relatively easy to detect if naturally contiguous patterns seem to stop abruptly along a given row. Most often there will appear to be a distinct line of demarcation between one portion of the map and another. Correcting the error is often a simple matter of downloading your grid data to a text processor and typing in the missing values, renumbering the rows as you go. Alternatively, you might be able to use the editing capabilities of the GIS to locate each grid cell and recode it. If the error occurs near the beginning or even the middle of the map, however, you will have to spend too much time editing to warrant using this method. It is best to use this method for correcting small numbers of grid cell values, or for correcting for missing rows that occur near the bottom of the map.

Missing attribute values in vector GIS are commonly caused by simply not including anything in the attribute tables for individual points, lines, or polygons. These can be identified by listing the tabular information and identifying missing attribute values in the tables, or by outputting the entities and their attributes as a video display or plot. The missing attribute values simply will not be included next to their respective entities. Commonly, if you use the second method, you can edit each entity by selecting it from the rest and inputting the appropriate attribute values. And, of course, you need to remember to save your edited work.

Incorrect attribute values are very difficult to detect, both in raster and in vector. In raster GIS, when they occur as individual cases, or short row or column segments, they are most often caused by typing errors when that form of input is used. When wholly incorrect attributes occur for large areas, they are most likely a result of inputting the wrong attribute value, either through digitizing or a block encoding method. In raster, an incorrectly coded grid cell is most likely to be identifiable as one individual that seems "out of place" among the surrounding grid cells. Unusually high or low attribute values show up easily. They normally appear as out-of-place grid cells or groups of grid cells that disrupt the natural organization of the map. If they occur as continuous strips of incorrect grid cell attributes, most software will allow you to use a run-length encoding strategy to edit these interactively. Individual grid cell errors can be selected individually and edited.

In raster maps that have few contiguous areas (e.g., maps of raw topographic data), incorrect values will not easily be seen on a two-dimensional display of the data because there are no large area patterns to disrupt. In such cases, a three-dimensional display of the raw data may show unusually high spikes or very low dips in the surface. While these abnormal features may be errors, be sure to check them out—they may be real attribute anomalies. Generally such incorrect values occur as individual grid cells and can be selected and edited interactively as before.

Incorrect attributes in raster systems also may be found along the margins of the areal patterns. In such cases, the typical culprit is either unreliable digitizing algorithms that do not use one of the four basic grid cell input methods developed by Berry and Tomlin (1984) or carelessness in the determination of the correct attribute codes along these margins. The pitfall here is that the incorrect values are most often identical to the neighboring area, thereby giving the impression that they are correct. You will need to compare the shapes of area patterns on the raster map against the original shape of the input map. Correcting such a problem usually means reevaluating each grid cell as to its correct attribute. In other words, which of the two adjacent areas does this grid cell really belong to? Once this has been determined, each can be selectively edited as before.

Incorrect attributes often are more difficult to identify in vector than in raster, since the former case generally calls for intimate familiarity with the source map and its attributes and attribute patterns before these can be compared to the digital version. If you are using a form of coding strategy that replaces the actual names of the items by, for example, a numeric value, there are many chances to enter an incorrect number. In such cases the codes will sometimes not correspond to linked tabular information included in other portions of your database or even in the data dictionary. The software should be able to flag such inconsistencies. Take, for example, a numerical coding strategy devised to represent the names of individual plant species for point entities in your coverage. While this releases the user from having to spell out the species names exactly during query operations, it increases the likelihood of input errors. If you have established an appropriate set of rules in your data dictionary—especially if it is an active data dictionary—it should be possible to flag any code you enter that does not represent a real species name stored in the tables. Many times, however, the code sequences become ingrained in your subconscious, resulting in plant codes that are real. They just aren't the correct ones. The only way to prevent such errors is to double-check each code as you enter it. Identifying this type of error requires you to display the coverage and compare it to the original. The work is tedious, but you should be able to detect most of the errors. By selecting the offending entities, you can easily edit their attribute values as before, interactively.

A common source of the foregoing type of attribute error, however, is failure to keep track of the attributes as you are typing them in: most of the time the problem is not simply that the wrong attributes are typed in but that attributes are misplaced—that is, matched with the wrong entities. Of course this mismatch is also a result of unconsciously using the wrong code, but we will treat it separately here, simply because it is distinctly different from placing a meaningless code in your tables. In many cases, these misplaced codes occur predictably. For example, you may tend to lose track of your place as you are typing codes in. This knowledge may provide a clue to finding these errors, particularly when the tables are printed and compared with the input data. Finding them with the use of graphic output is again a function of comparison to the input map. If the attributes were input using the label points on the entities themselves, this will be the best way to identify the errors. Alternatively, if your attribute tables were created separately and later linked to the attribute codes, systematic errors are more easily detected by comparing tabular data from the database with the original data from which the tables were produced. The systematic errors that are most likely to occur involve short periods during which incorrect codes are entered (again primarily by reason of losing your place), and then, suddenly they begin to assume the correct order of entry. As before, a good way to avoid this problem is to get assistance while typing in the data. Of course, if your software permits you to input your attribute data interactively, you are not likely to produce large systematic attribute misplacements, but only occasional ones.

DEALING WITH PROJECTION CHANGES

While the storage and editing subsystem has as a major function the correction of entity and attribute errors, it is also used to convert between the Cartesian (digitizer) coordinates and real-world coordinates based on a reference globe. Quite often the software will require you to identify the projection of the map upon input; but some programs, especially in the case of many (but not all) raster systems, do not. In any event, the conversion to a set of projected map coordinates is a necessity for vector systems for any analysis requiring real-world measurements. Moreover, because not all your input maps will have the same type of projection, it will be necessary to standardize them to permit comparison of the different coverages.

If you had to input the reference points using Cartesian as well as projected coordinates, you will use these as reference points for the coordinate transformations necessary to project your map. If your system operates on real-world coordinate systems but has you input the data strictly in a Cartesian coordinate structure, you must define the reference points as latitude/longitude coordinates to make the projection possible. Another important factor is the method by which the GIS stores and manipulates these coordinates. Some use latitude/longitude coordinates recorded in degrees, minutes, and seconds (**DMS**), while others will require you to convert these to decimal degrees (**DD**). The formula for converting from DMS to DD, which gives the numbers in degrees or fractions of degrees (Environmental Systems Research Institute, 1992), is

$$degrees + minutes/60 + seconds/3600$$

In many cases you will operate first on the reference coordinates by creating a separate coverage with these values only, recorded in digitizer inches. The values can be read directly from the associated tables, then edited by typing in their longitude and latitude equivalents. The reference points are now in geographic coordinates. Upon completion of these edits and after double-checking for accuracy, you will save the data and link them to the geographic coordinate system (latitude/longitude). This tells the computer that the numbers you edited are defined by a real-world (geographic) projection. That is, you are telling the computer that you have changed the Cartesian coordinates to the geographic projection.

As you remember, the map from which you input your data will likely have been produced through map projection. You must get your coordinates, which are now in a geographic projection, into the same type of projection as the original input document. Usually only a few commands, or even a single command, will be necessary to set the GIS about the transformation procedures of scale change, rotation, and translation that will produce the projection required for your reference points. Thus far we have been working with the reference points as a separate coverage, rather than with subsequent coverages. Since, in addition, these transformations require mathematical manipulation of your original geographical coordinates, errors will always be part of the procedure. No projection procedure is without error. Many software packages provide some measure of this error, and you should look at these numbers. High amounts of error resulting from this transformation are often a result of inaccurate digitizing, map sheet distortion, or incorrect recording of original reference coordinates (Environmental Systems Research Institute, 1992).

You now have a coverage of reference points, or ticks, that represent the projection of the original map input. Upon completion, the reference points can easily be linked to the input coverage from which you extracted the reference points to create the reference coverage. Whenever you use the original coverage, it will be connected to the reference coverage, which also will have correctly transformed coordinates. Thus measurements can be produced in real-world units rather than in digitizer inches.

JOINING ADJACENT COVERAGES: EDGE MATCHING

Now that you have created a coverage that contains all the pertinent coordinate transformation information, you must consider a closely related topic in the storage and editing subsystem—edge matching. In edge matching, two adjacent coverages, usually of the same theme, are physically linked to permit the analysis of a larger study area (Figure 6.9). There are two sources of difficulty when two adjacent coverages are input.

First, two maps that were input with the same projection but because they were put in separately are likely to display entity errors that are somewhat different. Keep in mind that the maps are registered to the digitizing tablet separately, the tick marks or reference points were input separately, and all the entities were input during a separate digitizing session. Therefore, while each coverage may be reasonably accurate with respect to itself, the differences in input errors between the two will most likely cause mismatches—sometimes subtle, sometimes obvious—between and among individual objects. You will need to link all the line and polygon entities that are supposed to be connected. For example, if a road digitized on one map sheet is supposed to be a straight line that runs across the two sheets, make sure that when the sheets are

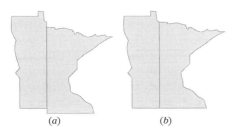

(a)   (b)

Figure 6.9 Edge matching. Two adjacent parts of a map *(a)* before and *(b)* after edge matching.

connected, you don't have a jagged edge or a slightly offset road. Both portions of the road will have to be connected so that they exist as a straight line.

The second situation that is likely to cause problems requiring edge matching arises either when two adjacent coverages are input from different projections (or with the same projection but based on different baselines or other starting points) or when the projection is applied specifically to that coverage without regard to its possible effects on the neighboring coverage.

Try taping any two adjacent USGS 7.5-minute quadrangle sheets together so that all line features are lined up. If they line up perfectly (a very unusual case), you are very fortunate, and all you have to worry about is errors introduced through digitizing. These errors will be minor if you prepared your map well and were careful during input. If the sheets do not perfectly line up, it is because the projection was applied to a single sheet at a time. You might still ask why the edges don't match up. The answer is simply that the mathematical procedures of map projection are imprecise, both because they are approximations of the three-dimensional reference globe and because of rounding error inside the computer.

Let's take a more obvious example. A match between two maps of the same area but produced with different projections is impossible. As any GIS vendor will tell you, however, the software can project the data into any projection you want. Here is another statement that ignores the two sources of imprecision: projection errors and computational rounding. Often you will find that the two maps, even when projected to the identical projection inside the GIS, will not completely line up. You may even see the Frankenstein effect mentioned in connection with the digitizing of adjacent polygons. After all, if the two maps have polygons, they too will be offset from each other.

Edge-matching problems can also occur in raster systems, at least for those that operate on projected surfaces, rather than simply in flat, Cartesian space. A common example of edge matching in raster systems is found in the use of remotely sensed data products, such as *LANDSAT* MSS (80-meter) images. Because horizontally adjacent scenes are sensed at different times, often days apart (Lillesand and Kiefer, 1995), there is the possibility that the satellite will not be located at exactly the same latitudinal coordinates. This often results in a skew between the two images of one to several pixels (grid cells). Most often this is corrected by moving one of the data sets until the pixels match. Finding these errors is not unlike our example for finding missing lines from raster data input. Even very complex satellite images usually contain areas that produce relatively similar responses that, when classified, would often correspond to a single area class. When this group of pixels is offset, even before classification, the condition is relatively easy to detect because the regular patterns themselves show the incongruity along this line. In addition, geographical coordinates are most often provided with such data, allowing the user to match the edges by matching the coordinates along both adjacent images.

CONFLATION AND RUBBER SHEETING

Another problem likely to need attention results when two coverages, perhaps representing the same study area but different time frames, are to be overlaid.

This situation is encountered primarily with vector data, although it can occur in raster as well. We will limit our discussion to vector, since many of these errors are suppressed because of the lack of spatial accuracy in the raster data structure. In fact, some prefer to use raster data for overlay operations because the measurement problems associated with registration of two vector coverages are avoided.

Let's assume that you are trying to examine the changes in land use for a given region. Your input methods rely on unrectified aerial photographs, all at the same general scale. After you input your first coverage, you establish a reference system by identifying latitude and longitude coordinates for known objects on the photos (perhaps by using coordinates for the same objects on maps or by using a GPS unit). You follow the same procedures for perhaps three more coverages, each representing a different time frame. Your first inclination is to go ahead and overlay the set of four, to begin analyzing changes in land use. When you display all four on the same screen, however, you notice a few bothersome anomalies. First, a stream that runs through the scene seems to show up in a different location in each coverage. Also, some identically shaped polygons seem to be doing a pirouette rather than staying in one place. Closer examination indicates that many objects have been moving around in the coverages as well. Yet you know that these entities do not physically move.

The problem you have reflects minor changes due to crabbing and yawing of the aircraft. In other words, because the platform was not stable, the view of the objects is altered (Avery, 1977). Without going into the details of transferring aerial photographic data into a GIS, it is at once obvious that you must find a way to georeference not just the map corners, but the internal objects themselves. No analysis of area change will be valid if the polygons themselves don't line up. You need to be able to physically tack down objects that are in the correct place while moving the remainder to more accurately indicate their locations in space. Such a process is sometimes called **rubber sheeting**, in comparison to stretching a map as if it were made of rubber, but the more technical term is **conflation**. Conflation is an interactive process that involves making decisions about which of the coverages you wish to adjust with respect to the others to induce them to coincide. Most often this is done by selecting the coverage that you believe is the most accurate in its representation of true coordinates. Then you must decide which objects to keep stable and which need to be adjusted. Objects that are to be "tacked down" (Figure 6.10) or made immobile with respect to the original coverage are individually selected. Sometimes the entire coverage is adjusted, but most often this approach produces unwanted distortions of objects that were originally quite representative of the correct locations and the correct shapes. It is best usually to spend the time identifying objects whose positions and shapes you wish to preserve and let the conflation algorithms modify the rest. Additionally, you may find that your first attempt was not as successful as you wished. This will require you to perform some additional adjustments using the same procedures. Most software will also allow you to modify or rubber-sheet individual objects to get the results you wish.

There are two things to remember when performing conflation on your coverages. First, the conflation is strictly a graphic manipulation. It does not assure that your graphic manipulations will yield the most accurate output. It simply assures you that the results are graphically reasonable based on your own faith

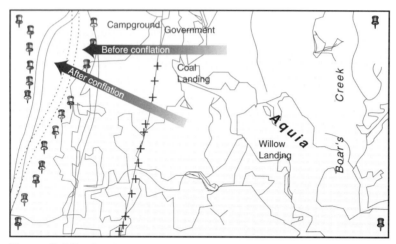

Figure 6.10   Conflation. The use of convolution to move objects in a map: some entities are tacked down to prevent them from moving, while the others are shifted to adjust to the same coordinates as the base coverage.

that the coverage you selected is the most accurate or truly representative of the real-world situation. Second, and certainly as important, there is a very good chance that the results of your initial GIS conflation run will turn out worse than the original map. Because of this, *under no circumstances should you delete the original coverages.* If the conflated coverage is indeed very poor, you may prefer to go back to the original and try it again, rather than modifying your first attempt. This lesson is one to keep in mind for the next chapter as well. Never get rid of a coverage until you are sure you don't need it. Your storage and editing subsystem is designed for such contingencies. If your system becomes cluttered, download some of your work to a backup unit. But keep all your files until they are no longer needed.

TEMPLATING

In the preceding section we mentioned the use of a single coverage as the most accurate among several for the same area. When you look at multiple coverages of the same theme for different time frames, you notice a number of graphical discrepancies, including one we have thus far neglected. Viewing the same coverages simultaneously, you also notice that, despite all your efforts to prevent it, the outside dimensions of all four study areas seem to differ slightly in shape. When you input these maps you chose certain data points as reference points and assigned them to true geographic coordinates, yet the four coverages are not identical. Probably, then, the differences in location of these reference points from coverage to coverage, combined with the nuances of the projection algorithms and computer rounding error, have produced slightly different results for each coverage.

If you must later perform an overlay of these four coverages, there will be numerous areas along the margins of some coverages that will not have associated areas for other coverages. Once again, you must select the coverage you trust most to be representative and use it as a **template** (some call it a cookie cutter). If the boundary of the template is within the boundaries of all the other coverages, you simply use this pattern to cut out the study area of the rest. However, if any of its boundaries extend beyond those of another coverage, you will need to select coordinates somewhat inside the margins of the template to ensure that all subsequent coverages will occupy all areas within the template. Once applied, the coverages should all have the same shape, coordinates, and size.

A note here about the statistics reported from multiple coverages is important. It is entirely possible that once you have applied a template to all your coverages, then produced output statistics, there will be slight differences in total area for each of the four. I've seen this result in a study of land use change in which every attempt to remove the discrepancy failed (Simpson et al., 1994). Much of the incongruence can be attributed to a combination of computational rounding and the algorithmic methods by which the GIS calculates area. For the most part, if the error is a minor percentage of the overall database, it should simply be accepted. No amount of editing is going to remove it. Your task is complete, and you are now ready to begin analysis.

Terms

tiling	edge matching	entity error
attribute error	entity–attribute	pseudo nodes
island pseudo node	agreement error	attribute pseudo node
dangling node	spatial pseudo node	overshoot
missing labels	undershoot	sliver polygons
weird polygons	too many labels	incorrect attribute values
DMS	missing attributes	rubber sheeting
conflation	DD	
	template	

Review Questions

1. What is the purpose of tiling in the storage and editing subsystem?

2. What is so important about editing a database? What problems can occur as a result of even a simple error in the database? Give an example other than one from the text.

3. What are the three basic types of error that need to be edited? Describe each. Which of these is (are) the most difficult to find? Why?

4. What are the six major areas of entity error that need to be addressed? Give an example of each.

5. What are pseudo nodes? How are they produced? How can they be avoided? Give an example or two of a pseudo node that is not an error.

6. What are dangling nodes? Overshoots? Undershoots? How are these caused? What can you do to avoid them?

7. Why is a missing polygon label point a problem? Why do we sometimes end up with missing or multiple label points? How can this result be avoided?

8. What are sliver polygons? Weird polygons? Describe methods of avoiding and correcting these conditions.

9. How do you find missing attributes in vector? Raster? How can you avoid creating a map that is incomplete in this report?

10. How do you find incorrect or displaced attributes in vector? Raster? What are the principal causes of each?

11. Describe the process of converting projections using a vector GIS.

12. What is edge matching? Why is it needed?

13. What is rubber sheeting (conflation)? Why is it needed? How is it done?

14. What is templating? What is it used for? How do you decide which coverage to use as the template?

References

Avery, T.E., 1977. *Interpretation of Aerial Photographs,* 3rd ed., Minneapolis, MN: Burgess Publishing Company.

Berry, J.K., and C.D. Tomlinson, 1984. Geographic Information Analysis Workshop Workbook. New Haven, CT: Yale School of Forestry.

Environmental Systems Research Institute, 1992. *Understanding GIS: The ARC/INFO Method.* ESR I: Redlands, CA.

Epstein, E.E. 1989. "Development of Spatial Information Systems in Public Agencies." *Computers, Environment and Urban Systems,* 13(3):141–154.

Laurini, R., 1994. "Multi-Source Updating and Fusion of Geographic Databases." *Computers, Environment and Urban Systems,* 18(4):243–256.

Lillesand, T.M., and R.W. Kiefer, 1995. *Remote Sensing and Image Interpretation.* New York: John Wiley & Sons.

Seipel, P., 1989. "Legal Aspects of Information Technology." *Computers, Environment and Urban Systems,* 13(3):201–205.

Simpson, J.W., R.E.J. Boerner, M.N. DeMers, L.A. Berns, F.J. Artigas, and A. Silva, 1994. "Forty-eight Years of Landscape Change on Two Contiguous Ohio Landscapes." *Landscape Ecology,* 9(4):261–270.

ANALYSIS: THE HEART OF THE GIS

CHAPTER 7

Elementary Spatial Analysis

Now we begin our journey into our digital world. Chapter 1 provided a conceptual and historical framework for GIS, setting the stage for our approach to this exciting field of study. In Chapter 2 we began to develop a geographical vocabulary, ensuring us a filter through which our geographic data will pass. Chapter 3 taught us how our world can be transferred to a cartographic framework, complete with symbols, levels of data measurement, and different scales and projections, all designed to model the phenomena we encounter in the real world. Chapter 4 described a computer framework within which we can represent our cartographic objects in an automated environment. In Chapter 5 we examined the techniques and technologies by which geographic data can be transferred to our digital environment, moving from map preparation to input. In Chapter 6, a consideration of data storage and editing, we saw how our work can be made easier and our results more meaningful by ensuring that our digital world correctly represents the earth features we wish to examine.

What we have done thus far is essential preparation for our journey. Some of you may find yourselves working in an environment that is mainly preparatory. Given that up to 80% of the effort for most commercial GIS operations is in preparing the digital database, it may seem that the rest is useful only to the analysts who will scrutinize the data you produce. For those who have a decided interest in analysis, the next seven chapters are what you have been waiting for. However, even if your final product is a completed database, I invite you to come along with us to see the analytical capabilities of the GIS and to watch how they operate. It is important to find out what your data can do, what their limitations might be, and the types of question the GIS analyst encounters.

There are two reasons for inviting everyone along on this journey, whether individual career goals involve database production and management, or exploration through the analysis subsystem. First, this subsystem is, for many, the most enjoyable part of the GIS world. While we may not look at work as fun, I believe you will find yourself spending far more hours browsing existing databases than you could have anticipated. For some of you who took this course because of your interest in analysis, that is probably a given, and your patience has been taxed as you awaited this opportunity. Our second, and perhaps most compelling reason for entering along these paths to database

179

exploration and analysis is that many operations possessing GIS analytical capabilities fail to use them at even the most rudimentary level. While this may not be a major drawback to the organization, depending on its actual needs, it often results in fairly limited use of a powerful technology. If the only requests are for cartographic output from existing databases, the GIS is not the correct tool. A computer assisted cartographic system could perform these tasks with greater ease, often at a fraction of the cost of a fully functional GIS. In other words, as we will see in Chapter 15 on design, the GIS becomes a $500 answer to a $5 question.

Most often, however, organizations having GIS capabilities really do need advanced analytics; it is limited knowledge of the power they possess that keeps them from asking the questions that would lead to good analysis. In such cases, and there are many, the GIS professional, whether database developer or analyst, must play the role of tutor. By taking the journey with us over the next few weeks you will see what the GIS is capable of. In turn, you will be able to share this exciting knowledge with your supervisor by suggesting analyses with which he or she may be unfamiliar. Beginning perhaps with simple queries and query combinations, then advancing to measurements and finally to comparative analysis, you can slowly introduce more and more advanced spatial thinking into your organization. This will result in an organization that is more competitive, more successful, and ultimately more profitable. And this, of course, will likely improve your own opportunities for advancement as well.

This chapter begins with a brief discussion of spatial analysis, proceeding to some of the more elementary, but extremely useful techniques involving simply locating objects based on what they are or how they are described. These simple tasks involve some additions to our spatial vocabulary—our geographic filter—that will assist us not only in the simplest identification and retrieval tasks, but also later on as we combine these with more advanced techniques. Finally you will learn some more complex identification and retrieval approaches based on higher level spatial objects that we will identify.

As before, our journey will be a conceptual one, and you will be spending more time learning to think spatially than on the commands you will need to perform analysis. Remember, the only way to decide which commands to use is to know what you are trying to do. It is a much shorter path from concepts to commands than from commands to concepts. So I invite you now to pick up your pith helmet and your machete. It is time to explore. Enjoy.

LEARNING OBJECTIVES

When you are finished with this chapter you should be able to:

1. Discuss why the analysis subsystem of the GIS is often the one most abused and why it is considered an incomplete set of geographical analysis tools.

2. Understand the importance of the analysis subsystem even for those who will primarily be building databases for others.

3. Explain the process and the importance of isolating, identifying, counting, and separately tabulating and displaying individual items.

4. Describe the differences in process for isolating, counting, and identifying the different object types (points, lines, areas) in a database.

5. Explain the use of measurable attributes in searching for lines and areas.

6. Define "higher level objects," giving examples for points, lines, and areas, and describing their utility.

7. Describe the approach to identifying higher level objects.

8. Be able to define and describe the different types of region.

9. Suggest how collections of points, lines, and areas could be considered to be communities.

INTRODUCTION TO GIS SPATIAL ANALYSIS

The analysis subsystem is the heart of the GIS: it is what the GIS is really all about. It is also, however, the most abused subsystem of the GIS. The abuses range from attempts to compare noncomparable nominal spatial data with highly precise ratio spatial data to statements about the causative nature of spatially corresponding phenomena made without testing for alternative causes. Much of the abuse of the analytical subsystem is a result of lack of understanding of the nature of the spatial data contained in the system. For example, some data are purposely ranked at the ordinal level, primarily on the basis of their perceived importance for a particular problem. Then, the same ordinal data are used for other, nonrelated analyses for which the ranking no longer applies beyond the original spectrum of values from which they were produced. This is not unlike correlating history grades to calculus grades to compare the intellectual capabilities of a student. In addition, it is not uncommon for the numerical values stored in a grid-based GIS, and meant to represent nominal categories such as land use type, to be multiplied or divided by ordinal, interval, or ratio data, yielding numerical values that essentially have no meaning. Yet the results of such analysis are frequently used for decision making. The list of possible offenses is quite large, mainly because the inherent power of the GIS is too great to be effectively applied without a firm understanding of fundamental geographic, numerical, and statistical concepts.

Beyond the problems of GIS abuse, there is a common belief that GIS is the panacea for all geographical problem solving. While the tool is quite powerful, most GIS systems are driven more by market demands than by academic requirements for solving very difficult geographic problems. In short, GIS is an incomplete set of spatial analytical tools. In many cases the user will be obliged to combine GIS tools with statistical analysis software, input/output modeling packages designed to model systems throughputs, mathematical modeling tools providing enhanced mathematical computations, geostatistical packages designed for advanced spatial analysis or subsurface modeling, or even advanced macro language packages (called GIS **applications development**), designed to simplify the GIS tool kit for a particular set of tasks. But before you get discouraged, remember that many different GIS packages are available, each with its own special strengths and weaknesses. Some packages provide direct links to other analytical software to enhance their utility; others provide data structures that allow the results of external analysis to be passed back and forth to the GIS. The limitations of the GIS you choose should be weighed against the kinds of analysis you most often perform. For this reason, and to be able to select the system or systems that best suit your needs, you should be aware of the myriad possibilities available in the larger field of automated geography.

Most GIS packages are heavily oriented toward the ability to overlay two or more map coverages to analyze corresponding patterns, or to operate on digital data obtained through remote sensing systems. While these two tasks are a small portion of the potential of a GIS, they do illustrate the market-driven nature of the GIS analysis capabilities. But even these systems are enlarging their capabilities. Systems designed primarily for cartographic overlay most often contain additional capabilities to operate on surface data, to compare data in a single coverage, to work with transportation-related databases, and the like. Those that were primarily designed as remote sensing image processing systems are now incorporating raster GIS capabilities and are coordinating data structures with other GIS vendors to allow data to be passed among different systems.

It is up to you as a GIS analyst to recognize the potential capabilities because such awareness will provide you with the conceptual framework to operate on the largest "superset" of geographic analysis capabilities you have at hand. In addition, should you be in a position to purchase a system or multiple systems, you will be able to determine the limitations of each system for your application before you spend money on one. Ultimately, the GIS should be able to automate geography. More specifically, the GIS should be able to automate most of the geographic analyses that have been developed over two and a half centuries of modern geographic thought.

In the following seven chapters I have limited the framework of geographic ideas to those that deal fundamentally and explicitly with maps. Although there are far more techniques that allow analysis of mapped data implicitly, these are more appropriately studied in courses in geographic analysis, statistical analysis, systems modeling, and geostatistics. You might consider enhancing your curriculum to include these courses.

A SIMPLE ANALYSIS FRAMEWORK

Chapters 7 through 13 are organized into a simple framework for analysis that should provide an ample survey of the existing capabilities of most GIS software. We begin in this chapter with the simplest operational capabilities of the GIS: counting and locating objects. This set of techniques most resembles the descriptive work of the early explorers, who often wanted to know what was located where and how many of them there were. The utility of this old approach should not be discounted or its importance minimized. It provides much of the basis for later analytical techniques.

In Chapter 8 I show how a GIS can be used to make measurements on maps.

We will be replacing devices like dot grids, distance wheels, and planimeters, traditionally used for measuring the sizes, lengths, and areas of objects, with computer algorithms that perform the same techniques. Combined with the ability to locate objects and count them, these techniques provide a full suite of initial capabilities necessary for most advanced techniques. Many uses of the GIS will require you to make simple comparisons of the sizes and amounts of objects on different parts of a map. More advanced techniques will require you to make some of the same comparisons on multiple maps.

While most geographic data are assigned names to identify them on the map, in many situations it is desired to reclassify them based on specific needs. This is the subject of Chapter 9. A common approach to the manipulation of classified data might be to group or aggregate them into larger categories that simplify later analysis, or stand alone to describe a particular phenomenon. By reclassifying spatial data, we are able to obtain patterns that often are more descriptive than those originally input to the GIS. For example, you may want to reclassify wheat, barley, and oats into "grain crops" to show the spatial distribution of that larger category. Or you might want to reduce the level of data measurement of ratio data to ranked (ordinal) to show categories of, say, earthquake hazard zones. There are many other cases, and we will look at each individually.

In Chapter 10 we turn specifically to surface data. We'll take some time to examine the unique data structures designed to operate on surfaces. Later, you will see how these surfaces can be used to work with classification manipulations to produce patterns of mutually homogeneous surfaces. For example, we will learn how surfaces can be used to determine what can be seen between two points on a map. Another technique allows us to produce areas of identical slope and aspect (e.g., all areas on a surface map that are facing south).

Patterns and arrangements of objects are the primary subject of Chapter 11. Here we will see how points, lines, and areas can form measurable patterns that provide us with additional insights about how the landscapes function. We will examine a variety of analytical techniques for characterizing point patterns. Then we will extend these measures to include approaches for characterizing arrangements of lines and areas. You will be introduced to a method that allows us to move from point objects to polygons, that allows us to examine the influence of one point on another. Because of their unique structures in the GIS we will isolate specific collections of linear objects called networks. We'll see how attributes attached to lines can be used to indicate specific relationships from one line segment to another. For example, we will see how speed limits, traffic counts, road types, intersections, stop signs, and so on can all be encoded and linked to the linear entities to give us a picture of transportation nets in the real world. Then we'll apply these data to analyses aimed at making decisions about transportation planning.

In Chapter 12 we will see how a GIS can compare variables from one coverage to another—a set of techniques most often called overlay. Given the frequency of the use of these techniques, we will place special attention on how the techniques are performed in a digital environment. Each overlay operation is described individually, and examples illustrate how and when each should be used. Because of the potential for error resulting from overlay, especially in vector, the potential pitfalls for the techniques are illustrated, and methods of avoiding them explained.

Finally, with a complete set of techniques readily in hand, Chapter 13 will provide an overall framework for developing more advanced combinations of individual techniques called cartographic models. After examining some of the history of modeling, we will proceed to show how modeling is performed with GIS. We will spend time on the process of isolating individual cartographic elements; then, using a systematic approach, we will piece these individual map elements together into a working cartographic model. Finally, we will present flowcharting techniques that can help assure that the models have been properly constructed and are likely to assist us in answering the appropriate questions.

THE FIRST TASK OF EXPLORATION: OBSERVATION

How the GIS Finds Objects

As we saw in Chapter 2, the trained explorer is one who knows what he or she is looking for. In each digital database there are likely to be many different coverages, each containing a separate theme. Within each theme, as within each focused field study, you are likely to be trying to isolate numerous features for study. You will often want to know which of the selected features occur most often, how often they occur, and where they are located. Let us say, for example, that you are interested in the number of individual trees (point features) located inside your coverage. Because these are point data, you are going to be counting objects that conceptually occupy no spatial dimension. A raster GIS offers numerous ways to find these items, but the simplest is to create a new coverage that eliminates all unnecessary data. To do this you mask all other portions of the database through a simple reclassification process (see Chapter 9), classifying all the trees of the selected species as your target points and reclassifying everything else in your coverage as background. Most raster GIS packages will then allow you to output the results in a table that allows you to count the target points (grid cells) and the background points (grid cells) (Figure 7.1, Table 7.1). Generally this will also allow you to output percentages directly as well, making possible comparisons of amounts of the coverage occupied by the selected tree species. Of course, the trees are only located within the grid cell, not wholly occupying it, so the percentage of coverage occupied by these target cells is rather skewed.

Identification of the absolute locations of point data in a raster GIS is, if you will forgive the pun, a pointless task, because the grid cells divide up or quantize space into spatially uniform packets. Still, you can easily identify the grid cell locations, most often by using an on-screen cursor device that allows you to point to each grid cell individually. This usually results in a readout giving the row and column numbers, as well as the attribute codes for the item selected or pointed to. Even in raster, a knowledge of the location of individual points is important. For example, you might be comparing the locations of trees to grid locations from remotely sensed data as a means of testing the sensor's ability to recognize the presence of trees at that location based on the electromagnetic properties returned to the sensor from that cell.

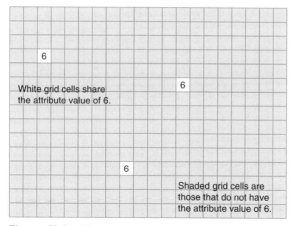

Figure 7.1   Finding raster attributes. Raster GIS showing the isolation of points by attribute. Each grid cell is identified by a color or shading pattern that represents its unique attribute values.

In raster, GIS lines are merely collections of grid cells that touch one another either along each side or diagonally in Cartesian space. Polygons are groups of grid cells that are connected in much the same way, or share attribute values (called **regions**). Because of the way attributes are linked in the simplest raster GIS software, to identify them as entities you must identify them by attribute, just as we did earlier when we isolated point data. And, as before, you can use a cursor to identify column and row locations and even attributes for each location. Tabulation of the results will again show the amount of each category and the percentages. Again, you need to keep in mind that for lines, percentages of the total database occupied by the various objects are most likely meaningless. For areas, the accuracy of the percentages is largely determined by the method used to encode the grid cell categories.

In vector systems the process of finding points is performed by displaying all points in the cartographic portion of the database. If you need to find points of a particular kind (e.g., points that have attributes indicating that these are

TABLE 7.1   Listing of Grid Cells Chosen by Attribute Value Codes

Category	Attribute Value	Cells[a]
Not coded	0	225
Prime land	1	642
Urban	2	201
Federal land	3	188
Grazing	4	981
Water	5	64

[a] Number of cells of each type found in the database.

TABLE 7.2 Listing of Entities Chosen by Attribute Value in a Vector System[a]

Landuse Code	Name	Number of Polygons
100	Row-crops	21
110	Fallow	18
120	Grains	65
130	Grazing	3
200	Residential	982
210	Commercial	124
220	Light Industry	192
230	Heavy industry	54

Landuse Code	Name	Polygon ID
130	Grazing	25
130	Grazing	28
130	Grazing	29

[a] The lower half of this table shows a list of polygons selected from the top half. Notice that three polygons matched landuse code 130.

telephone poles, bird nesting sites, etc.), you will need to access the attribute database and its tables (Table 7.2). Most often you will perform a search that identifies all table locations that exhibit the appropriate code for the objects you wish to see. Because these are linked to the entities, you will be able to selectively isolate these specific types of point object. Because vector data structures contain explicit spatial information, you can easily obtain the exact coordinates in projected space. These can be produced as tabular output or viewed directly on the screen by means of a pointing operation. In some cases you will be able to point to individual items, and in others you can surround several with a graphic window that will give the desired information for all items inside it. As with raster systems, you can also obtain attribute and co-ordinate information.

Because explicit coordinates are used in vector data structures, you can also obtain entity information about lines and polygons by selecting them individually. For example, you can identify all points in the coverage by simply displaying only points, or you can obtain a listing by accessing the attribute tables specific for points. This is true for lines of all kinds as well as for polygons. It is no difficult matter to identify all the polygons in a coverage, plus their attributes. To identify the absolute coordinates for one- and two-dimensional objects (i.e., for lines and areas), you will need multiple sets of coordinate pairs for each of the points that are used to identify all the line segments with which they are defined. Again, this can be done by pointing to individual objects or by surrounding them with a graphic window, depending on the software. And, as with point data, any line or polygonal object can also be selected based on a database query of the attribute tables.

Why We Need to Find and Locate Objects

While nearly any GIS can be subjected to find or locate queries, before we go any further, we might ask why we need to be able to find and locate objects in our GIS database. The twofold answer is blatantly obvious ("because it is a fundamental process comparable to traditional map reading"), and yet subtly obscure ("because of its relationship to more complex calculations").

In the first case, it is important to be able to isolate, count, and locate objects because these activities give us an understanding of the overall complexity of our map coverages. The objects in the map represent features on the ground. Based on our traditional communication paradigm, one primary purpose of the map is to display these objects, to facilitate the identification of the spatial relationships between and among themselves as well as to other objects on the landscape. The more objects of a particular type occur in the coverage, the greater their density in the real world. It is, for example, important to know whether there are many houses or just a few in a particular region. These numbers often relate directly to population densities, which may, in turn, be vitally important in determining the risk of a population to hazardous materials spills, flood events, earthquakes, and other disasters, to cite just a few examples. Likewise, the numbers of plants or animals in a coverage may very well be related to the environmental health of the region. Or the number of road networks might provide information important for routing sales staff or truck traffic. By extension, knowledge of the relative absence of these objects or features may be as useful under given circumstances. In other words, a quantitative measure of the amounts of these objects will allow us to make direct, analytical comparisons to other coverages or other variables in the same coverage.

As you can see, the actual existence of objects and their locations and distributions is often highly important. Comparisons to other features on the landscape might be used to determine causes for distributional patterns, or at least to suggest strong spatial relationships among cartographic variables. For example, an enumeration of the number of houses may be useful for examining their relationships to the amount of land available for new housing. Areas on a coverage with low numbers of housing units can be compared to a coverage of available land for house construction through a process called **overlay**, which we will cover in Chapter 12.

But the simple enumeration of objects and their locations also lets us examine their relationships to more prominent objects in the same coverage. Landscape ecologists, for example, are interested in the relationships between small, isolated polygons, which they call patches, to what they call the matrix, or overall background of the study area. By knowing how many patches occur in a study area, and how much area they occupy in the database landscape, ecologists obtain a measure of patchiness or fragmentation that is taken to be one indicator of overall landscape health. There are applications in economics, as well. The locations of shopping centers, for example, may be highly fragmented, requiring additional roads and public utility services to afford access to isolated polygons of retail activity.

As implied by the last two examples, among the most important aspects of being able to find and locate objects on the map is the ability to make further measurements and comparisons. From absolute numbers we can make extensions to relative numbers. For example, we can move from the total number of houses, roads, or polygons of a particular land use to relative numbers of houses per unit area, the relative number of road miles per square mile, or the relative number of polygons of land use type 1 to that of land use type 2. Each set of measurements or comparisons provides additional insights into our environment. But all require the initial ability to isolate, count, and locate the individual entities within the database.

DEFINING OBJECTS BASED ON THEIR ATTRIBUTES

Simply being able to find point, line, and area entities on a map is of little value. As you have seen, most of the objects analysts find, count, and locate in coordinate space are not selected based on entity type alone, but rather on what they represent in the real world. Just as we would not simply record points as points, lines as lines, and areas as areas in a field survey of an area, we likewise do not keep them stored as sterile entities in our GIS database. Instead, as we would in the field, we make a note of the types of point, line, and area objects, the amounts or values of each, and their categories, because our greatest interest is in their attributes or descriptors. For the same reason, we most often search for objects, count their numbers, and note their locations based on these attributes. We will examine individually the three entity types (points, lines, and areas), based on how we would search a database for their attributes and the insights we might gain from doing so. We will examine surfaces separately (see Chapter 10) because they are stored in the GIS in a fundamentally different way.

Defining Point Objects Based on Their Attributes

Point objects, like all entities, differ not only in their locations but, more importantly, in their attribute characteristics. Trees are different from houses, which are different from cars, which are different from businesses, and so on. These differences provide us with different but often related spatial patterns of each group of objects. Point objects can differ by nominal type, as in our trees, houses, cars, and businesses. These types may also be found within a given category. For example, trees can be identified as maples, oaks, pines, birches, and so on. Or houses can be separated as single-family slab houses, houses with basements, two-story houses, multiple-story houses, and the like. In each of these nominal categories, there is an inherent reason for segregating individuals from a larger category to isolate the patterns these individuals might show us.

If point objects can be separated by type, they can also be separated and classified by an ordinal categorization. We may, for example, want to retrieve trees by some measure of health or vigor, or even the ability to provide shade. For example, we might have poor, moderate, and good categories of trees based on the completeness of shade they could offer. A lumber company, on the other hand, might want to classify trees in ordinate categories based on the quality of timber they might produce upon maturity. In some cases we might need to know the relationships between actual tree species to be able to retrieve these objects; or, we may have already placed attribute data regarding some of these values as ordinal categories suitable for direct retrieval.

Of course, if we can classify point objects by nominal and ordinal categories, we can also separate out these objects by value in interval or ratio scales as well as by type. For example, houses could be selected on the basis of market value: houses that are below $50,000, between $50,001 and $100,000, and so on. These values would have to be stored in the database to permit retrieval of the corresponding points. Other examples might include the annual amount of sales for individual businesses, the amount of nitrogen fixation for selected plants, and the shoulder heights of deer located through telemetry. These quantitative measures will allow us to select a wide range of groups or classes of each object depending on our needs.

It is vitally important that a functioning GIS be able to define each group or category separately and tabulate the results of a search. In addition, we need to be able to produce a graphic coverage of the categories or values of point variables of interest, and only those categories or values. In other words, we need to be able to isolate these groups of point objects from those that are not important to our output or analysis objectives. Because GIS operates in a holistic cartographic paradigm, we should be able to retrieve all the raw data and then group the results in any fashion that suits our purpose.

Beyond simply isolating categories of data, the GIS must allow us to locate each item individually for each class of data and compare it with others of its own kind. That is, we need to be able to show the spatial relationships between individual items of a selected point object in a class of objects so that we can later perform analytical operations to quantify those relationships. Are all houses costing under $50,000 located within a given distance of one another? Are they distributed evenly or randomly? Are they clustered or dispersed in their distribution?

If we have a need to show spatial relationships between like-kind or like-value objects, it stands to reason that eventually we will need to compare them to other similar point objects of a different type or value. We will need to be able to show and later quantify the relationships, for example, between houses selling for $50,000 or less to those costing $50,001 and over. But even more, we are likely to want to show the relationships between and among point objects of a certain type, say houses, and those of another, say the locations of streetlights. And it follows that if we need to be able to compare the numbers and spatial locations of one group of point objects to another, we will soon need to show the relationships among entities of different types. For example, there may come a time when we will need to know the relationships between houses and available paved roads, sewerage lines, water and electric utilities, parks, and shopping centers. Of course, as we will soon see, each of these other entity types will also need to be separable based on type or magnitude.

Before we leave our discussion of points it is important to note that while most GIS software will allow us to isolate, select, quantify, and locate point objects, many, perhaps most, GIS packages are not specifically designed for determining the spatial relationships among these objects. Still, even though the software may not allow us to perform advanced spatial analytical techniques on these objects, we should be able to perform at least these basic functions; then we can use alternative analytical software to learn more about the spatial relationships that exist. Your GIS, at a minimum, should allow you to separate out the items you wish, and only those items, accompanied by their attributes and locational coordinates. Once these have been selected, you should also be able to produce a separate coverage of what you have chosen, to permit further analysis on them either within the GIS or using separate software.

Defining Line Objects Based on Their Attributes

As you remember, line objects are one-dimensional entities that are defined by two or more points with corresponding coordinate pairs. The line object can also contain nodes that are points specifically indicating either the beginning or the ending of a line, or some change in attribute along the line. As with point objects, line entities can be identified by attributes with varying levels of data measurement (i.e., nominal, ordinal, interval, or ratio). Examples of line item types include railroads, streets, fault lines, fencerows, or streams. Each of these is separable from the others because all differ in kind. In other words, each different object should be able to be identified, retrieved, and located separately. And, of course, they should also be able to be tabulated and displayed separately to identify the unique pattern each exhibits on the landscape.

Like point objects, lines should be separable based on ordinal rankings, or some measure of magnitude. Highway types such as single-lane, double-lane, three-lane, and interstate freeway are examples of line objects organized by ordinal ranks. These distinctly different road types can be compared only across the single spectrum of highway types; quantitative comparison to other nonhighway line features is not permitted. Line items that exemplify a measurable difference in magnitude might be exemplified by lines representing the actual stream flow of different tributaries of a river system, probably measured in cubic feet per second or cubic meters per second.

In some cases, a single line may experience a change in attribute type, rank, or magnitude along its overall length. For example, a road may change from single to double lane, a stream's flow may change to a higher value because of inputs from tributaries, or the measured traffic flow along a city street may change along a portion of its length where it connects to another street. By using nodes to indicate the changes, and by storing each segment between nodes with the appropriate attributes, we identify each segment as a separate, identifiable entity. The attributes can therefore be used to retrieve either whole lines or portions of a line.

Another attribute characteristic involves not just the attributes of the line itself, but a comparison of what falls on either side. For example, we may want to define a hedgerow or fencerow based not on type of vegetation but rather on a comparison of the land covers that fall on either side of the entity. We might want to be able to identify, for example, all hedgerows that have forest on one side and farm fields on the other, or all hedgerows that have only farm fields on both sides (DeMers et al., 1995). In raster systems, chores like this may be difficult, requiring us to perform elaborate categorical manipulations called **neighborhood functions**, about which we will talk in Chapter 9. In vector systems that employ a topological data model, a search could be performed to identify each attribute for each bordering polygon. Since, however, the relationships between the line and its neighboring polygons are explicitly encoded into the database during input, a simpler method for both raster and vector is to encode the attributes for the line that relate to the neighboring polygons. With raster this would mean either creating a separate coverage for the line entities to permit the attributes to be coded or, if your raster system employs a database management system, storing the attributes there as separate identifiers. In vector, the latter approach is most often used. Once again, you can see the importance of thoughtful planning before data entry, in this case with a view to having questions you might ask easily answered.

To locate the lines, it is necessary to be able to identify all the coordinate pairs that make up the line in vector, or all the grid cell column and row values in raster. This adds three other factors that might be used to retrieve the line—its length, its azimuthal direction, and its shape. Line entities may be straight with single orientations, or they may be jagged or sinuous with multiple orientations, like street or road patterns or meandering streams, where each straight-line segment indicates a unique orientation. Some line entities are simple, with a single line; others are complex, with a branching network that may form a hierarchy of lines, as in a branching stream network. All these are difficult for most raster GIS software to calculate, but vector GIS packages can easily perform length (see Chapter 8) and azimuthal calculations. Length calculations are merely a matter of performing the distance-between-points calculations using the Pythagorean theorem, if the data are in Cartesian coordinates, or great circle calculations for data that are projected (Robinson et al., 1995). Azimuthal calculations use the standard formulas for spherical trigonometry between each set of points (Robinson et al., 1995). Shape calculations for line entities are most often measures of **sinuosity**, or some ratio of the amount of overall length of a curvilinear line to its straight-line length. This merely requires two separate sets of calculations, one for the total length and one for the straight-line length. More complex analyses involving combinations of these calculations are possible but call for numerous sets of operations (Mark and Xia, 1994).

These three principal types of linear measure—length, azimuthal direction, and shape—can be defined as individual attributes for each line entity, or for each separable portion of a line entity. For example, you may want to find all hedgerows that are longer than 80 meters, or you might want to separate out all roads that have a north–south orientation, or you may want to isolate highly sinuous streams. To allow the lines or line portions to be selected, tabulated, and displayed separately, then, each should be separable with your GIS. In some cases, the line entities may extend beyond the map boundaries rather than being wholly contained inside the map. If you want to be able to eliminate these, it might be useful to include an attribute indicating that the line is incomplete. Analyses that are not able to separate out incomplete line entities are likely to be inaccurate even if the data are aggregated. Most of these operations are designed to obtain overall or aggregate statistics for your map, such as the average length of a certain line entity in a coverage or the overall average azimuthal direction or overall average sinuosity. At times, however, it may be necessary to combine operations to create regions of similarity in a coverage where the lengths, azimuths, and sinuosities vary from portion to portion on the map. We will look at these in Chapter 9.

Defining Area Objects Based on Their Attributes

As with point and line entities, area or polygon entities can be defined, separated, and retrieved based on category, class, or magnitude. As before, each of these attributes needs to be explicitly stored in the database, whether as grid cell attributes or as vector polygon attributes. Isolation and retrieval are performed in exactly the same manner as for points and lines. But, like lines, areas have added dimensionality that permits more attributes to be assigned to them based on measurements of their dimensions.

Among the more useful attributes for polygon entities would be a measure of their shapes. This shape could be strictly Euclidean in that it could be measured as a deviation from some known geometric shape with predefined properties, such as a circle or a square. Another group of shape measures uses a set of mathematics called fractal geometry, in which the irregularity of the outside of a polygon is measured. Related to shape is a measure of a polygon's elongation, or the ratio between its long and short axes. Although not generally a built-in function of vector GIS, measuring the long and short axes and expressing these values as a ratio is a relatively simple matter. The coordinates for each of the polygon points can be used to identify the points that are the farthest in a particular direction. But this requires the use of a higher level object, the centroid, against which to measure the rest of the polygon edges (see next section). Of course, elongation implies that one might also wish to determine the particular orientation of elongated polygons with respect to cardinal directions. While measures of shape are not frequently invoked by the GIS community, there is growing interest in the relationships between shapes of both human and natural area features and their functioning (Forman and Godron, 1987). At a minimum, some simple measures of shape would be useful, to facilitate the isolation, selection, tabulation, and display of polygons of a given category of shape and orientation. We will look more closely at measures of shape and orientation in Chapter 8. For the time being they should simply be thought of as measured attributes that can be used to locate and count polygons of a particular size category.

Another, somewhat more commonly applied measurable attribute of polygons is their size. In raster, size is determined by counting the number of grid cells of a given category. These categories do not need to be restricted to a single polygon; rather, they simply indicate the number of grid cells with a given value. Many times a reclassification process (see Chapter 9) will be necessary to isolate these polygons, and then their area can more readily be found by a simple count of grid cells. Of course, measures of size in raster are not going to be particularly accurate because of the quantized nature of the grid cell. In vector, the perimeter of the polygonal lines is easily calculated by adding the line lengths, while the area is calculated much as would be done manually, by a series of length-times-width calculations for portions of each polygon. The more complex the polygon, the more calculations must be performed. But, in general, any commercial GIS package should be able to provide reasonably rapid response times for area calculations. To isolate polygons based on area or perimeter, one need only select a set of category criteria, place the appropriately sized polygons in each, and perform a simple retrieval on those selected. Polygon size, like line length, is most often (but not necessarily) performed on each polygon for each category and is then averaged for tabulation. This indicates to the user the relationship of polygon size to total area of the database occupied by that category of polygon.

Two other measures regarding areas need to be mentioned as well, although they will be discussed in more detail in Chapter 8. The first is **contiguity**, a measure of the wholeness or amount of perforation of a polygon (Berry and Tomlin, 1984). A large polygon that contains many holes (small polygons contained within the larger polygon) demonstrates much less contiguity than one with only a few holes or none at all. Analyses may require the amount of contiguity for animal habitats or fire hazards for forest regions. Most raster GIS software contains some measures of contiguity, but vector GIS is less adept at performing this analysis. By determining the amount of contiguity, then classing the results into groups, the analyst can easily retrieve and display these groupings, or use them for further modeling.

The second additional measurable attribute that could prove useful for polygonal features is the **homogeneity** of an area, not necessarily defined as a single polygon (Berry and Tomlin, 1984). Homegeneity is a measure of how much area of a given portion of a map is directly in contact with polygonal features sharing the same attributes. For example, two polygons that have identical attributes may touch each other only along a small portion. In this respect homogeneity is very similar to a simple measure of size. However, homogeneity can also be defined to include the amount of internal **heterogeneity** that an area contains. One can, for example, group a number of polygons based not on the similarity of their attributes, but on some heterogeneous mix of attributes. Other homogeneity measures of attributes include minimum values, maximum values, averages, totals, and even diversity of attributes. Each of these can be grouped, with the result that the polygons contain a new set of identifiable attributes that describe them. For example, we may have a forest polygon that shows, when separated out based on species of trees, that there are 12 different species of tree contained within its borders. We could isolate all polygons that demonstrate 12 or more trees as a measure of the amount of diversity (in this case called "species diversity") within them. Thus we could map all areas that showed high diversity of tree species. Alternatively, we might be interested in polygons that, for example, show a high degree of human ethnic similarity—a lack of diversity in residential areas. We could then select all the polygons that contained fewer that three or four ethnic groups, to perhaps indicate some measure of ethnic segregation in a city.

Both raster and vector GIS typically have some capability to perform this type of analysis, although in some systems a direct function is lacking. Instead, a number of commands might be used to achieve the desired results. As always, these measured attributes should allow us to isolate, retrieve, and output the results individually. When we begin looking at classification inside the GIS, we will return to all these techniques as methods of reclassifying our existing coverage attributes.

WORKING WITH HIGHER LEVEL OBJECTS

Thus far we have worked with points, lines, and areas, based either on readily available attribute characteristics or on attributes that are measurable through sometimes simple, sometimes complex methods. All of these entities, however, have certain attributes that set them apart from the rest. Some attributes are a result of the methods of encoding—for example, nodes that are encoded during the digitizing process. Others may have to be determined—for example,

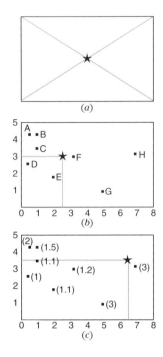

Figure 7.2   Types of Centroid. *(a)* Simple centroid, *(b)* center-of-gravity centroid (mean center), *(c)* and weighted mean center, where numbers in parentheses indicate assigned weights, *f* values in Table 7.4 [note how high weights for points G and H of *(b)* pull the mean center to the right].

centroids indicating the center of an area. We call these "higher level objects" for lack of a better term, simply because of their uniqueness and usefulness for later GIS analysis. We will separate these higher level objects into points, lines, and areas and examine each individually.

Higher Level Point Objects

Two primary types of higher level point object are centroids and nodes. A **centroid** is most commonly defined as the point that occurs at the exact geographic center of an area or polygon (Figure 7.2*a*). Its calculation is simple for simple polygonal shapes such as rectangles; when the polygons become quite complex, the complexity of the calculations needed to perform the task increases proportionately. Raster GIS is not well suited for this procedure. In many cases even vector GIS does not include this calculation as a stand-alone function. Simple or geographic centroids in vector are calculated by a rule called the **trapezoidal rule**, which separates the polygon into a number of overlapping trapezoids. Then each trapezoid's centroid, or central coordinates, are calculated and their weighted average calculated (Figure 7.3). The centroid may be important if you are attempting to produce a surface map from samples taken from within areas. For example, if you were to produce an **isoline** (defined as a line of equal value) or fishnet map of the population of the United States, but the data were sampled at the county level, you would have to place a centroid in each of the approximately 300 counties. The centroid would act as a point location, as if the data were actually calculated at this point. Then, through interpolation, the isolines or surfaces could be calculated based on these point locations.

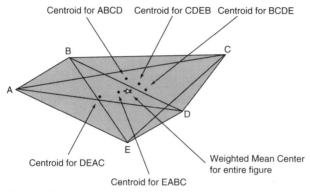

Centroid for ABCD   Centroid for CDEB   Centroid for BCDE

Centroid for DEAC

Weighted Mean Center
for entire figure

Centroid for EABC

Figure 7.3   Trapezoid rule. Diagram of the trapezoids defined in an irregular polygon. Each overlapping trapezoid has a calculated mean center from which the weighted mean center can further be determined. This weighted mean center is the centroid of the polygon.

Centroids can also be placed at the center of a distribution of some phenomenon, instead of at the absolute geographic center of a polygon (Clarke, 1990). For example, to continue our analogy of using county-level population data to produce an isoline map, we may find that for some counties, most of the population is clustered to one side or another. While we could still use the centroid of the county as the estimate of its population center, it would be more accurate if we placed it closer to the center of the distribution. This location, called the **mean center** or the **center of gravity**, requires us to separately average the X and the Y coordinates for all points in the coverage (McGrew and Monroe, 1993; Muehrcke and Muehrcke, 1992) (Table 7.3, Figure 7.2b). The final result would

TABLE 7.3   Table of Values Set Up to Allow Calculation on the Center of Gravity[a]

Point	X	Y
A	0.5	4.5
B	1.0	4.5
C	1.0	3.5
D	0.5	2.5
E	2.0	2.0
F	3.0	3.0
G	5.0	1.0
H	7.0	3.5
	20.0	24.5
Mean center	2.5	3.0625

[a] "Centers of gravity" (mean centers) are found by dividing the total values for X and Y by the number of points.

be a single pair of values representing the central point of this distribution of points. In addition, if these points had additional weights—for example, if the points indicated both the locations and the amount of shopping done at each store—we could further place our center based on this additional weighting factor. The procedure, called the **weighted mean center**, simply requires us to multiply each X and each Y by a weighting factor (amount of shopping in this case), then sum them up and divide by the number of points (McGrew and Monroe, 1993) (Table 7.4, Figure 7.2c). The result is a single pair of X and Y coordinates indicating the mean center of the distribution, modified by the weighting factor.

Such calculations are often applied in large-scale market analysis and economic placement analysis to define the center of the market. Having done this, the market analyst might choose to select areas near the market center for locating a new shopping center or other type of business. There are, of course, many other reasons for selecting either simple centroids or center-of-gravity centers; examples include studies of animal foraging activity (Koeppl et al., 1985) and analyses of the change in movement of population centers through time as a measure of large-scale migration (McGrew and Monroe, 1993). As before, many raster GIS systems do not have this capability built in, but the majority of commercial vector systems will allow the calculation of centroids and weighted mean centers.

Our second type of higher level point object, the node, was mentioned in Chapter 4. In this case the points are significant not as individuals, but as specific locators along line and area entities. Nodes do not occur as specific objects in raster GIS, so there is no need to comment on their calculation in those systems. Because nodes carry with them an indicator of a change in attribute, the ability to identify them is vital to many attribute selection and manipulation procedures. Nodes are generally encoded explicitly during input and should be easily separated or identified through simple search procedures in the GIS. There will be difficulty only when a node has been improperly coded as a simple point, rather than as a node. In such circumstances the ability to isolate line segments defined by nodes will be compromised. This is yet another

TABLE 7.4   Table of Values Set Up to Allow Calculations of the Weighted Mean Center Based on the Weights Assigned to Each Set of X and Y Coordinates

Point	X	Y	f	fX	fY
A	4.5	2.0	0.5	2.25	1.0
B	1.0	1.5	4.5	6.75	1.5
C	1.0	3.5	1.1	1.1	3.85
D	0.5	2.5	1.0	0.5	2.5
E	2.0	2.0	1.1	2.2	2.2
F	3.0	3.0	1.2	3.6	3.6
G	5.0	1.0	3.0	15.0	3.0
H	7.0	3.5	3.0	21.0	0.5
Total				52.4	28.5
Weighted mean center				6.55	3.512875

illustration of the importance of good planning and careful execution during the input phase of the GIS.

One other situation that is covered in later chapters needs brief mention here, namely, the use of the distribution of points to develop areas. For example, where large numbers of points are found to exist, whatever they might represent, their area is distinctly different from places where few points exist. You may, for example, find a great many weeds in particular parts of agricultural fields, indicating that these areas are different. What remains to be determined is the source of the difference: a former disturbance, a lack of pesticide, or something inherently different about the soil or how it is manipulated.

Other pattern characteristics such as uniformity of distribution of point objects, or randomness, or another identifiable pattern can also be used to define areas as specific **communities**, or areas sharing distributional patterns (Figure 7.4). Defining areas based on point distributional patterns is not a strong part of most GIS systems, but it can often be calculated in both raster and vector systems. We'll return to this topic later.

Higher Level Line Objects

Three different types of lines are particularly important and justify calling their objects higher level. The first was touched on when we discussed the relationships between line attributes and associated attributes for the adjacent polygons. These lines are most often called **borders**, and a major change in attribute value or even collections of attributes is recognized or assumed as one moves across them. In other words, borders are particularly important because of their locations relative to adjacent polygons. As in our hedgerow example, borders need to have associated with them a set of attributes that easily allow portions of the map to be separated out because all the objects on one side are substantially different from all those on the other. Take a simple example

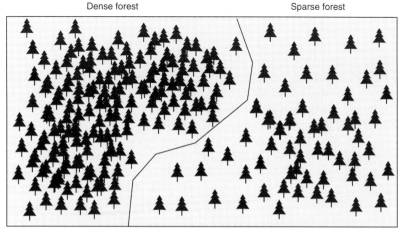

Dense forest   Sparse forest

Figure 7.4 Communities based on object density. Map illustrating two different communities of points based on density.

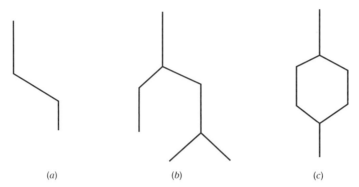

(a)   *(b)*   *(c)*

Figure 7.5 **Types of networks.** Networks composed of line segments with attribute codes indicating flow. The three major types are shown: *(a)* a straight-line network, *(b)* a branching network, and *(c)* a circuit.

of a national border between the United States and Canada. The line serving as the border between these two countries should allow one to identify all the states in the United States as fundamentally separate from all the provinces and territories of Canada on the north. While this may seem obvious, if the line separating these two areas on the map is not provided with attribute information clearly indicating its status as an international border, you may have to force the GIS to separate the United States from Canada, rather than simply performing a search based on the neighbors of the borderline.

Lines also can become higher level objects when they are located in a particular way relative to other lines, rather than to polygons as in our last example. In many cases lines are not simply indications of locations of linear objects, or of boundaries between polygons, but rather are connected by nodes to form **networks** (Figure 7.5). Networks can most appropriately be defined as a set of interconnected line entities whose attributes share some common theme primarily related to flow. As we will see in Chapter 11, networks allow us to perform modeling of a multitude of flow types from automobile and rail traffic to the flow of particular commodities or even the movement of animal species along corridors. In all these cases we need to be able to operate on networks, which means that the lines must share the attributes necessary for analyzing these flows (speed limits, frictions, etc.). Raster GIS is not particularly efficient at handling networks because there is no way to define them explicitly except to assign specific attribute values to the grid cells. Still, this can be done, although most GIS professionals will defer to vector GIS for working with networks.

Networks come in three major forms: straight lines, as one might find in a major interstate highway (Figure 7.5*a*), branching trees, as one might expect when looking at stream networks (Figure 7.5*b*), and **circuits** (Figure 7.5*c*), as one might find in street patterns whose lines lead directly back to the starting point (Muehrcke and Muehrcke, 1992). In addition, all these network types can be defined as directed or undirected. In a **directed network**, the flows are allowed to move in a single direction only (Figure 7.6*a*). Streams, for example, will flow downslope and will not, under normal circumstances, flow in the opposite direction. Likewise, one-way streets limit the flow of traffic to a single direction. In the event that one link in the network intersects another at an angle, there may be a change in flow direction, or it may be necessary to restrict

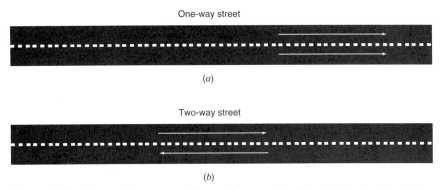

One-way street

(a)

Two-way street

(b)

Figure 7.6 Directed versus undirected networks. Directed networks *(a)* restrict the flow to a single direction, while undirected networks *(b)* allow flow in both directions.

the locations at which turns can be made in passing from one link to another. At the intersection of a two-way street and a one-way street, for example, you will not be allowed to turn from the two-way street into the oncoming flow of one-way traffic. But in networks called **undirected networks** (Figure 7.6b), the flows can go back and forth along the network in either direction.

Because networks have the capability of modeling flows in either a directed or an undirected fashion and because some network links will be connected to certain links, but not to others (e.g., a road overpassing another road), all these characteristic attributes must be explicitly encoded either during data entry or later (as edited attributes). Nearly all vector GIS software has the capability of storing these attributes and modeling flows with them. In fact some software has been specifically designed for work of this kind (see Chapter 11). A lack of attribute data for networks severely limits the use of linear features as higher level network objects. Lines connected to each other without attributes indicating that they provide a common pathway for flows offer no basis for network modeling.

A final characteristic related to lines is not unlike the last one mentioned for higher level point objects: that is, collections of lines can also act as identifiers for area patterns or communities. There may be areas of high density road networks, or areas low in hedgerows. Because of the density or lack of density of the line objects, we can define these areas as communities sharing these characteristics. And, as with point entities, additional patterning such as regularity, or randomness, can be used to identify communities of these line features. Most GIS software, however, does not contain algorithms specifically designed for this process. Most often you will need to use alternative software or modify the existing software through the use of on-board macro languages and other tool kits that may be obtainable from the vendor under certain conditions.

Higher Level Area Objects

As with points and lines, areas can also occur as higher level objects. The polygons themselves can be used to define regions of similar geographic attributes. In fact, among the more important aspects of geographic research in the past as today is the definition of regions: areas of the earth that exemplify some unity of describable attributes. For example, political regions are defined by national boundaries, ethnic regions by a similarity of origin, and biogeographic regions on the basis of organismic similarities. Inside the GIS these regional definitions can be based on the attributes defining each polygon or set of polygons. We might, for example, be able to define a region within our GIS by selecting all the polygons that show forest as the major vegetation component. This will give us a "forest" region. We will have to know beforehand which regions we are going to be looking for and how they are to be defined. Because defining regions is a major endeavor in itself, the chances are good that simply selecting the appropriate polygons or sets of grid cells will not suffice to produce definitions. Instead, we will most likely have to combine several sets of attributes from several different coverages to define our regions. In fact, the power to define regions based on a large variety of characteristic attributes put together in as many ways as we need is one of the nicest features of the GIS. For this reason we can define regions based on what we intend to do with the data. This leads us to another classic problem in geography, that of classification itself, and we will cover this in some detail in Chapter 9. In the meanwhile, the selection of regions can be thought of as isolating homogeneous sets or homogeneous combinations of factors. In some cases regions can also be thought of as areas containing a similar heterogeneous mix of attributes rather than sharing only a few common attributes.

But regions differ not only in their attributes and in the way the attributes are manipulated to define them, but also in the way they are configured in space. There are three basic types, based on spatial configuration: **contiguous regions**, **fragmented regions**, and **perforated regions** (Figure 7.7). Contiguous regions are wholly contained in a single polygon. While a contiguous region is wholly contained, its attributes can be defined as homogeneous or heterogeneous but with some similarity of heterogeneous mix, as we have seen. Fragmented regions, on the other hand, share some commonality of attributes; again they are either homogeneous or a heterogeneous mix containing identifiable similarities, but are composed of more than one polygonal form separated by intervening space that does not share the same mix of attributes. For example, one could define as a forest region a number of polygons, scattered throughout

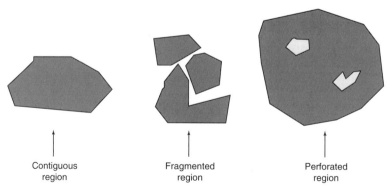

Contiguous region   Fragmented region   Perforated region

Figure 7.7 Types of region. Three major types of regions: contiguous fragmented, and perforated.

the coverage but having the same mix of trees or tree types. There are no limitations to the distance between polygons as long as the similarity of attributes is maintained. Likewise, for perforated regions, the defining criterion for the region remains the same—similarity of attributes or attribute mix. In perforated regions, however, the uniform polygon is interspersed with smaller polygons that do not share a mix of attributes with the surrounding polygon. As such, the region is defined as the surrounding matrix, while the smaller internal polygons are said to be the perforations. Clearly there is a possible relationship between perforated and fragmented regions. If the smaller polygons contained in the perforated region are found to share common attributes, they too may be considered to be a region and can easily be separated out from the background region.

All the objects we have identified, whether simple or higher order, must be identifiable by the user so that they can be operated on for further analysis. Each must be able to be isolated, separately tabulated, and displayed. The ability to perform analysis is strongly linked to the ability of the GIS to do this. In vector data models, and in raster data models that are linked to a database management system, most often these objects are selected through a formal search of the attribute tables. For simple raster model systems, the separation is once again relegated to some form of reclassification process—the topic of Chapter 9. But first, Chapter 8 will take us through some important measurements that also are essential to more advanced analysis.

Terms

applications development	regions	overlay
neighborhood functions	sinuosity	contiguity
homogeneity	heterogeneity	centroid
trapezoidal rule	isoline	mean center
center of gravity	weighted mean center	communities
borders	networks	circuits
directed network	undirected networks	contiguous regions
fragmented regions	perforated regions	

Review Questions

1. Explain why the analysis subsystem of the GIS is the most abused subsystem, and give examples to support your explanation.

2. Explain what we mean when we say that a GIS is an incomplete set of geographic analysis tools. What accounts for this?

3. If your career goals revolve around database creation rather than GIS analysis, why is it important to know about the analytical capabilities and limitations of GIS?

4. Why is it important to be able to isolate, identify, count, and separately tabulate and display individual items?

5. Describe the process of isolating, counting, and identifying point objects based on their attributes within a raster and a vector GIS. Why do we need to use attributes to find these objects?

6. What are the differences between finding point objects and line and area objects?

7. What measurable attributes can be used to search for lines? Areas?

8. What are higher level objects? How do they differ from simple objects?

9. Give some examples of higher level objects for points. Do the same for lines and areas.

10. What are centroids? How are they found in a vector GIS? What are the different types of centroid? How do they differ in how they are found in a vector GIS?

11. What are networks? How do they differ from simple lines? What forms can networks take? What needs to be done to identify networks, as opposed to simple lines, in a GIS?

12. What is the difference between a directed and an undirected network?

13. What is a region? What are the differences and similarities among contiguous, fragmented, and perforated regions?

14. Give some examples of how collections of points, lines, and areas might be considered to be communities.

References

Berry, J.K., and C.D. Tomlin, 1984. Geographic Information Analysis Workshop Workbook. New Haven, CT: Yale School of Forestry.

Clarke, K.C., 1990. *Analytical and Computer Cartography.* Englewood Cliffs, NJ: Prentice-Hall.

DeMers, M.N., R.E.J. Boerner, J.W. Simpson, A. Silva, F. Artigas, and L.A. Berns, 1995. "Fencerows, Edges, and Implications of Changing Connectivity Illustrated by Two Contiguous Ohio Landscapes." *Conservation Biology,* 9(5):1159–1168.

Forman, R.T.T., and M. Godron, 1987. *Landscape Ecology.* New York: John Wiley & Sons.

Koeppl, J.W., G.W. Korch, N.A. Slade, and J.P. Airoldi, 1985. "Robust Statistics for Spatial Analysis: The Center of Activity." *Occasional Papers,* 115:1–14, Museum of Natural History, University of Kansas, Lawrence.

Mark, D.M., and F.F. Xia, 1994. "Determining Spatial Relations Between Lines and Regions in ARC/INFO Using the 9-Intersection Model." In *Proceedings of the 14th Annual ESRI User Conference,* pp. 1207–1213.

McGrew, J.C., and C.B. Monroe, 1993. *Statistical Problem Solving in Geography.* Dubuque, IA: Wm. C. Brown Publishers.

Monmonier, M.S., 1982. *Computer-Assisted Cartography: Principles and Prospects.* Englewood Cliffs, NJ: Prentice-Hall.

Muehrcke, P.C., and J.O. Muehrcke, 1992. *Map Use: Reading, Analysis, Interpretation.* Madison, WI: J.P. Publications.

Robinson, A.H., J.L. Morrison, P.C. Muehrcke, A.J. Kimerling, and S.C. Guptill, 1995. *Elements of Cartography,* 6th ed. New York: John Wiley & Sons.

CHAPTER 8

Measurement

Our journey inside the world of GIS analysis has just begun. We've taken some of the same steps as our predecessors, although in much different terrain and with strikingly different tools. Still, the analogy is useful. We have started our journey by focusing on individual objects and collections of objects. We've counted them, grouped them, and noted where they were located so we could find them at any time. But a simple identification, enumeration, and locating of objects provides only a very rudimentary picture of our world.

Like the explorer of old, we need to describe not only what objects are, how many exist, and where they are, but how large they are, how far apart, and what the distance between them is like. Rather than focusing on location as we did in Chapter 7, we will begin measuring the sizes of objects, especially linear and areal objects. We will measure their lengths, areas, and perimeters. We'll also look at the relationships among some of these measures. Measurement will allow us to produce ratios of lengths to widths, and perimeters to areas. This will set the stage for more advanced analysis by providing a set of quantifiable feature attributes for each coverage that can later be compared within a single coverage or to those of other coverages.

Distances between objects will be measured as simple distances, showing the shortest physical space between and among features. But, as explorers, we soon see that we can't always travel directly from one point to another. Our travels may require us to move along trails blazed earlier. We will need to define distance from one place to another based on the paths we are forced to travel, the obstacles we must avoid, and the hills we must climb. In some cases we will define our distance incrementally, adding the distance of one leg of our journey to another to produce a total distance. In other cases we will search for better paths, less arduous terrain, seeking always the easiest, least-cost distances. These least-cost distances may be from point to point or from a single point to all other points in our terrain, thus creating a map for future travelers.

But our travels may take their toll on us, as each step drains more and more of our energy. We will record the cumulative effects of the journey, noting a progressive cost for movement through woods, swamps, and rugged topography. We will show impassable areas like cliffs and wide rivers. Sometimes we will find streams that can be forded, and we will note them to show others the way. We will be able to show future explorers which barriers are absolute and which are relative, allowing passage, but with difficulty. And we will provide measures of the relative difficulty of passable barriers.

As explorers through our digital world we are the trailblazers, measuring the sizes and shapes of the features we explore, showing paths to, around, and through them. Our digital maps, like their analog counterparts prepared during field excursions, will provide needed information for future explorers and for those who make decisions about the lands we travel. What we are about to do is not as difficult as it once was. But it is just as important as the original surveys of Lewis and Clark as they traveled the Louisiana Purchase. We provide a detailed survey of all we inspect for use in making plans for habitation, for natural resource planning, for transportation planning, and for decision making in many other areas.

When we get to Chapter 13 on cartographic modeling, you will easily see how simple measures of size, length, distances, and the like can be combined with many other analysis functions to develop far more complex models than can be provided without GIS. The measurements we can make so easily in our digital explorations set the stage for wiser decisions for future land use, supported by data of better quality. Therefore, to ensure meaningful results and successful explorations, we must understand how best to perform the measurements described. So, replacing tape measures, surveying instruments, compasses, and levels with the new tools of GIS, we take the next step in digital database exploration.

LEARNING OBJECTIVES

When you are finished with this chapter you should be able to:

1. Describe the process of measuring both straight and sinuous lines in vector and in raster, and discuss the advantages and disadvantages of each process.

2. Explain the process and the purpose for measuring polygons to determine long and short axe's, perimeter, and areas.

3. Explain some of the simpler measures of shape for lines and areas and discuss why they might be important.

4. Understand the concepts of spatial integrity and boundary configuration, describe how these might be measured, and state what they tell us about the real-world objects we are analyzing.

5. Explain the process of measuring both isotropic and functional distances among objects, through friction surfaces, and around barriers.

6. Explain the methods and potential problems associated with assigning friction and impedance values for modeling functional distance.

MEASURING LENGTH OF LINEAR OBJECTS

We know that aside from surfaces, which we will cover in depth later, there are three basic cartographic objects. Points are assumed to have zero dimensions, lines a single dimension of length, and areas two dimensions: length and width. Because points have zero dimensionality, there is no appropriate measure for them aside from magnitude attribute values, which are assigned to them and stored as grid cell attributes or saved in an attribute table linked to the entities themselves. While these point magnitude values are important for analysis, they require no formal calculations and we will separate them from the remaining discussion.

Lines, with a single dimension of length, will need to have this dimension calculated, again assuming that other magnitude attributes formally linked to the line entities can be simply retrieved from the database. But, as we have seen, the calculated lengths of lines can be used as defining attributes for isolating categories of line entities. Calculating line lengths in raster is a matter of adding the number of grid cells together to achieve a total. Let us begin by working with a straight-line entity in grid format that occurs as a set of vertical or horizontal grid cells. Knowing the resolution of each grid cell, generally assumed to be from side to side (**orthogonal**), rather than on the diagonal, we need only add up the number of grid cells and multiply that value by the grid cell resolution value. If a vertical line is composed of 15 grid cells, each with a resolution of 50 meters, our line length is calculated by multiplying 15 grid cells by 50 meters to obtain a total line length of 750 meters. The identical process would be performed for a horizontal line.

But suppose that a line is perfectly diagonal. What do we do when the grid cells in a line are connected to each other along a corner? A simple approach is to assume that the resolution is the same for the diagonal measure as for a side-to-side measure, even though we know that the diagonal measure across a grid cell is larger than the orthogonal by some fraction. Under these assumptions we could perform our calculations as before, by adding the number of grid cells and multiplying by the resolution value. The value we obtained is an approximation of the length along the diagonal line, but it is less than the actual line length by the ratio of orthogonal distance to horizontal distance of each grid cell. As it turns out, many simple raster GIS simply calculate the number of grid cells and use that number for the length, relegating the calculation of the actual distance measure to the user. This division of labor has the advantage of keeping the computational sophistication of the software to a minimum. More sophisticated raster systems are able to calculate the diagonal distance between each cell in a diagonal line of grid cells through the use of simple trigonometry (Environmental Systems Research Institute, 1993). Because a diagonal through a square grid cell produces a right triangle, the right angle sides of which are identical to the grid cell resolution, the hypotenuse is calculated through the Pythagorean theorem (see Chapter 2). You should note that the calculation of the hypotenuse will always give you a value approximating 1.414 times that of the horizontal or vertical distances. Therefore it is a simple task to multiply the resolution of each diagonally connected grid cell by 1.414 to obtain the correct value for distance. Some raster GIS systems are capable of this calculation.

A more difficult problem arises when the linear feature is more **sinuous** and winds erratically through the coverage. If we used the presence/absence method of encoding, it is entirely possible that the linear feature will produce a number of pairs, threes, or even larger groups of grid cells that represent the curved portions of the line. In such a case a simple count of grid cells may suffice for measuring distance because the exact location of the road network inside the grid cells is unknown. In fact, whether your GIS can perform trigonometric operations on the grid cells (or can multiply diagonally connected grid cell numbers by 1.414) matters relatively little in some cases because a linear object is not well represented by grid cells that quantize space anyway. In other words, the location of the line entity is not precisely known for any grid cell. Depending on the resolution and the irregularity of the line entity's path, it is possible for whole loops to be represented by a single grid cell, in which case the length of the object will be underrepresented regardless of the method of measurement. As indicated in Figure 8.1, a linear entity that is likely to cause this problem is a highly sinuous stream. For this reason, it is best, whenever possible, to use a vector data structure if your analysis relies heavily on measurements of linear objects.

The calculation of length along a vector line entity is computationally more exhaustive than simply adding the number of grid cells as in raster, but the results are decidedly more accurate, as is the representation of the line itself. For each straight-line segment in the linear object, the software will have stored a set of coordinate pairs. The distance between members of each coordinate pair can be calculated through the Pythagorean theorem. Simply adding line-segment lengths will produce a relatively accurate measure of the total or accumulated line length. Just remember that the vector representation of linear entities also employs a form of sampling, where changes in direction are represented as separate straight-line segments: the more line segments, the more accurately the linear object will be represented by the data structure, and the more accurate the measures of total line distance will be. This again reminds us of the importance of good map preparation and careful data input.

MEASURING POLYGONS

Polygonal features of course exhibit two dimensions, length and width. Because of the added dimensionality of polygons, we have more measurements that we can make upon them. We can, for example, measure the length of the short or

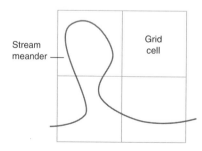

Figure 8.1   Sub resolution line object. A highly sinuous linear object, such as a meandering stream may have whole meanders self-contained within a single grid cell. This indicates the limitation of measuring length with raster data structures.

long axis of a polygon, calculate the lengths of the line surrounding the polygon (its perimeter), or measure the area occupied by this figure. As we learned in Chapter 7, these measures can be used as calculated attributes from which we can isolate areas within a particular set of size limits. We may choose to separate these out simply to create coverages of polygonal features of a particular size, or we may want to use the areas for additional analysis later on. For these reasons it is important that your GIS be able to perform such simple calculations.

Calculating Polygon Lengths

Because the orientation of the polygon is often linked to an environmental process, the orientation of polygonal features will at times be important to the GIS user (Muehrcke and Muehrcke, 1992). For example, asymmetrical forest patches oriented in a particular manner may be easily observed by birds as they fly over (Baker, 1989; Forman and Godron, 1981; LaGro, 1991; O'Neill et al., 1988). Or a scientist attempting to determine the movement history of glaciers may wish to know whether polygons representing certain glacial features have a particular direction. Conceptually, the idea of polygon orientation is quite simple. It is merely a matter of determining the direction of the longest axis for the feature. Most raster GIS software, however, lacks techniques for easily ascertaining the long axis. And because raster uses relative locations of grid cells, it is no easy task to determine orientation either.

In vector, the solution is one of calculating the lengths of each pair of opposing polygon vertices, much as we did with line entities. A comparison of these vertices will then show which line is the longest. That line is the long axis of the polygon, and its angular direction is calculated by means of spherical trigonometry (Robinson et al., 1995).

At times, however, the GIS analyst will be searching not for orientation but rather for the relationship between the major and minor axes. The latter gives a simple measure of shape that can also be used to separate out, for example, objects with a set ratio of short axis to long. Suppose a user wants to identify all long, skinny polygons, because of the unique way they operate on the landscape. Because long, skinny objects tend to have more surface area exposed to the surrounding matrix, if the region contains an object of this type that is a lake, there will be more potential beachfront property to interest a developer. Or an ecologist may want to study in small mammalian species that prefer to locate in long, skinny forest areas next to agricultural areas (Turner, 1991). In any event, it is obvious that being able to calculate the long/short axis ratio is a useful GIS function. Its initial calculation in vector GIS is performed as before, but first distances of all the opposing vertices must be found. Then, by comparing their lengths, the shortest and the longest axes can be identified.

When calculating long and short axes for polygons, it is notably easier to work with polygons that are wholly convex—that is, having no concave portions. If there are concave portions, or if the polygons are highly irregular, the calculations become considerably more difficult to describe, and the results are less useful. One approach, called the **least convex hull,** allows us to sort of shrink-wrap the polygon so that it essentially resembles a totally convex polygon. These calculations, while useful, are beyond the intent of this introductory work, and you are referred to any good book on computer graphics; or you may seek more detailed algorithmic descriptions provided by your GIS vendor.

Calculating Perimeter

Perimeters around polygons are created by connecting line segments in a **closed cartographic form,** where the beginning coordinates of the first line segment are identical to the ending coordinates of the last line segment. Therefore, calculating the perimeter is a matter of calculating the distance of each line segment using the Pythagorean theorem, then adding these individual values. Once again, raster data structures are not particularly well suited for this task, but, like measuring distances themselves, they can be made to produce a perimeter. To do so, each grid cell located at the outside perimeter of the collection of grid cells that make up a polygon must be separately identified and reclassified to produce a map of just these perimeter-forming grid cells. Quite often this task calls for extensive interaction with the software. Finally, however, the perimeter is defined by the sum total of grid cells, multiplied by the grid cell resolution. As before, the more complex the polygon, the more grid cells will be at a diagonal to their neighbors and the less accurate your results will be.

Measuring perimeter is a useful technique if it is necessary to surround an identified polygon with a linear object. If, for example, you wish to build a road around a lake (your polygon), you will need to know how long the road must be to run completely around the lake. Because the cost of building such a road will be, in large part, related to the amount of road itself, the value obtained by measuring the distance around the lake can be converted into total cost by multiplying the cost per unit length by the total number of road units needed. Similarly, if you are planning to fence in a large ranch, the perimeter of the property will immediately translate into the amount of fence you need to build. But while perimeter may be a useful measure, it is commonly associated with area as the ratio of perimeter to area. We will look at the calculation of areas first, then return to the perimeter-to-area ratio.

Calculating Areas of Polygonal Features

A frequently used measure in GIS is that of polygonal area. Area gives us some quantitative measure of the amount of each classified areal feature. Developers may want to know how much unsold land is available for purchase, whereas wildlife management specialists, knowing that most animals need a specific amount of territory in which to search for food, might be interested in the amount of natural habitat available for a particular animal. Finding area for a region, as defined in Chapter 7, is calculated in raster by selecting the grid cells that share common attributes (our region) and counting the number of grid cells that the region occupies. Actually, you don't even need to perform a calculation. Just tabulate the data in your coverage and read the number of grid cells sharing the assigned attributes from the resulting table. Of course, the table will also give you a percentage of grid cells in the database for that region, compared to the overall database. Other ratios can be performed by comparing the numbers of grid cells of one region to those of another.

If, on the other hand, you do not want to measure the grid cells of fragmented regions, the problem is a little more difficult. If you want to know only the area of a particular polygonal area, whether it constitutes a contiguous region or not, you need to isolate each polygon separately by selecting its grid cells and reclassifying them so that their attribute values are unique to the coverage. As we will see in the next chapter, there are generally functions available in raster that allow you to isolate such areas by size, providing an easy way to select the polygon you want to measure. If you have many polygons of approximately the same size within your coverage, however, you will need to isolate them either by their row and column locations or by individually recoding the grid cells of your target area using available editing functions. In any event, the process is not difficult; it simply requires forethought and planning.

In vector, the target polygon or polygons are selected either based on coordinate locations or by attribute values in the attribute tables. Then the area of the polygon is calculated. For simple cartographic forms, such as rectangles, triangles, circles, parallelograms, and trapezoids, the calculation is no different from what we learned in our youth. It is only when the polygons take very complex forms that the computations become more complex. The most common solution is to divide complex polygons into shapes that can be easily measured with readily available formulas (Clarke, 1990). In many vector GIS, the areas of polygons are calculated as individual portions of the object are digitized. As the operator digitizes a polygon, a special code is normally given to indicate that a polygon, not a line, is being digitized. Therefore, as the units in each set of line segments are digitized, the software determines the simple geometric shapes produced as the process continues, then calculates the areas of these shapes. Finally, the totals are computed to obtain the final polygonal area. Because the polygonal areas can be computed individually, their values are passed to the attribute tables immediately upon data input. Therefore, the area of a polygon can be obtained by making a simple selection from the attribute tables.

Frequently, there is a need to combine the values of area and perimeter in a ratio. This relationship, called the **perimeter/area ratio,** is the most compact shape available, and it provides a measure of the complexity of each polygon. Thus a circular object will have the smallest perimeter/area ratio. By contrast, long, skinny polygons tend to have larger perimeter/area ratios because their perimeters are so large and their areas so small. While the calculation of such a ratio might seem to be pointless, numerous tasks, especially in natural-resource-related operations require it. For example, the smaller the perimeter/area ratio of a forest stand, the more likely you are to have living within that forest stand animals that prefer interior environments for habitat. If you are a wildlife specialist, you probably have some idea of the requirements of the animals you are trying to preserve. Alternatively, if you are trying to develop lakefront property, the higher the perimeter/area ratio, the more beach area you can offer to prospective lot buyers. Thus, the higher the perimeter/area ratio, the more money you will make developing beachfront property.

MEASURING SHAPE

As you have just seen, there is a close relationship between shape and such
measures as perimeter and area for polygons and even length for linear objects.
Many times, the shapes of polygons or some measure of the sinuosity of linear
features will provide insights into the relationships between objects and their
environment (Boyce and Clark, 1964; Lee and Sallee, 1970). For example, the
sinuosity of a stream is related to such functions as stream sediment load,
slope, and the amount of water flowing through the stream. In turn, these
functional relationships have much to do with whether a stream is aggrading
(building up sediment load), at grade (in balance with stream inputs and outputs), or degrading (downcutting). Hydrologists, geomorphologists, and other
environmental scientists find these values useful in overall analyses of conditions in a region. And as we have seen, the relationship between perimeter and
area, as a simple measure of polygon geometry, has much to do with the functioning of anthropogenic and natural features. Therefore, it is important to have
at least a basic understanding of what these measures are and how they might
be performed in a GIS. In the next chapter we will see how these measures can
be used to reclassify portions of the landscape.

There are very few measures of shape readily available in GIS, although there
are some relatively sophisticated ones in the literature (Moellering and Rayner,
1982). This again indicates the nature of commercial GIS as a market/driven
technology: it is most often the scientific community that is interested in these
measures, but academics and researchers constitute a relatively small proportion of the GIS sales market. In addition, there is little available theory that
allows us to measure unusual shapes. Traditional Euclidean geometry limits us
to a rather small list of identifiable shapes whose dimensions have been documented and quantified. There is an emerging interest in nontraditional geom-
etries, such as fractal geometry, which uses imaginary numbers to define
shapes. But this approach, too, is atypical, so you or your instructor may
choose to ignore it. Interestingly, however, there have been at least two major
efforts to incorporate such measures, as well as additional measures of shape
and object interactions into both raster and vector GIS (Baker and Cai, 1992;
McGarigal and Marks, 1994). For the time being we will limit ourselves to simple
measures of shape that are commonly available in most GIS software.

Measuring Sinuosity

There are two simple measures of sinuosity that can be used as a measure of
linear shape. The first, which we have already examined to some degree, is the
simple measure of the overall length of a line object compared to the straight-
line distance between the beginning and ending points. This is found by measuring the straight-line distance between the beginning and ending coordinate
pairs in grid cells (in a raster GIS) or using the Pythagorean theorem (for vector
GIS). Then the distance along the line is measured, again in raster or in vector,
as appropriate, and the contributing values are summed up. Finally, a ratio is
formed between the straight-line distance (in the numerator) and the actual
linear distance (in the denominator). The closer this value is to one, the less

sinuous the line is. As we saw in the last chapter, we can now find linear objects of any selected sinuosity value and display or tabulate its values.

However, there is often a need to know more about the shape of a curve along a sinuous linear object. Very sharp curves in roads, for example, are likely to cause accidents. And in streams, sharp curves lead to a high degree of erosion along the outside of the formation and more sedimentation along the inside. For this reason, it is useful to be able to find the radius of each curve in the line. To do this we assume that the curves are essentially circular, although this may not always be the case (Strahler, 1975). Then we fit a semicircle to each curve and measure the radii. When streams are being represented as polygonal shapes rather than as lines, a ratio of the radius to the width of the stream gives yet another useful measure of shape.

Whether in raster or in vector, measuring the radius of each curve in a sinuous linear object is difficult without the intervention of the operator. Many GIS packages do not provide an adequate set of tools for such measurements, but many specialty packages will allow them to be made outside the GIS and the data passed back into the system for later analysis. To perform the operation in a GIS most often requires the user to define the circles for each bend manually, which can become tedious. Thus you need to examine the importance of such measures for your work, and the abilities of your software to perform them.

Measuring Polygon Shape

As we have seen, there are two fundamental aspects of measuring the shape of individual polygons. The first, based on the idea of perforated and fragmented regions, is given the general name of **spatial integrity.** A second measure is based on the boundary configuration, more closely related to our earlier discussion of perimeter and area relationships. The second measure is often used in conjunction with other commands that allow each individual contiguous polygon to be separated from polygons with the same attribute values. In other words, we can measure each polygon's boundary configuration in isolation from the rest of a possible fragmented region, rather than simply noting the overall size of the region made up of individual polygons.

Spatial integrity is a measure of the amount of perforation in a perforated region. The most common measure of spatial integrity is called the **Euler** (pronounced "oiler") **function** and named after the Swiss mathematician (Berry, 1993). Many organisms—for example, many bird species—prefer large, uninterrupted patches or polygons of a particular type of land cover, whereas others, like deer, seek out large forest areas broken up by intervening smaller patches of grassland or other land cover types. A patch of forest that is unbroken is said to be a contiguous region; if it completely surrounds smaller polygons, it is said to be perforated; and if it is completely separated from similar patches by some intervening type, we call it fragmented. The Euler function is a numerical measure of the degree of fragmentation as well as the amount of perforation. Let's take a conceptual look.

Figure 8.2 shows three possible configurations of polygons. The Euler function describes these functions with a single number, called the **Euler number.** Numerically the Euler number is defined by the following simple equation:

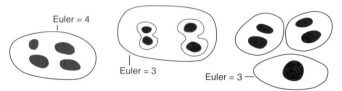

Figure 8.2   Example of Euler numbers. Three different land configurations and their associated Euler numbers: *(a)* four holes in a single contiguous region, *(b)* two fragments with two holes in each, *(c)* three fragments, two each with a pair of perforations and the third with a single perforation. It is important to note that the Euler numbers for *(b)* and *(c)* are identical even though the configurations are different.

$$\text{Euler number} = (\#\text{holes}) - (\#\text{fragments} - 1)$$

where #holes is the number of self-contained polygon perforations in the outside polygon and #fragments is the number of polygons in the fragmented region.

In Figure 8.2*a*, there are four holes or perforations in a single contiguous region. By inserting these numbers into the equation we get:

$$\begin{aligned} \text{Euler number} &= (4) - (1 - 1) \\ &= 4 - 0 \\ &= 4 \end{aligned}$$

In the second example (Figure 8.2*b*) we find two fragments in our region, each of which has associated with it two perforations or holes. Again substituting in our formula we find that our Euler number is $4 - (2 - 1)$, or $4 - 1 = 3$. The third example (Figure 8.2*c*) shows that with three fragments and a total of five holes, our equation gives $5 - (3 - 1)$ or 3, exactly the same as in Figure 8.2*b*. This shows that while the Euler number gives a measure of spatial integrity, care must be taken in the explanation of the results. You might want to examine a number of cases to see what results you get for, say, an area with three fragments but no holes: $0 - (3 - 1) = -2$. You should see, based on the last example, that at times the numbers will not be positive. In addition, if you tabulate some of your results, you may begin to see the relationships between Euler numbers for different combinations of holes and fragments (Table 8.1). This table will prove useful in determining the twin meanings of any given Euler number.

The second group of measures of polygonal shape, those related to the boundary configuration, are quite numerous. Among the more prevalent are those based on the axial ratio, those based solely on perimeter, those based on perimeter and area, those based only on areas, those based on areas and areal lengths, and others measuring circularity mean side and variance of side. The formulas for these, which can be found in Chapter 5 of a useful text on geology (Davis, 1986), can be referred to as your interests in shape analysis develop and your GIS skills improve. Some of the more esoteric measures are available in commercial GIS, while others are available in geostatistical software

TABLE 8-1   Matrix of Euler Numbers for Different Hole–Fragment Combinations[a]

	#fragments								
#holes	**1**	**2**	**3**	**4**	**5**	**6**	**7**	**8**	**9**
1	1	0	−1	−2	−3	−4	−5	−6	−7
2	2	1	0	−1	−2	−3	−4	−5	−6
3	3	2	1	0	−1	−2	−3	−4	−5
4	4	3	2	1	0	−1	−2	−3	−4
5	5	4	3	2	1	0	−1	−2	−3
6	6	5	4	3	2	1	0	−1	−2
7	7	6	5	4	3	2	1	0	−1
8	8	7	6	5	4	3	2	1	0
9	9	8	7	6	5	4	3	2	1

[a] Note the number of configurations that can produce the same Euler number. Not also the mirroring of Euler numbers on either side of the diagonal made with the zero Euler values.

or as macros for commercial GIS (Baker and Cai, 1992; McGarigal and Marks, 1994).

As we have discussed, most of these measures are strongly related to the ratio of perimeter to area. In fact, as you will soon see, the perimeter/area ratio itself can be considered to be a measure of polygonal shape. However, this doesn't necessarily describe the actual geometric shape of the object. Instead of a simple ratio, it would be really nice if we had a measure that was more shapelike. For this we most often compare the polygonal shapes we encounter to more familiar shapes we can easily describe (Muehrcke and Muehrcke, 1992). We might, for example, want to compare the polygonal shapes to such geometric forms as parallelograms, trapezoids, and triangles. However, these shapes come in a wide array of forms, each one of which would have to be described separately. And, of course, the simplest, most compact, and most easily defined shape is the circle. For this reason, the basic method of measuring shape is to compare it to the circle.

Because we use the circle as the comparative shape, we can say that this measure of shape is also a measure of the **convexity** or **concavity** of the polygon. A circle has perfect convexity because no portion of its surface is concave or indented. This is why the circle is the most compact geometric shape and, of course, why we use it to measure the shapes of other objects. As it turns out, all other geometric shapes have more perimeter than a circle. To compare our polygonal geometric forms to that of a circle is essentially the same as examining the amount of convexity of the polygon versus that of the circle. The general form of the convexity formula in vector GIS is

$$CI = \frac{kP}{A}$$

where

CI = convexity index

k = a constant

P = perimeter

A = area

So what we have is a perimeter/area ratio for each object, multiplied by the constant. The constant is based partly on the size of the circle that would **inscribe** the irregular polygon. In addition, it is designed to provide a range of positive values from 1 to 99, where 100 indicates 100% similarity to a circle. Stated differently, a 1 is as far from a circle as can be measured, and a 99 is as close to a circle as a shape can get without actually being a circle. A perfect circle thus has a value of 100.

In raster, the formula is based on the exact same concept, but the area is now recorded as the number of cells and its square root is used to provide the same 1–99 range of similarity values. Therefore, the general form of the formula for calculating convexity in a raster GIS is:

$$CI = \frac{P}{\sqrt{\#cells}}$$

where

$$CI = \text{convexity index}$$
$$P = \text{perimeter}$$
$$\sqrt{\#cells} = \text{area in raster format}$$

As before, a value of 1 is as far from a circular or totally convex shape as can be measured, and 99 is as close to perfect convexity as a shape can get without being a circle. Of course, in raster it is physically impossible to get a perfect circle.

As before, the important question again arises. What do we use measures of convexity for anyway? Ask yourself this question: When the American Plains Indians were attacking a wagon train, why did the wagon master have the wagons formed into a circle? Why not a trapezoid? Or a parallelogram? The answer is that there is something very important about the functionality of the circle, with its very compact shape. Specifically, it has the lowest amount of edge possible, given the area. Many creatures, including humans, like the security of a circular shape, with its readily defended perimeter (hence the term's use in the military). Other interior species take advantage of the lack of edge that offers protection from competition for needed resources like food and shelter. Yet, some species—for example, some small rodents—really like edges. They tend to use them as base areas from which to forage in the open along field crops, while the forest remains accessible for escape and shelter.

This scenario suggests another, quite simple measure of boundary configuration called **edginess** (Berry, 1993) that uses a device called a **roving window.** Also known as a **filter,** a roving window is a matrix of numbers of a preselected size that can be moved across a raster GIS or digital remote sensing database. The idea is to move the window across the database to either examine what is there or to modify the grid cell values. In remote sensing, two fundamental tasks for which the filter is used are **edge enhancement** and **smoothing.** In the former, edges and lines are detected, and their values are increased so that these features become more prominent. In smoothing, highly different values are suppressed, and their numbers are made more like their surroundings through some form of averaging. While these uses of filters could more closely be associated with classification, we will look at their use for edginess here as a measure of boundary configuration.

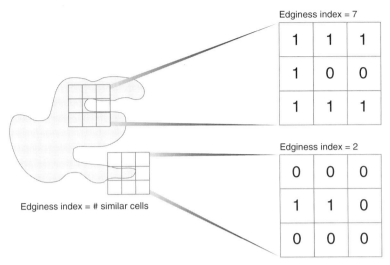

Figure 8.3 Raster measure of edginess. A moving filter used to determine the edginess of a polygonal edge in raster. Note how much more edge there is as the number of neighboring polygons with a single attribute goes down. *Source:* Modified from J.K. Berry, *Beyond Mapping. Issues and Concepts in GIS.* Fort Collins, CO, GIS World, Inc., © 1993.

Let's assume that we are using a 3 × 3 filter (a 9-cell window), which we will move along the respective boundaries of two raster-based polygonal structures to examine the numbers covered (Figure 8.3). We assign digital values to all the cells inside the filter: 1 for cells that are the same attribute value as the edge we are interested in and 0 if they are attribute values for any other object in the coverage. The amount of edginess is obtained simply by counting the number of grid cells that are 1s, or share the same attribute values as the polygon. The more 1s we get, the less edge we have, and the more interior. Thus the value of 7 in Figure 8.3 illustrates very little edginess: in the first portion of the matrix covered, nearly all the grid cells are connected to each other. Likewise, a 9 would be total interior, with all grid cells connected and no edge. By contrast, the edginess value of 2 in Figure 8.3 indicates that only a small number of the grid cells exist in the second portion of the map over which we have fitted the filter; the remainder are essentially background or matrix grid cells (0s). The 2 indicates a small, skinny protrusion into the matrix. Or stated differently, the lower filter covers a great deal of edge.

MEASURING DISTANCE

Although measures of shape are now becoming more important in GIS analysis, partially through an understanding of the relationships between form and function on landscapes, there has been a long-standing interest in the measurement of distance. The distance between features is important, not only in later analyses of the relationships between and among them, but more immediately because it provides a means of estimating travel to, from, and around them. As you will see, distance can be measured in simple terms where its calculation is the absolute physical distance between places in the coverage. In addition, however, distance can be measured to include the cost incurred while traveling rugged terrain, the cost of travel specifically along networks rather than a sort of off-road distance, or difficulties involving barriers that restrict or prevent movement. These latter approaches to distance measurement are collectively known as **functional distance.**

Simple Distance

As we have seen, calculating simple distance, known as **Euclidean distance,** is relatively easy in both raster and vector GIS. In raster, of course, the method is to add the column and row distances in grid cell units. And as we have learned, these grid cell distances can be converted to a standard measure of distance by multiplying the number of grid cells by the grid cell resolution. In addition, we know that diagonal grid cell resolutions can be calculated as the hypotenuse of a right triangle. Until now, we have been discussing distances between two point locations in a coverage, or along the grid cells of a linear object. However, this description does not always apply. We may, for example, want to know the distance between a single point in the coverage and all other possible locations. Again, this is done in raster by producing a series of concentric rings, one grid cell in diameter, around the starting grid cell. Each ring is essentially one grid cell away from the preceding ring. The result of such a distance measure is called an **isotropic surface** in as much as it is completely uniform in all directions (Figure 8.4a). If displayed in three dimensions, it will give the appearance of a cone whose center is the starting grid cell. An isotropic surface map from a point to all other locations is essentially a travel map showing distances to every other object in the database. It has the same effect as having a complete road map indicating the distance of travel in all directions from the starting point. Its advantage over a simple calculation of point-to-point distance is apparent when there are many to be performed: the isotropic surface has already

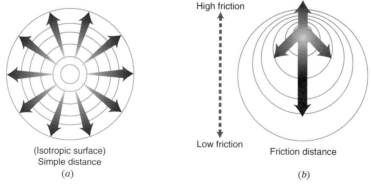

Figure 8.4 Simple versus functional distance. *(a)* Measuring simple distance from a central point produces an isotropic surface *(b)* Modification of an isotropic surface as movement is impeded through the friction encountered at the sites of obstructions.

calculated them. We will see later that we must modify this isotropic surface to account for differences in surface and the existence of barriers.

In one variation of measuring distances between a single grid cell and all other grid cell locations, we measure distance from a polygonal object to all other locations in the coverage. Suppose, for example, that you want to know the distance from the outside of a city, whose boundaries are represented by a large polygon, and any other location. The GIS begins at the margins of the polygon and counts the number of grid cells outward until it reaches the end of the coverage. The result again is a map of travel distance to all areas not in the city from any marginal location of the city.

In vector, once again, calculation of distance is simplicity itself, at least for calculating distances between points. By using the Pythagorean theorem for each connected line segment and then adding, we obtain a total Euclidean distance measure. The same operations are used to calculate distances between any two point objects that are not connected through line segments. However, the points must actually exist in the coverage. Because the vector data structure does not explicitly define the intervening spaces between existing objects, no calculations can be performed in these spaces unless objects are placed there.

The foregoing limitation also will apply to developing isotropic surfaces, whether from points to all other locations or from polygons to all other locations. To perform such calculations requires a different vector data structure, specifically designed for modeling surfaces. The most common of these vector surface data models, the triangulated irregular network (TIN), was described to some degree in Chapter 4 on cartographic and GIS data structures. We will return to the TIN model and describe its operation in more detail in Chapter 10. Distances can be calculated in the TIN data model, but at considerable computational expense. GIS users who work with surface models most often defer to the use of a raster system, especially when it is necessary to measure distances along surfaces.

Functional Distance

Although Euclidean distance is useful, obstructions or difficult terrain will frequently reduce our ability to travel "as the crow flies." For example, we may be restricted to the use of networks such as roads and rail lines, either because the intervening terrain is too rough, producing a **friction surface,** or because fences surround the intervening terrain, acting as a **barrier** to travel. Friction surfaces are areas that slow or impede progress, making travel to a given place slower than it would be if the surface were totally isotropic (Figure 8.4). Barriers come in two types (Figure 8.5). **Absolute barriers** (a cliff, a fenced area, a lake, etc.) prevent movement entirely. **Relative barriers** are identical to friction surfaces but occur only along limited portions of the coverage. Relative barriers to travel might be exemplified by narrow ridges of hilly terrain, shallow streams that are passable for off-road vehicles, or patches of forest that restrict but do not prevent completely the movement of animal herds.

Absolute barriers prevent or deflect movement, while relative barriers and friction surfaces incur a cost for movement, resulting in either slower movement or the expenditure of additional energy. You might imagine the slower move-

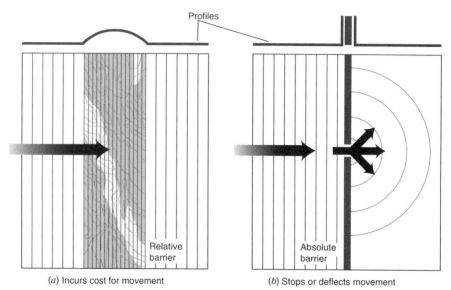

Figure 8.5 Relative versus absolute barriers. Relative barriers *(a)* incur costs for movement across them. Absolute barriers *(b)* either stop movement entirely, or deflect it to any available access points.

ment of large mammals through a swamp as opposed to rapid travel on an open prairie. Or you might consider an automobile that can climb a hill and cross a flat surface at the same speed but must use more gasoline per mile of travel in the uphill mode. Absolute barriers may have points of access through which travel is permitted, such as bridges over streams. In such cases, movement across the barrier is restricted to these access points, producing a bottleneck that results in a radial pattern of movement on its down-movement side. This is not unlike the pattern achieved as perpendicular onshore waves encounter the inlet to a bay, where the water that gets through produces a series of concentric waves inside the bay.

Raster modeling of barriers and friction surfaces is conceptually fairly simple. If, on an isotropic surface, the GIS adds one grid cell for each unit of travel, the resulting map will show that each additional ring of the isotropic surface is recorded as 1 + the row before it. So if we travel 10 units from our starting point or polygon, the outside ring will have the value of 10, the next inner ring will have the value 9, and so on, since the GIS counts grid cells as it moves along the surface. Let us say, however, that we want to place a relative barrier across our map from top to bottom. We want to show travel distance as a functional distance from left to right across the map. To create the barrier, we place values along the vertical line and will not allow the coding of another cell beyond it until it has counted beyond the highest value coded for the barrier. For example, if we used the **impedance value** of 5 for our barrier and began counting from left to right, it would continue to increase the next grid cell by a value of 1, until it reached the barrier. It then must add three additional counts before it can continue. Such a barrier is considered to be a relative barrier, because, once it has counted five additional numbers, it will continue to move along, adding one number to each additional grid cell (Figure 8.6*a*). In many

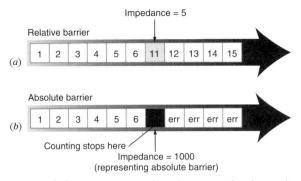

Figure 8.6 Impedence. Simple accrued values of grid cell counts as movement encounters a relative barrier *(a)* and an absolute barrier *(b)*.

cases, such relative barriers or friction surfaces may vary in value and extend across the entire surface of the coverage. For example, we might want to move through a coverage of land cover containing five or six different land cover types, each with its own friction value for travel. Most often this is done by reclassifying the land use attributes to reflect their friction value and naming this new coverage as the friction coverage (Figure 8.6*a*).

To make the barrier an absolute barrier, we assign it a weighting factor sufficiently high that it will either exceed some maximum number designated in the GIS program, or until it counts what should be the maximum number of grid cells in the GIS coverage. So, if we have a raster coverage that is 100 grid cells wide, we might put a barrier value or impedance value of 1,000 to assure that the process of moving and counting the grid cells stops before it passes the barrier (Figure 8.6*b*).

While the concept of assigning friction values and impedance values is a simple one, operationally it causes some difficulty. Just how much impedance does a forest cause to deer populations as they travel through it? How much more gasoline is used in moving up a 15% inclined surface than over a flat surface? How much longer does it take a large mammal to swim a stream 100 meters wide than it would to walk across a bridge 100 meters long? These questions are real questions, many of which will have to be answered to some extent before we can create our barriers and friction surfaces. Most often the answers will not only not be available but will depend on a variable as slippery as the agility of a deer, the gas efficiency of an automobile, or the swimming ability of a zebra. In other words, most often we lack absolute measures for impedance or friction values. Generally we tend to use a more ordinal ranking system to assign impedance values, based on a comparison of the relative difficulty of travel through each of the intervening friction surfaces or barriers. This is not always easy, and the results are not as precise as we would like. Caution is advisable in assigning these values to ensure that there is a reasonable explanation for the system used and that some justifiable comparison can be made among the friction surfaces. Above all, the results of analysis of distance based on barriers and friction surfaces should be viewed with some trepidation, especially if they are to be used for decision making.

Before we discuss the mechanics of raster approaches to non-Euclidean and functional distance, we must examine two additional characteristics of distance. Distance can be viewed not only as Euclidean or non-Euclidean, isotropic or functional, but also as **incremental distance** or **cumulative distance.** Incremental distance simply measures each unit, or increment, traveled, adding the second to the amount of distance traveled in the first, and so on. Each successive increment is added simply as a measure of distance, much like our procedure for working with the isotropic surface. In other words, the incremental distance measures the shortest path between two places without adding any friction or impedance values along the way. If incremental distance is measured for an entire surface, the result is a **shortest path surface,** while if the technique is restricted to lines or arcs (or linear arrangements of grid cells), we have a **shortest path** as a linear product, rather than a surface.

As we saw with friction and impedance, however, there is a continuous increase in cost as we encounter surfaces or obstacles. The purpose of calculating functional distance on a friction surface (e.g., a topographic surface) is to find the **least-cost distance,** or shortest path between two points in a coverage. Of course we may also want to calculate the **least-cost surface** for movement from one point location to all other locations. Let's take a look at these cases individually.

For this discussion let us assume that our surface is a real topographic surface, and the cost is related to the change in elevation from one grid cell to another. Conceptually, we are placing a drop of water at the top of our topographic surface and watching as it streams or drains down the topographic surface to lower elevations, finding the path of least resistance. Indeed, many GIS software packages use terms like "drain" or "stream" as commands. To create a least-cost distance (as opposed to a shortest path distance) in a 5×5 raster matrix, we begin at the top grid cell and perform a search of the eight nearest grid cells to evaluate which has the lowest friction or impedance value (in this case lowest elevation relative to the starting point). This grid cell, which is assigned a numeric value or a pointer to indicate that it has been selected, then becomes the starting point for the next iteration in a search of its eight neighbor cells. The process continues until the lowest elevation value in the database has been selected. What we have then is a selected route from the top of our hill to the bottom that incurs the least amount of effort.

Let's continue with our simple visual example. We want to find the easiest path down the mountain represented by the 5×5 matrix of Figure 8.7a. Starting at the top (column 5, row 5), we examine our neighbors (there are only three because we are at the corner of our coverage) to determine which of the connected grid cells is lowest relative to (5,5). We note that grid cell (4,4) has the lowest value, and we move there next. Proceeding to grid cell (3,3), we again perform a search of all neighbor cells, again choosing to move to the lowest available grid cell value (2,2). When we have reached the bottom of the hill at location (1,1), the trail left behind is our least-cost path. An analog version of this search is shown in Figure 8.7b.

While the least-cost route often suffices, it may prove useful to examine all possible routes from a particular place as we accumulate a cost for moving up or downhill through a friction surface. In this case, our friction values might be envisioned as topographic impedances, or friction impedances encountered during a trip through different types of land cover. Take a look at Figure 8.8.

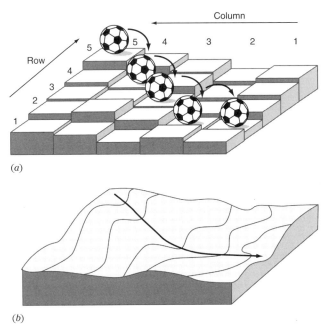

(a)

(b)

Figure 8.7   Least cost path. (a) Calculation of a least-cost path requires the computer to compare the starting grid cell to its immediate neighbors. A soccer ball bouncing down the steps (grid cells) of a raster surface is used to illustrate the idea. (b) The same function as it might be performed by a stream flowing down a mountainside.

Here we have two grids—the one indicating the friction or impedance factors and another indicating the calculated values of accumulated distance (weighted by the impedance factors). The process of computation is slightly more sophisticated than the shortest path model we just examined. In this case we see that the program doesn't simply search for the lowest value, but rather calculates a value for each adjacent cell based both on Euclidean distance as well as the impedance values. This example uses rational (decimal) numbers; diagonal distances between cells are measured when necessary.

We begin the process in the grid cell at column 0, row 0. The program then identifies all bounding grid cells and notes the friction factors (if any) for each of these bounding cells. The distance is calculated in half-step increments by multiplying each occupied grid cell (including the starting grid cell) by its friction factor and then by a value indicating its width (1 for orthogonal grid cells and 1.414 for each diagonal grid cell). Because we are moving from the center of the starting grid cell to the center of the next, we multiply each by 0.5 to indicate the half-step increments. Thus our formula for each step of half a grid cell is (Berry, 1993):

$$0.5 \text{ (grid distance} \times \text{friction factor)}$$

We then must add this to the previous friction value. So for a movement starting at column 6, row 6, moving to grid cells (5,5), (4,4), and (5,6), our resulting calculations and values are:

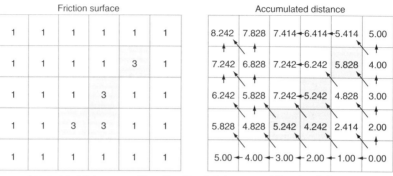

Accumulated distance = previous (functional) distance + next (functional) distance ...

Figure 8.8   Accumulated distance. Accumulated cost surface shown by indicating *(a)* the friction coverage and *(b)* the calculated values. Note how the calculation of diagonal grid cells uses a value of 1.414 as the distance rather than 1.0 for orthogonal grid cells. The shaded grid cells are those whose friction value is 3. To accumulate distance, the previous value must be added to the calculated value.   *Source:* Modified from J.K. Berry *Beyond Mapping. Issues and Concepts in GIS.* Fort Collins, CO, GIS World, Inc. © 1993.

5,5	$0.5 (1.414 \times 1.00)$	=	0.707
	$0.5 (1.414 \times 1.00)$	=	0.707
	Subtotal	=	1.414
	+ Previous value	=	0
	Total	=	1.414
4,4	$0.5 (1.414 \times 1.00)$	=	0.707
	$0.5 (1.414 \times 3.00)$	=	2.121
	Subtotal	=	2.828
	+ Previous value	=	1.414
	Total	=	4.242
5,6	$0.5 (1.000 \times 1.00)$	=	0.500
	$0.5 (1.000 \times 1.00)$	=	0.500
	Subtotal	=	1.000
	+ Previous value	=	0
	Total	=	1.000

The next step is to choose the smallest accumulated distance for each of the adjacent cells. Then the process is repeated. The friction distances for all steps are added together to yield an accumulated distance. As you can see, the process can be quite tedious. That, of course, is why we use computers. The result is a least-cost surface, rather than a least-cost path, and we can choose the least-cost path to and from anywhere on the coverage. This is done by simply selecting the adjacent cells with the lowest cost values.

In cases of vector coordinates describing points, lines, or areas available for calculating distances, a series of modifications of the standard Pythagorean theorem can be used to calculate the distances. We will examine these by first reminding ourselves of the original formula for Euclidean or straight-line distance between any two points.

$$d_{if} = \sqrt{(X_i - X_j)^2 + (Y_i - Y_j)^2}$$

In this equation the distance between points i and j is the square root of the sum of the square of the differences between the X and the Y coordinates.

If, however, we are unable to perform a straight-line distance—in other words, if there is an obstruction requiring us to deviate from a straight line, we can generalize the formula to a non-Euclidean form:

$$d_{if} = [(X_i - X_j)^k + (Y_i - Y_j)^k]^{1/k}$$

where the squares and the square root sign have all been replaced by k, a variable representing any of a range of possible values (McGrew and Monroe, 1993). Instead of squaring the differences between the X and the Y distances, we raise them to the kth power. And where the sums of the X and Y coordinate values were under the square root sign, we now raise them to the $1/k$th power (keeping in mind that the 1/2 power is the same as square root).

By inspection of the equation above, you will see that by replacing each variable k with 2, you return to the original equation for calculating Euclidean distance. So the generalized equation works just as well for straight-line dis-

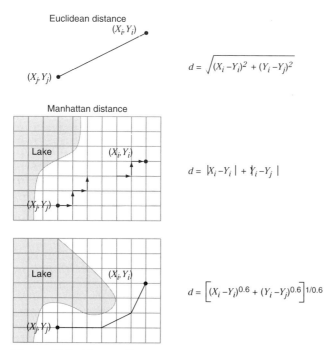

Figure 8.9 Vector distance measures. Two methods of calculating non-Euclidean distance in vector; each is a direct modification of the Pythagorean theorem formula. The exponent or k value is the only part of the equation that needs to be modified.

tances as it does for non-Euclidean distances. Let us assume, for example, that we are trying to find the distance between two points in the borough of Manhattan, where dozens of tall buildings and square blocks restrict our movement in square units. This measure, called the Manhattan distance, describes a rather restricted movement and changes our k factor to 1. Given that value our formula now converts to the following:

$$d_{if} = |X_i - X_j| + |Y_i - Y_j|$$

As you can see, the calculation of any distance in a vector database can be calculated by simply modifying the k factor to fit our needs (Figure 8.9). A k factor of 1.5, for example, simulates the distance modified by a combination of Manhattan and Euclidean. Even friction surfaces can be estimated through the use of these metrics by modifying the k factor. Take, for example, the use of a k factor of 0.6, which allows us to find the shortest distance around a barrier, such as a lake.

While these metrics are very useful, we will at times want to perform distance measures for networks. We have the capability to find shortest paths and least-cost paths, and to perform operations employing frictional impedance values as well as barriers. However, these are more closely related to connectivity than to distance, so we will examine them in Chapter 11 when we look at comparisons of variables within a single coverage.

Terms

orthogonal	sinuous	least-cost surface
perimeter/area ratio	least convex hull	closed cartographic form
Euler number	spatial integrity	Euler function
inscribe	convexity	concavity
filter	edginess	roving window
functional distance	edge enhancement	smoothing
friction surface	Euclidean distance	isotropic surface
relative barriers	barrier	absolute barriers
cumulative distance	impedance value	incremental distance
least-cost distance	shortest path surface	shortest path

Review Questions

1. Describe the simplest possible method of measuring a line in raster format. How can a result so obtained be converted to an actual measure of length when the grid cells are orthogonal? What do you do when they are lined up diagonally?

2. What are the possible pitfalls for measuring length with raster, especially with regard to the way grid cells capture linear data?

3. How do we measure the length of a sinuous linear object in vector? What is the relationship between sampling when digitizing a line in vector and the length of the line as it is measured?

4. Why would we be interested in measuring the long and short axes of a polygon?

5. Why is the relationship between perimeter and area of a polygon important? How might differences in this ratio be important in the natural world? In the anthropogenic world?

6. How is the perimeter of a polygon measured in vector? In raster?

7. How is the area of a polygon measured in vector? In raster?

8. Given the relationship between feature shapes and function, why are there so few operational methods of shape analysis in GIS?

9. Describe two methods of measuring sinuosity of linear objects. Give two examples of such measures that might be useful to the GIS user.

10. What is spatial integrity? How can we measure it? Why is a given Euler number not a definitive measure of the number of holes and the number of polygons?

11. Describe how polygon boundary configuration might be calculated. What is a measure of convexity? Why is the circle the most frequently used geometric object against which polygonal shape is compared?

12. How can a roving window be used as a measure of polygon boundary configuration in raster?

13. How is Euclidian distance from a point to all other points in a coverage measured in raster? What is an isotropic surface, and what is its significance?

14. What is a friction surface? A barrier? What are the two types of barrier? What is the relationship between a relative barrier and a friction surface?

15. Describe a simple raster method of measuring distance through a friction surface.

16. What is the difference between incremental and accumulated distance? How can we measure the path of least resistance (best path) through a raster coverage that includes surface data? What is a more accurate method of measuring accumulated distance (best path surface) through a raster GIS than simply adding up the friction numbers?

17. How can we modify the Pythagorean theorem to account for Manhattan distance, paths around barriers, and paths through friction surfaces in vector?

18. What are the potential problems of assigning friction or impedance values for modeling distance?

References

Baker, W.L., 1989. "A Review of Models of Landscape Change." *Landscape Ecology,* 2:111–133.

Baker, W.L., and Y. Cai, 1992. "The r. le Programs for Multiscale Analysis of Landscape Structure Using the GRASS Geographical Information System. *Landscape Ecology,* 7(4):291–301.

Berry, J.K., 1993. *Beyond Mapping: Concepts, Algorithms, and Issues in GIS.* Fort Collins, CO: GIS World. Boyce, R.R., and W.A.V. Clark, 1964. "The Concept of Shape in Geography." *Geographical Review,* 54:561–572.

Clarke, K.C., 1990. *Analytical and Computer Cartography.* Englewood Cliffs, NJ: Prentice-Hall.

Davis, J.C., 1986. *Statistics and Data Analysis in Geology,* 2nd ed. New York: John Wiley & Sons, Inc., New York.

Environmental Systems Research Institute, 1993. *Learning GIS, The ARC/INFO Method.* Redlands, CA: ESRI.

Forman, R.T.T., and M. Godron, 1981. "Patches and Structural Components for Landscape Ecology." *BioScience,* 31:733–740.

LaGro, J.A., 1991. "Assessing Patch Shape in Landscape Mosaics." *Photogrammetric Engineering and Remote Sensing,* 57:285–293.

Lee, D.R., and G.T. Sallee, 1970. "A Method of Measuring Shape." *Geographical Review,* 60(4):555–563.

McGarigal, K., and B.J. Marks, 1994. FRAGSTATS, Spatial Pattern Analysis Program for Quantifying Landscape Structure, Version Two. Forest Science Department, Oregon State University, Corvallis.

McGrew, J.C., and C.B. Monroe, 1993. *Statistical Problem Solving in Geography.* Dubuque, IA: Wm C. Brown Publishers.

Moellering, H.M., and J.N. Rayner, 1982. "The Dual Axis Fourier Shape Analysis of Closed Cartographic Forms." *Cartographic Journal,* 19(1):53–59.

Muehrcke, P.C., and J.O. Muehrcke, 1992. *Map Use: Reading, Analysis, Interpretation.* Madison, WI: J.P. Publications.

O'Neill, R.V. et al., 1988. "Indices of Landscape Pattern." *Landscape Ecology,* 1(3):153–162.

Robinson, A.H., J.L. Morrison, P.C. Muehrcke, A.J. Kimerling, and S.C. Guptill, 1995. *Elements of Cartography,* 6th ed. New York: John Wiley & Sons.

Strahler, A.N., 1975. *Exercises in Physical Geography* (2nd ed. New York: John Wiley & Sons.

Turner, M.G., 1991. "Landscape Ecology: The Effect of Pattern on Process." *Annual Review of Ecology and Systematics,* 20:171–197.

CHAPTER 9

Classification

We have taken only the smallest steps in our journey. We've discovered why it is important to measure the lengths, widths, perimeters, and areas of objects, and we've learned how this is done in the digital environment. In addition, we've seen how shapes as well as sizes can be important to understanding the functioning of our world and how we can use the power of GIS to measure some of these shapes. As we traveled we have kept track of our progress, and we have seen that our distances are not just purely Euclidean, but also functional. The functioning of our travel distance might be modified by topography, or by differences in land cover. We can measure these functional distances, not only in terms of how far apart things are when our concept of distance is modified by impedances encountered but also in terms of time, energy, or even costs. Now we know how to find our way through these different frictions or impedances, and we understand how the computer will allow us to model them. We have also encountered barriers, some of which were absolute. But all these objects, friction surfaces, and barriers need to be classified: when we have been able to categorize them, we will have a means of sharing what we have learned with the explorers who follow us into the digital world. That will be our next great task, learning the techniques of classification that will permit us to catalog what we see with the most useful measures possible.

Classification of the earth's features is an age-old problem. It is one that has engaged the minds of natural scientists of many types, each looking at the earth through different eyes and using a different vocabulary and a different intellectual filter. While the concept of classifying features dispersed throughout the earth's surface may seem quite simple, the practical aspects of the task are far from straightforward. How we classify what we see will be strongly reflected in how we analyze data and what our conclusions mean. Indeed, the way we classify data may even determine what our conclusions will be. The process cannot be taken lightly, for poor classification will often lead us astray.

Let's take a look backward, to the early nineteenth century, at travelers across the newly obtained Louisiana Purchase, who classified the grassland areas of the central plains when they first arrived. The term they used, and therefore the classification they employed, was "the great American desert." The word "desert" suggests that areas in question would not be useful for agriculture. As we know, however, the area is actually a prairie, not a desert, and its production of grain crops alone has led some to call it the "breadbasket of the world." Two quite different views of the same area.

Another example of a less historical, more GIS-oriented nature will also show how a misclassification can lead decision makers astray. Let's say we have a digital environment developed by an earlier GIS traveler. Among the many coverages is one called "land cover," which includes a classification of some areas called "wetlands." No problem, you say. If they are wetlands they are wetlands. That's it. But, before you go on, consider the following questions and statements. Suppose you have a large backyard, say a full acre. You have just watered it, and it now has many large areas of standing water. Are they wetlands? Are wetlands the opposite of uplands? Some well-established classification systems say they are. How big does a wet area have to be before we call it a wetland? Is it a wetland if it has waterfowl? What if the waterfowl only show up during migration periods at the changes of seasons?

As the foregoing not unreasonable questions indicate, how we classify things has much to do with how we view them, the scale or resolution at which we view them, and how we use the data later on. Our mode of classification may also influence the instruments we choose to employ: whether ground based or deployed from aircraft (aerial photography) or a satellite. The classifications may be simple, based on a single observed criterion such as ground cover, or complex, based on multiple sets of criteria (elevation, rainfall, ecological functioning, etc.). Some classifications can be created by combining many measurements of many different coverages. The ways we classify can change dramatically for a single set of coverages, based entirely on what questions we are trying to answer.

The journey gets a little more difficult when we begin to see how our geographic filter can be modified. In this chapter you will become acquainted with the general problem of land classification, which has often confounded the natural scientist. We will take a brief look at some classification systems and some basic methods of classification that can be used in the GIS environment. Finally, we will see how classifications themselves can be decomposed or combined with data from other coverages to reclassify what we originally had. This material is broken down into relatively simple reclassification procedures and more complex classifications.

The number of possible classifications and reclassification methodologies is infinite. All are highly dependent, as is much of GIS work itself, on the needs of the user. Don't get hung up on the classifications themselves; the arguments on the merits of one system over another are not relevant to you now, as a novice GIS traveler. Instead, try to spend the time on the general categories of classification and the performance-related concepts. This will allow you to adapt to any environment you travel and any mission you accept. So, back to the trail. We have more work to do.

LEARNING OBJECTIVES

When you are finished with this chapter you should be able to:

1. Understand the enhanced importance of classification afforded by new technology, as well as the common themes for the many classification systems available.

2. Know how to aggregate data in both vector and raster to achieve new classification systems.

3. Be able to reclassify data based on the manipulation of nominal, ordinal, interval, or ratio data.

4. Understand the types of filter available and know how to use both static and roving window filters for reclassification.

5. Have a thorough understanding of what neighborhoods are, what types are available, and how total and targeted neighborhood analysis can achieve new classifications.

6. Be able to describe the many ways neighborhoods can be created in GIS for two- and three-dimensional objects and surfaces.

7. Understand how intervisibility works to produce neighborhoods of viewability called viewsheds.

8. Describe how buffers are used for creating neighborhoods, and understand the four basic methods for determining the buffer distance.

CLASSIFICATION PRINCIPLES

Among the most natural things people do is classify or pigeonhole things. Even the most mundane items we encounter every day are classified. We wear different types of clothes, drive vehicles of different kinds, have different categories of employment. We are male or female, Republicans, Democrats, or independents, adults, teenagers, or children, and so on. The classifications we use are designed to put people, places, and things in a framework that allows us to understand how they function in similar fashion among members of the same group or differently from members of other groups or classifications. This is the essence of classification.

The tendency for humans to want to impose a classification framework has long been extended to the analysis of earth surface features. The centuries-old history of cartography shows us how people have grouped features by physical type (land versus water), by political subdivision, and by human endeavor (inhabited versus noninhabited). For centuries these classifications were placed on maps with little formal discussion about how they should be organized or what impact their meanings might have on decisions. Only since the importance of map classification as a means of decision making and scientific inquiry was recognized have the problems of areal classification been examined at length (Sauer, 1921). While the interest in examining the impact of land classification began to spread in the early part of the 1900s among organizations needing to make decisions about resources and land planning, a greater urgency seems to have emerged with the advent of two simultaneous technologies, satellite remote sensing and automated cartography/GIS. Because of the abilities of these systems to provide or manipulate vast amounts of spatial data, the time was ripe for the development of an interest in better classifying portions of the earth. Then, as now, it is vitally important that the sometimes mundane task of classification not be accepted on faith, but rather be viewed with a measured sense of cynicism because of the potential limitations these classification filters impose on our interpretations and decisions.

Land classification depends on the types of objects we are going to group. There are separate classifications for vegetation (Küchler, 1956), soils (Soil Survey Staff, 1975), geological formations, and wetlands (Cowardin et al., 1979; Klemas, et al., 1993), and for agriculture, land use, and land cover (Anderson et al., 1976; U.S. Geological Survey, 1992). These classifications may be simple, such as a classification of vegetation for mapping based strictly on the plants present, or they may be more functional, such as mapping ecosystems rather than vegetation alone (U.S. Department of Agriculture, 1993; Walter and Box, 1976). Variations in classification are dictated by scale when, for example, vegetation is mapped not for small regions but for the entire earth (United Nations Educational, Scientific and Cultural Organization, 1973). Or the system might be specifically oriented toward the mandates of a particular organization (Klemas et al., 1993; Wilen, 1990). Still other cases of classification are dictated more by the technology used to obtain the raw data, such as with the use of satellite remotely sensed data (Anderson et al., 1976). Yet others are designed specifically to address decisions based on known or suspected factor interactions, as exemplified by maps of biophysical units, land capability analysis, or land suitability, which have strong similarities but reflect subtle differences in the manner in which the classification is to be applied. Biophysical units use a combination of spatially arrayed co-occurrences of biological and physical phenomena related, for example, to soils and topography to evaluate the viability of the land units to support natural systems. Land capability most often applies specifically to the ability of the soil to support housing, septic systems, wildlife, agriculture, and other major categories. And land suitability maps areas based not only on soils, but also on other biological and physical properties, are used primarily to evaluate the utility of the respective regions to support a variety of human activities.

Collectively, all these classifications, and the thousands of others that have been and are still being produced, have one thing in common: they all have a target audience or end user in mind. The end users of some classified data sets, such as classifications of land suitability for a single land use type, will be very specific. In other instances the maps will be usable by a much larger audience, or even multiple audiences. The UNESCO (1973) vegetation classification system, developed before the advent of satellite technology, was designed to unify vegetation classifications around the world, to permit anyone interested in vegetation, anywhere on earth, to talk about plants using the same basic system. While this broad approach to classification may be practical as a general standard, its usefulness as a decision tool is restricted. In introducing some of the specific problems with selected map types in Chapter 3, we discussed the rigid nature of traditional maps developed under the communication paradigm. With today's GIS, we can store raw data before they are classified and manipulate them within the GIS depending on our needs. This is the most important point about classification in the GIS environment: the more closely your classification can be made to fit user needs, the more useful it becomes. GIS provides a wide variety of ways to classify and reclassify stored attribute data to achieve this result. Its functionality might actually be called one of reclassification, because the data that are input to the GIS are already classified. The GIS operator either displays the existing classification (a storage and retrieval operation) or manipulates the existing attributes to create a classification more appropriate to the questions being asked and the decisions to be made. In the remainder of the chapter, then, we will refer to the function as "reclassification."

ELEMENTS OF RECLASSIFICATION

On many occasions point, line, and area features, together with their attributes, are input to a GIS under a specific set of criteria. We have seen how points and lines can be reclassified by simply recoding the attributes in their tables, or by recoding the grid cell values to produce new point or line coverages. In this simple process, the user changes the attributes themselves, nothing more. The process is much the same when working on polygonal features with raster systems. By selecting the attributes for the areas in which you are interested, you merely change the numeric codes or attribute names for the grid cells. If the raster GIS is a simple one, without linked attribute tables, you must first change the numbers that represent attributes and then change the legend for the new coverage, to reflect the new situation.

While changing attribute code numbers in simple raster GIS is straightforward, the process can present some interesting problems, primarily because of the order in which the numbers are changed. For example, let us say that you have a coverage with attribute codes numbered 1 through 15, where each code number represents a type of agricultural crop. Let us also say that the numbers 1 through 5 and 13 all represent grain crops, while the remaining numbers (6 through 12, 14, and 15) all represent row crops. You are interested in producing a map with only two categories: grain crops, to which you will assign the value of 1, and row crops, which will have the attribute code value of 2. To do this, you need to give all values 2 through 5 and 13 the new attribute value 1 (you don't need to renumber 1 because it will remain the same) for the grain crops portion of the coverage, and you need to give all old values 6 through 12 and 14 and 15 the new attribute value of 2 for the row crops portion. Using the idea of first things first, you go ahead and change all the grain crop numbers, followed by all the row crops. So, numbers 1 through 5 and 13 become 1s and numbers 6 through 12 and 14 and 15 become 2s. After changing the legend, you will have a map of grain crops (1s) and row crops (2s).

Just to be different, let's see what happens when we renumber the row crops first and the grain crops second, beginning, that is, by changing all values 6 through 12 and 14 and 15 to 2s, and then changing all values 2 through 5 and 13 to the new value of 1. But upon inspection, you should notice a serious problem. All row crops have been given the value 2, but we will find ourselves changing these 2s to 1s in our second step of renumbering, in which all values 2 through 5 and 13 take the new value of 1. Thus our map will now contain only one attribute type (grain crops) with the single attribute value of 1. This is not what we had in mind when we started this process. The same kind of thing could have happened if the original attribute code of 1 was actually a row crop of some sort rather than a grain. By reclassifying the values 2 through 5 and 13 to a new value of 1 for grains, we have additional grid cells that were originally 1s and were meant to represent row crops but are now going to be representing grains. And if we reclassify the 1s (6 through 12, 14, and 15) as 2s to represent row crops, all the newly classified grain crops disappear and we are left with a single category of row crops.

While many more advanced raster GIS software products will help prevent problems of the type just described by keeping track of the attributes, simpler GIS that will not help us out will also force us to keep careful watch on how we perform our reclassification. To avoid improper reclassification with simple raster GIS, you can do one of two things. First, you can choose attribute numbers that are not already part of the coverage. This is a quite useful technique as well for avoiding confusion in the classification process. A second, less simple process is to keep track of how renumbering is going to affect your outcome. If you see that the order in which you renumber your coverage will create new values that would get renumbered in the next step, you may be able to avoid problems by reversing the order in which you proceed. Still, if you become confused, select the first option, or use a GIS that keeps track of the attribute names or manipulates the categories themselves as attribute table names, rather than displaying numbered codes for the attributes.

In vector, the process of reclassifying areas requires us to change both entities and attributes. The first task is to remove any lines that separate two classes that are going to be combined. This is called a **line dissolve** operation because it selects a particular line entity and dissolves or eliminates it. Then the attributes associated with the two polygons are rewritten for the new coverage as a single new attribute for both. Let's take a very simple example much like the one before but featuring only two polygons, one for wheat and the other for corn (Figure 9.1). Our purpose is to produce a single category called grain crops by "dissolving" (removing) the line separating the two original categories. We place the newly created category called grain crops in our attribute table and assign it explicitly to the new, larger polygon. We now have a new coverage with only a single categorical value. Of course, in most real applications, there will be far more line dissolves and far more attribute changes, but the process will be identical to the one we have just performed.

In both the raster and the vector reclassification of our polygons, an interesting pattern emerges. For both, we ended with fewer categories than we had originally. This result, called data **aggregation**, is a most useful and common simple reclassification. Imagine how you would go about trying to separate a large polygon into one portion with wheat and a second with corn. Unless you could compare this information to another coverage, it would be impossible to

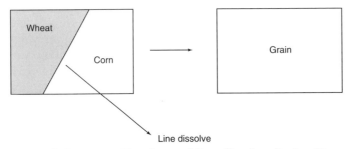

Line dissolve

Figure 9.1 Reclassification and line dissolve. Reclassification through data aggregation: the regrouping of corn and wheat into a larger class called "grains."

find the line separating them. And, of course, if you knew where it was, you wouldn't need to perform an analysis at all, you would simply use the other coverage. There are useful methods for separating out more detail from coarse polygonal information. These methods require the use of two or more coverages compared to each other through a process called **overlay**, which we will cover in Chapter 12.

The method of reclassification we have used thus far is based on nominal data only. It is not difficult to envision how we could perform the same process of reclassification on ordinal, interval, or ratio data. In cartographic methods this is done by creating range categories of data, often called **range graded** classifications. Such a scenario merely requires us to recode these data based on where they fit within the range classes. And, just as we did with nominal data, we simply recode our grid cells, or perform a line dissolve and attribute change in vector. We can also perform such operations as ordering the grid cell or polygon values, inverting them, or even using a process called **constrained math**, which permits the values of the target polygons to be operated on (multiplied, divided, etc.) by another number. But the process is fundamentally the same as what we found with nominal data.

All the methods thus far have one thing in common. The reclassification process is directed toward renaming the polygons based on the attribute values at their own location. It's a sort of local view [or what Dana Tomlin (1990) likes to call a worm's-eye view], in which each collection of grid cells or each polygon is seen as a distinct individual entity and the classification is constrained to the target area both in the use of initial values and in the reclassification itself. Reclassification based on **locational information** is only one of four basic methods; the others are based on **position**, **size**, and **shape information**. But these concepts do not separate out as individual techniques. Instead they are often combined with each other to provide a wide array of reclassification techniques. We will look at these, starting with a general description of neighborhood functions (based on the idea of a bird's-eye view), then proceed to examine several types including filters, two- and three-dimensional neighborhoods, and buffers. As you read along, notice the frequent similarities among the techniques as they are applied under different circumstances.

NEIGHBORHOOD FUNCTIONS

Reclassification based on locational attributes is very useful, but it limits us to the attributes within each object or feature. It would be nice if we could classify features based on a bird's-eye view. In other words, if we could fly over an area, we would be able to find out not only that a particular feature exists but where it is positioned within its surroundings. Such reclassification procedures are called **neighborhood functions** because the idea is to characterize each object as part of a larger neighborhood of objects based on some shared attribute. If you think about this for a moment, it is easy to recall growing up in a neighborhood. You might have classified that neighborhood based on size, or because you happened to interact with the people in it. After all, we tend to interact more frequently with people who live closer to us than those who live far away. In some cases, neighborhoods are formally defined by political or economic criteria.

Neighborhoods may also be defined in terms of a unifying attribute for an entire area (such a classification is called a **total analysis of neighborhood**), or the focus may be on smaller portions of the total area (a **targeted analysis**) (Star and Estes, 1990; Tomlin, 1990). A targeted analysis, also called **immediate neighborhoods**, includes only locations that are in immediate proximity (i.e., adjacent to) the target area or location. The total neighborhood analysis, also called **extended neighborhoods**, includes locations that are immediately in contact with the target area or location plus areas some distance beyond. While this separation of neighborhood functions is both intellectually stimulating and quite useful to the advanced GIS analyst, at the introductory level it can be somewhat confusing. As your GIS skills and modeling vocabulary grow, you may want to refer to the book by Tomlin (1990), who created this classification of neighborhood functions. In the meanwhile we will look at neighborhood functions operating on two- and three-dimensional objects. We will, however, be able to separate both these neighborhood function types as either **static neighborhood functions**, where the analysis takes place all at once for the selected target area, or **roving window neighborhood functions** (in raster), where the analysis takes place within the framework of a window that moves across the coverage. We saw an example of a windowing function in Chapter 8, when we characterized the edginess of the boundary for an area.

ROVING WINDOWS: FILTERS

As we saw in Chapter 8, there are functions that use a specified window to reclassify cells for a calculation of edginess. Such windowing functions are also called **filters**, especially when the window itself contains numbers against which the grid cells are to be compared. These can be used to reclassify whole raster coverages based in a wide range of alternative filter values and combinations of values. This technique is used quite frequently in image processing (Lillesand and Kiefer, 1995), but it has equal utility in raster GIS operations. In particular, the filter is used, much as in remote sensing, to isolate edges or linear features (an approach called **high pass filter**), to emphasize trends by eliminating small pockets of unusual values (often called **low pass filter**), or even give a measure of orientation (**directional filter**). While these filters can be statically placed over a target neighborhood to provide a neighborhood characterizing result (see next section), most often they are applied as roving windows to characterize an entire coverage.

The high pass filter is designed to separate out detail in a raster coverage that may be obscured by nearby grid cells that contain relatively similar attribute values. In remote sensing, these values most likely measure electromagnetic reflectance. However, we can use almost any surface-related data. Let's say that we are interested in finding small topographic ridges in our raster database. With each grid cell containing a value for elevation, we want to highlight the differences between the slightly higher values of a ridge and the slightly lower values surrounding this feature. The standard method for performing a high pass filter is to create a 3 × 3 filter with the weighting value of 9 at the center grid cell, in this case at location (2, 2), and a weight of -1 for all the remaining grid cells. This filter is placed over each 3 × 3 matrix of grid cells in your coverage, and the members of each corresponding grid cell pair are multiplied. That is, the elevation value of the center grid cell in your topographic coverage is multiplied by the weighting factor of 9, and all the remaining topographic values are multiplied by -1. Next, these nine newly created values are summed to obtain the final, high pass value, for the center grid cell (Figure 9.2). In other words, this single operation of nine grid cells operated on by the filter of nine grid cells produces a single value that is placed at the center of the new coverage.

The next step is to move the filter one grid cell to the right, so that the central cell in the topographic coverage is now (3, 2). The calculations are repeated as before, resulting in a new, high pass value for column 3, row 2. The procedure is repeated for the entire coverage, with the result that all lower values are somewhat suppressed, while all higher values are enhanced or made larger. What results is a coverage that shows the topographic ridge quite prominently. It can then be reclassified as you see fit, thus identifying as a specific topographic feature the region you have classified as a ridge.

Because edges and linear objects tend to come in different orientations (e.g., you may find a number of ridges related to some geological cycle of folding and erosion), it is sometimes desirable to "direct" your filter toward a particular orientation. If, for example, you are trying to highlight ridges that have an east–west orientation, you will want to enhance these features, leaving the others as low contrast objects in your topographic coverage. To do this you create a filter (we will stick to the 3×3 size) whose central column of numbers is positive and higher, while the numbers in the remainder are -1. For a northwest–southeast bias you will have the higher numbers going from upper left to lower right on your spatial filter. The same holds true for north–south and northeast–southwest bias, where the high numbers are oriented in the direction you are trying to enhance. In this way, the higher numbers will be multiplied by the higher positive values, and therefore enhanced or made larger, and the remainder will be suppressed by the operation with -1.

If you wish to suppress the higher values to remove meaningless topographic peaks, thus obtaining a coverage more closely resembling the overall trend in topography, you can use a filter in which there is less difference between values. The most common filter used for low pass operations is a 3×3 filter whose numbers are all the same, commonly 1/9 because there are nine grid cells in the filter. Since the idea now is to make all the numbers closer together in value, you move the filter as before, one grid cell at a time through the entire image, but instead of adding the values to obtain the new number for the central grid cell, you average them. Essentially, then, with each pass you are obtaining an average number for the grid cells under the filter.

While the normal method of filtering in raster databases employs the 3×3 filter, and set number combinations are programmed into standard software applications, there is no need to stick to either convention. Most software that contains the filtering capability will allow some flexibility in the size of filter and the numbers used. Smaller filters are more often used for edge enhancement, while larger ones are more common for low pass filtering or averaging. The larger the filter size, the more averaged the numbers will become because of the larger set of numbers used for averaging. It often takes experimentation to decide on the utility of abandoning the default filter size and number combinations.

	-1	-1	-1
	-1	9	-1
	-1	-1	-1

High-Pass Filter
Moving Window

	1	2	3	4	5	6	7	8	9	10	11	12	13
1	41	46	45	44	45	45	39	38	42	40	44	57	57
2	40	45	43	41	43	42	36	38	41	42	47	49	39
3	39	44	44	42	40	40	43	47	46	47	46	40	37
4	41	43	44	39	39	43	42	41	46	49	47	47	41
5	38	43	41	41	43	43	44	42	42	44	44	43	38
6	35	40	39	37	43	40	36	32	34	37	31	27	36
7	38	38	36	34	35	35	32	35	36	34	35	36	39
8	38	39	35	36	39	39	36	35	33	37	40	41	42
9	37	38	39	39	39	42	40	36	37	41	41	44	45
10	39	38	40	40	38	40	43	42	41	41	45	47	45
11	33	37	38	43	43	44	44	39	38	40	41	42	38
12	38	42	38	42	46	46	40	37	39	42	44	44	43
13	42	37	36	41	39	38	38	36	35	41	42	44	48

Original Brightness Values

High-pass Filtered Brightness Values

	1	2	3	4	5	6	7	8	9	10	11	12	13
1	31	60	53	45	56	71	30	27	59	18	20	119	145
2	26	64	37	23	48	47	-8	10	29	25	58	74	9
3	18	57	55	45	31	32	58	90	63	59	46	7	21
4	44	53	59	17	20	53	35	17	56	79	63	87	57
5	26	66	43	44	62	57	77	61	53	66	71	76	32
6	7	52	41	21	79	49	21	-13	4	33	-21	-59	15
7	41	42	26	6	11	15	0	41	47	23	32	33	40
8	39	52	16	28	52	53	30	30	6	36	51	47	48
9	27	37	46	45	38	64	47	17	27	54	33	50	60
10	59	41	48	41	12	27	60	60	55	45	68	82	72
11	-7	27	22	62	48	56	65	27	21	29	24	31	5
12	38	79	26	54	78	82	38	24	43	58	60	54	50
13	72	26	10	52	24	19	28	19	7	56	44	42	65

Note how some of the values are now much higher than before and others are much lower.

Figure 9.2   High pass filter. Operation of a high pass filter showing the use of a 3 × 3 matrix of numbers (the filter) designed to enhance the higher values and suppress the lower values.   *Source:* Robinson et al., *Elements of Cartography,* 6th Ed., John Wiley & Sons, Inc., New York, © 1995, modified from Figure 12.16, page 214. Used with permission.

POLYGONAL NEIGHBORHOODS

We know that a GIS should be capable of measuring the size of a polygon, or a fragmented region composed of numerous polygons. Suppose, however, that we are interested only in the identity of a region's polygons within a certain neighborhood or distance. For example, we are studying the spread of innovative farm practices to see whether a pattern of imitation emerges in which traditional farmers adopt new methods practiced by their neighbors. Let's say we are interested in neighborhoods of no-till agriculture. First we retrieve a coverage showing only locations in which no-till agriculture is being performed. Then we select a distance value as an estimate of where imitation is most likely to be observed. We might assume that no-till agriculture will be practiced only by connected neighbors, or we might assume that the idea will take root elsewhere, as well. In either case, the GIS takes this value, and spreads outward until it reaches the end of our selected search radius, all the while adding up the amount of no-till agriculture (essentially measuring the area of groups of polygons or polygons). The result will be small groups, medium groups, and large groups of fields that seem to indicate the sharing of this new practice. The GIS has treated the original region of no-till agriculture so that all the polygons or grid cells that fall within the specified distance of one another are given the same attribute number. The groups are numbered based on their rank compared to all other groups the software finds. Each group can later be reclassified by measuring its size. We might conclude that the farmers in the larger groups of no-till neighborhoods are more apt to talk to one another, or that the farms themselves are larger, or perhaps that these farmers knew some people at the nearby agricultural college. In any event, the values suggest that there is either a different mechanism for larger as opposed to smaller neighborhoods sharing these ideas, or some other functional reason for the difference in sizes of these clumps. The interpretation of the causes of these results will generally have to be further verified, but the neighborhood function of the GIS allowed us to recognize that differences exist.

In the preceding characterization of neighborhoods, we were looking at a single attribute to make our groups or clumps based on a specified distance. Many times, however, we are less interested in groups of identical polygons or grid cells, and more interested in defining the similarities and differences within a specified neighborhood distance. As an example, let's say that we want to determine the average age of people in a specified region based on census block data. By selecting a search radius, just as we did before, the software looks at the attributes for all the different census block polygons or grid cells, then performs a simple numerical average of these values. When it is done, it creates a new coverage for average age based on the values it calculated.

But average is not the only thing we can use to define new neighborhoods. Perhaps we are looking at a particular species of animal known to be particularly attracted to highly varied land cover types. The more diverse the cover types, the more these animals like the neighborhood, which provides a wide range of possible places to nest, to forage, and to take shelter from the sun and from predators. We may also know that this species needs a particular amount of land area in which to operate. We convert this known amount of area to a search radius and begin as before. In this case, the software looks at all the different land cover patterns within the search radius and counts them, returning the number of cover types as a measure of spatial or landscape diversity. Areas with the highest diversity would be the most likely place for the target species to nest. If we were trying to introduce this species into a region, we would most likely choose the areas of highest diversity.

As you might guess, if we can perform an average based on our surrounding neighborhood polygons or grid cells, we can also perform a wide range of other calculations. We might want to know some maximum value for a neighborhood—for example, the maximum number of crimes for the neighborhood for a given year. Or we might want to know the minimum value, such as the lowest price for homes in the neighborhood, to determine whether we could afford to live there. Other measures include total counts of all types in the neighborhood, the median, highest, and lowest frequencies, deviations from a central point to the mean of the surrounding values, and even the proportion of the neighborhood that shares attributes common to the central cell.

The measures we have identified in the last two paragraphs can operate in many different ways. If, for example, they return the value from analysis to the central point or target cell, that value (average, diversity, median, etc.) of the neighborhood is assigned to the central location in the neighborhood for a new coverage. We also do not necessarily need to assign a target distance at all, using instead a roving window approach to characterize the entire coverage based on the same measures as before (average, diversity, median, etc.). In this case the output values will not be returned to a central location but will be assigned throughout the new coverage, giving us an idea of the trends from one part of the coverage to another.

In some cases we can characterize a neighborhood based on the values contained within a single coverage, or we can use a separate coverage for target cells and another for the neighborhood measure. For example, if we want to know the neighborhood diversity for known bird nest locations, we would choose the bird locations as target cells and another coverage of land cover as our measure of neighborhood. In addition, measures of neighborhoods such as average, minimum, and maximum are comparisons of the surrounding polygons with each other, whereas other cases (e.g., measures of deviation from the average, the proportion of a neighborhood that contains the same value as the target location) require the target location to be compared with the surroundings.

But, using what we learned in Chapter 8, we can see that the ability of the GIS to measure size and shape offers other ways of characterizing neighborhoods. A measure of size, for example, is often combined with a measure of the clumps of a single polygonal attribute value. Most often this information is used to rank or order the results of analytical techniques that group data together into localized or regionalized clumps. The sizes of clumps may be very important in our analysis. Ecologists, for example, are aware of the sizes of range required by wolves and other large predators (Forman and Godron, 1987). They are also familiar with the minimum requirements of individual forest stands to support a diversity of animal species, or even to continue to exist as a forest patch. In some cases there may be a need to mix a neighborhood measure of diversity (where higher diversity indicates better habitat) with a measure of size (where larger areas are preferred). Many other combinations can also be employed. Let's take a look at an example. And while you read about this one, think of some others that might be of interest to you.

Jaguars are large predatory cats of a species that requires different types of habitat, especially jungle vegetation, for cover, and running water, because most of their diet consists of fish. For the sake of simplicity, let us say that jaguars require approximately 200 square miles of territory in which to live (a relatively accurate measure) and that they really need only two types of habitat—jungle forest and stream corridor. Let us further assume that researchers have indicated a requirement for territory that is at least 1 part stream corridor to 10 parts forest. We can identify the ratio between target points along the stream from a stream's coverage and the amount of surrounding (neighborhood) forest in the vegetation coverage that also includes stream corridor as a category. We can reclassify the vegetation coverage into stream corridor and forest vegetation to simplify our calculations. Then by using the neighborhood function that computes the ratio of the amount of neighboring area having the same category as the target location, we can identify the neighborhoods that have at least one-tenth stream corridor. Finally, now that we have created neighborhoods that suit the habitat-type needs of the jaguar, we can consider size. Neighborhoods that have at least 200 square miles of area with a stream corridor/forest ratio of 1:10 will be ideal sites for jaguar habitat.

Suppose, however, that the habitat neighborhood revealed by our analysis looks somewhat like an hourglass, having a narrow middle with the stream running through it. The areas outside our forest and stream corridor habitat are inhabited by humans, a species jaguars generally avoid. There may not be enough stream corridor in this bottleneck to support the animals' activities there. That is, as far as the jaguars are concerned, the two ends of the hourglass may be functionally separate habitat areas, neither of which is large enough for survival purposes. Using the ability of the GIS to measure distances across polygonal objects, we can perform an analysis that locates and classifies regions based on some value of narrowness, usually the narrowest portion of the polygon's shape. Thus we have employed another measure of neighborhood, this one based on shape.

Still other measures of shape introduced in Chapter 8 (the Euler function, measures of edginess, etc.) can be combined or used separately to characterize neighborhoods. In fact, the number of commands and command options available for classifying and reclassifying neighborhoods is quite large for most GIS systems. But we need not limit ourselves to an analysis of neighborhoods based on two-dimensional coverages. We can use our topographic data as a source of neighborhood characterization as well. Although we will go into greater detail specifically about surfaces and surface operations later, we can begin now to look at the utility of three-dimensional surfaces in reclassifying neighborhoods. Later, when we look more closely at surfaces and what we can do with them in isolation, you will be reminded of how we might use them in characterizing neighborhoods.

TERRAIN RECLASSIFICATION

Four basic characteristics of three-dimensional surfaces are commonly used for characterizing neighborhoods: steepness of slope, azimuth or orientation, shape or form, and intervisibility. To one degree or another, all can be performed with both raster and vector GIS, again depending on the sophistication of the software. In many cases, as with reclassifying neighborhoods with two-dimensional coverage data, these characteristics may very well be used in combination. We will look at each one individually and finish with a discussion of how they might be combined for more complex analyses.

Steepness of Slope

If you are planning on building a cabin on a mountain, you will likely want to know where the flattest parts lie, to be sure that your new home doesn't decide to relocate to the foot of the mountain. Or, if you intend to cut some timber along a mountainside, you may want to use the slopes to roll the freshly cut pieces downhill (not a generally accepted technique, by the way). Or maybe you are planning a ski resort and want to put up three different levels of slope: Snowplow Only, Extra Padding Needed, and Call Out the Rescue Squad. In all these scenarios you need to know something about slope, but you are not interested in the shortest route down, but rather in a general overview of the locations of steep, moderately steep, and relatively flat areas. The process is relatively simple conceptually: you need to know the relationship between the horizontal distance (measured in vector or raster) and the vertical change in elevation from bottom to top. A common way of expressing **slope** is rise over reach (rise/reach), where "rise" is the change in elevation and "reach" is the horizontal distance. To do this in vector you will need a data model similar to the TIN model we discussed earlier. Raster, of course, can readily manage this, although some minor errors due to grid cell quantization of space must be compensated for.

The general method of calculating slope is to compute a surface of best fit through neighboring points and measure the change in elevation per unit distance (Clarke, 1990). Specifically, the GIS will calculate the rise/reach value throughout the entire coverage, creating a set of categories of slope amount, much as we would do when defining class limits. If we wish fewer slope categories than are actually developed, we can reclassify the set produced by the GIS. While techniques designed to characterize different neighborhoods by the amount of slope on a topographic surface are in common use, the surface need not be a topographic one. As we will see in Chapter 10, our idea of a surface can be generalized to apply to any type of surface data that are measurable at the ordinal, interval, or ratio levels called a **statistical surface,** which is a surface representation of these spatially distributed statistical data. Thus we could analyze the steepness of slope of population change, or precipitation, or barometric pressure—any value that either is or can be assumed to be continuous across our coverage. By definition, however, it is clear that nominal data are not included. Imagine a raster database with attribute numbers representing land use types where 1 = agriculture, 2 = urban, 3 = manufacturing, and so on. Any surface produced from this set of attributes is meaningless because the numbers, although they themselves may seem to be ordinal, interval, or ratio, do not truly represent these levels of data. Instead they have been wrongly applied to noncomparable nominal categories. This point is made strenuously to emphasize that lack of understanding of basic levels of geographic data may lead to incredibly worthless results.

A couple of quick examples of the proper use of slope as a definition of neighborhood might prove instructive. Let's say we are trying to find slopes that are less than 25% (e.g., 25 meters change in elevation over 100 meters in horizontal distance), to define areas possibly suitable as sites for a cabin. In vector, using a TIN-like data model, we can perform any number of interpolation algorithms (more on interpolation in Chapter 10) to provide what essentially amounts to a contour map. To determine slope, the software simply compares the horizontal distance between the vertices each of the TIN facets with the respective horizontal coordinates. In fact, since the TIN model stores these calculated values in its attribute tables, no calculations need be performed once the TIN is in place. Each facet value can be retrieved, and the slopes can be grouped or classed as those with less than 25% slope and those with 25% or more slope. We can then rename these as "badslope" ($\geq$ 25% slope = too steep for building), and "goodslope" ($\leq$ 25% slope = acceptable for building). In this way we have reclassified neighborhoods based on slope, to facilitate decision making in the single area of suitability for building. Of course, there are numerous ways of manipulating the slope information, in particular those using nonlinear interpolation methods (more later) and those that generalize topography to indicate the trends in surface, called **trend surface analysis.** We will look at these to some degree in Chapter 10.

Both simple and complex methods of reclassifying neighborhoods based solely on slope can also be performed in raster GIS. The simplest method is to use a search of the eight immediate neighbor cells of each target cell. This is most often done by looking at all grid cells in the database and examining their neighbor cells, so that the slope values for the entire coverage can be performed. The software fits a plane through the eight immediate neighbor cells by finding either the greatest slope value for the neighborhood of grid cells or an average slope (normally the option is left to the user). For each group of cells, the software uses the grid cell resolution as the measure of distance, then compares the attribute values (elevations) from the central cell to all the surrounding cells (Figure 9.3). If, for example, we are building a ski resort and we

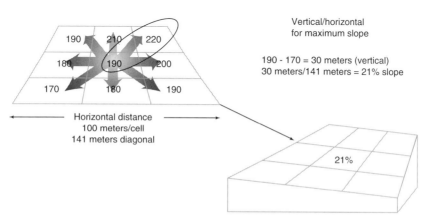

Figure 9.3 Trend surface in raster. Raster method of slope determination based on fitting a surface to the eight neighbors of a central target cell. This example shows how a raster-based GIS will indicate the maximum slope based on a comparison of a central target cell with its neighbors.

want slopes no greater than 15% for the use of inexperienced or novice skiers, between 16 and 25% for talented amateurs, and between 26 and 45% by professionals, we can select these three classes through a simple reclassification process. We would identify all neighborhoods whose maximum (or average) slopes are between 0 and 15%, 16 and 25%, and 26 and 45%. All slope values greater than 45% would be called "badslopes" because they are too steep for skiing.

As with vector, raster GIS software may also allow for nonlinear interpolation methods such as **kriging**, and surface-fitting processes. Many of these are based on different search methods employed while performing the analysis (Hodgson, 1989). If your software does not allow these as separate algorithms, there may be a way to produce a close approximation of some of these techniques by using multiple applications of the original technique. For example, it is possible to "smooth out" the slopes to obtain a trend surface of sorts by using whatever algorithm your software uses first, then repeating the process on the product of the first application. In other words, you perform a number of iterations of the slope calculation. The more iterations, the less rugged your surface will appear. If you keep it up, the final surface will appear as a flat plane. The technical jargon indicates that you will be moving toward a first-degree trend surface, from some higher order of surface.

Azimuth or Orientation (Aspect)

Because surfaces exhibit slopes, these features are, by definition, oriented in a particular direction, called the **aspect.** The two concepts of slope and aspect are inseparable from a physical as well as an analytical perspective. Without a slope, there is no topographic aspect. We have seen that azimuth or orientation can be a useful technique for classifying neighborhoods based on two-dimensional features. The same can be said for terrain features. There are numerous applications of this technique. For example, biogeographers and ecologists are aware that there is generally a noticeable difference between the vegetation on slopes that face north and slopes that face south (Brown and Gibson, 1983). The primary reasons for this differential entail the availability of sunlight to green plants, but our interest in the phenomenon is that GIS will allow us to separate out north- versus south-facing slopes for comparison to related coverages such as soil and vegetation. Another example of the utility of knowing slope orientation is found when we try to build wind generators. We want to locate these cheap alternative prducers of electricity high on the slopes of hills so they have ready access to the wind; but they need to be on slopes that face into the prevailing winds, rather than being sheltered from them. Geologists frequently want to know the prevailing slopes of fault blocks, or exposed folds, as a path to understanding the underlying subsurface processes. Or a grower may want to place his orchard on the sunny side of a hill to be able to take advantage of the maximum amount of sunshine. All these determinations and many more can be performed through the use of neighborhood functions that classify sloping surfaces based on their aspect.

In a vector GIS that uses a TIN-like data model, working with aspects is relatively simple. Each of the facets of the TIN model has a specific slope and aspect. The aspect is defined as the compass direction associated with each sloping surface (Evans, 1980). When the measurements we have seen are used for determining aspect, the TIN aspect values can be retrieved without additional steps. And, as before, we can group these aspects into classes. For example, if we are biogeographers interested in only north- and south-facing slopes, we can create three separate classes based on criteria selected earlier. Thus north-facing slopes could be classed within a single neighborhood if they range in azimuth from 345 degrees to 15 degrees, and slopes facing south could be classed into a single neighborhood if the range is between 165 and 195 degrees. This means that the classes have a 30-degree range, or 15 degrees on either side of the cardinal direction. All the remaining aspect values can then be classed as "badaspect" because they are not needed for analysis by the biogeographer. Of course, you may want to select your own class limits for these neighborhoods based on individual requirements.

In raster, the most common method is to again perform an analysis, throughout the entire coverage, that repeatedly compares a central cell to its neighbors. In this case the surface fitted through the nine-cell matrix evaluates either an average or a maximum direction by looking at the high and low elevation values within the matrix. If, for example, the highest value is located at the top center of the matrix and the lowest at the bottom center, the solution for this portion of the matrix is that the general aspect is south (assuming that your coverage is oriented north and south). Or you may find that the highest and lowest grid cell elevation values are located in the upper left and lower right, respectively. This will give you a southeast aspect to your sloping surface. The results of raster analysis of aspect can come either in degrees (0–360) or in a simpler set of vector values resembling the Freeman chain codes where, for example, north, south, east, and west would be 0, 2, 4, and 6, respectively, and northeast, southeast, southwest, and northwest would be 1, 3, 5, and 7, respectively. The actual numbers and methods will depend on the software.

Shape or Form

Another useful method of reclassifying statistical surfaces is to provide a measure of their form. The simplest method of visualizing surface form is to produce a **cross-sectional profile** of the surface. This is a common practice in many courses in map reading, geography, and geology, where students are asked to render the profile of a topographic surface along a line drawn between two points. This is done by transferring each elevational value to a sheet of graph paper where the horizontal is exactly the same width as the line between the points and the vertical axis is scaled to some **vertical exaggeration** (usually a whole-number power) of the original surface elevation values. The process is easy to perform in a vector system using a TIN model, where the line is drawn along some portion of the coverage (it need not be a straight line). The software then generates a profile identical to that produced with analog maps in the manual method. Keep in mind, however, that this visual is not a coverage, and the results are used merely as a means of interpreting what is there. We know, for example, that V shapes in profiles are most often associated with stream valleys, U shapes indicate glacial troughs. The more you know about the surface processes you are working with, the better your interpretation of the output is likely to be.

Your raster GIS may not have a method of producing a cross-sectional profile. Many, however, use an alternative method that creates a raster coverage by comparing a central target cell to two of its immediate neighbor cells. You can select the orientation of this search to be able to characterize a set of profiles for grid cells by looking at horizontal rows, vertical columns, or even diagonal sets of grid cells. The software will take each threesome and compare the central cell to its two immediate neighbors. Most often the software will describe each cell based on either its skyward angle (measured in degrees between each pair of surface elevational values) or by characterizing it as one of a number of form types. Form types are flat, rising, falling, monoclinal rising (flat then rising), monoclinal falling (flat then falling), rising then falling (convex as in a hill), and falling then rising (concave as in a valley), and are characterized by code numbers placed in the new coverage (Figure 9.4). The code numbers must then be interpreted manually.

Both surface form techniques, whether raster or vector, are designed to produce neighborhoods based on changes in surface value that can be interpreted by the user to represent specific features. Thus ridges, channels, peaks, watersheds, and so on may need to be identified as specific topographic features for later analysis. Watersheds, for example, are defined as all areas that drain into a stream network. More than simply draining into the stream channel,

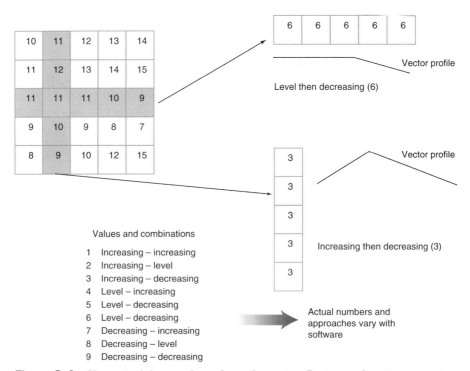

Figure 9.4 **Characterizing surface shape in raster.** Raster and vector output for characterizing a neighborhood based on form. In vector the output is a cross-sectional profile, while in raster the result is a coverage that shows the relationships between the target cell and its two neighbors. The two neighbors are selected as vertical, horizontal or diagonal neighbor cells.

watersheds tend to function ecologically as single, uniform regions. Ecologists, hydrologists, engineers, pollution and flood control experts, and many other specialists need to be able to define these areas precisely. We have only seen the simplest method, but other, more sophisticated methods have been applied to just this type of problem. As your skills improve, you may want to read up on these (Band, 1986, 1989; Douglas, 1986; White et al., 1992).

Intervisibility

While many neighborhood measures characterize all or part of the surface itself, another type of neighborhood measure uses the terrain in a more functional manner. The process called **intervisibility** recognizes that if you are located at a particular point on a topographic surface, there are portions of the terrain you can see (**viewshed**) and others you cannot see. Intervisibility analysis defines the regions that are visible from a particular point in the terrain. The many uses for this method include siting television, radio, and cellular telephone transmitters and receiving stations, locating towers for observing forest fires, routing highways that are not visible to nearby residents, and planning for artillery emplacements (Clarke, 1990). Any planning objective that requires anthropogenic features to be either visible or concealed will find utility in intervisibility analysis.

In vector, the simplest method is to connect a viewing location (observer location) to each possible target in the coverage. Next you perform **ray tracing**; that is, you follow the line (**ray**) from each target point back to the starting point, looking for elevations that are higher (Clarke, 1990). Higher points would obscure the observer's view of what is behind it (Figure 9.5). There are many possible ways to determine intervisibility in vector; the large number of calculations involved makes the ray tracing method a simpler and more useful technique, although perhaps less accurate. Intervisibility performed in vector requires the use of a TIN data model in which the surface is defined by the triangular vertices. A number of algorithms for this process have been provided for vector data structures, including TIN (DeFloriani et al., 1986; Sutherland et

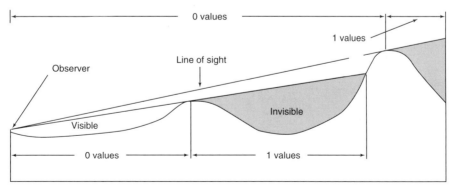

Figure 9.5   Visibility analysis. Ray tracing: a line (ray) is drawn from the observer to each other point in the coverage and the elevations compared. Areas beyond the observer that are lower in elevation are viewsheds, the remainder are not visible.

al., 1974). It is not our purpose here to go into these algorithms, but rather to show the concept. So, let's take a quick look at a possible application of intervisibility in vector to see what it does and how it might be applied.

Suppose you are planning to build a home in the foothills of a mountain range and you want to be able to see as much terrain as possible from your front porch. You have limited your candidate locations to three. For each location, identified in your GIS terrain coverage (TIN model), the software looks out in all directions at the vertices of the model and identifies them. It then retrieves the elevation values for all these points. Next it compares these elevations to the elevation of the potential building site. All areas that are higher in elevation (and all those beyond those locations) are not visible to you and must be reclassified as invisible; all the remaining areas are visible. The resulting polygon coverage shows you how much area is visible for each coverage tested. By simply tabulating the amount of visible area for each of the three tests, you can easily determine which site is best for building your home.

Raster methods of intervisibility operate in much the same way, but they are less elegant and more computationally expensive. The process begins by defining a viewer cell as a separate coverage against which the elevation coverage will be tested. Starting at the location of the viewer cell, the software evaluates the elevation (in the second coverage) that corresponds to that location. Then it moves out in all directions, one grid cell at a time, comparing the elevation values of each new grid cell it encounters with the elevation value of the viewer grid cell. Each time it encounters a grid cell with an elevation value higher than the viewer grid cell elevation, it reclassifies it as not visible and codes it to indicate a nonvisible grid cell. Each time it encounters a grid cell with an elevation value lower than the viewer elevation, it places a different code, indicating that the area represented by the cell is visible from the viewer's location. This, of course, is the simplest way to perform this method. There are others, and each yields different, but computationally valid results (Anderson, 1982; Dozier et al., 1981). You might want to check your software's documentation to evaluate the utility in your work of one type over another (Fisher, 1993).

Most applications of intervisibility are based solely on topographic surfaces, but in some cases the topographic surface will have forest cover with known individual heights or grouped heights associated with the trees. To perform intervisibility where the heights of these or other obstructing objects are known, the elevation coverage values must include the obstruction heights. These can be added in both vector and raster, usually by means of a mathematically based (addition) combination of the two coverages (see Chapter 12). Where topography itself is not important, the heights of the obstructions can be used alone to determine the degree of visibility of all points compared to those of the viewer. The applications of intervisibility analysis are endless, and the technique is quite common in surface neighborhood analysis.

BUFFERS

Another very common method of reclassification is a process called **buffering.** A buffer is a polygon created through reclassification at a specified distance

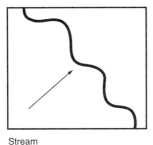
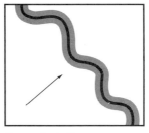

Stream   Stream plus buffer

Figure 9.6   Line buffers. Stream and its buffer based
solely on a distance measure selected by the user.

from a point, line, or area. Because it is based on the location, shape, and
orientation of an existing object, we could easily have classified buffering as a
method of reclassification based on position. However, a buffer can be more
than just a measured distance from any other two-dimensional object; it can
also be closely related and even controlled to some degree by the presence of
friction surfaces, topography, barriers, and so on. That is, while buffering is
based on position, it has other substantial components, as well. An area surrounding a stream that indicates something about the stream corridor offers
an example of a buffer. In Figure 9.6 the buffer was produced by reclassifying
the area on either side to differentiate it from the amorphous background.
Although the figure shows the stream plus the buffer, normally the buffer is
generated as a separate feature and is often stored as a separate coverage as
well. To produce this buffer merely requires that a specified distance be measured in all directions from the target object. We have seen how a GIS performs
a distance measure in both raster and vector; actually creating the buffer is
merely an extension of that procedure. But, because of the utility of the technique and the frequency with which it is applied, most GIS software includes
separate routines that allow the buffers to be produced.

Buffering is a matter of measuring a distance from another object, whether
a point, a line, or another polygon. In the case of point features we measure out
a uniform distance (in raster or vector) in all directions from that point (Figure
9.7). In some cases, we may want to produce a buffer around a linear feature,
as in Figure 9.6. Or, returning to Figure 9.7, we could begin with an area and
measure a set distance from its outer perimeter. There may even be a requirement for a buffer around a second buffer around still another buffer to produce
a **doughnut buffer,** so called because of its resemblance to the pastry and also
shown in Figure 9.7. This procedure is relatively simple in raster because each
buffer is simply a number of grid cell distances beyond the existing feature or
the previously created buffer.

In vector data structures, however, we must explicitly encode the topological
information for each polygon we produce. In particular, we are required to
provide topological information about the connections between polygons by
explicitly placing nodes at the beginning and end of each bordering line segment. A doughnut buffering procedure attempts to create an island polygon
that is not connected explicitly to the neighboring polygon. The difficulty of
producing a doughnut buffer in vector is largely a function of the data model
used, but by carefully following the software instructions, you can generally

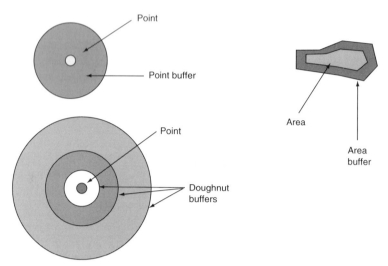

Figure 9.7 Point and area buffers. Point, area, and doughnut buffer.

obtain one. You may want to experiment on a test database before you try to produce a doughnut buffer on a real database.

Now we have seen that buffers are essentially distances measured in all directions from point, line, and area objects. But we have not mentioned one of the most pressing problems related to creating buffers—that of distance. How big should the buffer be? This question comes up often in software training seminars when students are asked to produce a buffer around an object in a coverage provided with the tutorial. Unfortunately, the most common response to this question in such a setting is "It doesn't matter; this is just to show you how to do a buffer." But, as we have seen, being able to build a buffer is not worth much if you have no idea what its purpose is or how large it should be. These basic conceptual questions are not really separate because the distance of our buffer is often dictated, or at least modified, by the purpose of the buffer. We'll look at these matters by focusing on the question of distance as the unifying factor.

How large should a buffer be? Some buffers are designed to indicate that around a given entity, for an unknown and perhaps unknowable distance, lies a region that needs to be protected, studied, guarded, or otherwise afforded special treatment. This scenario is not as uncommon as you might think. Many real-world buffers are just as arbitrary as the digital buffers we produce in our GIS. Yet in both cases, there is usually an underlying gut feeling behind our choice of a particular arbitrary distance. Contractors frequently build buffers around construction sites to protect passersby from falling debris and heavy machinery. The boundaries of areas believed to be contaminated areas by poison gas, nuclear disaster, and hazardous materials spills are generally imposed by governmental or law enforcement agencies. But, like the areas cordoned off around spaceships in the 1950s science fiction movies, quite often these danger zones are only guesses—that is **arbitrary buffers.** Most often the guesses are based on gut feelings or anecdotal information of unknown source.

They do share one characteristic, however, and it illustrates a time-honored approach—when in doubt, err on the side of conservatism. In other words, all these arbitrary buffers seem to be larger than they actually need to be. You probably notice this when you are forced to walk or drive around a construction buffer, or when a buffer surrounding a relatively small anomaly forces you to find another place to park. The extra distance that inconveniences the public is the added margin of safety for an arbitrary buffer.

Buffer distances can also be based on nearly any of the measures and re-classification procedures we have seen so far, whether two-dimensional or three-dimensional. For example, we could create a second type of buffer based on a functional rather than Euclidean distance from the target object. This would be a **causative buffer**—one based on a priori knowledge about the area within which the buffer is produced. Let's say, for example, that we are putting a buffer along a stream to indicate the potential for contaminating soil on either side of the stream. We know, however, that the soils on one side of the stream are heavily clay rich, while the soils on the other side are richer in sand. Because contaminants will pass through sand faster than through clay, the buffer must be based on the frictional or impedance quality of the clay soil. The result will be a buffer smaller in Euclidean diameter on the clay side and larger on the sand side, reflecting the difference in permeability of the soils. The use of friction surfaces and barriers is common when producing buffers, because they offer some justification for the buffer width we choose. However, as you remember from our discussion of friction surfaces, they tend to be poorly understood. Thus a buffer assigned based on a friction or barrier impedance value may be of little more use than an arbitrary buffer of simple design.

A buffer can also be based on intervisibility measures. In that case, the buffer is selected based not on an arbitrary value, nor on a poorly known friction value, but rather on a definable, measurable value—a **measurable buffer.** This is the third basic type of buffer we can employ. The measurements are not arbitrary, but are quite accurate, being based solely on measurable phenomena. Of course, there is always the option of combining the second and third methods of buffering, relying on a measurable phenomenon whose interaction with the buffer area is poorly understood. For example, we know that trees work quite well as filters of contaminating materials along stream corridors. We know also that the more trees, the more effective the filter will be. So, we could measure the density of vegetation along a stream corridor and then reclassify it to produce a number of neighborhoods, each representing a different class of tree density. Next, we could use these different tree densities as surrogates for friction surfaces, which we would then employ to modify the amount of buffering. However, we are not absolutely sure how directly the tree density is related to the amount of movement of contamination through the vegetation and into the streams. Thus we have some measurable values, and they can be logically applied; but we don't know the exact relationships among them. This is often the case when buffers are employed.

When the fourth and final method of buffering is employed, however, legal or otherwise mandated measures exist and must be adhered to—a **mandated buffer.** For example, if you build a home within the 100-year flood zone in some communities, you are not likely to be allowed to purchase flood insurance. While the 100-year flood zone is a measurable value, the insurance companies could just as easily have selected the 75-year flood zone or the 150-year flood zone. In other words, the value itself is measurable; its selection over other values is arbitrary. To develop a buffer of this kind usually requires a database of the terrain, and an ability to calculate the volume of water that would fill the stream floodplain if a flood of a size not likely to occur more than once in a century were actually to occur.

But other mandated buffers can be employed. We are told how close to a fire hydrant we can park, and how much of our frontage property actually belongs to the local community. Building codes specify distances around utility areas and between and among buildings; conservation organizations obtain easements on properties; legal ownership buffers are established around rail lines and power corridors, and so on. You might be able to think of a number of others. In each case, there is a legal reason for placing the buffer a set distance from other objects.

No matter which of the four basic buffer methods (arbitrary, causative, measurable, and legal mandate) is used, there is always the possibility that the buffer will not be the same along the entire length of linear objects or on all sides of a polygon. Such differences, exemplified by our different buffer widths along the stream based on soils, create a general class of buffers called variable buffers (Figure 9.8). The variable buffer can be based on friction, barriers, or any of our other neighborhood functions. It can be selected arbitrarily or on the basis of a measurable component of the landscape, or it can be mandated by law. In each case, special procedures must be invoked in the creation of the buffers. For vector, the nodes between line segments can most often be used to dictate buffer differences along the line segments. In raster, the grid cells need to be coded selectively so that a buffer for each group of grid cells can

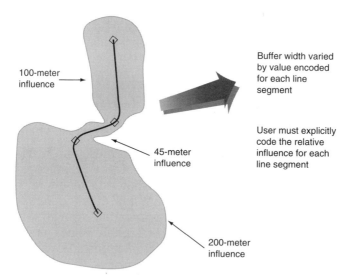

100-meter influence

45-meter influence

200-meter influence

Buffer width varied by value encoded for each line segment

User must explicitly code the relative influence for each line segment

Figure 9.8 **Variable buffer.** Variable buffers are created by using different impedance values on either side of a line, or by creating a separate buffer width for each line segment. In the latter case, each line segment must have identifying nodes on each end (in vector) or different distances must be measured for each set of uniquely identified grid cells (in raster).

Review Questions   **251**

be established; these buffers most likely will be combined later into a larger, buffer coverage.

Buffers are useful methods of reclassifying the landscape and are a common feature in many GIS analyses. The fundamental problem with buffers is that they frequently require us to know more about the interactions of our landscape's elements than we do. You should always attempt to overcome this obstacle by seeking out the best available knowledge about each situation before proceeding. The more you know, the more confident you will be about the distance you have buffered. If you have a legally mandated buffer distance, you need only be sure that your buffer corresponds to that mandate. But if your knowledge about the interactions of your landscape elements is weak, or if you have no knowledge on which to base selection of the buffer distance, your best bet is to make the buffer larger rather than smaller.

Terms

line dissolve
aggregation
overlay
range graded
constrained math
locational information
position measures
size measures
shape measures
neighborhood functions
total analysis of
  neighborhood
targeted analysis
immediate
  neighborhoods

extended neighborhoods
static neighborhood
  functions
roving window
  neighborhood functions
filters
high pass filter
low pass filter
directional filter
slope
statistical surface
trend surface analysis
kriging
aspect

cross-sectional profile
vertical exaggeration
intervisibility
viewshed
ray
ray tracing
buffering
doughnut buffer
arbitrary buffers
causative buffer
measurable buffer
mandated buffer

Review Questions

1. What two technological advances have heightened the importance of classifying spatial data? Why is it important that we view classification with a measured amount of cynicism?

2. Why are classification systems for earth-related data so varied? After stating some of the reasons, identify the common theme that runs through them all.

3. Give an example of a simple reclassification procedure that aggregates two or more classes together into a single class. How is this done in raster? In vector?

4. When reclassifying areas using a simple raster system, why is it important

to keep track of the order in which the grid cells are reclassified? Give an illustrative example.

5. What is a line dissolve operation? How is it used for reclassifying areas? How does it work?

6. What is the relationship between reclassifying areas based on nominal data types versus ordinal, interval, and ratio data types? What are range graded data? How are they used in the reclassification process?

7. What are filters? How are they used for reclassification processes? What's the difference between static and roving window filters?

8. What is the difference between a high pass and a low pass filter? Describe how both work and give their respective purposes. What is a directionally biased filter? How does it work to reclassify features?

9. What are neighborhoods? Give some real-world neighborhoods with which you are familiar. What is the difference between immediate and extended neighborhoods? What do neighborhoods have to do with reclassification? What is the difference between total and targeted neighborhood analysis?

10. Describe a process of defining neighborhoods based on a single attribute. What are some possible ways this could be used?

11. How could we use such mathematical procedures as average, maximum, and diversity to define a neighborhood? Create a table that shows a number of possible math-based measures of neighborhoods and an example of how each could be used. Describe how any one of these (e.g., diversity) actually works. Diagram the process.

12. Give examples of neighborhoods defined by a combination of possible methods (size, shape, diversity, etc.). Explain how this will be done: what steps are taken, how they are to be performed, and in what order. If you have access to a GIS, you may want to create a simple database to try this out.

13. What is slope? How is it implemented in a vector GIS? In raster? How can slope be used to define neighborhoods?

14. What is aspect? How is it used to create neighborhoods? Give an example of using aspect to define a neighborhood, describing both how it is done and what the output will tell you about your data.

15. How can the shape or form of surface data be displayed in vector? In raster? What is the general term for this output operation? What possible uses can be made from these techniques as they are applied to reclassifying areas?

16. What is intervisibility analysis? What is a viewshed? How does one calculate a viewshed in vector? In raster? Describe some uses of intervisibility analysis.

17. What are buffers? How are simple buffers based only on distance calculated? What is a doughnut buffer? Why is it so difficult to execute in vector GIS?

18. Describe how buffers can be produced based on functional distance, neighborhood classifications, and so on. What is a variable buffer? How is this done in vector? In raster? When would it be used?

19. What are the four basic methods of determining buffer distance? Describe them.

References

Anderson, D.P., 1982. "Hidden Line Elimination in Projected Grid Surfaces." *ACM Transactions, Graphics,* 1(4):274–291.

Anderson, J.R., E.E. Hardy, J.T. Roach, and R.F. Witmer, 1976. A Land Use and Land Cover Classification for Use with Remote Sensor Data. Geological Survey Professional Paper 964, U.S. Geological Survey. Washington, D.C.: U.S. Government Printing Office.

Band, L.E., 1986. "Topographic Partitioning of Watersheds with Digitial Elevation Models." *Water Resources Research,* 22(1):15–24.

Band, L.E. 1989. "Spatial Aggregation of Complex Terrain." *Geographical Analysis,* 21(4): 279–293.

Brown, J.H., and A.G. Gibson, 1983. *Biogeography.* St. Louis, MO: C.V. Mosby Company.

Clarke, K.C., 1990. *Analytical and Computer Cartography.* Englewood Cliffs, NJ: Prentice-Hall.

Cowardin, L.M., V. Carter, F.C. Golet, and E.T. LaRoe, 1979. A Classification of Wetlands and Deepwater Habitats of the United States. Office of Biological Services, Fish and Wildlife Service, FWS/OBS-79/31, stock number GPO 024-00524-6. Washington, D.C.: U.S. Government Printing Office

DeFloriani, L., B. Falcidieno, and C. Pienovi, 1986. "A Visibility-Based Model for Terrain Features." In *Proceedings of the Second International Symposium on Spatial Data Handling.* International Geographical Union Commission on Geographical Data Sensing and Processing and the International Cartographic Association, Seattle, WA, July 5–10, pp. 235–250.

Douglas, D.H. 1986. "Experiments to locate ridges and channels to create a new type of digital elevation model." *Cartographica* 23(4):29–61.

Dozier, J., J. Bruno, and P. Downey, 1981. "A Faster Solution to the Horizon Problem." *Computers and Geosciences,* 7(2):145–151.

Evans, I.S., 1980. "An integrated system of terrain analysis and slope mapping." *Zeitschrift fur Geomorphologie* (supplement) 36:274–295.

Fisher, P.F., 1993. "Algorithm and Implementation Uncertainty in Viewshed Analysis." *International Journal of Geographical Information Systems,* 7(4):331–347.

Forman, R.T.T., and M. Godron, 1987. *Landscape Ecology.* New York: John Wiley & Sons.

Hodgson, M.E., 1989. "Searching Methods for Rapid Grid Interpolation." *The Professional Geographer,* 41(1):51–61.

Klemas, V.V., J.E. Dobson, R.L. Ferguson, and K.D. Haddad, 1993. "A Coastal Land Cover Classification System for the NOAA Coastwatch Change Analysis Project." *Journal of Coastal Research,* 9(3):862–872.

Küchler, A.W., 1956. "Classification and Purpose in Vegetation Maps." *Geographical Review,* 46:155–167.

Lillesand, T.M., and R.W. Kiefer, 1995. *Remote Sensing and Image Interpretation.* New York: John Wiley & Sons.

Robinson, A.H., J.L. Morrison, P.C. Muehrcke, A.J. Kimerling, and S.C. Guptill, 1995. *Elements of Cartography,* 6th ed. New York: John Wiley & Sons.

Soil Survey Staff, 1975. Soil Taxonomy: A Basic System of Soil Classification for Making and Interpreting Soil Surveys. U.S. Department of Agriculture, Soil Conservation Service, Agriculture Handbook No. 436. Washington, DC: U.S. Government Printing Office.

Sauer, C.O., 1921. "The problem of land classification." *Annals, Association of American Geographers,* 11:3–16.

Star, J., and J. Estes, 1990. *Geographic Information Systems: An Introduction.* Englewood Cliffs, NJ: Prentice-Hall

Sutherland, I.E., R.F. Sproull, and R.A. Schumacker, 1974. "A Characterization of Ten Hidden-Surface Algorithms." *Computing Surveys,* 6(1):1–55.

Tomlin, C.D., 1990. *Geographic Information Systems and Cartographic Modeling.* Englewood Cliffs, NJ: Prentice-Hall.

United Nations Educational, Scientific and Cultural Organization, 1973. *International Classification and Mapping of Vegetation.* Paris: UNESCO.

U.S. Department of Agriculture, Forest Service, 1993. *National Hierarchical Framework of Ecological Units.* Washington, DC: U.S. Government Printing Office.

U.S. Geological Survey, National Mapping Program, 1992. Standards for Digital Line Graphs for Land Use and Land Cover, Technical Instructions, Referral ST0-1-2. Washington, DC: U.S. Government Printing Office.

Walter, H., and E. Box, 1976. "Global Classification of Natural Terrestrial Ecosystems." *Vegetatio,* 32:75–81.

White, D.A., R.A. Smith, C.V. Price, R.B. Alexander, and K.W. Robinson, 1992. "A Spatial Model to Aggregate Point-Source and Nonpoint-Source Water-Quality Data for Large Areas." *Computers and Geosciences,* 18(8):1055–1073.

Wilen, B.O., 1990. "The U.S. Fish and Wildlife Service's National Wetlands Inventory." *U.S. Fish and Wildlife Service, Biological Report,* 90(18):9–19.

CHAPTER 10

Statistical Surfaces

Our digital explorations now continue, but with new insights into what we can observe. Thus far we have concentrated on visible point, line, and area features, but we have made frequent reference to surface features as well. We've seen how the latter can be modeled in the computer environment, and we've just found that they can be used to modify our ideas of what a neighborhood is and how it operates. Surfaces are used frequently to model the friction or impedance that we encounter as we travel. They can be grouped into slopes and aspects, as well as viewsheds and specific features like valleys, hills, and watersheds. But the surfaces we will encounter will not always be topographic surfaces. As we will see, our geographic filter can distinguish surfaces that are continuous or discrete, smooth or rugged, physical or cultural. In other words our definition of surfaces will expand to include data of any type that either exist or can be assumed to exist as changing values throughout an area.

In our journey, of course, the topographic surface is the most readily understandable, and throughout this chapter we see frequent reference to topography as a classic type of surface. Topographical features are developed because at locations throughout the areas they occupy there are measurable objects, in this case elevation values. Because topographic elevation values occur everywhere, we say that the surface is a continuous surface. However, as we try to record this information with a view to producing reasonable and quantifiable descriptions, we find a major dilemma, not unlike that associated with trying to describe trees in a forest or blades of grass in a grassland—there are so many data that we simply cannot record the complete set. And so, as before, we must produce a meaningful sample of the elevation values from which we can reconstruct the essence of the topography. There is a strong similarity between the properties we sample for topographic surfaces, like barometric pressure, temperature, and humidity, and those of continuous surfaces. These also occur everywhere, but we cannot record data for them at every location. Instead we select locations to represent the distribution. We will learn how this is done, to ensure that our travels can be recorded properly.

But while continuous surfaces present a particular set of problems for the explorer, many other objects do not occur continuously everywhere; rather, they are found as discrete objects at specific locations. Because they occur with a very high frequency, or because we want to record them for very large regions, however, these discrete data must also be sampled. In some cases, as with continuous data, we will take samples at specific point locations, while in other cases, we must collect data for whole areas at a time. For example, we would surely find as we travel across the United States that there are differences in human population numbers for each county. Rather than taking the time to tour each county, locating individual people at every stop, we simply add up the number of people for each of the 3000 or so counties and use them as point samples. In both cases, we can produce maps that resemble the contours of a topographic map by assuming that they occur everywhere. In this way we can produce either a contourlike map or a fishnet map that models the trend or form of the distribution. And while we know that the objects we measure are not continuous, it is useful to employ these techniques for the sake of ease in communicating the shapes and patterns of distribution. In addition, we will see how these techniques can be used to predict distributions in places we have not sampled.

From time to time along our journey we will be able to record the locations of individual point objects but will refrain from recording any numerical attributes concerning them because such data represent not numerical values but only locations of points; hence we cannot appropriately present them in the form of a contourlike map or fishnet map. Still, there are ways to produce regions based on these point patterns, and we will see why this choice is made and how it is executed in a GIS.

Whether our data represented as surfaces are discrete or continuous, their importance cannot be oversold. Whole GIS systems have been developed based on data models that are designed for surface-related phenomena (e.g., pollution modeling, overland surface flows). Modern specialty software, often linked to GIS software, has been developed to perform a wide range of different methods of representing the data, modeling surfaces, and calculating volumes based on surfaces. Enormous volumes of surface data, especially related to topography and climate, are now available in GIS-compatible form to make our journey less tedious. With time, the amount of surface data available will increase at a heightened pace, as the importance of analyzing surfaces becomes more obvious.

Because of the increased importance and availability of surface data, it is important that we study this topic separately, despite its inclusion in earlier chapters. We will learn about the kinds of data that can be used in a surface modeling framework, the different surface models that can be used to display them, how to make predictions about missing data in a surface, and how such extrapolations might be employed in decision making. We'll also look at some commonly available surface data types and learn how they were derived, how we can use them in our work, and how they can be converted to data models of different types. And, we will see how surface data can be created from point observations of a wide variety of natural and human phenomena, to enhance their utility for modeling.

Once again, whether your interest in GIS is modeling, planning, decision making, or simply database development, it is essential to become familiar with the data and the techniques for creating, modifying, and modeling surfaces. If your interest is in using the data for your own work, the applicability of this chapter is possibly more obvious than for those who will work in database construction. For the latter group, it is important to understand what can be done with these materials as a means of recognizing when and why some data should be represented in this framework. So keep your eyes open as our geographic filter expands and our ability to describe and analyze our environment grows. As you finish this chapter, I hope you will become aware of far more surfaces along the way than you had been noticing before. The more you see, the more enjoyable your journey, and the more useful the results of your explorations.

LEARNING OBJECTIVES

When you are finished with this chapter you should be able to:

1. Define and describe a statistical surface and explain some nontopographic forms.

2. Describe how statistical surfaces can be represented using isarithms, and tell how the method of sampling of data changes how statistical surfaces are represented.

3. Explain how statistical surfaces can be represented by regular and irregular lattices and how these can be converted to other forms, such as discrete altitude matrices, digital elevation models, and TIN models.

4. Explain how a continuous surface is represented in raster format.

5. Understand and describe the different methods of interpolation, their uses, and their advantages and disadvantages.

6. Explain the four factors to be considered in any form of interpolation and understand how problems of each type can be solved.

7. Describe Z-value slicing and explain how it is used.

8. Explain the use of the method of ordinates for determining volumes.

9. Understand dot distribution maps as they apply to the input subsystem of the GIS.

10. Know the difference between traditional choropleth maps and dasymetric maps, and understand the utility of each.

WHAT ARE SURFACES?

Surfaces are features that are most often thought of as containing height values called "Z" values distributed throughout an area defined by sets of X and Y coordinate pairs. The Z values are frequently thought of as elevational values but need not be restricted to this one measure. Instead, any measurable values (i.e., in ordinal, interval, and ratio data scales) that can be thought of as occurring throughout a definable area can be thought of as comprising a surface; the term generally used is **statistical surface**, because the Z values constitute a statistical representation of the magnitude of the features under consideration (Robinson et al., 1995). Perhaps because the statistical surface extends our geographic filter to include such values as population, salary, animal densities, and barometric pressure, it is often considered to be among the most important cartographic concepts.

Behind the statistical surface is the idea that all the measured values (Z values) either occur continuously across the area of interest or can be assumed for the sake of mapping and modeling to occur continuously across it. Statistical surfaces that employ data that occur to some degree at every location inside the study area, such as elevation data, are called **continuous** (see Figure 2.3). Those that occur only as individuals, but with some difference in numbers per unit area, such as the number of houses per square mile in each neighborhood, are called **discrete** (see Figure 2.3). Conceptualizing the statistical surface for these two data types can be tricky, so we will examine them separately.

"Continuous" data are said to occur at every possible location in the area. That is, it is possible to obtain a measurement, no matter how small, for this variable anywhere within the area in question. We know, for example, that air temperature occurs everywhere. At every location we could obtain a measure of the temperature with a thermometer. If we did this, we would see that the values change gradually from place to place in a continuous stream. In some cases the temperature changes slowly, perhaps less than a tenth of a degree per 100 meters. We say that these continuous data demonstrate a **smooth surface,** with little change in statistical information per unit distance. At times, however—when we cross under the space between two totally different air masses along a frontal line, for example—the temperature values change very abruptly. We call the surface produced by rapidly changing values a **rough surface** because there is a major change in the statistical data with small changes in distance.

The obvious advantage of continuous data is the certainty that there are plenty of data to work with. In fact, in defining our statistical surface as continuous, we assume that there are an infinite number of data points from which we can make our measurements. This poses two closely related problems. First, it is physically impossible, even with the most advanced scientific instruments, to measure an infinite number of locations. Second, even if we could measure everywhere, no computer is capable of handling an infinite amount of data. So, defining a continuous surface based on an infinite number of data points must be replaced by a method of modeling that allows us to use samples taken at critical locations. These samples will serve to represent the most important changes in surface as a simplified representation.

Surface Mapping

The statistical surface can be represented by means of **dot mapping, choroplethic mapping, dasymetric mapping,** and **isarithmic mapping.** The first three methods deal most frequently with discrete surface data, and we will save these for later discussion. The fourth, isarithmic mapping, is the most common method of mapping continuous data, although we can use it for discrete data as well if we assume that the data are continuous. We can envision isarithmic mapping as a series of lines surrounding our topographic surface. Each line, commonly called a **contour line** in topographical contexts, represents all points

Figure 10.1 Perspective view of surface with contours. Perspective of a topographic surface showing the contour lines that will be used to represent it as a contour map when viewed from above.

that occur at the same elevation. The general term for lines connecting points of equal statistical value is **isarithm.** Figure 10.1 shows a perspective view of a topographic feature with the contour lines drawn in.

Viewed from the top, isarithms appear as a series of semiparallel lines. If they surround a topographic form, such as a hill, they will be closed and will encircle the feature; if not, they continue until they hit the edges of the map. What we are seeing, then, is line symbols used to display surficial data. Moreover, they allow us to observe specific patterns and shapes based on the type of topography encountered. We see, for example, that the isarithms that cross stream valleys tend to look like the letter V pointing uphill or upstream. We see that steep slopes show more isarithms spaced closer together horizontally than gently sloping surfaces. We talked earlier about neighborhoods being reclassified based on slope. You can now see how the traditional contour lines can be used to represent these neighborhoods of slope as well: steep slopes with closely spaced contour lines; gentle slopes with widely spaced contour lines (Figure 10.2).

We can also observe another important aspect of these isarithms by looking at Figure 10.1 again. It is no accident that the vertical distance between isarithms is the same for any two pairs of lines. The vertical distance between

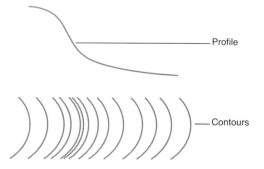

Profile

Contours

Figure 10.2 Contour spacing and surface configuration. Steep slopes are shown by closely spaced isarithms and gentle slopes shown by widely spaced isarithms. The distance between lines (the contour interval or class interval) is based on elevation differences, not horizontal differences.

lines was preselected to allow the reader to more easily understand what the lines represent. Each contour map has its own vertical distance between isarithms, depending on whether there is a rapid or a slow change in elevation. This preselected value, called the **contour interval,** is used to divide or quantize the change in Z value by choosing a uniform amount of elevational change that will continue to be used at every step. Thus the person viewing the map knows that each contour line indicates a set difference in elevation, rather than one that changes all the time. Imagine the isarithms that would result if we used a different, randomly selected distance value for each contour line drawn: if we had a steep mountain in our topographic surface, the lines would not necessarily be close together; instead the distance could change arbitrarily, thus preventing us from interpreting closely spaced lines as steep slopes and widely spaced lines as gradual surfaces.

It is important to understand how coutours are produced, especially within a computerized GIS. We do this in vector, as you have seen, by recording samples of Z values throughout our surface. Then, by constructing triangular groups of elevation values that somewhat abstractly represent flat facets of the surface, we create a TIN model. With the TIN model we can represent the slopes, the azimuths, and surface changes for any type of surface. This is a method of sampling based on a subset of points from our infinite number of possibilities. But where do we get these samples? How do we select them?

SAMPLING THE STATISTICAL SURFACE

It turns out there are two major ways in which we can obtain samples of Z surface data. The first employs selected point values, in which case the isarithmic map produced is said to be **isometric.** This is the method most commonly applied to elevational data for the development of contour lines. It is also the method used in developing maps of barometric pressure, temperature, and a host of other data from selected weather stations sited around the globe at point locations. However, we may also be able to work from data that are aggregated by small areas. We will stick to our straightforward example of population by county. We can gather data on the number of people who live in each of the approximately 3000 counties in the United States. Although we know that these data are discrete, if we choose to, we can treat them as if they were continuous. So, by assuming that each of these areas is a point, we can produce the same kind of isarithmic map that would result if the data had been in fact gathered at point locations. This type of isarithmic map, called an **isoplethic map,** requires us to determine where we are to place the points. We have already studied how centroids or centers of gravity can be calculated and used in determining such placement.

In the case of isoplethic maps, we choose the data collection point from within each area based on its central point or on a central point weighted by the clustering of the data. But in isoplethic mapping we have to decide where we are going to select the data sampling points in the first place. There are two general approaches. The first is with the use of a regular sampling procedure or a **regular grid.** Often the method based on a regular grid is called **regular lattice** sampling because the points are later connected to form a triangular lattice structure much like trellises used in gardens to support climbing plants (Figure 10.3a). The regular lattice approach to sampling has the advantage of simplicity in that there is no need to spend a large amount of time deciding where the points are to be placed. A distance for both X and Y directions is selected, and the points are measured where the X and Y lines cross. A major problem arises, however, when it is assumed that by sampling regularly throughout a surface, we can represent both smooth and rough surface features. This assumption is not often valid. A rough surface implies a rapid change in Z value with distance. We say that there is more information per unit area because every change in elevation is thought of as a potential piece of information regarding the form of the surface. When there is an increase in the amount of surface information with distance, there is an identical need to sample that area more than an area exhibiting little change in Z value. For this reason we will obtain a better understanding of surface shape if we use an alternative method of sampling called the **irregular lattice** (Figure 10.3b).

In an irregular lattice we can determine the amount of sampling to be done in any part of the surface based on prior knowledge of how rugged or smooth it is. We are not restricted in either the X or the Y direction. Nor are we restricted in the number of point samples we can make within any given area. It might be assumed that in the absence of restrictions we are likely to make more samples using this method. It turns out, however, that this is not necessarily the case.

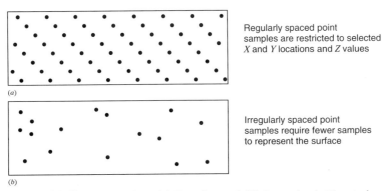

(a)   Regularly spaced point samples are restricted to selected X and Y locations and Z values

(b)   Irregularly spaced point samples require fewer samples to represent the surface

Figure 10.3   The lattice. (a) Regular and (b) irregular lattices of points. The points can be sample locations themselves or points as they are represented in the GIS database.

Most often we can obtain a good model of the surface with fewer data samples because we can reduce the number of samples for relatively smooth surfaces well below the number that would be needed for a regular lattice. In other words, while we increase sample density for areas with increased surface information, we can decrease the sample density for areas with decreased surface information. Increased sample density for smooth surfaces gives no relevant information, it just adds unnecessary time and effort as well as increased data storage later on.

THE DIGITAL ELEVATION MODEL (DEM)

Now that we have a procedure for the selection of sample data points for creating a map of a statistical surface, we next need to understand how a surface could be represented inside the computer, both in raster and in vector. We have introduced the TIN model, the basic vector data structure for representing surfaces in the computer. However, the TIN model is one of a number of methods of storing Z-value information, creating a group of products collectively called **digital elevation models (DEMs).** Such methods are based either on mathematical models or on image models (including TIN) designed to more closely approximate how they are normally sampled in the field or represented on the paper map (Mark, 1978). While mathematical models are quite useful, the currently available DEMs are most often of image models of some description, so we will focus on the latter, leaving the more advanced student to pursue the mathematical model representations. Image models come in two general types: those based on points and image models based on lines.

Image models of Z surfaces based on lines are nearly the graphical equivalent of the traditional method of isarithmic mapping we've just seen. In many cases such models are produced by scanning or digitizing existing contour lines or other isarithms. The purpose is to extract the form of the surface from the lines that most commonly depict or describe that form. Once input, the data are stored either as line entities (each line with a common elevation attribute, and each line segment connected by points) or as polygons of a particular elevational value. Since it is not particularly efficient to calculate slopes and aspects and to produce shaded relief output from such data models, it is more common to convert them to point model form, treating each point connecting each line segment as a sample location with an individual elevation value. This produces what is known as a **discrete altitude matrix.**

The discrete altitude matrix is a point image method that represents the surface by a number of points, each containing a single elevational value. This is not unlike the procedures for sampling continuous surfaces with point observations. Much like the methods of sampling the topography in the field, the altitude matrix uses a number of data points sampled from stereo pairs of aerial photographs, usually using a device called an analytical stereo-plotter that locates and measures elevational data based on the displacement of the two photographs (Kelly et al., 1977). Again paralleling the field sampling procedures, the analyst can use a regular lattice or an irregular lattice. Since the regular lattice produces a large amount of data redundancy where the topographic information is minimal and undersamples in areas with a great deal of topographic information, the irregular lattice is preferred. Progressive sampling using automatic scanning stereo-plotters that increase the number of samples in highly changing topographic regions offers an easy route to execution (Markarovic, 1973).

Irregular lattices can be converted to a TIN model under two different approaches. The first is to use the data points in the irregular lattice itself as the basis for the triangular facets in the TIN. This has the merits of requiring little additional input, instead allowing the GIS to create the triangular facets. Alternatively, a clear plastic grid can be placed over the points and the distances between data points and their elevational values used to predict values for all the vertices of a regular matrix. This process, called **interpolation**, is also an analytical technique that we will look at more closely in the next section. The product of such an interpolation approach is a predictive set of data points that are used to create the triangular facets of the TIN. While the use of interpolation allows the addition of data points to the altitude matrix and, ultimately, to the TIN, predicted values are not likely to be as accurate as measured data points. Thus, any model produced from such a technique has an added amount of elevational error.

DEMs are readily available for many parts of the world as altitude matrices with grids of 63.5 meters obtained from 1:250,000 scale topographic maps; they are becoming available in larger scale formats such as can be obtained from 1:25,000 maps and aerial photography (Appendix A). Among the advantages of using DEMs derived from interpolation techniques that create a regular matrix is ease of input to raster GIS. DEMs based entirely on an irregular lattice of point values and input to a raster GIS will have to be interpolated using raster techniques. We will look next at the representation of continuous surface data in a raster GIS, followed by a closer examination of how the interpolation process works.

RASTER SURFACES

Because raster data models divide the surface of the earth into distinct units, each unit or grid cell can contain only a single absolute elevational value. This in effect takes a continuous data variable and converts it to a discrete representation (Figure 10.4). While each grid cell contains a single absolute elevational value, it also occupies space. To more accurately represent elevation in raster, you may have to select a relatively small grid cell size, to increase the accuracy with which the point elevational values for each grid cell represent an appropriate elevation throughout the entire space occupied by the grid cell.

Although grid cell size is important in coding continuously varying data into a raster GIS, it is equally important to decide where within the grid cell area you want the actual elevation point to be located. You have options for indicating the exact location of the elevation point: at the center of the cell or in one of the four corners (Figure 10.4). For analyses that require the use of topography, however, the location will have an impact on the results of the calculation. For example, when we measured least-cost paths in Chapter 8, we began in the center of each grid cell because we were assuming that the elevational value was stored in that location. If your grid cells are coded so that

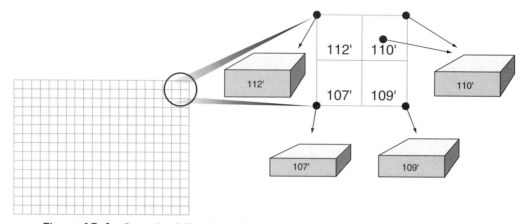

Figure 10.4 **Quantized Z surface.** Raster representation of a continuous surface. Note that for each grid cell, an absolute elevation is assigned. To ensure accurate analysis later on, it is important to decide specifically where the point elevation is located within the grid cell space.

the elevational value is at one of the four corners, your calculations of distance will be off by at least half a grid cell distance. Thus in deciding where to place the elevational value within each grid cell, you should first examine how the raster GIS you are using actually calculates such measures as functional distance, slope, and aspect.

In many instances data will be available for only a sample of grid locations, whether you are using raster or vector. In raster, you are most likely to obtain your elevational data as an altitude matrix in one of the two forms we discussed—regular or irregular lattice. If the regular lattice is small enough to fit your grid cell size, you can easily convert directly from the elevation points at each lattice vertex to a grid value for each of these points (again deciding beforehand where you want the elevational data to be located). When the data are provided in an irregular lattice form, you are going to have to estimate or predict all the missing values. This process, called interpolation, is needed because all the grid cells in your coverage must have elevational values. But, as we will see next, interpolation is a useful analytical tool for modeling in its own right, as well as a good technique to combine with other analysis methods for more complex models.

INTERPOLATION

Because Z values for continuous surfaces are samples, whether they are topographic, economic, demographic, or climatic, we need to be able to create models that depict with relative accuracy the features we observe. In traditional cartography, for example, we need to be able to convert from point samples of elevation or other statistical surfaces to a form of display that uses isarithmic lines to show the form of the surface. But we also need to be able to produce visual representations of other types, such as fishnet maps and shaded relief maps. And, of course, we need to be able to calculate slopes, aspects, and cross sections, and to predict unknown elevations for objects that occur at places for which we do not have elevational data. Interpolation provides much of what is needed to perform these operations.

The process of interpolation can be conceptually very simple, but it requires an a priori assumption. Before we begin working with surface data, let's look at a mathematical examination of interpolation based on the idea of a mathematical progression. Progressions are numbers that occur in some identifiable sequence. Thus the mathematical sequence

$$1\ 2\ 3\ 4\ 5\ 6\ 7\ 8\ 9\ 10$$

is based on starting at 1 and adding 1 at each successive step. It is called a linear or arithmetic sequence because it increases by the same amount each time, and each increment is developed by simple arithmetic. Other arithmetic series could be

$$10\ 20\ 30\ 40\ 50\ 60\ 70\ 80\ 90\ 100$$

which uses 10 as the addition factor, or

$$1000\ 900\ 800\ 700\ 600\ 500\ 400\ 300\ 200\ 100$$

in which 100 is subtracted at each step. Still other arithmetic series may be created using multiplication or division.

Linear Interpolation

Within these simple series we can easily identify the a priori assumption mentioned earlier: namely, that the numbers at each successive step can be determined by a simple mathematical procedure. If we can discern the mathematical procedure, we can also insert any missing values. So, for example, if we are told that the following simple series is missing two values:

$$30\ 40\ 60\ 70\ 80\ 100$$

we can infer that the series is based on addition and that the missing values lie between 40 and 60 and between 80 and 100. Because this is a linear series, we can see that the missing numbers are 50 (40 + 10) and 90 (80 + 10). This is nearly identical to **linear interpolation,** the method of assigning values between points of known elevation spread over an area.

Take the simple example illustrated in Figure 10.5. We are looking at a single line transect of data points that range between 100 feet in elevation and 150 feet in elevation. If we assume that the surface changes in a linear fashion, just as in a simple series, have a linear progression, it is obvious that four numbers, spaced equal distances apart, can be interpolated between 100 feet and 150 feet. By segmenting the distances between these two points into five equal units, we can treat the distances as surrogates for changes in elevation. Therefore, at each unnumbered segment we need only insert a 10-foot elevation to

Figure 10.5 **Linear interpolation.** Linear interpolation to find missing values between 100 and 150 feet.

obtain the missing values. If we do this for an entire area, rather than just for a single line transect, we can obtain values for all the 10-foot contour interval locations. By drawing smooth lines to connect these segments, we can create contours of 100, 110, 120, 130, 140, and 150 feet. In other words, we are able to create an **isarithmic map** (contour map in this case) that allows us to visualize the elevational features.

Thus far, of course, we have worked with linear progressions, assuming that the surface changes in this linear fashion. At times, however, a series of surface values does not conform to such a linear relationship. In some cases, the series is more logarithmic; in others it is predictable only for small portions of the surface. Under such circumstances the results of a linear interpolation method will not accurately depict the surface. And there are other needs for surface information that may require us to determine the overall trend in surface, rather than trying for the most accurate depiction. Some of these techniques can get complicated mathematically, and it is not my purpose to examine them in painful detail. Instead we will stick to the conceptual level in examining a few of the nonlinear methods of interpolation, to get an idea of how they might best be used in GIS.

Methods of Nonlinear Interpolation

Nonlinear interpolation techniques are designed to eliminate the assumption of linearity called for in linear methods. We will look at three basic types of nonlinear interpolation method: weighting, trend surfaces, and kriging. There are advanced texts that cover the many additional approaches in great depth (Burrough, 1983; Davis, 1986). Here we look only at the general types.

Weighting methods assume that the closer together slope sample values are, the more likely they are to be affected by one another. For example, as we go up a hill, we note that there is a much greater similarity in the general trend in elevation values close to you than there would be if we were to try to compare your local elevation to points far away. Likewise, as we go downhill, there will be a similar change in elevation values for neighboring points. Nearing the bottom of the hill, however, we quickly notice that the elevation values change rather quickly at the base of the hill, while the plain beyond the hill once again takes on a certain similarity in elevational change. To more accurately depict the topography, we need to select points within a neighborhood that demonstrate this surface similarity. This is done by a number of search techniques including defining neighborhood by a predefined distance or radius from each point, predetermining the number of sample data points, or selecting a certain number of points in quadrants or even octants (i.e., one point for each quadrant is used for the interpolation) (Clarke, 1990).

Whatever method is used, the computer will have to measure the distance between each pair of points and from every kernel or starting point. The elevation value at each point is then weighted by the square of the distance so that closer values will lend more weight to the calculation of the new elevation than closer distances (Figure 10.6). There are many modifications of this approach. Some reduce the amount of distance calculations by employing a "learned search" approach (Hodgson, 1989), others modify the distance by weighting factors other than the square (e.g., cubed or higher powers), and some include barriers that simulate coastlines, cliffs, or other impassable features likely to affect the output of an interpolation (Shepard, 1968). The barrier method is especially useful in the development of surface models that can account for these local objects. And, as in the use of barriers for other modeling tasks, the interpolation cannot pass through the barrier in its search for neighboring weights and distances.

Under some circumstances we are more interested in general trends in Z surface rather than in the exact modeling of individual undulations and minor surface changes. For example, we might want to know the general trend in population across a country, to support demographic research, or whether a buried seam of coal trends toward the surface, to indicate how much overlying material needs to be removed for surface mining operations. The most common approach to this type of surface characterization is called "trend surface." As

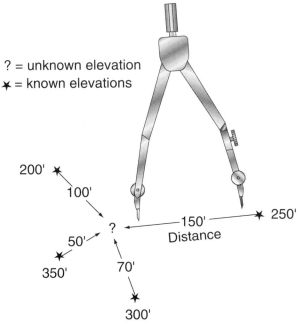

Figure 10.6 Distance-weighted interpolation method. Note how the closer Z values have more weight than those far away. For example, the missing value is more likely to be closer to 350 feet in elevation because of its proximity to that elevation.

in the weighting method, in trend surfaces we use sets of points identified within a specified region. The region is based on any of the methods already discussed for weighted methods. Within each region, a surface of best fit is applied based on mathematical equations such as **polynomials** or **splines**. These equations are best described as nonlinear progressions that approximate curves or other forms of numeric series. To develop the trend, each of the values in the region is examined and fitted to the mathematical equation. From the equation used to estimate the surface of best fit, a single value is estimated and assigned to the kernel. As we have seen, the process continues for other target points or kernels, and the trend surface can then be extended for the entire coverage.

The number assigned to the target cell or kernel may be a simple average of the overall surface values within the region, or it may be weighted based on the particular direction in which the trend is moving. Trend surfaces, as we saw in Chapter 9, can be relatively flat, showing an overall trend for the entire coverage, or they can be relatively complex. The type of equation used will determine the amount of undulation in the surface. The simpler the trend surface looks, the lower the degree it is said to have. For example, a first-degree trend surface will show a single plane that slopes across the coverage—that is, it trends in a single direction. If it trends in two directions it is said to be a second-degree trend surface (Figure 10.7).

The final method of interpolation, known as **kriging**, optimizes the interpolation procedure based on the statistical nature of the surface (Oliver and Oliver, 1990). Kriging uses the idea of the regionalized variable (Blais and Carlier, 1967; Matheron, 1967), which varies from place to place with some apparent continuity but cannot be modeled with a single smooth mathematical equation. As it turns out, many topographic surfaces fit this description, as do changes in the grade of ores, variations in soil qualities, and even a number of vegetative

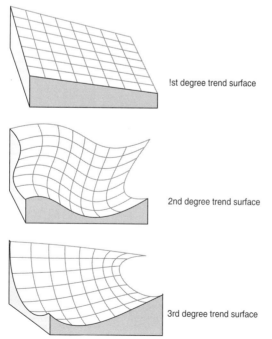

!st degree trend surface

2nd degree trend surface

3rd degree trend surface

Figure 10.7 Degrees of trend surface. First-, second-, and third degree trend surfaces based on the complexity of the polynomial equation used to represent the surface.

variables. Kriging treats each of these surfaces as if it was composed of three separate values. The first, called the **drift** or **structure** of the surface, treats the surface as a general trend in a particular direction. Next, kriging assumes that there will be small variations from this general trend, such as small peaks and depressions in the overall surface that are random but still related to one another spatially (we say they are spatially autocorrelated). Finally, we have **random noise** that is neither associated with the overall trend nor spatially autocorrelated. Clarke (1990) aptly illustrates this set of values by means of an analogy: if we are hiking up a mountain, the topography changes in an upward direction between the starting point and the summit; this is the drift. But along the way we find local drops denting the surface and accompanied by random but correlated elevations. Also along the way we find boulders that must be stepped over, which can be thought of as elevation noise because they are not directly related to the underlying surface structure causing the elevational change in the first place (Figure 10.8).

Now that we have three different variables to work with, each must be operated on individually. The drift is estimated using a mathematical equation that most closely resembles the overall change in the surface, much like a trend surface. An expected elevational distance is measured with the use of a statistical graphing technique called the **semivariogram** (Figure 10.9), which plots the distance between samples, called the **lag,** on the horizontal axis, the vertical axis gives the **semivariance,** which is defined as half the variance (square of the standard deviation) between each elevational value and each of its neighbors. Thus the semivariance is a measure of the interdependency of the elevational values based on how close together they are. A curve of best fit is then

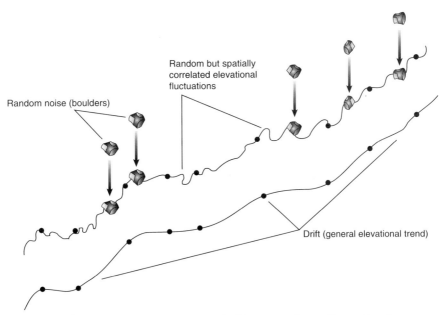

Figure 10.8 Elements of kriging. Drift (the general trend), random but spatially correlated elevational fluctuations (the small deviations from the trend), and random noise (boulders) exemplified by a hiking trail up a mountainside.

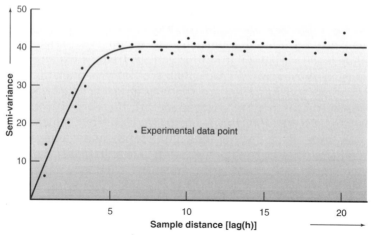

Figure 10.9 Example of a semivariogram. Semivariogram showing the relationship between sample points (dots) and the fitted line. Note how within the lag the Z values are related to one another and beyond the lag there is no relationship at all because the dots are too far apart. *Source:* Adapted from P. A. Burrough, *Principles of Geographical Information Systems for Land Resources Assessment,* Oxford University Press, Oxford Science Publications, Monographs on Soil and Resources Survey, No. 12, © 1986; Figure 8.11.

placed through the data points to approximate their locations (giving us a measure of the spatially correlated random component). If you look closely at the semivariogram, you will notice that when the distance between samples is small, the semivariance is also small. This means that the elevational values are very similar and are therefore highly dependent on one another because of their close spatial proximity. As the distance (lag) between points increases, there is a rapid increase in the semivariance, meaning that the spatial dependency of values drops rapidly. Eventually a critical value of lag known as the **range** occurs, at which point the variance levels off and stays essentially flat. Within the range (i.e., from 0 lag to the point at which the curve levels off), the closer together the samples are, the more similar they probably will be. Beyond the range the distance between sites makes no difference; the sites are totally unrelated at any of these larger distances. This information gives us a measure of what neighborhood needs to be applied (e.g., with weighted interpolation techniques) to encompass all points whose elevational values are going to be related.

A third feature of importance in the semivariogram is that the fitted curve does not run directly through the origin. Both mathematically and conceptually, if you have no distance between samples, there should be no variance because the items would essentially be the same point sample. But remember, the curve is an estimate of the locations. The difference between an expected variance value of 0 at 0 lag and the predicted positive value is an estimate of the residual, spatially uncorrelated "noise" variance and is called the **nugget variance.** As Burrough (1986) puts it, this nugget variance "combines the residual variations of measurement errors together with spatial variations that occur over distances much shorter than the sample spacing, and that consequently cannot be resolved."

Now that we have defined the three important components of the regionalized variable with the semivariogram, we can determine the weights needed to perform interpolation for local regions. Unlike the weighting methods employed before, however, the weights for interpolation within neighborhoods are chosen to minimize the estimation variance for any linear combination of elevation samples. This variance can be obtained directly from the model that created the semivariogram in the first place.

Kriging comes in two general forms. The general form, universal kriging, is most often used when the surface is estimated from irregularly distributed samples where trends exist (a condition called nonstationarity). Punctate kriging is the elementary form and assumes that the data exhibit stationarity (lack of a trend), are isotropic, and are collected at equally spaced point locations (Davis, 1986). Most often punctate kriging is used specifically for finding point estimates from other point estimates, rather than for defining surfaces.

Because kriging is an exact method of interpolation, it frequently gives a relatively accurate measure of the elevations of missing values. This exactness often comes at a cost in time and computing resources, however. Even so, kriging has another advantage over other interpolation methods in that it provides not only interpolated values but also an estimate of the potential amount of error for the output. One might conclude that given its ability to supply an error measure and the exactness of the computed surface, one might be well advised to use the method all the time. This is not the case, however. When there is a great deal of local noise due to measurement error or large variations in elevation between sample points, it is difficult for the technique to produce a semivariogram curve. Under such circumstances the results of kriging are not likely to be substantially better than those of any other method.

No matter which interpolation technique you use, most GIS software will provide at least a minimum number of approaches for your use. In vector data models, most often TIN, the process of interpolation is most easily performed by isolating individual points and their associated elevational values and converting them to an altitude point matrix (sometimes called a point coverage). From that point coverage, the interpolation procedures can easily operate on the selected algorithms, just as described. Actually, the TIN model itself can also perform interpolation (McCullagh and Ross, 1980), but the method is somewhat more involved, and we leave such techniques for an advanced GIS course. In raster coverages with missing values, the grid cells that contain elevational values are treated like points located inside each grid cell area (e.g., in the center). Then the distance between grid cells can be measured as described earlier, and the elevational values are again interpolated based on the techniques described. In this case, the grid cells receiving new elevational values are assumed to be placed in the same locations as the original target grid cells. When your GIS does not provide the desired interpolation technique, there is usually a method for converting the point coverages to a form compatible with specialty software designed expressly for working with surface data. They can be converted back to your original software for later processing. For some basic overviews of interpolation methods, you might consult Lam (1983) and Flowerdew and Green (1992).

USES OF INTERPOLATION

Interpolation is a useful technique for creating isolines that describe the surfaces with which you are working. Or the technique can be used to display the surface as a fishnet map or a shaded relief map if your software offers these options. But beyond creating a useful output, what would we most often use interpolation for? Let's say that you are planning a residential development and you do not want it to be located within the 100-year flood zone. And for some reason there is no map available that shows the boundaries of this area. However, you know that the elevation of the 100-year flood is 185 feet above sea level. You also have the surveyor's notes for a number of earlier subdivisions, and these notes include elevation data for each house built. By plotting the data on a local map, you can use interpolation to determine an estimate of the elevations of your proposed subdivision. From that you can draw in an isopleth indicating the 100-year flood zone and by simply comparing your location to the isopleth, you will know whether you need to choose another site.

Now suppose that you are building a highway through an unmapped area and you cannot begin without knowing the average grade or slope tendency. You may have to produce a trend surface map to show where the general slope is. Or suppose you are a mining engineer, trying to determine the general trend in an ore body from a large number of subsurface cores that show both the top and the bottom of the ore. Again a trend surface interpolation technique will provide information about the thickness of the ore body as it slopes across the subterranean surface. In addition, you may want to know about the quality of the ore seam. Here a kriging technique would prove useful because it is the nature of ore bodies to exist as regionalized variables.

In fact there are quite a number of uses for predicting surface values in different areas based on samples. If you are trying to predict changes in soil nutrients along a sloping surface, if you want to examine trends in vegetation cover at a distance from a source of water, or if you want to examine trends in population change across a large region based on sample data from previous decades—all these activities require some form of interpolation. What you should remember is that interpolation is in essence a predictive model of Z value. Whenever you want to determine the Z value, no matter what it might be, and you have other Z values on either side, interpolation gives you the needed capability.

PROBLEMS IN INTERPOLATION

We have now seen a number of methods of interpolation, some more exact than others. When performing any of them, however, four factors need to be considered:

1. The number of control points
2. The location of control points
3. The problem of saddle points
4. The area containing data points

Generally it is safe to say that the more sample points we have, the more accurate the interpolation will be, and the more like the surface will be our model. However, there is a limit to the number of samples that can be made for any surface. Eventually one reaches a point of diminishing returns: having more points doesn't substantially improve the quality of the output and indeed quite often increases the computation time and the data volume. In some cases too much data will tend to produce unusual results because clusters of points in areas where the data are easy to collect are likely to yield a surface representation that is unevenly generalized, and therefore unevenly accurate. In other words, having more data points does not always improve accuracy; Figure 10.10 shows that at a certain number of points, the accuracy actually falls.

Of course, the number of control or target points is frequently a function of the nature of the surface. We have seen that the more complex the surface, the more data points you need. But for important features of particular interest, such as depressions and stream valleys, we should also place more data points to be sure of capturing the necessary detail. Additionally, although the location of sample points relative to one another has an impact on the accuracy of the interpolation routine, the relationship is not perfectly linear (Figure 10.11).

The problem of sample placement is even more severe when we consider interpolation from data collected by area to produce an isoplethic map. We have seen that our GIS offers the capability of determining the centroid of each polygon or the center of distribution of each polygon. When the data points are relatively evenly distributed, it is easiest to use the centroid-of-cell method. And the center-of-gravity method is most useful when the sample points are clustered or unevenly distributed. With both methods, however, there is always the chance that the center will occur outside the sample polygon, especially if the polygons are of unusual shape. When this occurs, the easiest solution usually is to pull the centroid or center of gravity just inside the polygon at its

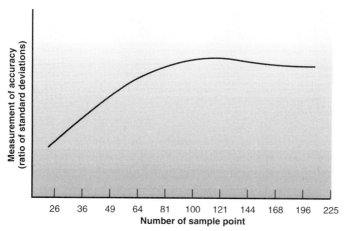

Figure 10.10 Accuracy of isarithmic map versus number of data points. Characteristic curve of a hypothetical relationship between the number of points and the accuracy of an isarithmic map. *Source:* A. H. Robinson et al., *Elements of Cartography*, 6th Ed., John Wiley & Sons, Inc., New York, © 1995. Adapted from Figure 26.29, page 513. Used with permission.

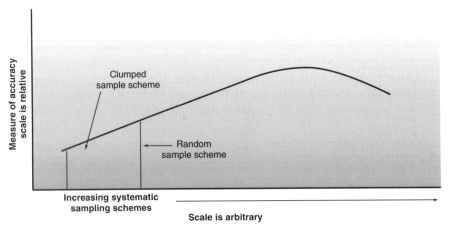

Figure 10.11 Sample point scatter and isarithm accuracy. Characteristic curve of a hypothetical relationship between the distance between data points and the accuracy of an isarithmic map.   *Source:* A. H. Robinson et al., *Elements of Cartography,* 6th Ed., John Wiley & Sons, Inc., New York, © 1995. Adapted from Figure 26.30, page 513. Used with permission.

nearest possible location. This will probably have to be done interactively within the GIS environment.

The **saddle-point problem,** sometimes called the alternative choice problem, arises when both members of one pair of diagonally opposite Z values forming the corners of a rectangle are located below and both members of the second pair lie above the value the interpolation algorithm is attempting to solve (Figure 10.12*a*). This generally occurs only in linear interpolation because any distance weighting will likely solve the problem. When it does occur, however, the computer software is presented with two probable solutions to the same question of where to assign the contour line (Figure 10.12). A simple way to handle this problem is to average the interpolated values produced from the diagonally placed control points, then place this average value at the center of the diagonal. Then the interpolation can proceed with a single solution to the

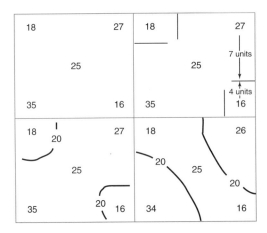

Figure 10.12 Saddle point problem. The saddle-point problem, where Z values are arranged rectangularly. Note the two possible solutions to the same interpolation problem. *Source:* A. H. Robinson et al., *Elements of Cartography,* 6th Ed., John Wiley & Sons, Inc., New York, © 1995. Adapted from Figure 26.27, page 512. Used with permission.

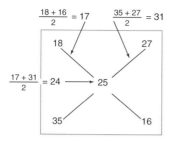
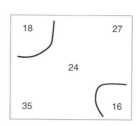

Figure 10.13 Solution to saddle point problem.
The solution uses an average value placed at the exact center. *Source:* A. H. Robinson et al., *Elements of Cartography*, 6th Ed., John Wiley & Sons, Inc., New York, © 1995. Adapted from Figure 26.28, page 512. Used with permission.

problem because it now has additional information with which to calculate the interpolated values (Figure 10.13).

The final problem that must be considered in interpolation is a common one in GIS operations involving the area within which the data points are collected. More specifically, for the interpolation to work properly, the data points that are to be estimated through a process of interpolation must have control points on all sides. If, as is often the case, we select a study area for analysis and use that same study area to perform interpolation, we will soon need to interpolate points near the margins of the study area. When we approach the map margin, we see that the interpolation routine is faced with control points on two or three sides of our unknown elevation points because the map border precludes any data points beyond the margin. As we have seen, best interpolation results are obtained when we are able to search a neighborhood in all directions for selection of control points and determination of weights. In the absence of these surrounding data points, the algorithm will use whatever is available, biasing the results away from the border.

As an example, let's try to perform interpolations along the left margin of a map. We will assume that this margin cuts directly through a hill that rises from the center of the map toward the left margin. The adjacent map sheet indicates that the hill continues to climb for some distance beyond the left margin of our study area. But, because there are no data points to the left of the map border, the algorithm will search for its neighbors to the right, above, and below the point to be estimated, and the interpolated values will be influenced in these three directions. That is, the matrix generated by the algorithm probably will use elevational values that are lower than those that in fact occur on the adjacent map sheet. Therefore, when the interpolation is performed it will tend to underestimate the missing elevations, producing a surface map that is of little use at the margins. Any slope calculations, azimuthal calculations, line-of-sight, least-cost path, or surface calculations will be in error near the margins, as well.

Despite the obvious problems encountered when performing surface calculations under such circumstances, even experienced GIS users occasionally fall into the trap of selecting surface data points from within the original study area only. Sometimes this procedural error occurs because surface data were not part of the original design, sometimes because the study area was selected on

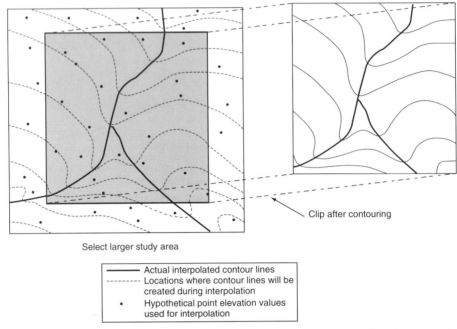

Clip after contouring

Select larger study area

———	Actual interpolated contour lines
--------	Locations where contour lines will be created during interpolation
•	Hypothetical point elevation values used for interpolation

Figure 10.14 Avoiding interpolation error at map margins. Solution to the problem of missing data for interpolation at the margins of a study area: We extend the study area and perform interpolation or other surface procedures; then the completed coverage can be clipped and subjected to operations by additional coverages, with accurate results.

the basis of the confines of a single map (even, perhaps, a topographic map), and sometimes because of time limitations. No doubt there are other reasons, but the result is the same—a compromised database and poor analytical results.

The solution to this problem is as obvious as the problem itself. Simply extend the borders of the elevation coverage beyond the boundaries of the initial study area (Figure 10.14). It is not necessary to expand the entire study area to accommodate the need for increased surface control points. Instead, we create a surface coverage whose outside margins are larger than the original study area, then perform the interpolation to produce the necessary surface configuration. Once done, the outside margins of the original study area can be used as a sort of cookie cutter, to shave off the edges (of the elevation model, slope model, viewshed, etc.). The idea is to operate on the larger elevational coverage to obtain whatever coverage will be used in conjunction with the other, smaller coverages. Determining the additional size of the surface coverage needed to produce reasonable interpolation results is difficult. In theory it would be useful to perform some type of neighborhood analysis, perhaps like that used in the kriging operation, to determine how far apart data points must be to ensure that they do not influence their neighboring elevational values. However, in practice, this is time-consuming and, ultimately, impractical. The best advice is to err on the conservative side by expanding the area sufficiently to assure good calculations. Most often extending the margins 10% on all sides will guarantee good results, but the complexity of the surface must be factored into this decision. The more rapidly elevation changes along the margins, the wider the margins should be, to compensate for the weights the neighboring control points will put on the calculation of missing values when weighting or kriging techniques are used. No standard guidelines exist here. You are left to use your own experience and firsthand knowledge of your data. Any additional border will increase the quality of your analysis, and is always preferred to no additional borders at all.

SLICING THE STATISTICAL SURFACE

As we learned earlier in the chapter, a common method of displaying surface information makes use of isarithms. And we also saw that the isarithms are commonly drawn at set intervals (a contour interval in topographic maps). The interval chosen allows us to depict the shape of the surface, but we always assume that the elevational values between the contour lines exhibit some continuously changing value because we assume that the surface itself is continuous. We also assume that the interval was selected to best display the surface features. Most GIS software, whether raster or vector, offers the ability to change the isopleth interval, or even to convert the area under each individual isopleth interval to a flat surface, eliminating the need to assume a continuous surface. For simplicity we will call this general group of techniques **slicing**, and envision their execution as applying a sharp knife horizontally through the surface.

Slicing can be simply a matter of selecting a different isoplethic class interval set to permit us to look at surface features differently (Figure 10.15). We might, for example, want to make the vertical distance between isopleths larger, to bring out overall shapes of features without great amounts of confusing detail. This useful visualization technique gives an impression of trends without necessitating the actual computation of a trend surface for the area. Alternatively, to see more of the detail in the surface, we can decrease the vertical distance between the isopleths. Of course this latter approach assumes that the sample data are sufficiently detailed in the first place.

A more common application of slicing could more appropriately be called a neighborhood function, as discussed in Chapter 9. We look at it here because in its performance it relates more closely to the interpolation procedure. This approach assumes that by placing contour lines at individual intervals, we have effectively reduced a continuous surface to a discrete surface more closely resembling a stair-step feature. Why would we want to degrade our data from continuous to discrete? Perhaps the following example problem requiring neighborhoods will help explain.

Let's say you are working for a consulting firm under contract to a foreign government. You have been asked to define appropriate land uses for a large portion of land based on a combination of soil capabilities and elevation classes. Elevation classes have been chosen rather than slope classes partly because most soil surveys incorporate slope into the soil capability classes and partly because the client's agricultural advisers know that some plants grow better at certain elevations than at others. Based on each crop's elevational requirements (let's say there are five crops), you can divide the surface into five elevational

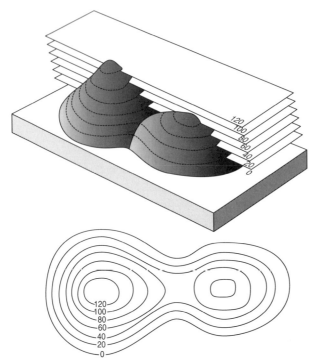

Figure 10.15 Slicing the *Z* surface. Slicing through a *Z* surface to produce different contour intervals.

regions by slicing through it in four places. Once you have performed the slice, the groups can be reclassified by their influence on the five crops. So you now have five groups of areas, based on elevational classes as they affect the five crops, and called "crop1elevations," "crop2elevations," and so on. Thus you have converted the ratio data of the surface to nominal data based on the interactions between crops and elevations. These, as we will see in Chapter 12, can now be combined with other coverages to make decisions about where the crops should be planted.

CUT AND FILL

While interpolation is useful for predicting surfaces, and even for selecting appropriate elevational ranges for crops, it can also potentially be applied to another set of problems related to the volume under surface features, or the volume of material related to cutting and filling (as in removing ore and back-filling to produce a flat surface during reclamation activities). Both types of procedure require a knowledge of two separate surfaces, a top and a bottom. By producing difference values for both surfaces, and applying contour slicing or integral calculus techniques, we can determine the volumes. Let's look at how this is done.

Suppose our backyard pool has vertical sides and covers an area of 150 square meters. The pool is 2 meters deep across the entire surface area. By multiplying the depth by the area, we obtain the overall volume of the pool: $2 \times 150 = 300$ square meters.

But many things that are measured across a large area are not uniformly distributed. For example, consider precipitation. Let's say that we have 15 rain gauges located throughout a county of approximately 100 square kilometers (1,000,000 square meters), and after collecting rainfall data we want to find the total amount of precipitation at every location for the entire year. Because of the variability of precipitation, we are likely to find amounts ranging in value from perhaps 50 centimeters (0.5 meter) to as much as 150 centimeters (1.5 meters). To simplify our calculations of the total volume of rainfall throughout the area we can use a technique called the **method of ordinates** that totals all these values (Muehrcke and Muehrcke, 1992). By summing all the values, we create a flat surface similar to a backyard pool. Suppose, then, that our average precipitation throughout the county is 100 centimeters (1 meter). We can then multiply 1 meter by 1,000,000 square meters to obtain a value of 1,000,000 cubic meters of water.

But suppose we want a more exact calculation of the countywide volume of precipitation. Figure 10.16 outlines the steps. We first record the coordinates for each of our rain gauges and produce an altitude matrix, just as we would do for elevation data. Then we use an appropriate method of interpolation to generate a set of contours, which we slice into equal parts, guided by the contour intervals. If we then assume that all slices are of equal elevation (discrete), we can perform the same simple volume calculation for each of the separate contour slices. By simply adding these together, we obtain a somewhat more accurate measure of the overall volume than if we assumed the entire county received an average amount of precipitation. Because both these techniques conceive of our volumes as having vertical surfaces, the calculation of volume is very simple in either vector or raster. For each slice of the volume, we simply multiply the area attribute value by the depth value. You might try this out on a simple example database with your own GIS system.

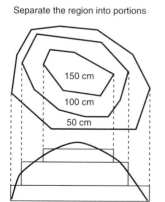

Separate the region into portions

150 cm

100 cm

50 cm

Calculate the area of each portion

Multiply that by the amount of precipitation

Add these products to find the total volume of precipitation

Figure 10.16   Method of ordinates. The method of ordinates used to determine the volume of water in an area.

Of course, we would really like to be able to be as accurate as possible in calculating volumes. Thus it is useful to know that by increasing the number of slices, you radically increase the accuracy of your calculations. As you increase the steps, you approach a point at which the number of stepped surfaces is nearly infinite and you are essentially subtracting the elevational values of one surface, the elevation values, from those of the base or ground level. These values can include the raw data plus any number of interpolated values you might see as useful for your calculations. Integral calculus can then easily be applied. Fortunately we seldom have to do this by hand. Instead the algorithms are already present in the software to do the necessary integrations.

Keep in mind that calculations of volume are essentially the same for a hill or a lake in that one of the surfaces is essentially flat. This of course simplifies the problem somewhat. But if your purpose is to determine how much ore there is in a subterranean deposit, both the upper and lower surfaces will be irregular and in all likelihood will have to be interpolated as well. Still, you can use the same approaches as before, except that you can no longer assume that one of the surfaces is flat. By simply subtracting the upper elevational values from the lower, you can obtain a variable height value. Then, by noting the area and allowing the integration equations inside the software to do most of the work, you can get the needed result.

While the conceptual model of volume determination is a simple one, you should be aware that it is not always implemented into commercial GIS. Some commercial products specialize in this sort of software, usually packages related to geological sciences and mining. If you either don't have these techniques available or don't want to piece them together with a number of different steps, you most often will find that the available volumetric software data can be converted to the GIS and back again at will. This practice is becoming more prevalent as time goes on.

OTHER SURFACE ANALYSES

We have already seen how continuous surface data can be applied to a number of different analytical techniques such as distance measures, intervisibility, and neighborhood characterization. At this point you may wish to review Chapters 8 and 9 to refresh your memory about how surface data can be used to create friction surfaces or barriers, and how neighborhoods can be reclassified based on slope, aspect, and intervisibility (in addition to elevation ranges), and by means of buffers. Most analysis with surfaces is performed as a combination of a variety of these techniques.

A wide range of displays and additional coverages can also be produced with the use of surface data. For example, most GIS software allows additional coverages to be visually displayed over three-dimensional displays of the surface itself. You might find this capability useful for showing the relationships between elevation and soils, vegetation, or land use distribution prior to slicing your topography. Most software can also produce coverages of surface area occupied by elevational features such as hills. Or you might find it useful to produce a measure of actual road miles, as opposed to horizontal road miles, as you travel up topographic surfaces. The latter technique is especially useful if you own a road construction company and are being paid by the mile for producing a road: if you are paid by map distance rather than road distance, you could find yourself seriously underpaid. In all these approaches you need to consult your software documentation to determine whether the desired techniques are available. You will also want to know whether the operation will produce a coverage that can be further analyzed, or merely a display from which you will need to make decisions almost from scratch.

DISCRETE SURFACES

As we have noted earlier in this chapter, statistical surfaces can be mapped in any of four different ways. We've already examined the primary method of mapping continuous data by the use of altitude matrices, TIN models, isarithms, and their derivatives. But statistical surface data can also occur as discrete objects. For these we need to look at some additional methods that can be applied to data of these types, for display as well as for analysis.

Dot Distribution Maps

Sometimes in the course of our explorations we will want to record the counts and locations of objects by points on the kind of map called a **dot distribution map.** This approach most often uses specifically defined areas in which to count up the objects encountered (number of birds by county, bushels of grain harvested by farm, etc.). Another common form of dot distribution map does not employ preselected areas, but rather records each individual by a single dot. A quick look at maps of plant and animal inventories, available through research articles or books, or surveys produced in a number of biological survey offices, will provide a representative sample of the latter approach.

When a dot, called the **unit value,** indicates more than a single observation, three interrelated questions must be asked (Robinson et al., 1995). First, how many objects do we represent by a single dot? Next, how big do we make the dot based on the unit value? And third, where do we locate the dots once the first two questions have been answered?

While the foregoing questions are important for cartographic representation, their importance to us does not appear unless we are asked to input such maps into a GIS. Most often, on the rare occasion when we must do this, the dots are counted by enumeration area and converted to a single value for each polygon. They can then be analyzed as value by area data in a choropleth map context. We'll look at choropleth maps in a little more detail later. In addition, the dots can be assumed to be continuous data, in which case we can decide whether to use a centroid-of-cell or center-of-gravity technique to represent each area's value as a single point and operate on the entire set in an isoplethic context. In this way we can work much as we did with continuous data, by interpolation or other surface approximation techniques, slope and aspect analysis, and so on.

When each dot indicates a single count of the objects—say, for example, the locations of bird species—we must assume that the locations are precise, and only the dot size is of concern to us if we are to input the counted objects into a GIS. We see, however, that the vast majority of single-value dot distribution maps, especially those produced before the advent of modern GPS and telemetry technologies, display the dots in a wide range of dot sizes. It is common to see maps at scales of 1:100,000 with dots that would cover hundreds or even thousands of meters on the ground. Unless we have specific locations, such as those that can be obtained through advanced technologies, we are forced to assume that the point locations are most accurately represented by the center of the dot itself. While this may be somewhat incorrect, we are left with little alternative.

In some cases we are interested in the distances and groupings of individual point distributions. Most GIS software is not particularly well adapted for performing such analyses, but we will look at these aspects further in the next chapter. Sometimes we also find that each individual point location is accompanied by additional data, recorded at any of the four levels of data measurement. Within the context of the statistical surface, however, we are concerned only with those that occur at ordinal, interval, or ratio scales. If we can assume that there is a continuity available for such values as the weights of birds collected or the size of brood for mammal species at points locations, we can once again operate on these point data as if they were point samples of a topographic surface, especially if they are recorded at interval or ratio scales. However, the assumption of continuity in most of these cases is invalid, and we will be forced to look at these as discrete events and compare them individually, either statistically or through comparisons of their spatial interactions. Again, we will look at this type of analysis in the next chapter.

We have also seen reference to the conversion of individual point distribution maps into some form of area distribution (Chapter 3), with special reference to zoological mapping. As we observed, there is little guidance as to how to make these conversions, and simple graphical techniques that are not related to the functions of the organisms or other objects in their environment may give erroneous results. Most often if you are asked to perform an analysis of such distributions, you need to know as much about the environment and the nature of the point phenomena as possible, either through your own study or by detailed consultations with your clients.

Choropleth Maps

Quite often, of course, statistical data are recorded for a specific area without regard for specific locations. We've seen that we can assume these data to be continuous and can operate on them as isoplethic maps. While this works well when we are trying to observe patterns across larger surfaces, most often the data are collected to observe highs and lows within each area unit, or to compare area units to other units in other coverages (more on that in Chapter 12). In Chapter 3 we looked at the idea of value by area mapping which we call choropleth mapping. And we saw how maps by area can be made by using selected classes, or by using the raw data, in a **classless choropleth mapping** approach. In the GIS environment, which features software capable of producing a wide range of output maps, we are most likely to want to obtain the original data for each area and to perform the many GIS functions based on these. Here, however, we need to reconsider the **classed choropleth map** because of the frequency with which we will be asked to input such documents for later analysis.

Traditional classed choropleth maps, based as we have seen on the communication paradigm, are of limited usefulness in the GIS context because of the design methods that are employed to produce them. They can be input to the GIS for analysis, but we should be aware of the design methods used to create them. Generally three design elements are required to determine how the polygonal categories are placed on the map: size and shape of areas, number of classes, and method of class limit determination. As you might guess, the larger the sizes of choropleth units, the more generalized the data are. So, for example, a map of the Canadian provinces and territories that assigns only two classes will produce a highly generalized set of data, probably of little use for further analysis inside a GIS. By contrast, if individual counties were employed, and there were 15 categories, the groupings of data will provide more useful information that might well be input to the GIS for analysis. Traditional cartographers also attempt to keep the shapes of the areas as similar as possible, a practice that results in modifications of the data that will affect what we can do to analyze them. In addition, a wide range of numerical classing techniques can be applied to the statistical data. These include constant series or equal steps, systematically unequal stepped class limits, and irregular stepped class limits. A knowledge of the approach used in the choropleth map you are using inside the GIS will afford insights into how the data are organized and may prove useful in later analyses. Most often, however, classed choropleth maps are not well suited for GIS analysis, unless we can obtain more detail about the functioning of the areas relative to other variables—a technique called dasymetric mapping.

Dasymetric Mapping

To improve the quality of choropleth maps for use in a GIS, the technique called dasymetric mapping is sometimes employed. Because choropleth maps, whether classed or classless, reflect the structure of the data collection areas, they are often poorly matched to the distributions themselves. Dasymetric mapping attempts to break down this artificial structure to unmask the hidden distributions. Rather than being restricted to artificial collection units, dasymetric maps assume areas of relative (i.e., compared to other locations) homogeneity of statistical data (Robinson et al., 1995). They employ the same data, but compare this material to other supporting data or information from other sources. These comparisons make it possible to improve the quality of the original choropleth maps by dissolving the boundaries imposed by the primary collection units. Because the work requires the use of supporting information, however, we will delay a detailed discussion of this highly useful technique until Chapter 12, where we compare different coverages. In the GIS domain, dasymetry is most often performed through the process of overlay of data contained in other coverages, rather than simply by adding data to the original map.

Terms

statistical surface	rough surface	dot mapping
smooth surface	dasymetric mapping	isarithmic mapping
choropleth mapping	isarithm	contour interval
contour line	isoplethic map	regular grid
isometric	irregular lattice	digital elevation models (DEMs)
regular lattice	discrete altitude matrix	nonlinear interpolation
linear interpolation	isarithmic map	splines
weighting methods	polynomials	structure
kriging	drift	lag
random noise	semivariogram	nugget variance
semivariance	range	method of ordinates
saddle-point problem	slicing	classless choropleth mapping
dot distribution map	unit value	
classed choropleth map	discrete	
continuous		

Review Questions

1. Define a statistical surface. Describe a nontopographic statistical surface. What is the difference between a continuous and a discrete statistical surface?

2. What is the difference between a smooth and a rough statistical surface? What is the importance of each in terms of point sampling to represent the surface configuration?

3. What is an isarithm? How do the inferences we can make from closely spaced isarithms differ from what widely spaced isarithms tell us? What is the specific term assigned to isarithms that measure topographic elevation?

4. What is a contour interval? Why must a contour interval remain the same throughout a map?

5. What is the difference between an isarithmic map and an isoplethic map? What assumptions must be made before we can use isolines for isoplethic maps?

6. What is the difference between a regular and an irregular lattice in terms of sampling for surface data? What are the advantages and disadvantages of each?

7. What are digital elevation models? What is the relationship between a discrete altitude matrix and a TIN model?

8. Describe and diagram the method of representing continuous surface data with a raster data model. Why is it important to know where in each grid cell the actual point locations for topographic data are intended to lie?

9. Describe and diagram linear interpolation. What are the major drawbacks to linear interpolation?

10. Why might we be more inclined to use nonlinear interpolation instead of linear interpolation? What are the three basic types of nonlinear interpolation routine?

11. Describe and diagram the use of weighted methods of nonlinear interpolation. Why might we want to include barriers in weighted interpolation methods?

12. What are trend surfaces? When might we want to use this interpolation approach rather than weighted methods? Give an example showing its utility.

13. What is the difference between kriging and weighted interpolation methods? What is a semivariogram, and what does it tell us about the drift, random spatially correlated variations in the surface, and random noise?

14. Provide a diagram of a semivariogram based on hypothetical data. Explain the relationship between lag and semivariance. What can we say about data within and beyond the range? What does this tell us about selecting the neighborhood for interpolation?

15. What is the difference between universal and punctate kriging? When is kriging not likely to give better results than any other form of interpolation?

16. Give some concrete examples of the use of interpolation beyond simply creating an isarithmic map.

17. What four things do you need to look out for in any interpolation routine? Describe the problem of sample placement in isoplethic mapping. What is the saddle-point problem? How can it be rectified?

18. Describe the problem of no data points at the map margin as it applies to interpolation results. How can such inadequate results be avoided?

19. What is *Z*-value slicing? Other than changing the class interval (e.g., the contour interval), how can slicing be used for creating neighborhoods? Give an example.

20. Describe the method of ordinates for determining the volume of a *Z* surface feature. Explain how this could be done in raster and vector GIS.

21. Given that dot distribution maps are not commonly used in GIS output,

why might we need to know about them? How can we perform an analysis on point data to which magnitude attributes are attached?

22.   What is the difference between choropleth maps and dasymetric maps? How are dasymetric maps generally produced from traditional choropleth maps?

References

Blais, R.A., and P.A. Carlier, 1967. "Applications of Geostatistics in Ore Evaluation." *Ore Reserve Estimation and Grade Control,* 9:41–68.

Burrough, P.A., 1983. *Geographical Information Systems for Natural Resources Assessment.* New York: Oxford University Press.

Burrough, P.A., 1986. *Principles of Geographical Information Systems for Land Resources Assessment.* Oxford Science Publications, Monographs on Soil and Resources Survey. New York: Oxford University Press.

Clarke, K.C.,1990. *Analytical and Computer Cartography.* Englewood Cliffs, NJ: Prentice-Hall.

Davis, J.C., 1986. *Statistics and Data Analysis in Geology,* 2nd ed. New York: John Wiley & Sons.

Flowerdew, R., and M. Green, 1992. "Developments in Areal Interpolation Methods and GIS." *Annals of Regional Science,* 26:67–78.

Hodgson, M.E., 1989. "Searching Methods for Rapid Grid Interpolation." *Professional Geographer,* 41(1):51–61.

Kelly, R.E., P.R.H. McConnell, and S.J. Midenberger, 1977. "The Gestalt Photomapping System." *Photogrammetric Engineering and Remote Sensing,* 43(11):1407–1417.

Lam, N.S., 1983. "Spatial Interpolation Methods: A Review." *American Cartographer,* 10:129–149.

Mark, D.M., 1978. "Concepts of Data Structure for Digital Terrain Models." In *Proceedings of the DTM Symposium,* American Society of Photogrammetry, American Congress on Survey and Mapping, St. Louis, MO, pp. 24–31.

Markarovic, B., 1973. "Progressive Sampling for Digital Terrain Models." *ITCJ* 1873-3:397–416.

Matheron, G., 1967. "Principles of Geostatistics." *Economic Geology,* 58:1246–1266.

McCullagh, M.J., and C.G. Ross, 1980. "Delaunay Triangulation of a Random Data Set for Isarithmic Mapping." *Cartographic Journal,* 17(2):93–99.

Muehrcke, P.C., and J.O. Muehrcke, 1992. *Map Use: Reading, Analysis, Interpretation.* Madison, WI: J.P. Publications.

Oliver, M.A., and R.W. Oliver, 1990. "Kriging: A Method of Interpolation for Geographic Information Systems." *International Journal of Geographical Information Systems,* 4(3):313–332.

Robinson, A.H., J.L. Morrison, P.C. Muehrcke, A.J. Kimerling, and S.C. Guptill, 1995. *Elements of Cartography,* 6th ed. New York: John Wiley & Sons.

Shepard, D., 1968. "A Two-dimensional Interpolation Function for Irregularly Spaced Data." In *Proceedings of the Twenty-third National Conference of the Association for Computing Machinery,* pp. 517–524.

CHAPTER 11

Spatial Arrangement

Our journey thus far has focused on characterizing the objects we have seen. But to evaluate our environment properly we also need to know the relationships among the individual items we see and the intervening space among them. Rather than looking at the amount of space occupied by a single object, or how the object's space is configured (i.e., its shape), we will now concern ourselves with the overall arrangement of objects in space. Such arrangements can be characterized by how many there are in a particular area, or whether they are distributed evenly or in groups. We will look at the distance relationships between objects themselves and how they are related to the overall size of the area they occupy.

Such distributional patterns can be seen in many situations. We know, for example, that some human population distributions are widely dispersed, as in farm households in a rural area. Other population distributions are more concentrated in larger collections we call cities. Plants and animals may occur individually in evenly distributed arrangements, or they too may form more concentrated groups. Even physical features such as sediment types and landscape features—streams and hills, mountains and valleys—may occur as widely spaced individuals or as larger groups. Anthropogenic objects like roads, fences, and houses can occur in patterns, as well. As our geographic filter increases, we will see still more. Observing that there are differences in spatial arrangements of objects allows us to ask questions about what these patterns are, how they can be classified, and how they might provide insights into the processes that made them.

If we revisit the areas where we first observe arrangements of objects, either in the real world or, as we would do in GIS, through multitemporal coverages of the same objects, we see that the observed patterns change. Regions that once appeared to show diffused and scattered patterns may now show evidence of grouping and coalescence. Objects that were once randomly organized in space may now begin to occur in regular, repeating patterns. Some distributional patterns may show growth and expansion, while others may be shrinking or vanishing. Patches or areas may coalesce, single linear objects may begin to connect to form networks, stabilized dunes may begin moving and diffusing. In all these cases time is an important component in our understanding of spatial arrangement. And we soon wonder what the processes are that cause the change from one arrangement to another. We can ask many questions: what the directions of change might be, whether there are driving forces that we can understand, what the upper or lower limits of these driving forces might be, and so on.

As before, the study of arrangements and patterns may not be our primary focus, but even those who will later enter the workforce as GIS technicians need to understand the approaches to modeling. Before any modeling can take place, we need to be able to construct the databases that will be analyzed. We need to know what the analytical capabilities are so we can decide which software will best suit our needs. Or we may need to understand what alternative software can be applied and what compatible data structures can be employed to move the necessary information to and from the GIS.

For those who are interested in patterns on the earth, the analysis of arrangement is an exciting one. Yet, it can also be somewhat frustrating because the GIS tool kit is not often well designed to accommodate such analytical techniques. In fact, there are only a few conceptual models available from which software could be written. If you find yourself frustrated, try to keep in mind that where there is frustration there is also room for more exploration: room for digging deeper into your digital universe than you might have thought possible. You may find that while the GIS does not provide a ready set of techniques for analysis of arrangement, it frequently is helpful to begin by dissecting the problem into its component parts, then applying the GIS tool kit to each part of the problem individually.

In this chapter we will be looking primarily at the arrangements of objects in a single coverage. This in itself is somewhat limiting in scope. It has been my experience that you will begin to see techniques that employ data from other coverages. That is exactly what you should be thinking. If you are patient for a short while, you will soon see how the arrangements of individual objects can best be compared to one another. At that point you will start thinking about how the GIS will allow you to compare objects in individual coverages. This is the subject of Chapter 12, but it is useful to stretch the ideas of this chapter as a preparation for the next.

While you work through this chapter, try to envision objects you might encounter in your work. The examples given here are based on one person's experiences. By sharing your ideas with your classmates and with your instructor, you can expand your collection of potential objects and patterns. While you are doing that you might ask an even more important question: What possible mechanisms might account for the patterns of trees, cars, farms, people, animals, and so on? The purpose of the GIS is to assist us in making decisions. Those decisions may very well be based on the knowledge of the relationship between pattern and process. The GIS will help us analyze the patterns. But you will most often be responsible for determining the potential processes and finding ways to link the two inside your digital world. Let your creativity work its magic as you continue on your journey.

LEARNING OBJECTIVES

When you are done with this chapter you should be able to:

1. Clarify what we mean when we say we are interested in arrangements of objects. Why is this knowledge useful to us?

2. Define a uniform distribution. Tell what makes a uniform distribution regular versus random. Define a clustered distribution.

3. State the significance of regular, random, and clustered distributions in terms of the potential processes that caused them. Give some examples of each distributional pattern and give an hypothesis that might describe the causes.

4. Tell the purpose of quadrat analysis. Describe how it is done, and name the statistical test we use to confirm or reject our results.

5. Tell the purpose of variance–mean ratio in point distributions and state how it differs from traditional quadrat analysis.

6. State what nearest neighbor analysis tells us that quadrat analysis does not and describe how the two techniques are similar.

7. Give the purpose of Thiessen polygons. Diagram and describe how they are created.

8. Describe the join count statistic, answering the following questions: What does it tell us? How does it differ from a simple measure of contiguity? What is the difference between free sampling and nonfree sampling as these apply to testing the results of the join count statistic?

9. Describe the process of performing a nearest neighbor statistic on line patterns, and state what this quantity tells us about the distribution of our lines. Describe some situations in which the nearest neighbor statistic for lines might give misleading results.

10. Describe the use of line intersect methods for analyzing line pattern distributions. Define a random walk.

11. Define a vector resultant, and state what it tells us about the patterns of linear objects. Define resultant length and discuss how it compares to resultant force problems in physics.

12. Define the mean resultant length, explain how it differs from the raw resultant length, and tell when we would use it. Answer questions such as the following: What would a large mean resultant length tell us? What is the difference between circular variance and mean resultant length?

13. Describe how we can adjust our measurements for mean direction, mean resultant length, and circular variance to account for orientations that can be measured in either of two different directions.

14. Describe how the gamma index is performed, and spell out what it tells us about a given network. What does a gamma value of .48 tell us in terms of the amount of connectedness in our network?

15. Tell what the alpha index is, how it is different from the gamma index, and how it is the same. Give some non-road-network examples of how both these indices might be useful.

16. Describe, in general terms, the gravity model. Explain how distance affects the interactions of point objects and how the magnitude of the nodes

changes the interactions between them. Give an example of how the gravity model might be applied.

17. Describe the types of attributes that might need to be assigned to the nodes and the arcs between nodes in work with routing and allocation problems.

POINT, AREA, AND LINE ARRANGEMENTS

Spatial arrangement is the placement, ordering, concentration, connectedness, or dispersion of multiple objects within a confined geographic space. Until now, most of the analytical techniques we have examined have dealt either with individual objects, or collections of objects when they might be defined as regions or neighborhoods or conceived of as statistical surfaces. We have even touched very briefly on the possible interactions of objects, regions, surfaces, and neighborhoods with those of other coverages. However most of the objects we encounter in a single coverage also have a definable spatial patterning that might indicate possible mechanisms for producing them.

In a purist approach, the term "spatial arrangement" generally refers to the simple cartographic display of spatially distributed objects. In such a "map as communication" mode, we would say that the map shows us where the objects are located and illustrates the shape of this distribution. That is enough, we might say. But we intuitively recognize that there is more that can be known. There is more that can be used to describe the interactions of each individual object to its neighbors and the relationship of all these objects to the whole space in which they reside. And, of course, if we can find ways to measure these relationships, we can find ways to isolate and understand the possible mechanisms that produce such patterns.

The pages that follow reveal a number of possible ways to analyze and evaluate spatially distributed phenomena. We will examine some of the potential approaches that are specific to points, lines, and areas (and surface features covering areas) because of their respective changes in dimensionality. As you examine each, remember that one of the primary differences among points, lines, and areas is the scale at which we perceive them. If you keep this point in mind, you will see that some of the techniques will begin to sound very much alike from one type to another. Look for this pattern. It will help you formulate your own approaches to analysis inside the GIS.

Some of the techniques we will see are relatively simple to compute, while others are very difficult, requiring a great deal of computer time. Some of these techniques are very frequently applied; others, while extremely useful, have received little recognition because they are seldom heard of except in graduate-level courses. It would be easy to ignore this subject, but I feel it is my obligation to let you examine early on the types of analysis that can be applied, within a GIS environment to study the arrangement of objects in a single coverage. If they are of use to you, you will find yourself digging deeper; if not, at least you know they exist and you can prepare your digital database to accommodate them if need be. While completeness is my goal, every effort is made to keep the discussion at a conceptual level; whenever possible, I will use graphic devices to make these concepts clear.

POINT PATTERNS

Perhaps the most common techniques for analyzing spatial distributions are applied to point patterns. Point objects can be individual trees, houses, animals, street lights, and even cities, depending on the scale (Figure 11.1a). As we will see later, point objects can be represented in linear, or area form, as well (Figure 11.1b).

The simplest measure of point patterns is a matter of determining the **density** of the distribution. This is done by dividing the number of points by the total area in which the points exist. Of course, this may yield a useful descriptor of the numbers within the distribution, but it does little to characterize itself. Population density, housing density, tree density, and so on are commonly used to provide a measure of the compactness of points. With that information we can compare the densities to those of comparable point objects in other areas to make comparisons of the different mechanisms that might be operating. Or we might compare the points in the same location but at different times to give us an idea of change in density through time. We might find, for example, that population density is increasing in urban areas through time, or that housing density is increasing, or that tree density decreases as the trees mature and compete for space and sunlight. Even this simple statistic, easily calculated in either vector or raster, can give us many useful insights into our data.

No matter what the overall density of the distribution, we may be interested in the form of the distribution in addition to the numbers per unit area. Point patterns are said to occur in one of four possible arrangements, each with a specific set of criteria. A pattern is **uniform** if the number of points per unit area in one small subarea is the same as the number per unit area in each other subarea. If the points occur on a grid, separated by exactly the same distance throughout the entire area, the uniform pattern is said to be **regular,** as we saw with a regular grid in sampling surface data points. In other cases uniformly distributed points may occur in **random** locations scattered throughout the

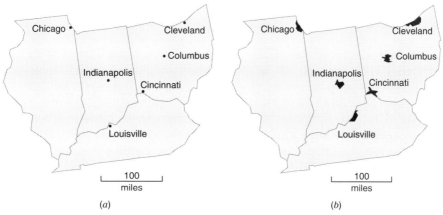

Figure 11.1   **Point and area representation of cities.** Area objects, in this case cities, can be viewed (a) as points or (b) as lines, depending on the scale at which they are presented. This shows the close association between the analytical techniques that can be applied.

study area. Or, in other cases, the points are grouped in tight arrangements in a pattern called **clustered** (see Figure 2.8).

Quadrat Analysis

Uniform point patterns are defined based on the relationships among uniform subareas called **quadrats** of the larger area. This is a very common method of analyzing discretely occurring biological data. Each quadrat contains a number of points defined by, for example, individual herbaceous plants, ant mounds, or any other type of point phenomenon. If each uniform quadrat contains the same number of point objects, we say the distribution is uniform for the overall study area. Uniform distribution is not likely to be found among biological phenomena, since living organisms tend to migrate toward specific locations that are rich in nutrients, have good drainage, contain certain types of soil texture, and so on. If distribution is in fact uniform, we could suppose that there are no major mechanisms controlling the locations of the objects. Thus we need to be able to test to see whether there is uniform distribution.

In the standard method of distribution testing, called **quadrat analysis,** we assume that roughly the same number of objects will be found in each subarea. We determine how many objects are likely to be in each subarea by dividing the total number of data points by the number of subareas. The number we get will be our "expected distribution" if it can be characterized as uniformly distributed. In a fairly simple statistical test that can be applied to this set of data, called the **chi-square** (χ^2) test, we subtract the expected number of points for each quadrat from the observed numbers O, square the result, divide by the expected number, and take the summation. This translates into a simple formula:

$$\chi^2 = \Sigma \frac{(O - E)^2}{E}$$

The results of this mathematical manipulation give us a value that we can test against predefined critical values in a standard table. If the number we get is not substantially different from what we would expect, the distribution is uniform; a notable difference, however, demonstrates some nonuniform pattern that might suggest an underlying process. Although this technique might be considered to be purely statistical, it can be performed in some GIS software, especially when the data model is raster. Many specialty software packages can perform this analysis, as well. If you are unfamiliar with the use of statistics, especially as they apply to spatial data, you might want to consult a simple text on the use of statistics in spatial data analysis, such as McGrew and Monroe (1993). However, for simplicity's sake, it is easiest to simply remember that the larger the chi-square value, the less likely the data occur in a uniform pattern.

Although most of you have probably assumed that the output of GIS analysis would be another map, in this case the output is simply a single value that must be evaluated to determine whether a uniform distribution can be seen. At this point, you might be asking yourself this question: "If my distribution is not uniform, what mechanism might contribute to the anomalies?" Most often the point distributions we see relate to other mapped variables (coverages) for the same study area. In other words, additional relationships may exist among coverages. These potentially related coverages can be additional points, but often they are maps of areal distributions of phenomena. In our biological examples these might be soils variables. We will then need to compare the points in the single coverage to the polygons in a second coverage. This useful technique is covered in the next chapter, but you might want to think about it now.

While quadrat analysis can help us decide whether a uniform distribution exists in a given situation, it can be used to provide us with more information, as well. A special technique, based on quadrats, is called the **variance–mean ratio (VMR).** VMR is an index that shows the relationships between the frequency of subarea variability and the average number of points in each subarea (McGrew and Monroe, 1993). It is calculated by dividing the variance of the subarea point frequencies (defined by the weighted standard deviation: see McGrew and Monroe, 1993) by the mean for each subarea (VMR = var/mean). As before, the test statistic is chi-square and is calculated by multiplying the variance–mean ratio by one less than the total number of subareas. High chi-square values indicate that our distribution is clustered (McGrew and Monroe, 1993); that is, the variability between the number of points in each subarea and the average for the entire area is very high. Alternatively, small chi-square values mean that the distribution is more evenly dispersed or uniform. Intermediate values indicate a distribution more closely associated with a random process, where some quadrats seem to have slightly more and some slightly less than an average value.

As before, the results of the analysis suggest that if the distribution is not statistically random (i.e., if it is either a uniform or clustered), you may be able to determine a possible causative mechanism by wisely selecting a set of variables to compare to your point coverage. Uniform distributions, for example, tell us that point objects are sprinkled more or less evenly within each subarea; but uniform distributions can occur in regular patterns as might typify trees in an orchard, or randomly, as is more characteristic of trees in a forest. In randomly distributed points we will not find the same number of points in each subarea, while in regularly distributed points we generally will.

Nearest Neighbor Analysis

Thus far we have confined our characterization of point distribution patterns to the numbers of points occurring within subareas. In other words, we are studying the distribution of points by comparing the areas they occupy. However, it is also instructive to examine the locational relationships from one point to another. To do this we most often rely on an alternate method of point pattern analysis, **nearest neighbor analysis,** a common procedure for determining the distance of each point to its nearest neighbor and comparing that value to an average between-neighbor distance. Calculating this statistic involves simply determining the average or mean nearest neighbor distance between each possible pair of near neighbor points (near neighbors are defined as the closest point to each test point you select). The mean nearest neighbor distance provides a measure or index of the spacing between points in the distribution. This in itself is useful because obviously point objects will conflict if they are too close together. For example, we know that many animals have requirements for space expressed as "range." When one animal's range overlaps that of another of the same species, conflict is likely. You have probably seen this many times in your backyard, where members of the same species of bird routinely attack one another. This is much the same for lions as it is for some species of ant.

But, as before, it would be useful to be able to extract additional information from the statistic. As it turns out, just as in quadrat analysis, we can compare the nearest neighbor spacing index to three possible patterns that might occur: regular, random, and clustered. This technique, described in detail by McGrew and Monroe (1993), can be generally explained for each of these cases by creating an index against which to compare your results as follows. For an index of random distribution, simply divide 1 by twice the square root of the density of points (number of points/area). If you want a test of maximum dispersion (regular distribution), divide 1.07453 by the square root of the point density. Finally, for a test of maximum clustering, where the points would normally be on top of one another, we can simply assume the value is of the divisor 0. Any value you obtain for your index that is not 0 will be a positive number. A simple comparison of your average nearest neighbor distance to that of these three index values will give you an idea where along the continuum they fall.

Let's take a quick look at how this works with the set of data in Table 11.1 and the associated plot (Figure 11.2). We have six data points, each with a single X,Y coordinate pair, all falling within an area of 25 square units. The average nearest neighborhood value from these data turns out to be approximately 1.4. For randomly distributed points, the index is found from 1 divided by 2 times the square root of the density (in this case 6 data points per 25 square units or 0.24). Since the square root of 0.24 is 0.4895 the final result for our random nearest neighbor distance is 1.02. Our average nearest neighbor of 1.4 is somewhat higher than the random value, 1.02. Our test index for perfectly dispersed

TABLE 11.1 Calculating Nearest Neighbor Distance*

Point	Coordinates		Nearest Neighbor	Nearest Neighbor Distance (NND)
	X	**Y**		
A	0.7	1.0	B	1.6
B	1.25	3.0	C	1.4
C	2.5	3.7	D	1.3
D	3.3	2.75	C	1.3
E	4.0	4.0	C	1.34
F	3.8	1.0	D	1.5
Average NND				8.44
Random average NND				1.4
				1.02

* Nearest neighbor distance calculated by the Pythagorean theorem.

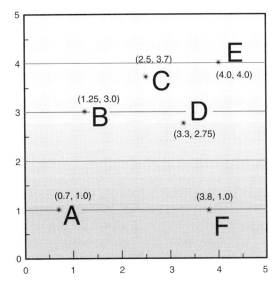

Figure 11.2   Nearest neighbor point coordinates. Locations of data points for determining nearest neighbor distance (see Table 11.1). Each point (e.g. point A) has an associated nearest neighbor (in this case point B). The distances are measured using the Pythagorean theorem.

points is 1.07453 divided by 0.489 (the square root of the density), to give a rounded value of 2.19. This would be the value if our point arrangement was perfectly uniform. Our average nearest neighbor is much smaller than that. And, of course, a perfectly clustered distribution is 0, and our average nearest neighbor is much larger than that. So, what we find is that our average nearest neighbor is slightly more dispersed than random, or somewhat between truly uniform and random. In other words it is starting to take on a more regular configuration, but as yet is still rather randomly distributed.

Again, these values give only a measure of the patterning of these point features. Most often we will need to decide how we might later compare these to other coverages or simply use the statistic as a descriptor for further statistical analysis. The nearest neighbor statistic is an absolute measure, hence, is not immediately amenable to comparison with nearest neighbor statistics for other point distributional patterns. The nearest neighbor index can be standardized to permit such comparisons (McGrew and Monroe, 1993), but the techniques are beyond normal use in a GIS setting. In addition, there are other methods of determining the degree of clustering based on other statistics (Davis, 1986; Griffiths, 1962, 1966; Ripley, 1981), but again these are more detailed than we need for an introduction to the use of nearest neighbor analysis in GIS.

THIESSEN POLYGONS

Points can be organized into a more regional context without relying on the use of other coverages for comparison by means of **Thiessen polygons** (also referred to as **Dirichlet diagrams** and **Voronoi diagrams**). Rather than simply looking at the arrangement of points, we can instead grow polygons around each point to illustrate the "region of influence" it might have on other points in the coverage. For example, as we will see later when working with the gravity model, we assume that the distances between points impose an attraction on the neighbors. In addition, the magnitude of the node—for example, the size of a city—is often directly related to the degree of influence. We will restrict ourselves to the case of all points considered to be of equal magnitude. This simplifies the description, allowing us to understand it in this simple case. When your skills increase you can carry the technique to its logical conclusion by weighting the points.

Creating Thiessen polygons is relatively simple conceptually, but the process can become quite involved as the number of points increases. For insight as to how to proceed to develop the polygons, let's first take a look at what these shapes are to represent. If we have a number of point objects, such as towns (again, assuming for simplicity that all are the same size), we can imagine that each point is surrounded by a single, irregular polygon. But the polygon has an important property—every location within it is closer to the point encircled than to any other point in the coverage. By contrast, then, every location outside each individual polygon is closer to some other point than to the enclosed point. In other words, the boundaries of each polygon give to the point it surrounds the most compact possible area of influence. In most cases, each point in the coverage will have its own Thiessen polygon, sometimes called a **proximal region,** to show its exclusive area of influence (Clarke, 1990). Now, think about how we might be able to do this.

Let's take a small sample of points to show this (Figure 11.3). With five points as a starting place we might conceptualize how the Thiessen polygons are derived by envisioning a bubble growing outward from each point, much like the bubbles formed from bubble pipes. If you look carefully, you will see that each of the interfaces between bubbles forms a straight line (if viewed orthogonally). This straight line is oriented perpendicular to a line that connects each pair of neighboring points. If you measure the distance between the two end points of the perpendicular line, you will see that the distances are identical on

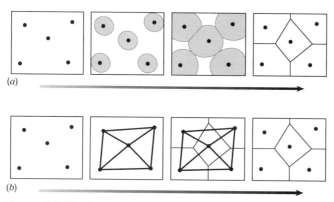

Figure 11.3 Creating Theissen polygons. Construction of a Thiessen polygon from five points: (*a*) arrangement of the data points, (*b*) construction of the related Thiessen polygons. Note the similarity between the form of the Thiessen polygons and the pattern we might get from a bubble pipe. *Source:* Figure derived from Environmental Systems Research Institute, Inc. (ESRI) drawings.

either side of the line forming the interface. In other words, the edges of the polygons are formed by an equidistant (bisector) line perpendicular to the line connecting each pair of points. Algorithms for producing Thiessen polygons have been implemented in both CAC and GIS systems for decades, both in vector systems (Brassel and Reif, 1979), and even in quadtree data structures (Mark, 1987). Those of you inclined to examine details of algorithms might consult those references.

For the rest of us, perhaps the most important question is: What do we do with Thiessen polygons?" We can gain some initial ideas by looking at the reason for their development in the first place. The Thiessen polygon was named after the climatologist A.H. Thiessen, who wanted to be able to interpolate climate data from highly uneven distributions of weather station data. In other words, he was trying to describe and analyze point data with area-based symbols and analytical techniques. Thus if we have only a few scattered points of data, but we want to characterize regions based on these points, we use Thiessen polygons to divide our space into a polygonal form. Because we assume that in each polygon, the influence of the enclosed point is absolute, we can treat the data as a polygonal coverage.

Most uses for Thiessen polygons involve a determination of the influence of point data representing shopping centers, industries, or other economically based activities. If we modify the position of our perpendicular line based on the size or other magnitude of each point, our Thiessen polygons become even more representative of the actual influence of industries or shopping on the surrounding space. With such information, the economic placement specialist can determine, for example, how much of a city's population (based on proximity) will likely frequent a planned shopping center. Thiessen polygons are not restricted to economic geography tasks, however: Hutchings and Discombe (1986) found that this device offered a practical approach to the detection of spatial patterns in vegetation. Actually, the use of this technique is likely to increase as the functional capabilities of the GIS become more well known to the users.

AREA PATTERNS

Until now we have looked only at the distribution of points as they are arranged in a defined space. But, we know that points are not the only objects that can be distributed in a spatial context. If we change our scale somewhat, our points become areas, and their patterns can then be analyzed by similar methods. We can begin to analyze area patterns much as we did with points—by determining the density of polygons per unit area of our study area. In performing a density measure of polygons, however, we must first measure the area of each class of polygons in which we are interested. We can do this in raster or vector, just as we learned in Chapter 4. Then, knowing the area of each category of polygons, we divide the area of each polygon type (i.e., each region) by the overall area of the coverage. This gives a percentage rather than a number per unit area as in point density. It is possible, of course, to calculate the number of polygons (or clumps of grid cells) per unit area, but because of the potential for widely varying polygon sizes, this approach is not likely to be of much use.

Again, with areas, we might be interested in more than just a measure of the density of the polygons. We may very well be interested in the arrangement and pattern shapes created by groups of polygons. Groups of polygons arranged differently in space may suggest causes for such arrangements, much as they do for point patterns. Examples of potentially interacting polygons could be new plowing innovations at certain farms, cities, and towns and the movement of goods and services among and within them, or even water sources distributed within an area that might provide good spots for overwintering birds. But before we can examine the interactions of polygonal objects with one another, we must know something about how they might be arranged. As with point patterns, areas can be clustered, dispersed (regular), and randomly spaced relative to one another (see Figure 2.8). In addition, area patterns can be connected to one another, or separated by some definable distance. We will look at some methods of analyzing these pattern metrics.

Extending Contiguity Measures: The Join Count Statistic

As we work with polygonal coverages we will create many **binary maps,** or maps with only two categories of polygons—most often to help us define particular neighborhoods that characterize a variable as good or bad for a given decision. For example, we can have goodsoils and badsoils for row crops, goodslopes and badslopes for housing, or goodaspects and badaspects for setting up solar energy panels. It might be useful to be able to define the distributional patterns of some of these variables, perhaps because we need to place our homes, crops, or solar panels in a single large group (characteristic of clustered distributions), rather than scattered about or totally dispersed. Or we might be trying to determine the pattern of certain area features such as eroded surfaces, weedy vegetation, or settlement patterns to determine whether some underlying cause might be assigned to the observed patterns.

We have already seen that we have measures of immediate neighborhoods based on their **contiguity,** defined as the condition of polygonal entities in contact with one another. We have seen how contiguity might easily be defined in terms of the amount of area occupied by connected polygons or grid cells. And we saw how these could be ranked to produce new neighborhood classifications (Chapter 9). But, while a simple measure of contiguity might be useful for examining the sizes of connected polygons of a single type, it tells us little about the pattern formed by these regional polygons. For that, we will use the **join count statistic.** This statistic is not limited to binary maps, but because working with binary maps affords a better feel for what it does, and because it is relatively simple (and common) to move from multicategory maps to binary maps (McGrew and Monroe, 1993), we will restrict our discussion to the analysis of binary polygon maps.

By definition, a join (not related to the relational join) is two polygons sharing a common edge or boundary. The join count statistic counts the number of joins in a polygonal pattern and characterizes the join structure for each coverage (McGrew and Monroe, 1993). Take a look at Figure 11.4a, which shows an area with 15 polygons and indicates all the possible neighboring line segments (joins); associated with each polygon as a specific number of joins. We will

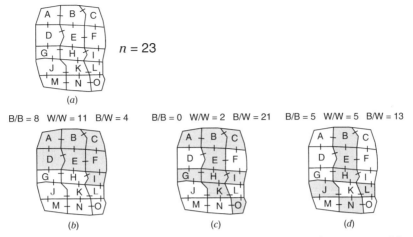

Figure 11.4   Join count statistic. Join count statistics for an area with 15 polygons. (*a*) the 23 possible joins, (*b*) clustered distribution, (*c*) dispersed distribution, and (*d*) random distribution.

conform to accepted practice and identify each of the binary polygons as either black (shaded) or white (unshaded), so we can observe the numbers of connections between each category for each of the three cases: clustered, dispersed, and random.

There are 23 possible joins (i.e., adjacent lines between polygons in each of the polygons in Figure 11.4*a*. In Figure 11.4*b* we find 8 black-to-black joins, 11 white-to-white joins, and 4 black-to-white joins. This count of joins, indicating that there are very few joins between the black and white units, means that most of the white polygons are connected to each other and most of the black polygons are connected to each other. In other words, the polygons are clustered, just as we saw with our points before. Figure 11.4*c* shows a totally different set of numbers; here most of the joins (21 out of 23 possible) are between differently classed polygons—a very dispersed array. Figure 11.4*d* is an intermediate case: both black-to-black and white-to-white joins are low, but not nearly as low as they were in Figure 11.4*c*. The number of like polygons that are connected to each other is high (13 of 23 possible), but again not nearly as high as in the dispersed case. This arrangement clusters some small groups of polygons but disperses others. In other words, it demonstrates a condition of randomly dispersed polygons.

This approach is likely to be most easily performed in a vector GIS built on a topological data model. Each shared boundary is easily identified and could be counted very readily. If, however, you anticipate doing a join count analysis, it might be useful to code the neighboring attributes along each line segment explicitly during the input phase of the GIS. This will expedite the analysis later on, allowing the joins to be enumerated by simply retrieving the line attribute codes and tabulating them (DeMers et al., 1996). Failure to prepare the database in this manner will require substantial manipulation to extract these values during analysis.

Now that we have seen how the join count statistic works, it is important to ask what we might do with the results. We have determined the observed number of like and unlike joins, and we can even conceptualize the differences among three extreme cases. But how would we actually compare the results of a single analysis for one database to what might be expected under clustered, dispersed, or random conditions? Our primary concern is whether the pattern of polygons can be described as having occurred through a random process. Random patterns most likely tell us that the polygons are where they are for reasons not discernible from the patterns themselves. Either of the other two conditions would tell us quite the opposite—that something is causing or controlling the distribution of polygonal patterns.

As with point distributional patterns, the easiest way to determine whether the patterns are random is to compare the observed versus the expected distribution patterns. In the analysis of point patterns we most often deferred to the use of the chi-square statistic. But that statistic assumed that we knew what the expected pattern should be under random conditions. If we knew the expected distributional patterns for polygons (based on the number of joins), we could compare them in much the same way. If we then found the observed count to be significantly different from the expected count, we could call the pattern nonrandom (either regular or clustered).

But how do we know the expected, random pattern of joins against which our observed values could be tested? It turns out there are two general approaches to this problem. The first, called **free sampling,** assumes that we can determine the expected frequency of within- and between-category joins either based on a theory about the types of polygon we are modeling or with respect to the known patterns for a larger study area. In the former case we might, for example, know that because of particular zoning ordinances in a city, shopping centers or industrial parks will occur with predictable regularity compared to other types of land use. We could then compare these patterns to the regularity of shopping areas in another city to see whether common zoning rules are shared or whether the rules are different, resulting in widely different patterns of shopping centers or industrial parks to other land uses. In the second case, where we use a known pattern from a larger area, a similar comparison can be made. Let's say that for our home county we have an established pattern of polygonal joins developed from mapping and analyzing different crop types. Then, later on, we want to scrutinize the patterns for a single township located near an urban area. We can compare the number of joins found for the subarea to that of the larger area to see whether a similar pattern emerges.

The second approach to testing the patterns of polygons, called **nonfree sampling,** is the one that is more often applicable. It makes no assumptions about known theory and does not rely on a comparison of numbers of joins in a subarea and in a larger area. Instead of comparing a known (expected) pattern to an unknown pattern (observed), we instead compare the number of joins to estimates of a random pattern of joins for the polygons that exist. In other words, we create a random pattern of blacks and whites based only on the polygons themselves. Then we can compare our results to our estimated random pattern, with a view to spotting any significant nonrandom distribution that might suggest a causative mechanism.

Both free and nonfree sampling will provide insights into area patterns, but the calculations are more likely to be performed by the advanced GIS user than by the novice. Thus we shall not go into the details of calculating the statistics of comparison. You may, however, wish to consult the studies showing how this was done with respect to Republican and Democratic gubernatorial seats in the eastern United States (McGrew and Monroe, 1993).

Other Polygonal Arrangement Measures

Analyses of polygon arrangements can become quite sophisticated, and software links to a GIS have provided opportunities to perform them (Baker and Cai, 1992; McGarigal and Marks, 1994). The landscape ecology community is a frequent consumer of these techniques, usually treating polygons as patches, especially with respect to a larger, more uniform background, a so-called matrix. You might want to consult Forman and Godron (1986) for an overview of some of these methods as you acquire sophistication in GIS analysis. In general you will find measures of polygonal **isolation,** measures of **accessibility** (connectedness of polygons to lines), polygon **interactions,** and **dispersion.** Since many of these measures are borrowed from the literature in geography, biogeography, ecology, forestry, and other disciplines, the examples will be varied enough to give you a feel for how these additional measures might be used.

LINEAR PATTERNS

Just as we have seen with points and areas, lines occupy space and occur in particular configurations and patterns on the land as well. We see patterns of lines all the time, but often we miss them. Streets and highways form lines with particular definable patterns that we recognize as being related to the networks produced by humans to move people and things between the spatially arranged points called towns. We see fencelines, also occurring in particular configurations and numbers depending on field size, lot size, and shapes of the polygons they trace (Simpson et al., 1994). Striations on exposed bedrock show parallel lines indicating the dragging of rock below the glaciers as they scoured the land thousands of years ago. The mechanisms that caused each of these line patterns can best be understood if we first define the specific patterns and concentrations before us. Next we will look at some of the more common measurements of line patterns available to the GIS professional.

Line Densities

Because of the added dimensionality of lines (as with polygons), especially lengths and orientations, linear analysis is somewhat more complicated than the study of points. Some researchers have tried to examine the distributions of line lengths (Aitchison and Brown, 1969), and some have looked at line spacing (Dacey, 1967; Miles, 1964), much akin to the analysis of nearest neighbors in point distributions (Davis, 1986). We will examine these as well as the larger body of knowledge of line orientation in the next sections. In the meanwhile we need to introduce the simplest measure of line patterning—line density.

Like point and polygon arrangements in an area, lines can occur frequently or infrequently with regard to the space they occupy. To measure this density value for points, which are considered to be dimensionless, we merely divided the number of points by the area. When we worked with polygons, however, we accounted for the area occupied by each polygon by adding all the areas and dividing this sum by the area of the entire map. In the same way, to measure the density of lines (with one dimension), we divide the sum of all the line lengths by the area of the map coverage. Our measure of line density will then be measured in units such as meters per hectare or even miles per square mile. Except for comparing with other values for different regions or for the same region at different time periods, there is not much we can do with this information. We need to know more about the distributional patterns of the lines, just as we did when we were working with points and polygons. Next we will examine the nearest neighbor statistic as it is applied to lines rather than points. You will see that while the calculations are somewhat different because of the added dimensionality nature of lines, the results provide a useful characterization of line patterns.

Nearest Neighbors and Line Intercepts

We have seen how the nearest neighbor statistic can be used to characterize the spatial arrangements of points. The distribution of pairs of lines can be determined in much the same way, although our computations are somewhat more complex because unlike points, lines have dimension. It might seem simple to adjust for this added dimension by simply selecting the center of each line and performing a nearest neighbor on that point alone. Because the linear objects are often of different lengths, however, this procedure is not going to give us a true picture of the arrangement of the lines themselves. From a statistical standpoint it is often considered useful to sample in a random fashion. Following that approach, our first task in measuring nearest neighbors between line objects is to select a random point on each line on the map (or on each line segment if the lines are not straight). Next, the distance between each point and its nearest neighbor is drawn perpendicular to that line (Davis, 1986) (Figure 11.5). Then we measure these distances and calculate the mean nearest neighbor distance for all the distances. As with all nearest neighbor statistics, we need to be able to test this value against a random distribution. Dacey (1967) determined values for expected nearest neighbor distance, expected variance, and standard error in a random distribution of lines. These values allow us to compare the expected and the observed and to generate a statistic against which to test our hypothesis of randomness (Davis, 1986). Should you find such a measure useful, you are referred to Davis (1986) for a description of the equations.

This test will work for most line patterns, whether the features are straight lines or curved lines, but it has some limitations. If lines in your coverage are very sinuous, and especially if they change directions frequently, this approach

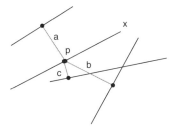

Figure 11.5 Nearest neighbor distances between lines. Finding the nearest neighbor between lines using a randomly selected point along each line as the sample location. *Source:* J. C. Davis, *Statistics and Data Analysis in Geology,* 2nd Ed., John Wiley & Sons, Inc., New York, © 1986. Adapted from Figure 5.16, page 313. Used with permission.

is less than successful. In addition, for the test to be useful the lines should be at least 1.5 times the length of the mean distance between the lines. If the number of lines on the coverage is small, the estimate of density used in nearest neighbor analysis should be adjusted by a weighting factor of $(n - 1)/n$, where n is the number of lines in the pattern. So rather than simply dividing the sum of the lengths by the area, we use the formula

$$\frac{(n - 1)L}{nA}$$

where L is the sum of the lengths and A is the area. This adjusted line density value will improve the quality of the nearest neighbor statistic.

   Line intersect methods are alternative methods for analyzing distributions of lines. One simple approach is to convert the two-dimensional pattern into a one-dimensional sequence by drawing a sample line across the map and noting where the sample line intersects the coverage line objects. There are at least two basic methods of producing sample lines (Getis and Boots, 1978). The first is to randomly select a pair of coordinates and connect them with lines. A second method is to draw a radius at a randomly chosen angle and starting point, then measure a random distance from the center, finally constructing a perpendicular to that radial line (Davis, 1986). Once the random test line has been drawn over the coverage lines, the distribution of intersection intervals can be tested through the use of simple statistics like the **runs test** or other tests of sequences of data. An alternative to a single line is to use a zigzag line that crosses the coverage two or three times. The zigzag path (often called a **random walk**) will again produce a series of line intersects, the distances of which can be tested with any simple statistical test for sequences of data (Figure 11.6).

   If, after performing our nearest neighbor statistic or line intersect method, we find that our distribution is not a random one, there is reason to believe that a process other than random chance contributed to the observed distribution. Here, again, it is useful to consider how the linear patterns might be compared to patterns of other objects in difference coverages (Chapter 12).

DIRECTIONALITY OF LINEAR AND AREAL OBJECTS

Linear objects can exhibit more than just distributional patterns on the landscape. Often we find that such features as sedimentary bedding planes, glacial

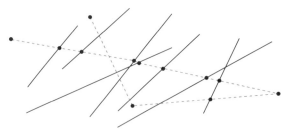

Figure 11.6 Random walk method to evaluate line patterns. Modification of the line intersect using a zigzag line called a random walk to identify sample points. *Source:* J. C. Davis, *Statistics and Data Analysis in Geology,* 2nd Ed., John Wiley & Sons, Inc., New York, © 1986. Adapted from Figure 5.17, page 314. Used with permission.

striations, water-deposited pebbles, glacial boulder trains, fencerows, street grids, and forest blowdowns exhibit a preferred orientation. These features and many more are oriented in a particular direction, often strongly related to a functional force. We know, for example, that tree shelterbelts in agricultural regions are often placed in directions opposite the prevailing winds, that trees felled from hurricanes give evidence of the wind directions occurring within the storm, and that glacial striations are indicative of the movement of the glacier. But when we analyze orientation, we see that if lines are oriented in, say, a predominantly north–south direction, we have two directions to choose from. If these linear objects are one-way streets, however, the orientation doesn't tell us the single direction in which traffic must flow. So, besides orientation we also need to consider directionality. We can also think of the distributions of linear objects as either two-dimensional (i.e., distributed purely in azimuthal directions), or as three-dimensional (i.e., distributed in an azimuthal direction combined with an angular direction in a sphere). While both are important, we will concentrate on the simpler two-dimensional forms and leave the spherical directional and orientation measures for more advanced treatments (Davis, 1986).

In traditional statistical analysis of orientation, the data from maps of linear objects are transferred to a circular graph called a **rose diagram,** which plots all linear features starting at the center of a circle and draws each observation as a single line. In some rose diagrams the length of the line also indicates the magnitude (such as wind speed) or length of the feature (as in the lengths of hedgerows or fences). But while rose diagrams are a useful framework for the visual inspection of directions, the measurements themselves obtained directly from our coverage are much more useful for numerical analysis. Our first analytical technique is to figure out what the **vector resultant** is. This is the procedure your high school text used to describe the usual example of three people pulling in different directions on a large object. In that instance the vector gives us an idea of where the object is likely to move and with what speed.

Let's take an example of directional analysis in two dimensions by considering a large number of trees that were blown down by straight-line winds associated with frontal activity. Each blown-down tree might be plotted as a line feature in a coverage; the positions of the top and bottom of each tree

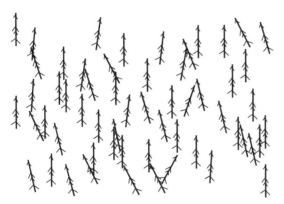

Figure 11.7 Line patterns of felled trees. Map of tree blowdowns showing the general trend in orientation as well as some deviations from the trend.

would be recorded as well, giving us both the orientation and the direction of every downed tree (Figure 11.7). The meteorologists want to deduce the overall direction of the winds on the basis of the felled trees, but the trees are not all perfectly oriented in a single direction. Our first task is to analyze the vector resultant of all the blown-down trees.

The vector for each tree is defined by an angle θ (theta) measured from the base of a tree to its top. We multiply the X coordinate of each tree by the cosine of theta and each Y coordinate by the sine of theta. To find the resultant vector, we sum all these values for both coordinates, and resultant vector values X_r and Y_r show the dominant direction of the end points of all the trees in the blowdown. Figure 11.8 shows a resultant vector of **R** obtained from three values of theta, the vectors **A**, **B**, and **C**.

We may want to go further, however, and determine a mean direction $\bar{\theta}$ based on the resultant vector. As an average of the directions for all the downed trees, it is analogous to the mean of any other set of data. The formula for mean is:

$$\bar{\theta} = \tan^{-1}\left(\frac{Y_r}{X_r}\right)$$

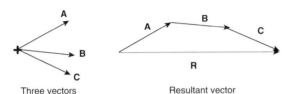

Three vectors   Resultant vector

Figure 11.8 Resultant vector. Resultant vector for three individual linear features. *Source:* J. C. Davis, *Statistics and Data Analysis in Geology,* 2nd Ed., John Wiley & Sons, Inc., New York, © 1986. Adapted from Figure 5.20, page 317. Used with permission.

Because the mean direction of our vectors depends not only on the dispersion of the trees, but also on the number of observations, we can standardize these values by dividing the coordinates of each resultant vector by the number of line objects in the coverage. This will allow us to compare two different areas. For example, by comparing the average vectors for downed trees from two study areas, we could determine whether the winds were generally coming from the same direction.

As with any sample of points in which the mean serves as a measure of the central tendency of the data, or the trend of the data to group around some central point, we can use the average to develop other statistics that define the amount of spread away from the mean. Figure 11.9 shows two cases of three vectors being used to determine the resultant vector **R**. When the vectors are fairly closely spaced, the resultant vector is very long, while the widely dispersed vectors give a resultant vector that is considerably shorter. Think about this in terms of three people pulling on ropes to move an object. The closer the people are and the more nearly they are pulling in the same direction, the more force results, to pull the object along. On the other hand, if the people are pulling in quite different directions, the object may move, but the amount of force applied is considerably less. In other words, the resultant vector is smaller.

We can determine the length of the resultant force by using the Pythagorean theorem on our resultants (X_r and Y_r). The formula is simply:

$$R = \sqrt{X_r^2 + Y_r^2}$$

where R is the **resultant length.** Thus, not only do we know the average direction in which the trees lie, but we also have a measure of the compactness of the distribution. The more compact the distribution, the longer the line. Again, this is analogous to three ropes pulling an object. The closer together the ropes (in other words, the more closely spaced the vectors), the stronger the force (or the greater the length of the resulting vector).

To compare the resultant length of our vector to that of another site, we again need to standardize the data. We find the **mean resultant length** $\bar{R}$ by dividing the resultant length R, found with the Pythagorean theorem, by the number of observations, n. This value will range from zero to one. The mean

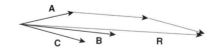

Closely spaced around a single direction

Widely dispersed vectors

Figure 11.9 Vector dispersal. Resultant vectors determined when individual lines are close together and widely spread. *Source:* J. C. Davis, *Statistics and Data Analysis in Geology,* 2nd Ed., John Wiley & Sons, Inc., New York, © 1986. Adapted from Figure 5.21, page 318. Used with permission.

resultant length resembles variance in linear statistics inasmuch as it is a measure of spatial dispersion about a mean value. The value of mean resultant length, however, is expressed in the opposite direction: that is, large values indicate that the lines are closely spaced and small values mean they are more dispersed in direction. Thus a large mean resultant value in our tree example would mean that the winds were nearly perfectly unidirectional, while a small value would suggest eddies in the overall wind pattern. If the inverse nature of the mean resultant length seems counterintuitive, you can use its complement, defined by 1 minus the mean resultant length. This value, called the **circular variance,** will give increasing values as the spread of vectors increases. If you are statistically inclined, you might see some possibilities for directional analogs to standard deviation, mode, and median. Gaile and Burt (1980) provide equations for such measures.

One final problem remains with respect to orientation. As we saw earlier, the orientation of any linear object gives two possible (and opposite) directions. Unlike our trees, however, some features can be expressed in either of two directions. Let us say, for example, that two researchers are looking at different portions of a set of fencerows. One researcher states that most of the fences are oriented toward the north. The second, however, says they are oriented toward the south. Both are correct: that is, if we took one vector that was oriented directly north and the other directly south and used our equations for mean vector length, the vectors would tend to cancel each other.

Fortunately, Krumbein (1939) found a simple way to resolve this problem. He observed that when he doubled the angles, no matter which direction was originally used to record the data, the same angle was recorded. Suppose that a fencerow is oriented northwest to southeast (that is, 315 degrees to 135 degrees). If we multiply each of these values by 2 we obtain the same number: $315 \times 2 = 630$ degrees (-360 degrees $= 270$ degrees) and 135 degrees $\times 2 = 270$ degrees. The calculations for mean direction, mean resultant length, and circular variance will now be doubled. To get our true values for these measures, all we need to do is divide by 2.

These simple measures of directionality and dispersion can be tested for randomness (Batschelet, 1965; Gumbel et al., 1953) and for specific trends (Stephens, 1969) through standard processes of hypothesis testing. Other sets of directional data can be compared to those for our original coverage (Gaile and Burt, 1980; Mardia, 1972). These tests are beyond the scope of this book; the works just cited offer detailed discussion of these more complex techniques.

In the context of GIS, we are most likely to use the measures simply to help us characterize the distributions inside our coverage. Or, more likely, we will use them to discover useful comparisons to other coverages to help us understand the causes. Commercial GIS frequently provide tools that will allow us either to calculate these directional statistics directly or to pass preliminary data from the GIS to other software designed specifically for such analyses. You should be aware, however, that an increasing number of software systems, especially among the geosciences, are designed for handling directional statistics. Raster GIS systems are not particularly well suited to this type of analysis, but most topologically based vector GIS systems will at least allow you to compile some of the preliminary values (e.g., angles for line segments), which can be stored in the database system directly as attribute values that can be passed to other software systems.

CONNECTIVITY OF LINEAR OBJECTS

An important aspect of the spatial arrangement of lines is their ability to form networks. Networks occur in many different forms and are both natural and anthropogenic. Roads and rail lines are networks that serve to move and transport people and materials from place to place. Telephone lines are networks to allow the movement of information and communication. Rivers and streams are networks for aquatic creatures, and even hedgerows may act as networks to allow the movement and migration of small mammals through a landscape. The list of systems that may be defined as networks is quite large. And while the density of such features and their orientation may be of interest to us, we need to be able to analyze the actual connections made by these features and the amount of connectivity provided from place to place. Most of you have probably encountered the frustration of traveling in a large city or across country and finding that a direct path to a location you wish to visit does not exist. Even a quick look at a road map of North America shows a great many places that can be visited only through very long and indirect paths. That is, the areas are characterized by a relative lack of **connectivity.**

Connectivity then is a measure of the complexity of a network. There are several devices for calculating this value (Haggett et al., 1977; Lowe and Moryadas, 1975; Sugihara, 1983; Taaffe and Gauthier, 1973). The two most common ones are the **gamma index** and the **alpha index.**

The gamma (γ) index compares the number of links L in a given network to the maximum possible number of links between nodes. To calculate this, we simply produce a ratio of the two. First we count the number of links (line segments between nodes) that actually exist in the coverage. As you might guess, this is very difficult in raster GIS, but easily accomplished in vector. Once we have the number of actual links, we need to determine the number of possible links in the coverage. This is not as difficult as it might appear because there are limits to the number of links, determined entirely by the number of nodes. If, for example, we have three nodes present, only three links are possible (Figure 11.10), but if we add another node we see that three additional links are possible, for a total of six (Forman and Godron, 1987). So if we assume that no new intersections (line crossings) are formed, the maximum number of possible links is increased by three each time (Figure 11.10). In other words, the maximum number of links L_{max} is always $3(V - 2)$, where V is the number of nodes.

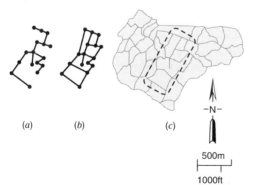

(a)   (b)   (c)

-N-

500m

1000ft

Figure 11.10 Gamma index. Two different networks of hedgerows based on the same set of nodes. The network on the left (a) is minimally connected and has no circuits, while the network on the right has higher connectivity and has circuits that provide alternate routes for travel. (From Forman and Godron, 1987.) *Source:* R. T. T. Forman and M. Godron, *Landscape Ecology,* John Wiley & Sons, Inc., New York, © 1986. Adapted from Figure 11.12, page 418. Used with permission.

To find the gamma index, we simply divide the number of links L by the maximum number of links L_{max}.

$$\gamma = \frac{L}{L_{max}} = \frac{L}{3(V-2)}$$

The gamma index will range from a low of 0, where none of the nodes is connected, to 1.0, where all possible links between all nodes are present (Forman and Godron, 1987).

Figure 11.10, illustrating a network with 16 possible nodes, reveals the difference in connectivity for two different cases. The case on the left shows 15 links among the 16 nodes. This gives us a connectivity of

$$\gamma = \frac{L}{3(V-2)} = \frac{15}{3(16-2)} = \frac{15}{42} = 0.36$$

While the case on the right has 20 links for the 16 nodes, yielding a connectivity of

$$\gamma = \frac{L}{3(V-2)} = \frac{20}{3(16-2)} = \frac{20}{42} = 0.48$$

So the first network is about one-third connected (36%), while the second is almost half-connected (48%).

In Figure 11.10 we were looking at hedgerows, which might be used as corridors for small mammals. The same arrangement of features could easily have been a road net developed for human transportation. The results would have been the same. In the first network the route from the starting point at R to the ending point at S would be relatively long, compared to that for the second network with its increased connectivity. If we were transportation planners, this difference should suggest that increased connectivity also will improve the movement of goods and services between nodes.

However, connectivity is not enough to fully characterize networks. Recall our discussion of the formation within networks of circuits—loops providing alternate paths for travel from node to node. Perhaps the best example of a circuit would be a beltway, one of the circular highways surrounding urban areas in North America. As we travel across the country, we must decide whether to travel through or around each city. If we choose to go around, thus avoiding the urban traffic, we have two possible routes: we can turn either left or right, to go around the city counterclockwise or clockwise. Thus, the loop feature offers choices for travel, allowing us to select the shortest route to a destination on the other side of the city.

The index designed to measure **circuitry,** the degree to which nodes are connected by circuits of alternative routes, is called the alpha (α) index. Similar to the gamma index, this measure is a ratio of the existing number of circuits to the maximum possible number. We see that a network with no circuits present contains one link fewer than the number of nodes: ($L = V - 1$). If you look again at Figure 11.10, you see that this is true where the first network (a) has 15 links and 16 nodes. This figure is minimally connected, in that it has the fewest number of links possible, given the number of nodes, but each is still connected to at least one link. By adding a link to this network, we create a circuit or loop. As such, when a circuit is present $L > (V - 1)$. Therefore, the number of circuits present in the network (i.e., the number of links present minus the number of links in our minimally connected network) can be given by $(L - V) + 1$. Thus we know the actual number of the circuits in the network (Forman and Godron, 1987).

An alpha index for the amount of circuitry is given by the ratio of the actual number of circuits present to the maximum number possible in the network. Therefore, since the number of possible links is determined by $3(V - 2)$ and the minimum number of links in a minimally connected network is expressed as $V - 1$, we can determine the maximum possible number of circuits (by combining these two simple formulas). That is, we subtract the number of links in the minimally connected network, $V - 1$, from $3(V - 2)$, the maximum number of links, and from $3(V - 2) - (V - 1) = 2V - 5$, we obtain the final formula for the alpha index:

$$\alpha = \frac{\text{actual number of circuits}}{\text{maximum number of circuits}} = \frac{(L - V) + 1}{2V - 5}$$

As before, the values for the alpha index range between 0 (a network without any circuits) and 1.0 (a network with the maximum number of circuits present).

If we again use the network from Figure 11.10, the alpha value for Figure 11.10a is

$$\alpha = \frac{(L - V) + 1}{2V - 5} = \frac{(15 - 16) + 1}{(2 \times 16) - 5} = \frac{0}{27} = 0$$

showing us that there are no circuits present. And the alpha value for Figure 11.10b is

$$\alpha = \frac{(L - V) + 1}{2V - 5} = \frac{(20 - 16) + 1}{(2 \times 16) - 5} = \frac{5}{27} = 0.19$$

showing that there is 19% circuitry out of a possible 100%. So in Figure 11.10a there are no options for travel; movement from point R to point S has only a single option. For Figure 11.10b, on the other hand, we have numerous alternative routes, some longer than others, to get from R to S.

Because of the addition of links needed to produce circuits, it is entirely possible to consider the alpha index as an alternative method of measuring connectedness. However, because the alpha and gamma indices give us a different look at the overall network patterning, they might more appropriately be combined to provide an overall measure of the amount of **network complexity.** To perform the alpha or the gamma index in a GIS requires that the GIS be a vector system. This relationship between these two indices and the vector GIS model is particularly strong because, like the vector GIS model itself, the indices are based on topological properties. In this case, the primary topological foundations are based on **graph theory,** which, as we saw when we looked at the DIME topological model, is concerned less with the lengths of lines or their shapes than with the degree of connection between them.

Of course, for modeling transportation, we need to know more than just how connected the network is—for example, the lengths of the lines between nodes, something about the direction in which travel can proceed along these lines, and impedance values for these lines. In addition, there are numerous other simple indices, derived mostly from transportation and communication theory, for characterizing the connectivity of networks. We can, for example, calculate the linkage intensity per node or the number of alternate routes between nodes; we can determine which node has the most links (**central place**), and we can ascertain delineations of regions based on connectivity and accessibility (Haggett et al., 1977; Lowe and Moryadas, 1975). And, of course, all these could be combined with each other and with the measures of linear dispersion, arrangement, and orientation to give a more detailed view of networks. As you work more with your GIS, you will see how these approaches might be applied, and you can consult the references at the end of this chapter to learn more. In the meantime we will look at one last measure that is in common use for modeling relationships among linear objects, based somewhat on the connectivity we have just examined.

GRAVITY MODEL

In our discussion of connectivity we made the simplifying assumption that all the nodes were of equal size or importance. This allowed us to concentrate on the degree to which the links connected these nodes or the number of circuits and alternate paths available. However, think about the nature of towns and the nature of cities. Large cities are obviously more likely to provide more opportunities for shopping, more access to the arts, and more occasions to attend professional sporting events. This is why we tend to go to cities for shopping and entertainment. But towns and cities are not the only pairs of nodes that can exhibit different amounts of attraction. Take, for example, a small pond versus a lake. The lake is likely to attract more waterfowl than will be found near the pond.

Regarding the two examples of contrasting nodes, you should note that we have defined a magnitude for each that we can use to separate them out in a hierarchical fashion. Both large cities and lakes have a greater likelihood of drawing activity, whether it is performed by ducks or people, than their smaller counterparts. The magnitude of the attraction can be thought of much as astronomers envision the gravitational attraction between two celestial bodies. The larger the body, the greater the gravitational attraction between itself and neighboring bodies. We know this is true because astronauts who landed on the lunar surface were experiencing a gravitational attraction of only about one-sixth that of earth.

Converting the concept of gravitational attraction to use in a two-dimensional setting provides the measure of the interaction between two nodes in a GIS coverage that we call the **gravity model.** In its simplest form, the gravity model takes the form of the following formula:

$$L_{ij} = K\frac{P_iP_j}{d^2}$$

where L_{ij} is the interaction between nodes i and j, P_i is the magnitude of node i, P_j is the magnitude at node j, d is the distance between the two nodes, and the constant K relates the equation to the types of object being studied (population size, animals, etc.). The values of P can be represented by a force of interaction between the nodes such as the demand for products, the amount of retail sales for city shopping centers, or the amount of wetland habitat for waterfowl.

As with gravity, the larger the magnitude of the nodes, the greater the interaction between them is likely to be. Conversely, the greater the distance between the two nodes, the less attraction there is between them. You might again think about the draw to a large city for shopping as an example. The larger the city, the more shopping it is likely to offer, but the farther away you are from a given city, the less likely you are to visit it regardless of the bargains available.

Some GIS software systems, both raster and vector, have the capability to develop gravity models between points. There are actually quite a number of variations on the simple model you have seen here, but they are best studied in more advanced texts on spatial analysis. And while most gravity models are used for economic placement analysis, much like Thiessen polygons, there are quite a number of possible applications. Investigators have used gravity models to describe the flow of passengers by air and land between cities, the volume of telephone calls, flows of birds, and the seeds the birds disperse among woodlots (Buell et al., 1971; Carkin et al., 1978; McDonnell, 1984; McQuilkin, 1940; Whitcomb, 1977). As long as your concern is the movement of objects between nodes of different magnitude, the gravity model may be of use to you.

ROUTING AND ALLOCATION

Among the most useful applications of networks in GIS are the related tasks of **routing** and **allocation.** Routing involves finding the shortest path between any two (in the simplest case) nodes in a network (Figure 11.11). Because the nodes can be assigned weighting factors as in the gravity model, a route might be between one point and the nearest point with the highest weight (as in the amount of demand for a product). (See Lupien et al., 1987, for an excellent discussion of routing and allocation.)

Each link in the net can also be assigned an impedance value, much like a friction surface, but imposed only on the line itself. The impedance value might be related to a speed limit along a street. By using an accumulated distance (see Chapter 8 for a discussion), based on a combination of calculated distance and the impedance factor, the most efficient route can be found, rather than just the shortest. Nodes can also be coded with stops (indicative of traffic signals or stop signs), turning impedances based on the difficulty of turning left or right at an intersection, and even barriers preventing movement and forcing traffic along another path. As with calculating distances along surfaces, all these measures require prior knowledge of the nature of the streets, intersections, and other nodes. Frequently the weights and impedances are somewhat arbitrary or are based on intuitive knowledge, rather than absolute certainty of how they will affect travel.

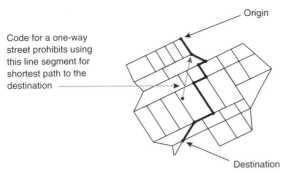

Code for a one-way
street prohibits using
this line segment for
shortest path to the
destination

Origin

Destination

Figure 11.11   Shortest path along a network. A
simple road network showing the solution for the
shortest path algorithm: notice the code entered
along the diagonal road to indicate this one-way
street is going in the wrong direction for it to be
usable for the shortest path solution.

Although routing can be done in raster, it is much easier if performed in a
vector system. Because of the close relationship of graph theoretic topology
and the topological vector data model, the two perform well together. You
should also be aware that given the many possible routes, especially where
circuits are involved, there are likely to be a number of possible ways for finding
your route. A discussion of these could fill a volume on its own.

Allocation is a process that can be use to define, for example, the location
of a market, the areal extent of a water treatment center, or the boundaries of
a series of fire station service areas. A network structure within a vector GIS is
most often used. The idea here is that the capacity of a given service is distributed throughout the net. Each link in the network has a specified number of
items that must be served. For example, each street segment contains a number
of homes that would need to receive water service from the water treatment
center. Or each house could be thought of as a fire-prone structure that may
need the services of the nearby fire station (Lupien, et al., 1987). In addition,
each service or market center is reasonably capable of operating at a certain
maximum capacity. So the maximum number of units that each service area
can accommodate is compared to the number of service sites that occur along
the network links. Say, for example, that a fire station can service only 100
homes with a reasonable level of safety. Beginning at the service location, the
GIS will travel outward, following the nearest roads, counting the number of
houses that occur along those links until it reaches its capacity of 100. If each
street contains 10, the GIS will designate the 10 nearest streets (each with 10
homes) to ensure efficient allocation of the maximum capacity for the service
center (Figure 11.12). The specified amount of route might include the number
of people to service (i.e., number of residents per house), the road miles that
can be traveled, and a travel time limit that must not be exceeded. If our road
net were completely uniform (no impedances, no stops, no changes in speed
limit, etc.), calculating this allocation would be a simple matter of deciding on
the criteria for allocation and pushing our boundary outward until they are
met. For example, if we wanted to allocate the routes for newspaper carriers

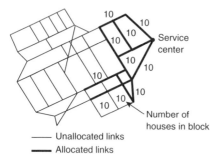

Unallocated links
Allocated links

Figure 11.12 Allocations of networks. A simple road network showing how links containing 10 homes each are allocated to a service center (say a fire station) that can service only 100 homes safely.

so that each person's vehicle would have to travel only a certain number of road miles, the software would simply add up the road miles as each route spreads outward from the starting point, until that value is reached, whereupon the links in the road that were allocated would be assigned attribute codes to correspond to the different carriers.

As you might guess, most real allocation problems are quite complicated. In some cases our primary interest might be the number of people we can service in a given amount of time. If you were planning to allocate newspaper routes, you would need, for example, to be able to distribute newspapers to the subscribers by a specified time of day, say 9 A.M. And, of course, no road network is totally uniform. Some roads have higher speed limits than others, some have stop signs, some include turns that are slower than others, and so on. And, of course, you need to know that individual addresses are somehow connected to the streets you intend to service (the process of linking addresses to streets is called **address matching**). Most of the work of determining these allocations requires us to encode all the appropriate attributes, impedance values, turning impedances, and other variables. Once this has been done, the software is quite capable of providing a reasonable result. You must, of course, watch out for missing or incorrectly coded attributes along the lines, or incorrect address matches between individual homes and the streets. The time required to explain the possible methods, applications, pitfalls, and remedies would probably fill a book the size of this one, so, I leave you to experience these on your own when you begin working with linear objects in a transportational setting.

THE MISSING VARIABLE: WHY WE NEED TO USE OTHER COVERAGES

Only a single, simple reminder is necessary at this point. We have seen how points, areas, and lines can be examined for their arrangements, connectedness, and orientations. And we have seen how we might be able to characterize neighborhoods of these collections of objects. Your ability to use the information you have teased out of these analyses is often only as good as the comparable data available in other coverages. This is especially true if you are trying to determine the relationships of objects to a causative mechanism. In the next chapter we will learn how comparisons among different coverages can be made. There are far more than you might imagine, and their variety should provide ample opportunity to test your abilities to relate variables to one another.

Terms

spatial arrangement
density
uniform
regular
random
clustered
quadrats
quadrat analysis chi-
  square
variance–mean ratio
  (VMR)
nearest neighbor analysis
Thiessen polygons
Dirichlet diagrams
Voronoi diagrams
proximal region

binary maps
contiguity
join count statistic
free sampling
nonfree sampling
isolation
accessibility
interactions
dispersion
line intersect methods
runs test
random walk
rose diagram
vector resultant
resultant length

mean resultant length
circular variance
connectivity
gamma index
alpha index
circuitry
network complexity
graph theory
central place
gravity model
routing
allocation
address matching

Review Questions

1. What do we mean when we speak of an interest in arrangements of objects? Why is this knowledge useful to us?

2. How do we define a uniform distribution? What makes a uniform distribution regular versus random? What is a clustered distribution?

3. What is the significance of regular, random, and clustered distributions in terms of the processes that could have caused them? Give some examples of each distributional pattern and give a hypothesis that might describe the causes.

4. What is the purpose of quadrat analysis? Describe how it is done. What statistical test do we use to confirm or reject our results?

5. What is the purpose of variance–mean ratio in point distributions? How does VMR differ from traditional quadrat analysis?

6. What does nearest neighbor analysis tell us that quadrat analysis does not? How are the two techniques similar?

7. What is the purpose of Thiessen polygons? Diagram some samples and describe how they are created.

8. Describe the join count statistic. What does it tell us? How does it differ from a simple measure of contiguity? What is the difference between free

sampling and nonfree sampling as these apply to testing the results of the join count statistic?

9. Describe the process of extracting a nearest neighbor statistic from line patterns. What does this quantity tell us about the distribution of our lines? Describe some situations in which the nearest neighbor statistic for lines might give misleading results.

10. Describe the use of line intersect methods for analyzing line pattern distributions. What is a random walk?

11. What is a vector resultant? What does it tell us about the patterns of linear objects? What is the resultant length? How does it compare to resultant force problems in physics?

12. What is the mean resultant length? How does it differ from the resultant length? When would we use it? What would a large mean resultant length tell us? What is the difference between circular variance and mean resultant length?

13. How can we adjust our measurements for mean direction, mean resultant length, and circular variance to account for orientations that can be measured in either of two directions?

14. Describe how the gamma index is performed. What does it tell us about our network? What does a gamma value of .48 tell us in terms of the amount of connectedness in our network?

15. What is the alpha index? How is it different from the gamma index? How is it the same? Give some nonroad–network examples of how both of these indices might be useful.

16. Describe, in general terms, what the gravity model is. How does distance affect the interactions of point objects? How does the magnitude of the nodes change the interactions between them? Give an example of how the gravity model might be applied.

17. Describe the types of attribute that might need to be assigned to the nodes and the arcs between nodes in routing and allocation problems.

References

Aitchison, J., and J.A.C. Brown, 1969. *The Lognormal Distribution, with Special Reference to Its Uses in Economics.* Cambridge: Cambridge University Press.

Baker, W.L., and Y. Cai, 1992. "The r. le Programs for Multiscale Analysis of Landscape Structure Using the GRASS Geographical Information System." *Landscape Ecology,* 7(4):291–301.

Batschelet, E., 1965. Statistical Methods for the Analysis of Problems in Animal Orientation and Certain Biological Rhythms. American Institute of Biological Sciences Monograph. Washington, DC: AIBS.

Brassel, K.E., and D. Reif, 1979. "A Procedure to Generate Thiessen Polygons." *Geographical Analysis,* 11(3):289–303.

Buell, M.F., H.F. Buell, J.A. Small, and T.G. Siccama, 1971. "Invasion of Trees in Secondary

Succession on the New Jersey Piedmont." *Bulletin of the Torrey Botanical Club,* 98:67–74.

Carkin, R.E., J.F. Franklin, J. Booth, and C.E. Smith, 1978. Seeding Habits of Upper-Slope Tree Species. IV. Seed Flight of Noble Fir and Pacific Silver Fir. U.S. Forest Service Research Note PNW-312. Portland, OR: USFS.

Clarke, K.C., 1990. *Analytical and Computer Cartography,* Englewood Cliffs, NJ: Prentice Hall.

Dacey, M.F., 1967. "Description of Line Patterns." *Northwestern Studies in Geography,* 13:277–287.

Davis, J.C., 1986. *Statistics and Data Analysis in Geology,* 2nd ed. New York: John Wiley & Sons.

DeMers, M.N., J.W. Simpson, R.E.J. Boerner, A. Silva, L.A. Berns, and F.J. Artigas, 1996. "Fencerows, Edges, and Implications of Changing Connectivity: A Prototype on Two Contiguous Ohio Landscapes." *Conservation Biology,* 9(5):1159–1168.

Forman, R.T.T., and M. Godron, 1986. *Landscape Ecology.* New York: John Wiley & Sons.

Getis, A., and B. Boots, 1978. *Models of Spatial Processes, An Approach to the Study of Point, Line and Area Patterns.* Cambridge: Cambridge University Press.

Gaile, G.L., and J.E. Burt, 1980. Directional Statistics: Concepts and Techniques in Modern Geography. Geo Abstracts, No. 25, University of East Anglia, Norwich, England.

Griffiths, J.C., 1962. "Frequency Distributions of Some Natural Resource Materials." *Pennsylvania State University, Mineral Industries Experiment Station Circular,* 63:174–198.

Griffiths, J.C., 1966. "Exploration for Natural Resources." *Journal of the Operations Research Society of America,* 14(2):189–209.

Gumbel, E.J., J.A. Greenwood, and D. Durand, 1953. "The Circular Normal Distribution: Tables and Theory." *Journal of the American Statistical Society,* 48:131–152.

Haggett, P., A.D. Cliff, and A. Frey, 1977. *Locational Analysis in Human Geography.* New York: John Wiley & Sons.

Hutchings, M.J., and R.J. Discombe, 1986. "The detection of spatial pattern in plant populations," *Journal of Biogeography,* 13:225–236.

Krumbein, W.C., 1939. "Preferred Orientation of Pebbles in Sedimentary Deposits." *Journal of Geology,* 47:673–706.

Lowe, J.C., and S. Moryadas, 1975. *The Geography of Movement.* Boston: Houghton Mifflin.

Lupien, A.E., W. H. Moreland, and J. Dangermond, 1987. "Network Analysis in Geographic Information Systems," *Photogrammetric Engineering and Remote Sensing,* 53(10):1417–1421.

Mardia, K.V., 1972. *Statistics of Directional Data. London: Academic Press Ltd.*

Mark, D., 1987. "Recursive Algorithm for Determination of Proximal (Thiessen) Polygons in Any Metric Space." *Geographical Analysis,* 19(3):264–272.

McDonnell, M.J., 1984. "Interactions Between Landscape Elements: Dispersal of Bird-disseminated Plants in Post-agricultural Landscapes." In *Proceedings of the First International Seminar on Methodology in Landscape Ecological Research and Planning,* Vol. 2, J. Brandt and P. Agger, eds. Roskilde, Denmark: Roskilde Universitetsforlag GeoRuc, pp. 47–58.

McGarigal, K., and B.J. Marks, 1994. FRAGSTATS, Spatial Pattern Analysis Program for Quantifying Landscape Structure, Version Two. Forest Science Department, Oregon State University, Corvallis.

McGrew, J.C., and C.B. Monroe, 1993. *Statistical Problem Solving in Geography.*, Dubuque, IA: Wm. C. Brown Publishers.

McQuilkin, W.E., 1940. "The Natural Establishment of Pine in Abandoned Fields in the Piedmont Plateau Region." *Ecology,* 21:135–149.

Miles, R.E., 1964. "Random Polygons Determined by Lines in a Plane, I and II." *Proceedings of the National Academy of Sciences,* 52:901–907, 1157–1160.

Ripley, B.D., 1981. *Spatial Statistics.* New York: John Wiley & Sons.

Simpson, J.W., R.E.J. Boerner, M.N. DeMers, L.A. Berns, F.J. Artigas, and A. Silva, 1994. "48

Years of Landscape Change on Two Contiguous Ohio Landscapes." *Landscape Ecology,* 9(4):261–270.

Sugihara, G., 1983. "Peeling Apart Nature." *Nature,* 304:94.

Stephens, M.A., 1967. "Tests for the Dispersion and for the Model Vector of a Distribution on a Sphere." *Biometrika,* 54:211–223.

Stephens, M.A., 1969. "Tests for Randomness of Directions Against Two Circular Alternatives." *Journal of the American Statistical Association,* 64(325):280–289.

Taaffe, E.J., and H.J. Gauthier Jr., 1973. *Geography of Transportation.* Englewood Cliffs, NJ: Prentice-Hall.

Whitcomb, R.F., 1977. "Island Biogeography and 'Habitat Islands' of Eastern Forest." *American Birds,* 31:3–5.

CHAPTER 12

Comparing Variables Among Coverages

The process of overlay requires both graphic and attribute comparisons. Some of the details can become quite complicated, especially the computer graphics associated with vector overlay. The following graphic descriptions of the overlay process should give you a feel for how the computer accomplishes vector overlay. Rather than detailing what is essentially computer graphics rather than geography, the chapter inputs a general, conceptual understanding. You are encouraged to consult the numerous texts on computer graphics; or you should consider taking a course or two if your interests lie in this direction.

As you approach this chapter, imagine variations on the examples that are included. Based on your own interests or needs, try to see relationships between these examples and the kinds of overlay typically used in your area. Seek to determine which general types of overlay operation might be useful to you for your own field of study. Keep in mind that this chapter does not detail all possible methods of logically or mathematically combining coverages. Instead, it focuses on a few that are readily understandable.

Experiment! If you are using software for your course, try out different methods on the same set of data. Relate such trials to specific questions you are asking, to see which method gives you the best answer. There is no substitute for experience. And experience gained on even the simplest GIS packages is easily transferable to the larger, more powerful commercial software. You will find that the amount of time required to learn the largest system will be greatly reduced if you learn how overlay is done conceptually. After that, it is merely a matter of finding the documentation that says something like "OVERLAY" and determining which buttons will execute the correct approach for a particular type of comparison among cartographic variables.

Upon completing this chapter you should be able to:

1. Understand the relationship between manual and automated overlay approaches.

319

2. Describe the overall advantages of overlay as a GIS analytical technique, and describe the advantages and disadvantages of raster versus vector overlay approaches.

3. Describe examples of the use of point-in-polygon and line-in-polygon overlay operations.

4. Understand the difference between CAD-type overlay operations and GIS-type overlay operations.

5. Describe how dasymetric mapping could be used in GIS.

6. Understand the problems associated with overlay operations and be able to discuss the types of error associated with them.

THE CARTOGRAPHIC OVERLAY

Among the most powerful features of the modern geographic information system is its ability to place the cartographic representation of thematic information of a selected theme over that of another. This process, commonly called **overlay,** is so intuitive that its application long preceded the advent of modern electronic geographic information systems. A very early application of the analog overlay technique dates back to the days of the American Revolution. During the Battle of Yorktown in 1781, George Washington engaged the services of the French cartographer Louis-Alexandre Berthier, who employed the use of map overlay to examine the relationship between Washington's troops and those of the British general, Charles Cornwallis (Wikle, 1991). This was certainly not the first use of the manual overlay operation, but by citing it I can easily document that the idea has been around for many years.

A readily understandable military application also illustrates the inherent usefulness of graphically displaying functionally related mapped information. It was natural for Washington to think in terms of the functional relationship between the locations of his troops and those of his adversary. The process of overlay showed the spatial correspondence between the two groups, and knowledge of the relative locations of American and British troops was recognized to be essential to making decisions about troop movements. In other words, the spatial correspondence could be directly related to cause and effect. However, this is not always the case in showing mapped correspondences. Take the following example.

A novice GIS user sits down at a computer workstation containing dozens of map coverages for his county. The coverages include land values, zoning, soil types, soil engineering properties, land use, roads and other transportation types, natural vegetation, hospitals, fire departments, and schools. After being shown how to display and overlay maps the novice wants to try out the capabilities of the system. He peruses the maps and discovers that there appears to be a nearly total spatial correspondence between land that is very low in value and the distribution of old growth oak/hickory forest vegetation. One conclusion this overlay operation might foster is that oak/hickory forest contributes to poor quality soil which, in turn, results in low land values. While this might conceivably be true, at least true in the sense that the existence of oak/hickory forest on those land parcels makes that acreage costly to develop because the trees have to be removed, it also is possible that other factors contributed to the correspondence observed. Thus when our novice overlays a land value map with a map of land use, he is rewarded with a particular land use—in this case, forestry—that corresponds nearly exactly to the region occupied by the old growth forest. Not surprisingly, although this land would be worth money to a lumber company, its cost per acre is much lower than that associated with nearby industrial parks, shopping malls, multiple family housing, and the like.

Many other correspondences are certainly possible from the example coverages above. Beyond the close relationship between forestry and oak/hickory forest there are also likely to be correspondences between land values and transportation, between land use and zoning regulations, even between major road intersections and the locations of certain service industry outlets such as the gas stations. Many of these data show at least some spatial correspondences, and many will be related to yet other factors by means of some causal mechanism. It is important, however, to be aware of the enormous potential to make cause-and-effect judgments based on visual correspondence only. Because the map is so likely to be interpreted as a true picture of reality, it is doubly important to obtain evidence that a true cause-and-effect relationship exists before proceeding on that assumption.

The conceptual framework behind spatial correspondence of different phenomena has been available for some categories of mapped data, but certainly not all. Sauer (1925), for example, established a model for the interactions of general categories of earth-related data in his work on the morphology of landscape. His research established the existence of strong correlations among human activity, landform, and other physical parameters. Although Sauer did not formalize these correlations for use with maps, it was clear that he saw connections among distributions of these phenomena on the surface of the earth. Later researchers formalized this design into a wide variety of approaches named, for example, sieve mapping (Tyrwhitt, 1950) and biophysical mapping (Hills et al., 1967).

Perhaps among the most influential practitioners of this early developing approach was Ian McHarg, whose work relating environmental phenomena has spawned an entire school of thought among today's landscape architects, allowing considerably more work to be performed using the computer than was originally possible from field observation and single coverage cartography alone (Simpson, 1989). McHarg (1971) retained an analog method, applying clear acetate overlays of selected mapped environmental phenomena to allow him to evaluate the environmental sensitivity of an area. The darker the area of each coverage, the higher the sensitivity. As more coverages were placed on top of one another, the areas of high environmental sensitivity became progressively darker. The area was then remapped showing a gradation from higher to lower sensitivity, and decision makers could use the map product in assessing available alternatives (Figure 12.1; Steinitz et al., 1976).

There is still much to learn about the causal relationships among spatial phenomena. This subject comprises a large part of what geographers do. Geographic information systems are now providing easily available map overlay procedures that may result in the development of new hypotheses, new theories, perhaps even some new laws about these pattern similarities. Many spe-

SLOPE
ZONE 1 Areas with slopes in excess of 10%.
ZONE 2 Areas with slopes less than 10% but in excess of 2½%.
ZONE 3 Areas with slopes less than 2½%.

SURFACE DRAINAGE
ZONE 1 Surface-water features—streams, lakes, and ponds.
ZONE 2 Natural drainage channels and areas of constricted drainage.
ZONE 3 Absence of surface water or pronounced drainage channels.

SOIL DRAINAGE
ZONE 1 Salt marshes, brackish marshes, swamps, and other low-lying areas with poor drainage.
ZONE 2 Areas with high-water table.
ZONE 3 Areas with good internal drainage.

BEDROCK FOUNDATION
ZONE 1 Areas identified as marshlands are the most obstructive to the highway; they have an extremely low compressive strength.
ZONE 2 The Cretaceous sediments: sands, clays, gravels; and shale.
ZONE 3 The most suitable foundation conditions are available on crystalline rocks: serpentine and diabase.

SOIL FOUNDATION
ZONE 1 Silts and clays are a major obstruction to the highway; they have poor stability and low compressive strngth.
ZONE 2 Sandy loams and gravelly sandy to fine sandy loams.
ZONE 3 Gravelly sand or silt loams and gravelly to stony sandy loams.

SUSCEPTIBILITY TO EROSION
ZONE 1 All slopes in excess of 10% and gravelly sandy to fine sandy loam soils.
ZONE 2 Gravelly sand or silt loam soils and areas with slopes in excess of 2½% on gravelly to stony sandy loams.
ZONE 3 Other soils with finer texture and flat topography.

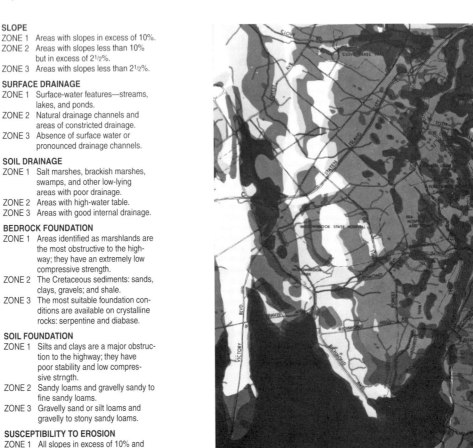

COMPOSITE: PHYSIOGRAPHIC OBSTRUCTIONS

Figure 12.1   Acetate overlay to determine environmental sensitivity.
Example of Ian McHarg's use of manual acetate overlay operations to illustrate increased sensitivity as a number of categories of physical parameters overlap. This map shows the overlapping of slope, surface drainage, bedrock foundation, soil foundation, and susceptibility to erosion for the Richmond Parkway study area in New Jersey. *Source:* Ian McHarg, *Design With Nature,* John Wiley & Sons, Inc., New York, © 1995. Reprinted with permission.

cialists in other fields are also learning more about corresponding spatial patterns that had, until recently, been unobserved, perhaps even unobservable without automation. As more specialists in more fields learn about unique spatial patterns, we will have an increasingly detailed set of rules about what coverages can legitimately be overlayed and which ones can reasonably permit statements about cause and effect. Until then, we would be wise to stay alert, especially because the modern GIS provides not just a single method for quickly and easily overlaying maps, but rather dozens of ways. This power to overlay can easily be translated into a large number of opportunities to make mistakes and to overlay nonassociated coverages, subject to false conclusions. It can also increase our ability to knowingly lie with maps well beyond what was possible before GIS could be used to compare spatial phenomena (Monmonier, 1991). In the following pages we will look closely at our options for overlay operations, considering what our limitations might be, what we can reasonably overlay, when we should overlay, and what problems we may run into along the way.

POINT-IN-POLYGON AND LINE-IN-POLYGON

It is traditional to look at the overlay operation in either its analog form or its digital counterpart as a method of comparing one polygon-based map with another. This approach works well if all data are polygonal. Often, however, point and line data must be compared either to each other or to the locations of polygons. Thus a means of comparing these disparate cartographic objects might prove useful to the GIS practitioner. Let us look at a few examples.

We begin with a wildlife biologist who wants to employ **biophysical mapping** to discover whether there are functional similarities between the locations of brown-headed cowbirds and certain types or arrangements of forest vegetation. The biologist begins by plotting the known locations of these birds on an aerial photograph on which she has classified the forest vegetation. Next she manually tabulates the number of cowbirds in each forest area. Based on these records, she is able to determine whether the birds have a preference for certain types of vegetation. She finds that the cowbirds seem to occur in most of the forest patches on her map, but they tend to be located almost exclusively toward the edges of these patches. This operation has enabled the biologist to reject her premise that the birds are selective for vegetation type and to modify it to indicate the nature of this species as one requiring forest edges.

Our wildlife biologist has just performed a **point-in-polygon** operation on her analog map. Beginning with a polygonal vegetation map (the classified aerial photograph) she prepared another map of point locations of cowbirds, a nest parasite (egg robber) that has a serious effect on the songbird populations in the eastern United States. In doing so she created a third map showing the locations of points (the cowbirds) inside the polygons (the vegetation patches). This map made it possible to instantly visualize and tabulate the relationships between the attributes of these two entities.

Another example might prove useful as well. Suppose a detective is trying to determine whether there are any spatial relationships between certain neighborhoods and the rising incidence of purse snatchings in his city. The detective is not going to be able to go into the field to collect data on the incidence of purse snatchings. Instead, he must rely on years of records compiled by police officers and detectives who encounter these crimes on a one-by-one basis. Similarly, he will take the crime records from the existing files and relate the addresses of these incidents to the city street map. Outlined on the street map are the existing neighborhoods, so when he places the purse-snatching events on the map by address, he will be placing these points into the neighborhood polygons as well. In addition, he will plot these crime statistics, creating a map for each month of the year for each year for which there are statistics. This tedious process would have proven much easier if the data had already been incorporated into a GIS.

Let's look at what our detective finds. Upon examining all the point locations together, he notices that there are far more purse snatchings in the Pleasantfield neighborhood than anywhere else. The pattern is striking because Pleasantfield is not a poor neighborhood, where one might expect to find many people in desperate need of small sums of money. Nor is it a particularly affluent neighborhood, whose residents perhaps carry larger sums. Instead, it is a middle-class neighborhood, made up mostly of people with average salaries, modest homes, and domestic automobiles. Apparently Pleasantfield is an area that needs to be patrolled more closely, but thus far the detective has not been able to pinpoint the reason for the large concentration of purse snatchings there.

His next move is to look at the data on a monthly basis. A comparison of point locations of purse snatchings for the month of December alone reveals nearly the same pattern shown overall. The pattern breaks down for the other months, however. Knowing that the holiday shopping season peaks in December, our detective begins to formulate an idea about the possible causes of the seasonal peak of purse snatching. Next he maps only the December crime data for each year starting with the most recent year for which data are available. He goes back, year by year, and observes the same relationship occurring until 1983. At that point, the purse-snatching incidents seem to be more widely dispersed throughout the different neighborhoods, as opposed to being rather concentrated in Pleasantfield. The 1983 data correspond to the year before a shopping mall was built in Pleasantfield. He's getting closer. As he compares a map of point locations of purse snatching with a detailed map of the Pleasantfield neighborhood, he sees not only that a spatial relationship exists between these crime data and the Pleasantfield polygon, but that certain high traffic streets leading to and from the mall and its parking lot seem to experience the most purse-snatching events in the neighborhood. This latter comparison shows the importance of certain lines (in this case streets) occurring inside polygons (in this case neighborhoods).

Our detective has demonstrated the strong relationship between points and polygons, and we can see how useful comparing these entities can be. He has also demonstrated the close relationship between comparing variables within a single coverage, as we did in Chapter 11, with comparing those between coverages. In examining his maps, he discovered not only that there were more points in a certain polygon, but that the points themselves were grouped in close proximity to themselves and to linear features in the coverage. This is also true of our wildlife biologist, who discovered the proximity of the point locations of cowbirds to forest edges. In each case, it would have been most useful to have had the capability to perform the analyses automatically.

A third example will illustrate the possibility of correspondences between line phenomena and polygons or **line-in-polygon** procedure. Suppose you are a historical geographer in the Minneapolis/St. Paul metropolitan area. You are interested in the influences on city growth, and you know that the Twin Cities area has experienced several growth spurt periods. You also know that the architecture during different time periods is easily identifiable and quite distinctive. Your first step is to map the regions identifiable by different architectural periods as separate polygons. The mapped pattern immediately shows three major zones, in which residential expansion seems to have proceeded amoeba-like, with long, fat tendrils spreading out from a more central core area. These tendrils seem to have taken unique, period-dependent directions. Because of the linear nature of the polygons, you begin to suspect an underlying linear feature as the cause for the patterns.

Returning to your laboratory, you begin to compile a series of maps based on the most prominent linear patterns in any urban environment—the transportation nets. You compile a map of rail lines, another of old streetcar rail lines, and another of major highway patterns. After transferring these maps to Mylar at the same scale as the maps of architectural periods, you overlay the patterns to view possible correspondences. To your delight, you see that the polygons showing the earliest growth spurt have running through them the major rail lines that existed in the area at that time. You next overlay the streetcar lines and immediately see a similar pattern for the second growth spurt. Finally, after overlaying the major highways, primarily the interstate system begun in the 1950s, you see once again the strong spatial relationships between transportation networks and urban expansion (Adams, 1970) (Figure 12.2).

The results of the application of map overlay techniques to urban history scenarios, as above, allow the scientist to formulate hypotheses concerning the fundamental causes of urban expansion patterns. They further facilitate in-depth analysis to prove that visual pattern similarities indicate true cause-and-effect relationships. As with the two preceding examples, however, a substantial amount of time is necessary to compile the maps and to examine multiple coverages. In addition, if you wanted to compare additional variables in this manner, each set would have to be compiled and overlaid. If you had already placed large numbers of mapped variables for the Minneapolis/St. Paul area into an automated GIS, you could overlay any number of these variables to examine at will alternative hypotheses about the spatial patterns of urban change. In addition, once you had been able to verify that certain patterns have functional relationships, this knowledge would be available for use in planning city expansion under controlled growth conditions. The automation of these

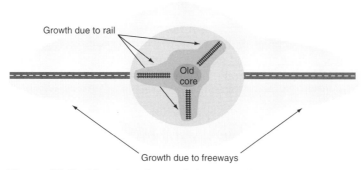

Figure 12.2   Line in polygon. Schematic showing how a line-in-polygon overlay for the Minneapolis/St. Paul area might appear. The periods of urban expansion are illustrated by polygons, while the lines within the polygons are illustrative of the strong spatial relationship between growth periods and major period transportation nets.   *Source:* Adapted from John S. Adams, 1970. "Residential Structure of Midwestern Cities." *Annals, Association of American Geographers,* 60(1):37–62. Blackwell Publishers. Used with permission.

simple geographic analytical techniques then has both conceptual and practical benefits.

POLYGON OVERLAY

As we have seen, the traditional approach to overlay is one of comparing polygons on one coverage to polygons on another coverage. The predominance of polygon overlay procedures among GIS systems today derives from the practical considerations for the early GIS work carried on under the auspices of the Canadian government during the development of CGIS. Canada's needs were primarily to examine spatial relationships of large regions on an area-by-area basis. This is still a major thrust among GIS practitioners, and as a result, vendors have developed systems whose strengths in overlay operations are still fundamentally tied to polygon overlays of these types. Thus there are many varied approaches to performing polygon overlay operations, each answering specific user needs. Before we explain the more technical aspects of overlay in general and polygon overlay specifically, let us continue to identify applications by looking at case studies.

A regional planner wishes to prepare a plan for controlled expansion of population in a rural farming county expected to experience massive urbanization over the next 20 years. Her mandate is to allow growth in areas that have soils that are capable of supporting houses with basements, yet to preserve the highest quality soils for agriculture whenever possible. She also must plan to prevent development of lands that are owned by the federal government, are already being used as farmland, contain known archaeological sites, or comprise the habitats of threatened or endangered species. The state constitution permits the county government to develop zoning regulations that prohibit urbanization in areas it wishes to preserve for other uses.

The first task of the regional planner is to determine what areas should be zoned against urban land uses. After researching the foregoing requirements, our planner creates maps detailing the quality of soils for building houses with basements, areas of high agricultural potential soils, existing farmland areas, known archaeological sites, habitats of threatened or endangered species, and areas presently owned by the federal government. To produce a map showing where urbanization will be zoned against, she must block out all areas that cannot support houses because of the existing restrictions. She prepares a Mylar overlay map of areas having soils that cannot support houses with basements **or** high quality soils capable of agriculture **or** existing farmland **or** archaeological sites **or** habitats of threatened or endangered species **or** federal ownership. Each of these factors could be shaded, giving a map with two primary categories: open, or suitable for expansion, and shaded, or under some restriction (Figure 12-3). The shaded areas then are land parcels that will be zoned against urban land uses by the county government.

You will notice that the connectors between the restrictions above were all the same: the word **or**, in boldface type. This convention indicates that there is a logical mathematical function operating. In set theory, the word "or" corresponds to a "union" operation that includes all these factors. We will look at the **or** function in more detail when we consider how one would automate the procedure just performed.

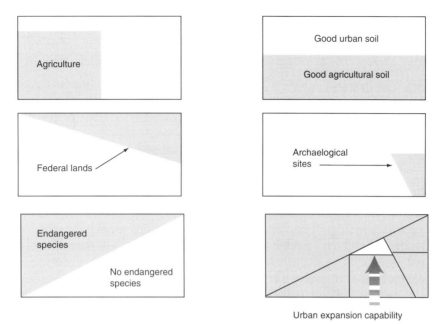

Figure 12.3 Logical "or" overlay. Shaded polygons showing relationship among polygons that are to be zoned against urban expansion. This illustrates the use of the logical "or" search in raster overlay operations.

There is an alternative approach to the overlay operation. Let us say that our planner decided to come at the problem from the opposite direction. That is, she chose to indicate areas that will allow urbanization. It is easy to see that the open areas in Figure 12.3 would answer this question quite nicely, but for illustration we will have our planner create from scratch a map showing areas that will allow urbanization—that is, a map indicating all areas that do not carry restrictions. Stated differently, she must find all areas that have soils capable of supporting basements **and** have no soils capable of supporting farmland **and** have no existing farmland **and** have no archaeological sites **and** have no threatened or endangered species **and** are not owned by the federal government. By preparing Mylar overlays of her maps, where areas not having urbanization restrictions exist are shaded dark, she will be able to see areas that are capable of supporting urbanization as white (or open) areas, just as before. The operational difference is that this time the planner is looking only for areas that fit all the criteria. Such an "**and**" approach is a method of using Boolean logic for set theory associated with intersection. Union and intersection are frequently used operations in automated GIS software, and we will revisit them when we come to the automation of overlay operations.

In the preceding **or** and **and** examples, all the variables in each coverage were weighted equally and could formally be called **exclusionary variables;** that is, they were designed to prevent the performance of a particular type of activity within zones that they occupy. Although this method of overlay comparison is very common, it is rather limiting to users because of the Boolean logic and the nominal level of the data applied to the variables. In many applications of overlay we find that the binary responses generated by Boolean logic are unrealistically restrictive. Take the coverage used to indicate the quality of soils in which houses could be built in our zoning scenario. If you look closely at a normal soil survey report, the engineering properties of soils as they are related to building houses exist as a set of ranked classes ranging from severe limitations to moderate to slight and even no limitations. This ordinal ranking of areas gives planners flexibility by allowing them to decide whether houses should be built in areas with moderate soil limitations. After all, if modern technology can overcome the limitations defined in an earlier statute, planners certainly could consider for urbanization areas that do not conflict with other variables at all but have moderate soil limitations as their only limiting factor.

We could modify our analysis by assigning a higher importance to areas on the map that exhibit severe soils limitations for building houses with basements, a medium level of importance for those with only moderate limitations, and a low level of importance to areas with slight or no soils limitations. This approach, which we will call a **mathematically based overlay,** allows us to assign weights to individual variables to account for the differential impacts they might have on our decision. To do this manually, each factor would be assigned a shading pattern indicating its degree of importance as a factor in the model. For our zoning example, this would mean that soils poorly suited for building houses because of severe limitations might be rendered in black, while the other categories would have systematically lighter shades of gray, with white indicating no limitations.

As we look at our model we see that other factors might also be classed as ordinal variables rather than Boolean, depending on how they are interpreted. Capability of soil for agriculture is another variable that obviously could be assigned ordinal value based on its range of limitations. But some of the others could also be raised to the ordinal measurement scale. Take the existence of archaeological sites. By creating around an existing archaeological site a series of circular regions or buffers as we did in Chapter 9, we could easily change an entire land parcel from one on which urban expansion is totally banned to one with a variety of levels of limitation. Thus we could create a first buffer of, say, 200 meters around the site, on which no housing at all would be permitted. The second buffer, of another 200 meters, might be classed as "possible use for limited housing." And still a third buffer of another 200 meters could indicate that the area could be used for most types of single-family dwelling, but not for high density housing. The area outside the outermost ring in our doughnut buffer could then allow the building of homes of any kind, including apartment complexes.

Once again, this series of distinctions could be imposed manually, by shading each category darker as the degree of restriction becomes more severe. As you might guess, this would require our planner to measure outward from each archaeological site, draw in each successive polygon, and shade it appropriately. Although this procedure is certainly possible, it is very tedious and quite time consuming. As we will see soon, the desired effect can be obtained easily in either a raster or vector GIS.

But even in an automated environment, a problem remains. If our planner is to give a realistic appraisal of areas that should not undergo urbanization, all the factors should be scrutinized and weighted to reflect what is really important to the planning mission. We have thus far assumed, incorrectly, that all the mapped coverages are of equal importance to the problem of land planning. This is clearly not true. Take the factor concerning the present existence of agriculture. In the current version of the overlay operation the existence of agriculture could, by itself, prohibit the zoning of land for nonagricultural use because the importance of the presence of crops or livestock is considered to be of equal weight with all the other factors. We could easily modify this perspective by assigning weights to each coverage based on its importance to the overall planning mission. This, of course, requires awareness of what the ranks should be. The assigning of weights to each coverage as well as to the individual factors is conceptually very simple within a mathematically based map overlay operation. The task is difficult manually, however, and we now must use the power of the computer to assign weights and to combine the factors through a simple addition of the factors.

Addition is, as you have probably guessed, only one of a large number of mathematical operations that could reasonably be applied to a combination of coverages. Multiplication, division, subtraction, squares, square roots, minimum, maximum, average, and many more traditional mathematical operations could be applied, as well. The mathematical manipulations that potentially could be employed are quite large, and few are easily performed manually. Not many undauntable souls would be interested in trying to combine these coverages by assigning numbers and individually calculating the staggering number of values that would be needed to determine the outcome from physically overlaying Mylar versions of these coverages. In fact, there is another set of techniques, called dasymetry, for combining coverages based on mathematical manipulations of coverages that act as statistical surfaces. Because dasymetry is potentially much more complex, however, we will postpone a discussion of this powerful array of methods until we have discussed the automation of the overlay process.

A final general category of overlay is still available, even in the analog environment. We'll call this approach to overlay **selective** rather than mathematical. In fact, **selective overlay** is not necessarily a separate category, but rather a combination and an extension of all of the techniques we have discussed thus far. In many cases we will have a set of rules that allow us to decide which factors we might want to use to cover other factors as we compare two coverages. This approach, sometimes called **rules-of-combination overlay** (Chrisman, 1995), allows us to use exclusionary rules, weightings, mathematical manipulations, and Boolean logic all at the same time. Such a set of rules might resemble a formal computer algorithm with its if-then-else command structure. Using our planning and zoning example, we could create a selective-overlay-based coverage that follows this simple set of rules:

    If moderate or less soil limitations for housing
        and
                moderate to no soil limitations for agriculture
        or
                greater than 200 meters from an archaeological site
    Then
        Minimize coverage values
    Else
        Maximize coverage values

This simple programlike structure adds a level of control to the overlay operation by imposing multiple rules to determine how the overlay is to be performed. To develop this routine, our planner may have to know beforehand what the likely combinations are going to be. In fact, a tabulation for each coverage of the quantities of each value for each factor would be most useful. Such a tabulation would require another form of overlay, called **identity overlay**, that can isolate any of these numbers and preserve all their attribute information. Because it is normally employed within such overlay procedures, I include identity overlay within the selective or rules-of-combination category, although other authors could find ample justification for argument. In any event, the identity overlay, like the selective overlay operation itself, is far too complex to be performed manually. In fact few of the operations discussed thus far, beyond the most simple approaches, are realistically possible without some form of automation. Therefore, we will now revisit these basic operations to see how they might be automated and how automation will enhance their capabilities.

AUTOMATING THE OVERLAY PROCESS

As we have noted many times in this chapter, the process of overlay, whether for point-in-polygon, line-in-polygon, or polygon overlay, can become very complex and nearly or totally impossible to perform with analog maps. Variable factor weights, differences in the importance of coverages, complex mathematical procedures, and iterative rules of combination all contribute to these difficulties. We will now see how the computer can assist us in the performance of these often used techniques; we will examine the basics of computer automation, and we will learn still more techniques for overlay that are seldom undertaken without the assistance of computers.

Automating Point-in-Polygon and Line-in-Polygon in Raster

The automation of the point-in-polygon and line-in-polygon procedures that we discussed using our wildlife biologist and police detective examples is conceptually a relatively simple matter. In a grid-cell-based GIS, the implicit locations of grid cells representing crime statistics, bird locations, or any point activity can be compared by assigning to these points, as well as to the polygons, unique and easily separable numbers or category names; thus upon comparison it becomes immediately evident which numbers represent the grid cells that occur in the same location space as the grid cells of the areas or polygons they occur within. If we had, for example, a coverage with a single polygon, say grassland, it might be represented by a numeric value of 1. The surrounding background might be coded with a value of 0, indicating "area not classified." Another coverage shows points, expressed as grid cells, that represent the location of large weeds. In this case the weeds are assigned a numeric value of 1, while nonweed areas are given a value of 0. If we were to overlay these two maps by a process of simple addition, we would find that for some areas of the grass coverage, the grid cells are numerically different from the surrounding grass (Figure 12.4). These numerically highlighted grid cells occur because the addition of the 1s from the grass area grid cells to the 1s from the weeds in the second coverage would now be equal to 2.

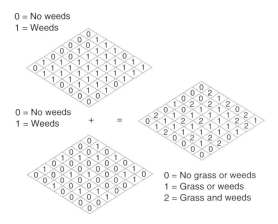

0 = No weeds
1 = Weeds

0 = No weeds
1 = Weeds   +   =

0 = No grass or weeds
1 = Grass or weeds
2 = Grass and weeds

Figure 12.4   Raster overlay using addition. The results indicate the differences between grid cells with weeds and grass and those with just grass.

As you might have noted, we used the same number for grass in the first coverage and for weeds in the second coverage. This choice resulted in a number of grid cells not originally coded for grass that now have a 1, which could represent grass or the weeds. We know that the 2s are weeds (points) within grass (polygons), but the 1s outside the original grass polygon could be either grass or weeds because their numbers are identical. Had we assigned the weeds a much higher number, say 10, we would find it very easy to separate the categories. Those with 10s would be weeds on an unclassed background; those with 11s would be weeds in grass—our point-in-polygon operation. Certainly, if our GIS software is capable of storing additional attribute data, or if it carries the legend information with it, we would be able to identify these points correctly. In fact most operational GIS software has numerous methods of assuring this. Still, it is important to keep this simple example in mind, because any mathematical manipulation we might want to perform on this coverage later could cause us serious difficulty. It is also important to keep the separability of the category numbers obvious because we will encounter the same problem when we perform raster polygon overlay operations.

As you can see, the point-in-polygon overlay operation requires no explicit record keeping of the exact locations of either the points or the polygons. Because most simple grid-based systems can handle only a single grid cell value at any location, the numeric combination will be easily identifiable as unique numbers, at least if we plan ahead. For more sophisticated grid systems we will be able to identify the locations by grid cells whose labels are combinations of individual labels (such as weeds + grass). Grid-based systems that are linked with database management systems will have modified attribute records for each of these grid cells to indicate the co-occurrence of two or more attributes at one grid cell location.

In our last example, the method of creating the point-in-polygon overlay strongly resembled our earlier discussion of mathematically based polygon operations. It turns out that the process of point-in-polygon overlay is identical to polygon overlay because, as you remember, each grid cell is a uniformly quantized portion of the geographic space. As such, each grid cell is essentially a tiny, uniform polygon, and the overlay operation is therefore a polygon operation. This conception carries forward for grid-cell-based operations of the line-in-polygon type, where the lines consist simply of a series of small, connected grid cells sharing identical numbers and labels.

Automating Polygon Overlay in Raster

We have seen how automated grid-cell-based overlay operations are performed for point-in-polygon and line-in-polygon. By extension, the process of grid-cell-based polygon overlay is a simple matter because the polygonal areas would be irregular blocks of grid cells also sharing an identical set of numbers and associated labels. Unsurprisingly, grid-cell-based polygon overlay tends to lack spatial accuracy where grid cells are large, but, like its simpler point-in-polygon and line-in-polygon counterparts, gains an unusually high degree of flexibility and speed because of its simplicity. It is well recognized that raster overlay operations in general are preferred because of the computational ease with which they are performed (Burrough, 1983). Because each grid cell in a coverage is, by default, coregistered with the same cell for a different coverage, there is no need for the computer to calculate composite reference locations for each overlayed coverage. Instead, all the computer resources are used to compare the attribute data, thus greatly improving the speed of this operation. As we will see later in the chapter on cartographic modeling, this is much akin to a modified version of matrix algebra, often referred to as map algebra (Tomlin and Berry, 1979).

The simplicity of the raster overlay process greatly improves the flexibility of the overlay operation as well. Each grid cell has associated with it a series of numerical values, plus their attributes and labels. These numerical values can therefore be compared to other coregistered variables through a nearly unlimited set of logical, conditional, or mathematical operations. We have seen how the grid cell values can be reclassified through simple decision rules, measures of size, shape, and contiguity, slope and aspect categorizations, and many more. Each of the resulting neighborhoods is a polygon, and each of these polygons can easily be compared to polygons on other coverages through many of the same operations.

Let us consider one or two of the simpler operations to show how the process takes place and to examine how even simple operations can give us a great deal of flexibility. We have two coverages, one showing an intervisibility surface indicating lowland locations that can be observed from the hills surrounding a potential nature preserve. Our purpose is to gain community support for a nature preserve by choosing a location that will allow residents in the surrounding hills to view the wildlife with binoculars. There are, however, other conditions that must be met. For example, the park must be located in areas that already have existing wildlife. Our task, then, is to compare a polygonal map showing the locations of existing undisturbed areas to our intervisibility areas map. In raster this operation can be performed in many different ways. Let's assume that the map of intervisibility has the numerical values 10 for visible and 1 for nonvisible. Likewise, our coverage of existing undisturbed areas has values of 10 for undisturbed and 1 for disturbed. By simply adding the values for these two coverages, we see that areas that are both undisturbed and visible will have values of 20, while all other areas will be 11s or 2s, all of which could be renumbered to the same value (Figure 12.5. This operation is identical to our weed and grass example for point-in-polygon.

Alternatively, assuming the same grid cells represented undisturbed and visible, we could have multiplied the two coverages to get different values. In this case the areas sharing these conditions would achieve values of 100, while

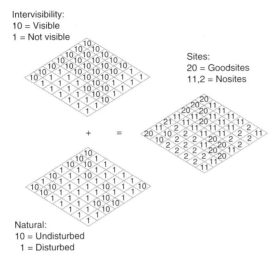

Intervisibility:
10 = Visible
1 = Not visible

Sites:
20 = Goodsites
11,2 = Nosites

Natural:
10 = Undisturbed
 1 = Disturbed

Figure 12.5 Confusion using raster overlay by addition. Although addition will allow separation, it can also result in some confusion. Note that the result of this additive overlay produces values of 20, 11, and 2. The only values of interest in this case would be 20, and the rest of the numbers could be reclassified as a single category.

all other areas would be 10s or 1s, which could then be renumbered to a single value, as before (Figure 12.6). As you can see, both types of polygon overlay operation achieve the result of highlighting the same set of areas. Either operation could be used to achieve basically the same result. By using the multiplication operation, however, we have made the numerical differences more evident, hence the categories are easier to separate. With operational raster GIS software this is not a major problem because the category labels are carried forward to show the resultant categories. Forethought exercised prior to the overlay operation, however, will reduce confusion. By preprocessing the coverages to obtain specific results, an operator can both simplify the ultimate answers and formalize his or her thought processes to be sure of getting the desired product.

Let's preprocess our existing coverages a little to simplify our answers. For our intervisibility coverage we could renumber the nonvisible areas to 0. If we do the same for the natural areas coverage, we will wind up assigning 0 to nonnatural areas. We see that all we have left are 10s and 0s. If we then add the two coverages, we will get a coverage with 0s, 10s, and 20s. The 0s indicate no natural areas and no visible areas. The 10s could indicate either natural areas or areas visible from the hills. Finally, values of 20 indicate correspondence between natural areas and visible areas. Again, not an insurmountable number of categories to deal with, especially if our software carries the labels with the coverages. If we multiply the two coverages together, however, the 10s multiplied by 10s will become 100s, but any number that is multiplied by 0 will become 0. Thus, we have created a binary map of locations: areas that are both natural and visible are represented by 100, and the rest are represented by 0. This is a much more elegant approach.

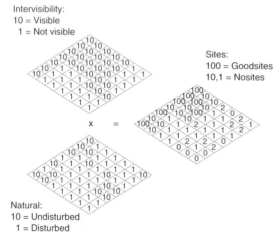

Figure 12.6 Raster overlay by multiplication.
This approach reduces the categorical output
by one, thus lessening potential confusion.

As we have seen then, raster polygon overlay provides a rapid, simple, but extremely flexible environment in which to compare variables between two coverages or among many coverages. These attributes of raster overlay allow us to build a wide array of complex mathematical comparisons among multiple coverages. This further allows us to develop an equally large number of cartographic models. We will cover cartographic modeling in the next chapter. Until then, it is a good idea to keep the preceding lesson in mind, if for no other reason than to simplify the accounting necessary to keep track of the results of multiple compared coverages.

Automating Vector Overlay

Having reviewed the simpler raster method of automating the overlay process, we turn to an examination of vector overlay operations. Vector overlay operations offer the same advantage as vector-based automated cartography—they produce map products that resemble the traditional hand-drawn map far more than do raster maps. This similarity brings us back to the hand-drawn overlay process because the maps resulting from vector overlay offer potential solutions to some of the graphical problems associated with manual overlays. We have seen how the process of polygon overlay can become very complex, especially where multiple coverages are used to produce a single map product. To perform even a simple polygon overlay using Mylar sheets limits us to the use of very discrete patterns and a minimum of categories. Let us assume, for example, that we have 15 different environmental sensitivity factors in a project such as one originally implemented by Ian McHarg. Using gray-shaded maps means that it is possible to obtain up to 15 classes of environmental sensitivity. Although each of the categories might be important, our visual sense does not allow us to discriminate among more than 8–10 categories at a time. The limitations of analog maps also severely restrict our ability to examine corresponding spatial patterns, both because of the length of time it takes to produce them and because categories with many corresponding dark gray shades are likely to look identical to the viewer. Today's vector geographic information systems provided a wide array of tools for performing such an analysis and producing readable maps with the power of multiple colors, shades, and intensities.

Given its utility, and the frequency with which vector overlay is employed in commercial GIS packages, it is not a far stretch to question whether a system that cannot produce new coverages through, for example, polygon overlay is really a GIS at all. While such an argument may seem to be purely academic, in reality any limitations in the vector overlay process also limit the utility of the software for the user. To become familiar with all the possibilities, we will look at overlay both as a graphic artifact, where the maps produced cannot be saved as separate coverages, and as a method of producing new, retrievable coverages. Then we will look at how point-in-polygon, line-in-polygon, and polygon overlays operate in vector. This knowledge will be very useful to those who use geographic information systems as well as to those who act as consultants to potential GIS customers about selecting systems.

TYPES OF OVERLAY

CAD-Type Overlay

The first and simplest method of automated vector overlay is very similar to the analog method in that we simply apply the display symbols of precategorized data to a single Cartesian surface. Such data might include a range-graded choropleth map of population, a single value-by-area map showing the locations of wild plant species, land use polygons, line symbols indicating road networks, or point symbols showing archaeological sites. The process of graphic overlay is akin to the method of thematic combination that is used to produce more general maps, like those in a regional atlas. Such a map allows the reader to see a considerable variety of factors laid out on the same graphic at the same time and can be used to show graphically how these different factors might or might not be spatially associated. To perform this relatively simple operation with the computer requires only that the software be able to keep track of the locations of each point, line, and polygon to ensure that all the coordinates can be represented on the screen.

The result of this operation is a visual graphic, not a coverage. The software is not responsible for combining the attributes for each object displayed, since the attributes are most often merely labels attached to the graphic elements. There are no tables that associate the labels with other attributes. Nor is there any topology. The image is merely a composite artifact of superimposing the individual images to make a single, unified image. Such products can be printed or plotted as hard-copy output, or, in some cases, saved as graphic computer files. So why are these documents not coverages?

The question is an important one, and it continues to confuse even some experienced GIS users. Let's use an example from computer aided design (CAD) software. Say that we have digitized a map of roads using **CAD** software. On another layer of the software we digitize the locations of water bodies from the identical map. Next we combine the two layers into a composite that shows both roads and lakes. Visually we can locate the roads that would get us to the lake so we can go boating. This is useful information, so we save the composite map for another time. Later another user wants to ask the software which roads enter the area and run parallel to and within 100 yards of the lakefront. Although the solution to this simple query can readily be observed by a person who has found your map and is looking at it, there is no way for the computer to answer the question because it does not know explicitly what names might be associated with what roads. It has no associated database; instead, the names are also graphic devices placed on or near the entities. The computer also lacks the capability to determine any measure of proximity or connectedness because it does not know how to determine these measurements. It is not a GIS coverage. With a little work, it would be possible to import this map to a functional GIS system, to build the database for road names, to create topological structures, and then to answer the question asked. Thus enhanced, our original composite has become more than a graphic device designed specifically for the communication paradigm. It is now a bona fide coverage suitable for analysis under the analytical or holistic paradigm.

CAD-type overlays are very useful within their analytical limits. It is not always necessary to perform computer-intensive queries to determine neighbors or locate the spatial congruity between two objects. These tasks can easily be done by hand. Much of the utility of this type of overlay lies in its simplicity and the rapidity with which it can be produced. Many **AM/FM** (automated mapping/facilities management) systems operate on the same structure and provide exactly what the users want. That is, they produce maps of selected items reasonably fast for users to take into the field or job site, where the graphics can be read and interpreted manually. It is only when the complexity of the databases or the complexity of the queries is very large that the limitations of manual interpretation of automatically produced maps become a problem. Let us say, for example, that you have a map of underground gas lines, overlaid with a map of streets (including street names), overlaid with a map of housing, overlaid with a map of population information. Someone in the area of the coverage reports a smell of gas. You look at your map to see where the leak might be located, but the lines are rather dense because of the high number of symbols representing residences. It is also difficult to separate the lines that represent roads from those that represent gas lines. You are aware that there are several above-ground shutoff valves, but the complexity of the map obscures the tiny symbols that show them. Such uses require overlay operations (as well as others) that can keep track of multiple attributes for each entity and can highlight the shutoff valves, as well as the locations of residences that might need to be evacuated. All this could be done by hand, given enough time. It might, however, prove easier to use an individual map for each theme, to prevent the graphic symbols from interfering with the map interpretation. In short, you need something more for this emergency situation than a basic CAD-type overlay is capable of producing.

Topological Vector Overlay

The idea of topological data structures introduced in Chapter 4 outlined how the modern GIS can explicitly relate points, lines, and areas in a single coverage.

We further saw how these relationships allow us to determine functional relationships among the entities, such as defining left and right polygons associated with a specific line, defining the connections among line segments to examine traffic flows, or searching for selected entity combinations based on individual or related attributes. It also established a method for overlaying multiple polygon coverages to ensure that the attributes associated with each of the entities could be accounted for and so the resultant polygons with multiple attribute combinations would be maintained. This topological result, known as the **least common geographic unit (LCGU),** was introduced by Chrisman and Puecker (1975) specifically to show how the changes in polygonal entities can eventually reach a point at which no further divisions are possible. Associated with these smallest divisions, then, is a specific set of attributes that also cannot be broken down into any further categories. In addition, the attributes associated with the LCGU would effectively emulate the "darkest" or "most sensitive" area polygons found through using the McHarg (1969) method of manual cartographic overlay. For each of the remaining polygons, there will be a number of attribute combinations with lesser "darkness."

In the subsections that follow we will see how topology allows us to create coverages that can continue to be analyzed and operated on in a vector GIS. We will also look at a few examples. Then we will compare the vector and the raster overlay processes.

Topological Vector Point-in-Polygon and Line-in-Polygon Overlay   As we have discovered, most operational vector geographic information systems are designed around the polygon-on-polygon overlay operation. Still, we can take certain steps to automate the point- and line-in-polygon operations in vector. We need software that is able to locate the coordinates of each point, each line, and each polygon (and the associated line segments). In addition, the software must be able to calculate the locations of points or lines relative to the polygon segments with which we are to compare them. If, for example, a point is located at coordinates $X = 2.300$ and $Y = 3.450$ in Figure 12.7, the program must be able to determine whether this point is located within the polygon whose

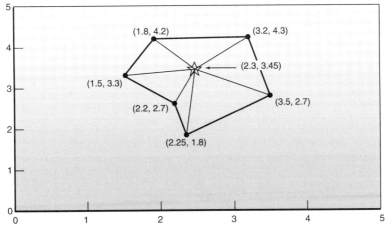

Figure 12.7   Point in polygon. Illustration of the graphical nature of point-in-polygon in vector. Example distance calculations are shown for reference.

outside coordinates match those in the figure. As you can see, in this case they do. In an earlier example, we attempted to show the locations of cowbirds inside forest polygons; we will no doubt want to do this for a number of these points. A polygon overlay search might be performed to create a coverage that shows only the polygons that contain cowbird locations. This will require the software to perform a number of calculations to determine which of the polygon line segment coordinates surround each of these points (Figure 12.8).

To fulfill the objectives of our study, we might want to know what types of land cover polygon do contain these point coordinates. In this way we will be able to determine whether there is a preference for cowbirds in forest polygons as opposed to urban polygons or row crop polygons. Therefore, the software must be able to produce table (in addition to map) output that shows how many points are located in each of the polygon types. The search will require the computer to know not only whether our points occur within polygons but specifically which attributes are associated with each polygon. Software that does not explicitly include topology will be required to check each point in each polygon line segment to determine the coordinate limits of the polygon. This information will have to be checked against attributes to define what the polygons represent. Then the software must be able to compare these data against the locations of our cowbird point data. As you might guess, for very large databases this can be a time-consuming process. If, instead, our polygons were already defined by their line segments, with each line segment explicitly related to each other line segment, we could save computation time in comparing our points with the polygons and their attributes.

One problem practitioners will encounter with some vector GIS software goes beyond the mere computation time necessary to compare a point coverage with a polygon coverage. Some software, although it is capable of comparing these sets of entities and their attributes, is not capable of preserving the map output as a separate coverage. Although the tabular output is useful, the ability to operate on the final map product is missing. This can be a problem if your intent is to, for example, create variable buffers around the point locations

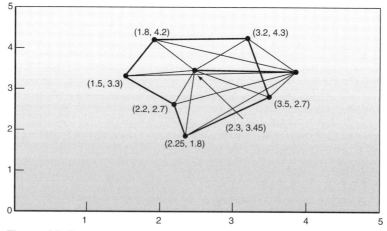

Figure 12.8   Line-in-polygon calculation. Example distance calculations are shown for reference. Note that the addition of numerous line segments (and their associated points) greatly complicates the calculations.

based on a value assigned to each polygon attribute type. This limitation is generally a function of how the program stores its associated table information. Exactly why this occurs is beyond the scope of an introductory course, and often the reasons are of little importance to the user. A simple method of working around this software limitation is to create very tiny buffers (too small to be seen on the output map), which become minute polygons. You can save this coverage, and the points will be represented by the tiny polygons.

By extension, line-in-polygon is a process of comparing the coordinates of lines with those of the polygons to evaluate when the lines occur within the graphic space of the polygons. The computations are somewhat more complex but are essentially comparable because each line segment begins and ends with explicit point coordinate pairs. The only complicating factor in this case crops up when a line begins within a particular polygon and ends outside it. As long as any point coordinate pair falls within the polygon, we can assume that the line also falls within it. This, however, oversimplifies the real situation. It is more reasonable for the computer to calculate where the line segment of each linear feature intersects a line segment for each polygon. Once this is done, a node is placed at this point and the line attributes and each type of polygon attribute can be kept separate. In this way, if a fencerow encoded as a line feature crosses a polygon of agriculture and a polygon of grassland, we can determine how much of the fencerow occurs in each. We can also create a table showing this relationship.

As you can see, the point- or line-in-polygon overlay in vector is not simply a matter of showing the graphical relationships; the attribute relationships must be revealed, as well. After all, our entities are meant to represent some portion of the real world. You will remember that many real-world features can be encoded as points. And, of course, each point not only could represent the location of a single object but might also include a large set of associated attributes. In the case of our cowbirds, we might have associated with each location such information as gender, tag number (if known), physical condition, size, and whether the bird was observed singly or as one of a mated pair. Each of these attributes could be selected and compared to preselected polygon attributes. Likewise, the lines could show roads and could include road type, road condition, number of accidents per segment, when the road was last surfaced, and many other attributes associated with transportation. Multiple attributes can be compared in numerous ways. And, as we have seen before, they will yield fundamentally different results. Which we use will depend on which attributes we wish to compare, how we want them compared, and ultimately what we are looking for in these comparisons. We'll return to this subject when we discuss vector polygon overlay operations because they are essentially of the same general type. Which of these techniques is available depends largely on the software you are using.

Vector Polygon Overlay   The main problem accompanying the use of point-in-polygon overlay and line-in-polygon overlay is that the lines do not always occur entirely within the polygons. Our most powerful solution to this problem was to let the software determine the points at which each set of lines intersected, so that nodes could be placed at these points. For vector polygon overlay your task is essentially the same, except that you must now be able to account for the overlapping intersections among lines defining polygons and for the attributes associated with each of these. Each polygon line segment on one coverage that crosses each polygon line segment on subsequent coverages must have an associated node that shows where the segments intersect. In addition, the software must keep track of the attributes for each original polygon as well as for each newly created polygon in the final coverage.

There are a number of ways to perform vector polygon overlay and an increasing body of how-to literature, but such details are beyond the scope of our discussion. Our concern is to compare mapped variables, not to study the specifics of how the computer program performs the graphics or how it keeps track of the variables. For the following discussion we will assume that the vector-based software incorporates a tabular method of keeping track of variables. This is most often done by using a database management system either tied specifically to the graphics or linked externally with them.

Because many vector-based geographic information systems are linked to a database management system, it should be no surprise that database queries based on Boolean logic are used for spatial database queries as well. In fact the **Boolean overlay** is among the more commonly used approaches. They are also very easy to understand. You remember back in your youth when a teacher drew diagrams showing the relationships between two sets of numbers, two sets of dishes, and so on. Diagrams of that type, called Venn diagrams, are graphic "mappings" of the relationships between two sets. Venn diagrams closely resemble the graphic devices we commonly call maps. Let's take a closer look at this.

Consider the following. You have two sets of data, each one representing a number of types of land use. Set A, using the list notation, would be A = row crops, grain cropping, ranching, dairying, forestry, single family housing, multiple-family housing, retail, manufacturing, and office space. For the second set, B, we have: B = dairying, forestry, retail, and office space. To find out what land uses these two sets have in common, we simply perform an intersection on the two sets (Figure 12.9), just as in our nonspatial example earlier.

In the case of Boolean vector overlay, we do not compare the attributes themselves directly, but rather the space occupied by any two sets of attributes. Thus if we had a simple map showing only two land use polygons, one might show cropland and the other an urban setting. Now for the same portion of the earth's surface, we have another coverage showing two other attribute variables, such as owned and rented land. As you can easily see, when we try to find

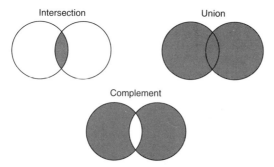

Figure 12.9   Boolean operators. Venn diagram showing intersection, union, and complement. The same two sets (lists) of land uses could be used in each.

out which portions of land are owned farmland and which are owned urban, we create a new map through the process of intersection that shows the designated portions only. The graphics portion of the program determined the polygon intersections and created specific polygons based on these line intersections. Then the attributes associated with the polygons that fit both criteria are selected from the database, while those that do not fit both criteria are not selected. In other words, performing an intersection of the two maps is equivalent to drawing a Venn diagram for the two spatial locations of the two sets of attributes. By doing so we are now able to locate both the owners of farms and the owners of urban property.

Again, going back to elementary mathematics, you remember that the union operation is the total of all the elements in both sets and can be illustrated with a Venn diagram (Figure 12.9). Extending this idea to our earlier map coverages, we could produce an overlay of the two maps to show all locations that were owned or rented or were agriculture or urban, obtaining a map with a single category that included all these attributes. In this case our map would cover the entire study area (Figure 12.10). A third method would be to use the complement of these two coverages, analogous to the Boolean "not" operation, which also can be illustrated by a Venn diagram (Figure 12.9). In this case we might want to cover these two maps to show all areas that are owned or rented and have some type of human activity, but not human habitation. The effect would be a map showing owned and rented farmland.

Our last simple example was selected for two reasons. First, it clearly shows how easy it is to overlay variables through the logical complement procedure. It also illustrates, however, that it is not always necessary to overlay maps to achieve the desired output. To determine where owned and rented farmlands lie would simply be a matter of reclassifying our land use coverage ("farmland" would become "owned or rented farmland"), and we would not select from the same coverage the attributes called "urban." This is an important lesson because it shows that simple solutions, requiring no overlay, are often readily

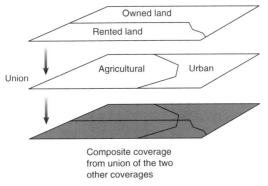

Composite coverage
from union of the two
other coverages

Figure 12.10 Union overlay. The union operation is used in a vector overlay of two coverages: The first is a map of land ownership versus renting and the second is a land use map showing agricultural versus urban. The result of this operation is a map of the entire study area, indicating agricultural and urban as well as owned versus rented land.

available. Because overlay of large polygon coverages can be a lengthy process, whereas solutions to problems are often needed quickly, it is prudent to think about a problem before you attempt to solve it. This is especially true when working with Boolean overlay operations, which are relatively simple and, in addition, offer more cases in which reclassification operation would suffice. Still, many other types of overlay operation could be avoided if the reclassification is employed by itself. Or, the complexity of the overlay procedure could be greatly reduced by preprocessing the overlays, as we did in raster, by reclassifying the polygons that are not of interest in the subsequent overlay operation. Our purpose is to whittle down the number of categories, thereby reducing the number of polygons. This process makes the results of the overlay simpler and easier to understand. It also has the fortunate side effect of eliminating some computer operations and speeding up the overlay process.

Before we leave this discussion it is well to revisit the idea of identity overlay. Although it is generally unnecessary to formally define identity overlay as a method, the use of the term will allow us to at least have a moniker to hang on it. In identity overlay, we can achieve results similar to our earlier examples except that while we are using Boolean operators to perform the actual cartographic overlay, we retain the attributes; thus our resulting polygons show combinations of all those that intersect. In other words, we overlay the graphic entities and tabulate all the category combinations. So, for example, if we have four different categories (say land use, soil type, animal communities, and vegetation type) that intersect in space, all the categories are tabulated. Therefore, we may have a polygon that contains orchards (the land use), mollisols (the soil type), field voles (the animal community), and pecan trees (the vegetation type). By maintaining all these categories, we can easily cross-tabulate the amount of each category that corresponds to each other. Many overlay operations are performed so that these statistics can be obtained. In fact, your vector GIS may use identity overlay to perform this operation, thereby requiring you to reclassify the categories once they have been overlaid.

A Note About Error in Overlay

As in our raster examples, the operational use of cartographic overlay tends to involve numerical data more often than simple nominal data. Overlay data may be ordinal, interval, or ratio; they may be range graded to indicate groups of values for a given set of polygons. The values in turn place constraints on the use of simple Boolean logic and identity overlay operations that are designed more for handling nominal data than for mathematical manipulation. The more precise mathematical values give us other powerful options for overlay. Instead of being restricted to Boolean logic and set theory, we now have available the same set of mathematical operators we used in mathematically based raster overlay. Any fully operational database management system has the capability of allowing mathematical manipulations of the data in the attribute tables associated or linked with the graphics of the GIS. Because the vector GIS normally contains the attributes and their labels and can link these with the entities, the process is even easier than it would be in the simplest types of raster system, and nearly identical to the raster systems that also contain a database management system tool kit.

The operations of mathematically based overlay in vector are performed much the same on the attributes in the database tables and in each individual georeferenced set of grid cells. Except for computational difficulties due to the graphical procedures, the user will see little difference in the results of such operations. This similarity in operational capabilities will prove very important when we consider cartographic modeling in the next chapter. Because of the nature of the output from vector GIS, the visual results of mathematically based overlay will look quite different in vector and in raster. One would expect the cartographic output to resemble a hand-drawn map. The result, however, can be both surprising and annoying, as dozens or even hundreds of tiny, unexpected polygons of interacting polygons show up, especially around the margins of intersecting polygons. This apparently innocuous visual difference can have quite profound impacts on how we interpret the results of our analysis.

As an example, let's make a simple comparison of multiple land use coverages representing separate time periods for identical pieces of land, digitized from aerial photographs classified on the same scale. When we compare a single set of polygons of our urban land use category for just two of these time periods, we notice that the polygons looked identical when mapped separately (Figure 12.11). But, when combined, they do not exactly match. Instead, there are a number of sliver polygons at the margins of the overlaid/intersecting polygons (Figure 12.11). Now we must decide: Has there in fact been no change from time 1 to time 2? Have there been slight changes in the land use locations? Was there a problem with the digital input of the data in the first place? Instinctively we know that changes as minor as the slivers in the composite are unlikely, especially because the polygons look so much alike. We can probably eliminate our second possibility, as well, again because the polygons are nearly identical in shape and even minor changes in land use would have imposed greater modifications.

Ignoring the possibility of land use changes of a minuscule percentage, we are left with the answer that the polygons probably are identical, and we try to account for variations caused by digitizing. Or, if the polygons are essentially

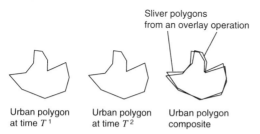

Sliver polygons
from an overlay operation

Urban polygon
at time T^1

Urban polygon
at time T^2

Urban polygon
composite

Figure 12.11   Silver polygons created by vector overlay. These sliver polygons result from minor differences in two marginally intersecting polygons. Do the discrepancies indicate real change or error? Note that most of the slivers occur on the periphery of the two otherwise identical polygons. Care must be taken in interpreting the results of vector overlay, especially if small amounts of change are important.

the same shape, but one is slightly rotated relative to the other, the slivers might be due to a minor rotation or distortion in one of the aerial photographs from which the polygons were digitized. In this case we might be able to perform a computerized form of rubber-sheeting. This remedy is readily available in GIS software but rarely gives 100% co-occurrence even if done with care. The process also requires that we anchor a number of points that we know should be located at exactly the same place on both coverages. While these are held in place, the rest of the map can be "conflated" or rubber-sheeted to move the polygons until correspondence is achieved. Often, however, when we do this, other polygons that were well aligned move out of register. In short, there is no easy solution to this problem, which is much more common in vector GIS overlay operations than one would like to admit. Some users even prefer to use raster, thus giving up some locational accuracy, to avoid the tedious adjustments frequently required. Even so, conflation is a valid approach and one that might be considered if a process must be repeated many times, or if polygons are very small and the sliver polygons are likely to be tiny.

As you might guess, problems are compounded when multiple coverages are overlaid, especially if the coverages encompass fundamentally different themes. For example, if we overlay a coverage of soils with one of vegetation, we no longer have the luxury of assuming that any two polygons might be the same shape. It is not necessarily true that there will be a perfect one-to-one relationship between soils and vegetation. In fact, the purpose of the overlay might be to see where the categories deviate. Our dilemma here, then, is to separate the error from the actual polygon differences.

If, as in our last example, each map has its own unique amounts, types, and sources of error for both entities and attributes, the question is, How do we deal with error in multiple coverages? Unfortunately, despite ongoing research attempting to identify how much error is propagated with multiple coverages, there are few general principles and even fewer answers, especially if the coverages are from widely different sources, (Chrisman, 1987). Most of the time you will be left to your own devices. Your decisions must be based on your own knowledge about the data, about data quality, about the quality of the input, and even about field work. Your response to error is heavily dependent on how precise your output must be for the problem at hand. It is logical to assume that if you have many different coverages, the results of overlay will be only as good as the worst coverage employed, but it is very important to note that this is not often the case. This "weakest link" approach, although appealing, is based on the assumption that all coverages are equally important. As we have seen, however, we have the capability, in either raster or vector, of weighting each coverage through the overlay process. In addition, even if we do not explicitly weight each coverage, some will ultimately be more important to our model results than others. Before you overlay your coverages, you should become familiar enough with your data to be able to decide what types and amounts of error you can accommodate.

DASYMETRIC MAPPING

Before we leave the topic of polygon overlay, I believe it is important to discuss an old cartographic technique for detailing polygonal information based on other variables. The technique, called **dasymetric mapping,** is based on the idea of the choropleth map discussed in Chapter 3, and it gives us insights into some of the approaches currently taken inside GIS, as well as others that have not yet been considered by the GIS community. Its use requires that the data be in ordinal, interval, or ratio form and that they exist as a statistical surface (see Chapter 3). The first documented use of dasymetric mapping was employed to improve the categorization of population densities on Cape Cod through a method called **density zone outlining** (McCleary, 1969). The technique, also called density of parts, was employed to obtain greater detail of individual densities of unresampled areas based on a more detailed knowledge of some smaller subareas that had been resampled (Wright, 1936). An excellent example of the application of density of parts reported in Robinson et al. (1995) closely associates the example with Wright's original 1936 method. The student of GIS should become familiar with this technique because of its power to improve the quality of quantitative data contained in polygons by comparison with more detailed data for another coverage.

We have already employed one method of dasymetric mapping without identifying it as such: pure dasymetry, first used by Hammond (1964) in his work on landforms, entails the delineation or classification of areas of geomorphic type based on a recategorization of topographic data. Topographic surfaces perfectly fit this approach, which calls for continuous distributions of an infinite number of points. When we studied reclassification of continuous surfaces, we created neighborhoods for "intervisibility," "south-facing slopes," "steep-slopes," and so on by grouping preselected ranges of our data sets. This was a modified form of pure dasymetry. Once again we see that the modern techniques we use so often have roots in a time well before the advent of the computer.

Several more forms of dasymetric mapping could prove potentially useful; among the most powerful is the type called "use of other regions with the assumption of correlation," with its strong implication of a form of cartographic overlay. Thus we might look at our example of density of parts and suggest that this was also a comparison of variables between or among maps. The assumption-of-correlation approach is strikingly different, however. Instead of isolating portions of an area for which detailed study improves the information content, then using this information to improve what we know about the remainder, we use either limiting variables or related variables contained in other coverages.

We used the **limiting variables** form of dasymetry when we discussed exclusionary factors within overlay. As an example of how this might be applied to improve the quality of our polygonal data and the models we produce from them, let's say that we have a map of Minnesota that shows the population by county. We also have a map giving as a polygonal value the area in square miles for each county. By overlaying the two coverages, having divided the population coverage by the area coverage, we obtain a map of population density for each county in the state. However, Minnesota is the "land of 10,000 lakes," which implies a substantial surface on which people do not reside unless they have houseboats. To improve our population density coverage result, we should have "excluded" from the area coverage the amount of water contained in each county. If we had used this "limiting variable" to create a coverage of land area by county and employed in producing our population density coverage, the population density of counties with large water bodies would have gone up.

Such techniques can be used frequently in conjunction with cartographic overlay to isolate and exclude areas that bias quantitative polygonal data.

Our final example of dasymetric mapping, called **related variables,** corresponds strongly to our use of mathematically based cartographic overlay methods as in both raster and vector GIS. We have seen that variables contained in our polygon coverages often interact in ways more complex than simple exclusion. Statistical tests such as correlation and regression are often used to show how geographically dispersed variables are related to one another and how these relationships can allow us to predict variations in one based on the changes in the other. This is no less true when using GIS. If we, for example, know that there is a high correlation between percentage of cropland and percentage slope on the land, we can predict the amount of cropland based on this correlation and its associated regression line. With this information, we can develop a detailed predictive coverage of "% cropland" based on the slope alone. Alternatively, we might have coverages of existing percentages of cropland and slope. By cartographically overlaying these, we would create a coverage showing the true relationships between these two variables in a particular area, or we could overlay our composite on a map of "predicted % cropland." The differences between the two would give a visual display, as well as a quantifiable difference between the actual and the predicted, which could be used to evaluate the predictive model. The areas that do not fit the predictive model could then be compared to other coverages to develop hypotheses concerning variations in the model. This is an excellent example of how GIS can be used to develop hypotheses for scientific applications of GIS or to create predictive GIS models for decision making in commercial applications of GIS. We will look at the use of this powerful application in the next chapter.

As we have seen, dasymetric mapping has a great deal of potential to improve the use of GIS in both academic and commercial settings. Still, the subject is relatively untapped within the GIS literature. Raster GIS lends itself readily to many forms of dasymetry, and several methods are being employed every day without the knowledge of the users. Vector GIS is also capable of performing some, perhaps all of these techniques, but few have attempted in a serious, systematic manner to evaluate its potential (Gerth, 1993). An awareness of dasymetry as a potential set of cartographic overlay methods will likely result in vast improvements in the GIS overlay tool kit currently available to users.

SOME FINAL NOTES ON OVERLAY

Because of the visual appeal and intuitive nature of cartographic overlay, this set of techniques is often considered to be what GIS is. This perception limits the number of potential GIS users and may even result in reduced sales of commercial products. In addition, it retards many users from evaluating other powerful techniques already available in GIS. The student should be aware that despite the power of cartographic overlay, there are numerous alternative methods for solving GIS-related problems. Overlay is frequently found to be far more powerful when combined with other methods of spatial analysis than when used in isolation. In the next chapter we begin to develop complex cartographic models that will force us to select appropriate techniques and to combine them in a rational manner to obtain useful results.

Constantly remind yourself that cartographic overlay is powerful but mindless. No GIS software is capable of evaluating whether the coverages you are using are functionally related. For personal applications, you should take care to know which spatial variables are likely to be functionally related before you begin blindly overlaying them. If you are operating a GIS for a client, it is even more important to be vigilant because the client will turn to you for answers. Think about your variables before you select those that you will overlay. Determine why they might be related. You might even consider pretesting statistically a sample of the variables to help you make this determination.

A statistical approach that is often useful employs the **integrated terrain unit mapping** approach (Dangermond, 1976). Very much like biophysical mapping units, **integrated terrain units (ITUs)** generally reflect the assumptions that all the data are integrated on the ground and that you employ a method, such as digitizing from aerial photographs, that requires you to extract these interacting variables from the same source of information. This assures that the variables you include are related spatially if not in a logical sense. If you do not have the capability of employing ITUs and can't find a logical reason, or at least a statistical reason, that would indicate a relationship among your variables, perhaps you should reevaluate the solution to your problem. This is no different from testing the relationships between the surrogate radiometric responses gathered from a satellite and the land features they are meant to represent.

TERMS

overlay
line-in-polygon
selective overlay
CAD
Boolean overlay
limiting variables
ITUs
biophysical mapping

exclusionary variables
rules-of-combination
  overlay
AM/FM
dasymetric mapping
related variables
point-in-polygon

mathematically
  based overlay
identity overlay
LCGU
density zone
outlining
integrated terrain unit
  (ITU) mapping

Review Questions

1. What are some obvious limitations of the manual overlay process? What advantages have been offered through automation?

2. What are the advantages of raster-based overlay operations over their vector-based counterparts? What are the disadvantages?

3. Why is overlay still among the strongest operations in most modern commercial geographic information systems?

4. From your own studies give examples of how point-in-polygon and line-in-polygon could be useful to you.

5. What are the limitations of CAD-type overlay operations? Give an example of how CAD overlays would prohibit subsequent analyses on the product of the overlay process.

6. What is dasymetric mapping? How can it be employed in vector GIS? Raster?

7. What is the difference between related variables and limiting variables in dasymetric mapping? Give an example of the use of each variable type.

8. What problems or difficulties might you encounter when comparing multiple coverages with vector overlay techniques? Are there any approaches you might consider to solve these problems? What should you do before you overlay to lessen this?

9. Give some examples of how portions of dasymetric mapping might readily be applied with existing raster or vector GIS. Why do you suppose there has been so little attention paid to the use of dasymetric mapping in GIS overlay?

References

Adams, John S., 1970. "Residential Structure of Midwestern Cities." *Annals of the Association of American Geographers,* 60(1):37–62.

Burrough, P.A., 1983. *Geographical Information Systems for Natural Resources Assessment.* New York: Oxford University Press.

Chrisman, N., 1987. "The Accuracy of Map Overlays: A Reassessment." *Landscape and Urban Planning,* 14:427–439.

Chrisman, N., 1996. Exploring Geographic Information. New York: John Wiley & Sons.

Chrisman, N., and T.R. Peucker, 1975. "Cartographic Data Structures." *American Cartographer,* 2(1):55–69.

Dangermond, J., 1976. Integrated Terrain Unit Mapping (ITUM)—An Approach for Automation of Polygon Natural Resource Information. Publications of the Environmental Systems Research Institute, no. 160, 11 pages. Redlands, CA: ESRI.

Gerth, J.D., 1993. Towards Improved Spatial Analysis with Areal Units: The Use of GIS to Facilitate the Creation of Dasymetric Maps. Unpublished MA paper, Ohio State University, Columbus.

Hammond, E.H., 1964. "Analysis of Properties in Landform Geography: An Application to Broad-Scale Landform Mapping." *Annals of the Association of American Geographers,* 54:11–19.

Hills, G.A., P.H. Lewis, and I. McHarg, 1967. *Three Approaches to Environmental Resource Analysis.* Cambridge, MA: Conservation Foundation.

McCleary, G.F., Jr., 1969. The Dasymetric Method in Thematic Cartography. Ph.D. dissertation, University of Wisconsin, Madison.

McHarg, I.L., 1971. *Design with Nature,* Garden City, NY: Natural History Press.

Monmonier, M., 1991. *How to Lie with Maps.* Chicago: University of Chicago Press.

Robinson, A.H., J.L. Morrison, P.C. Muehrcke, A.J. Kimerling, and S.C. Guptill, 1995. *Elements of Cartography,* 6th ed. New York: John Wiley & Sons.

Sauer, C.O., 1925. "Morphology of Landscape." In: *Land & Life,* J. Leighly, Ed. Berkeley: University of California Press, 1963, pp. 315–350.

Simpson, J.W., 1989. "A Conceptual and Historical Basis for Spatial Analysis." *Landscape and Urban Planning,* 17:313–321.

Steinitz, C., P. Parker, and L. Jordan, 1976. "Hand-Drawn Overlays: Their History and Prospective Uses." *Landscape Architecture,* 56(4):146–157.

Tomlin, C.D., and J.K. Berry, 1979. "A Mathematical Structure for Cartographic Modeling in Environmental Analysis." In *Proceedings of the 39th Annual Meeting, American Congress on Surveying and Mapping,* pp. 269–284, Washington, DC.

Tyrwhitt, J., 1950, "Surveys for Planning." In *Town and Country Planning Textbook.* London: Architectural Press.

Wikle, T.A., 1991. "Computers, Maps and Geographic Information Systems." *National Forum,* Summer, pp. 37–39.

Wright, J.K., 1936, "A Method of Mapping Densities of Population: With Cape Cod as an Example." *Geographical Review,* 26:104–115.

CHAPTER 13

Cartographic Modeling

Thus far in our discussion of the analysis subsystem of a GIS we have taken a step-by-step approach to the types of modeling capability available in a functioning GIS. The classification I have used for these techniques is simple and, I hope, straightforward, but not every book on GIS will use the same one, nor will every GIS vendor. Some will not even attempt such a classification of techniques; some will offer more detail, whereas others will take a more general tack. But whatever software system you use, or whatever advanced GIS texts you read, you should find that most of the approaches can be matched fairly easily with the one here. Don't be overly concerned about what the techniques are called, or how they are grouped. Most commands are given names that are quite descriptive. Evaluate the nature of your problem and then match the conceptual needs with the appropriate techniques and command sequences available for the software at hand.

As you become more familiar with GIS you will quickly see the limitations of using only a small subset of techniques for solving advanced problems. In the following pages we will view the analysis subsystem as a set of interacting, systematic, and ordered map operations that, together, can be used to perform some very complex modeling tasks. The secret is to see these complex models as being composed of much smaller, simpler model components. Often the individual components can be solved with relatively few analytical operations. Each one can then be combined with others to create larger modules of your more complex model, and these can be combined further until your entire model is constructed. Like any system, a spatial model can almost always be broken down into these component parts as long as you are familiar with how the model works.

It is easy to become intimidated by the complexity of some models available in the literature or found in GIS poster sessions and demonstrations. Keep in mind that the people who developed these models were once where you are now—just starting out, and perhaps unsure of their abilities. In fact some of them did not have the advantage of classroom instruction in GIS, so they learned on the job how to model with GIS. While this approach is effective, it is not very efficient. Take the time to learn from their successes by examining whatever models might be available to you through GIS conferences, poster sessions, and classroom or vendor demonstrations. There is no substitute for experience, but you can take advantage of the experiences of others by looking systematically at the modeling process in the following pages and comparing what you learn with what has been done with operational GIS software. Again and again you will be amazed at how the apparent complexity diminishes and the utter simplicity of the overall models becomes evident. You may find yourself using portions of canned models to create your own, or you may soon introduce your own components to enhance existing models.

This chapter emphasizes the development of appropriate flowcharts for each model you require. While flowcharting may sound like very tedious work, it forces you to think carefully about what coverages and data elements you need to obtain the answers for which you are searching. It is easier to detect missing coverages, or to spot coverages that will be used many times, by examining a flowchart during model development than to deal with a model that is finished or nearly finished. I believe the flowcharting of cartographic models should be a required preliminary step to the development of any model. During this process you may discover that you do not have a clearly defined objective, whereupon you will have to rethink your goals for building the model in the first place. In addition, even if your primary interest is not modeling, but database development, you cannot properly develop a database without first specifying the final product that is to be obtained from the modeling process.

Examine the models and flowcharts given in the text. Compare the flowcharts to the descriptions of these models. If you look closely, you will probably find dozens of ways to improve on the approach or to enhance the model by including additional information. The models given should not be considered to be absolutely complete. Nor should you assume that they represent the only solution to the respective problems. Instead, you should envision alternative solutions for each problem—perhaps even doing an alternative flowchart for each. Then make a comparison—it may appear that your approach is more easily accomplished (i.e., a more elegant solution) or more realistic (i.e., more likely to give an accurate representation of what is being modeled). Practice is essential to good flowcharting, as it is to good modeling.

LEARNING OBJECTIVES

When you are finished with this chapter you should be able to:

1. Understand how commands are assembled to produce a cartographic model.

2. Illustrate the cyclical movement from subsystem to subsystem inside a GIS.

3. Know the difference between deductive and inductive cartographic modeling and the advantage of deductive over inductive approaches.

4. Determine the difference between explicitly and implicitly spatial variables and discuss the role of spatial surrogates for nonspatial variables.

5. Produce and explain a simple model flowchart.

6. Explain the problem of cartographic modeling constraints that are too severe or too loose and define methods of circumventing these problems.

7. Describe the idea of weighting of coverages as they impact the outcome of cartographic models, especially with regard to how these weights might affect the error component of the output.

8. List and discuss three basic areas of cartographic model verification.

9. Describe alternatives to cartographic output for GIS results.

10. Discuss the role of prototyping and preplanning in the production of a successful cartographic model.

MODEL COMPONENTS

In our last few chapters we saw a wide variety of possible individual techniques for working with points, lines, areas, and surfaces. Each of these simple techniques alone can be powerfully applied to the solution of geographic problems. And, of course, each has a wide array of possible options and subtle variations that could radically change the result of an analysis. It is not too difficult to envision a nearly astronomical number of possible combinations and permutations from even a simple geographic information system. Let's say, for example, that our software has only 30 commands with no options available for any command. The possible combinations of commands, calculated as 30! (30 factorial), would be equal to 2.65^{32}, a very large number of possibilities. From this simple exercise in factorial mathematics, imagine how many possible command combinations there might be with a piece of software that puts 7000 commands at the user's disposal. As a practical matter, the number of combinations could be described as infinite.

You have seen, however, that there are limitations to the ways in which the options can be applied. To develop a neighborhood function that shows the number of land uses within a specified area, for example, you would have to use a measure of diversity. You would ignore all the other options because "diversity" correctly answers the question at hand. This constraint on the numbers and types of option used to answer specific questions greatly simplifies GIS analysis. It also focuses your attention on the techniques or options that are absolutely necessary. From our discussion of overlay commands in Chapter 12, you will recall that there are many options for overlaying one coverage with another. You must first decide what result you are looking for and what is the best way to achieve it. For example, an overlay operation using complement may very well produce a map that has the same polygonal entity shapes as a map based on union. One approach, however, is going to require you to perform a second operation to reclassify the result because the attributes are the opposite of what you want. Both achieve the same answer, but one is more efficient than the other. Efficiency is very important in working with large models, both to prevent you from making mistakes and to simplify the overall model as you begin combining many commands.

As with the options for each command, the potential number of command combinations is daunting. Fortunately the potential for interaction of commands is also limited by the nature of the models you are creating and the types of object on which you are operating. While it is common to reclassify polygonal data prior to polygon overlay operations, this is not generally done prior to an analysis of intervisibility for a topographic surface. Each command combination is contingent on the nature of the data, the type of operation being performed on the data, and the kinds of answer sought. It would be very useful to be able to create a set of dos and don'ts for which commands will be combined with which others and which should not. Unfortunately, such a cookbook would be too large to be practical. Instead, you will have to experiment, learning from your own experiences and those of others around you. But you are not going to be left alone and unaided in your search for proper command combinations. Formalized procedures for combining GIS functions to analyze complex spatial systems are now available, and we will examine them in some detail.

THE CARTOGRAPHIC MODEL

"**Cartographic modeling**" is the term coined by C. Dana Tomlin and Joseph K. Berry (1979) to designate the process of using combinations of commands to answer questions about spatial phenomena. More formally, a cartographic model is a set of interacting, ordered map operations that act on raw data, as well as derived and intermediate map data, to simulate a spatial decision-making process (Tomlin, 1990). Let's examine this definition in greater detail.

The first condition of the definition is that the map operations must interact with one another. Each individual operation performed on a coverage must have as a purpose the creation of a result, usually another coverage, that can be used by the next operation. The next operation may be a further manipulation of the same coverage, say to isolate certain map variables, or it may operate on a second coverage, perhaps as a map overlay operation.

Let us say that you have a map of "landcover" for a study area, and you want to determine whether there are sufficient large, contiguous, federally owned rangelands to make it worth your while to petition the government for permission to graze your cattle there. Your first operation would be to reclassify the coverage to create a new coverage called "landuse," to isolate the land uses related to the land covers, as in reclassifying all the grassland polygons as either pasture or natural grasslands. You also would create a coverage called "whosegrass," to show pastures that are owned by the federal government and those that are privately owned. This may require you to perform another reclassification, although you could have done this work during your initial reclassification. Your next map operation would be another reclassification to create a coverage called "bigfed" that isolates federally owned grasslands that contain over 25 hectares of continuous grass. This reclassification will call for a clumping function that isolates polygons of 25 hectares or larger (Figure 13.1), and you will need to combine this clumping operation with a neighborhood function that separates out contiguous polygons.

Thus you have the opportunity to operate either on the raw data, as in our original ("landcover") map, or on coverages derived from this initial operation, as in the reclassified "bigfed" coverage used to isolate large, federally owned pastures. A cartographic model employs each analytical operation on either raw or intermediate map data. You can also see how each step is combined to add information content to the data by giving the data a context about which

1	0	0	1	0	0	0	2	3	2	2	1	1	1	1	2
1	1	0	1	0	0	2	2	2	2	2	2	1	1	2	2
1	1	1	1	1	0	2	2	2	2	2	2	2	1	2	2
1	1	1	1	1	1	2	2	2	2	2	2	1	1	2	0
1	1	1	1	1	1	1	2	2	2	2	0	0	1	1	3
1	1	1	1	1	1	0	0	2	2	0	0	0	0	3	3
1	1	1	1	1	0	0	0	0	3	3	0	0	3	3	0
1	1	1	1	0	0	0	0	3	3	3	3	0	3	0	0
1	1	1	0	0	0	0	0	0	3	3	0	0	2	2	2
1	1	0	0	0	0	0	0	0	0	2	2	2	2	2	2
1	0	0	0	0	0	0	0	0	2	2	2	2	2	2	2
0	0	0	0	0	0	3	3	3	3	2	2	2	2	2	0
1	1	0	0	0	3	3	3	2	2	2	2	2	2	0	0
1	1	1	0	3	3	3	3	2	2	2	2	2	0	0	0
1	1	1	1	3	3	3	2	2	2	2	2	0	0	0	0
1	1	1	1	1	1	3	2	2	2	2	0	0	0	0	0

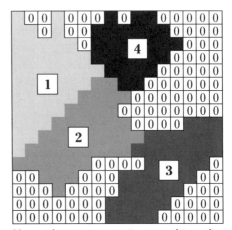

Original values representing different types of rangeland. Each grid cell contains one hectare of land.

New values representing a ranking of contiguous rangeland parcels based on size. 0's represent those under 25 hectares. 1's are for areas with 44 hectares, 2's for areas with 42 hectares, 3's for 41 hectares and 4's for 28 hectares.

Figure 13.1   Reclassifying neighborhoods by size. Clumping to isolate the polygons of 25 hectares or larger.

it relates. Our original coverage, "landuse," was not particularly specific. It related to all possible land cover in the study area, without regard to what the question might have been. As we manipulated this coverage we created "value-added" coverages of specific attributes that could be used to decide whether the coverage contained federally owned pastures large enough to graze a certain number of cattle.

This brings us to the last portion of the definition. Our manipulation of raw and intermediate data was designed to simulate the process of deciding whether the coverage in question contained "large enough" contiguous areas owned by the federal government. A negative result will require you to look elsewhere for such grasslands, while a positive result will prompt you to begin seeking the permits necessary for your grazing operations. A positive result means simply that there are sites available. But suppose you want the option of increasing the size of your herd. You would then have to perform yet another operation on the derived coverage "bigfed." By asking the computer to rank by size the federally owned pasture polygons that exceed 25 hectares, you will be in a position to determine which combinations would allow you to increase your grazing herd. These steps add information to each preceding coverage in an incremental fashion.

Now that you have completed your simple decision model, think carefully about the GIS subsystems involved. The first step was to input the "landcover" coverage, employing the input subsystem of the GIS. Next you had to store the map and edit any mistakes. When you were satisfied that it was an accurate model of present land use, you had to retrieve it for analysis, calling on the storage and editing subsystem of the GIS. Having retrieved the coverage, you reclassified it based on attributes more closely associated with anthropogenic use of the land. The reclassification process used the analysis subsystem. The product of the reclassification is a stored map coverage, representing the use of the storage and editing subsystem as well as the output subsystem, even though the output may never have been printed as a hard-copy map. Then, to solve your problem, you retrieved the new coverage to continue to analyze it. And at any point, you might have wanted to view the map.

Clearly the foregoing process is not linear—we did not go from input to storage to retrieval to output, as in computer-assisted cartography. Instead, we went from input to storage to retrieval to analysis to storage and output of our first intermediate coverage. From there we retrieved the first intermediate coverage and output the map on the monitor to determine what analysis to perform next to create a second intermediate coverage, and so on. In short, the process of cartographic modeling is cyclical rather than linear (Figure 13.2). The cyclical nature of cartographic modeling allows you the greatest flexibility to move from subsystem to subsystem in a series of data transformations all designed to produce a final spatial information product.

MODELS IN GEOGRAPHY

Before we begin an in-depth examination of cartographic modeling, it is important to review the conceptual roots of geographic modeling that ultimately led to the development of GIS. A fundamental understanding of historical context will give us many useful insights into how, when, what and, more importantly, why we model spatial phenomena related to the surface of the earth. Spatial models have been a mainstay of geographic research and applications for many years, well before the quantitative revolution of the 1950s and 1960s. As explained in Chapter 2, the map itself is a model of reality that allows us to view and evaluate at a glance the spatial relationships among variables. With the advent of quantification, the original communication paradigm was replaced by an analytical or holistic paradigm. Still, the map as communication device has had long and important service as an hypothesis-formulating tool. Its primary function in this regard is to prompt the viewer to ask questions about the patterns displayed, especially questions regarding the relationships between observed patterns and the functional processes that may have formed them.

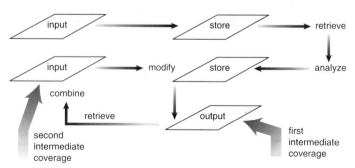

Figure 13.2   Cyclical modeling process. The illustration shows the movement from one GIS subsystem to another to produce new coverages.

Questions regarding pattern and process have been employed historically to explain, for example, the relationships between agricultural activity and transportation costs. The most famous of these models, developed by Heinrich von Thünen in 1910, is now called the **isolated state model** (Isard, 1956). The original model explained areal agricultural activity patterns as a series of concentric circles from a central market. Modifications of this model that include transportation route availability, multiple market availability, land value, soil quality, production costs, and market prices are still used today to help predict the viability of certain types of agricultural activity. At approximately the same time, another set of geographic models was being developed by Alfred Weber (1909). Originally designed to predict the spatial patterns of point locations of industry, Weber's models have been modified to allow practitioners to find optimum locations for businesses and services. Such models, today known as **location–allocation models,** comprise a frequently used set of GIS techniques available in many commercial systems.

Although operationally complex, the location–allocation models are simple in concepts. Allocation modeling is designed to minimize the distance traveled from an existing business or service to the intended public. Take the example of school district allocation. Say that a small city has two high schools of roughly the same size and classroom capacity, located on opposite sides of the city. The district school superintendent must decide which students will attend which school. To allocate an equal number of students to each school, he must start by determining how many students will be eligible to attend. But it is not enough to say that since there are 400 students, 200 will attend each high school. The superintendent must also assign each student to the school that is closer to his or her home. This means finding out where the students live and calculating how far each residence is from both schools.

The superintendent's problem becomes more difficult if only one school exists because the other has not yet been built. The new school should be located with a view to minimizing the overall Euclidean distance between it and the majority of the students who will be attending. This is normally done by a repetitive or iterative approach: the school is located at a specific hypothetical location, the student numbers and distances are calculated, and the results examined to determine the degree to which each location satisfies the established criterion. This process continues until the modelers are satisfied that the best answer has been found. One would not want to perform these calculations by hand. In fact, the mathematical complexity of the problem most often requires an approximate solution even with modern computer technology. Many of today's operational GIS have the capability to provide this approximate solution.

In our example problem for location–allocation modeling, we have assumed that students within a specified distance will attend a given school. In the business environment, however, each store or business has a clientele that is strictly voluntary and must be attracted, one customer at a time. The decision to patronize a business is partially a function of distance, but other factors may be present, as well. A group of economic geography models called **gravity models** have been developed to determine the best location for a business based on the idea that draw is inversely proportional to functional distance (see Chapter 8) from the potential market (Abler et al., 1971). As you may remember from a basic course in physics, the concept of distance decay is very much like the physical law that the gravitational attraction of a body is inversely proportional to the square of the distance from the center of a neighboring body. The major differences are that we are now looking at the interactions of people and services, that people make decisions based on functional distance and a desire to patronize a service, and that these interactions take place on a two-dimensional surface. Add to this the possibility of competing businesses located in close proximity to the proposed site. Gravity models are commonplace in economic location theory and are now well established in commercial GIS software.

Numerous other geographic models have found their way into GIS software, while some are still awaiting implementation. Some have been designed to examine the changes in population densities and locations within cities (cf. Casetti, 1969); others have shown that cities themselves operate as regional hierarchies, exerting differential attractions on regional businesses (Christaller, 1966; Brush, 1953). Still others have been developed to plan transportation networks based on speeds, road types, and other factors. Still other models examine the movement of ideas through space (Hägerstrand, 1967). Such **innovation diffusion models** are now used extensively in the ecological community to track movements of plants and animals through geographic spaces.

Perhaps among the more sophisticated spatial models are those recently employed by landscape ecologists to examine the relationships between structure and function of large portions of the biosphere (Forman and Godron, 1987). The variables covered by these models include patch (polygon) size, landscape connectivity, patch diversity, contagion, and shape. The belief among these scientists is that a knowledge of the structure of landscapes is directly related to their functional capabilities and to overall environmental quality. The roots of many of these techniques are predominantly geographic, whether explicitly or implicitly, and their use today shows the increasing need to examine spatial relationships. The landscape ecology community has already implemented many of these inside existing commercial and public domain GIS software (Baker and Cai, 1992; McGarigal and Marks, 1994). Far more models could be mentioned. My purpose is not to provide an exhaustive listing of all spatial models, however, but to illustrate that many solutions have already been developed, tested, and employed. As you begin modeling complex systems, you are not alone in the cartographic world. No matter what your field of endeavor, as you continue to use GIS, you will continue to experience different modeling needs and seemingly unique problem sets. It is generally useful to examine the literature to see what others have done when confronted with a similar problem. By doing a little research before you begin your modeling, you can often save many hours and avoid a great deal of frustration. In turn, you will likely develop your own solutions to difficult problems that the next generation of cartographic modelers will employ in their applications.

TYPES OF CARTOGRAPHIC MODEL

There are strong similarities between the categorization of statistical techniques into descriptive and inferential types and the major types of cartographic models we will encounter. Tomlin (1990) separates cartographic models into descriptive (like the statistical counterpart) and prescriptive (somewhat related to the inferential statistical techniques) types.

Descriptive cartographic models are those that describe and perhaps, under certain circumstances, explain particular patterns and pattern associations produced from analysis. The simplest **descriptive models** are those that merely illustrate existing conditions by isolating preselected phenomena and presenting the results in a form that allows the user to see at a glance what features are located in specific locations and how they are associated. This is not grossly different from the original communications paradigm in cartography except that the entity and attribute data have been preserved in a computer database. Despite its simplicity, this type of model is still among the more often used because it offers a relatively straightforward way of obtaining easily recognized patterns of spatial phenomena. A common complaint among well-trained GIS professionals working for non-GIS-trained personnel is that the new users never do any modeling. Rather, they ask the GIS initiate to "make me a map of" some phenomenon. This function could, of course, be just as easily performed with a computer-assisted cartographic system, or even with a CAD system. Still, its utility should not be diminished. Instead, the GIS professional should see this as the starting point for more complex models that might be suggested to the impatient user.

It would be natural to go from pure descriptions of existing conditions to a model that predicts how these existing conditions might affect the location of an industry, how these conditions might indicate a change in natural vegetation through time, or where an optimum habitat for waterfowl might already exist. Such a model may, for example, describe the results of damming a stream by showing the areas above the dam that would be inundated. It could also show the areas currently experiencing flooding downstream that would benefit from the construction of the proposed dam. A city planner might be able to predict locations of potential urban expansion by using knowledge of where expansion has taken place in the past to identify spatial phenomena that might be used as predictors.

The latter case should be examined more carefully because it shows the predictive potential of descriptive cartographic models. **Predictive models** allow the user to determine what factors are important in the functioning of the study area. They also permit the user to determine how these factors are associated with each other spatially. Prediction based on these associations can, of course, be very tricky. It requires that the variables have a clear and verifiable causal relationship. As we saw in Chapter 12, a spatial association among different mapped variables does not, of itself, dictate that there are cause-and-effect relationships, only that the variables occupy more or less the same space. A knowledge of the environment being modeled is as vital in predictive cartographic modeling as it is in inferential statistics such as regression analysis.

Predictive modeling is most often associated with the second major type of modeling defined by Tomlin (1990), prescriptive modeling. However, as we will soon see, there is no clear separation between descriptive and **prescriptive models.** Rather, the two might be thought of as two ends of the cartographic modeling spectrum, with prediction and prescription increasing as we approach the prescriptive end. Take the following example of prediction based purely on descriptive modeling techniques. The owner of a very large ranch, say in the southwestern United States, wishes to evaluate the carrying potential (i.e., the number of animals that can survive on the land) of her property holdings with respect to grassland for cattle and habitat suitability for native wildlife such as quail. The model requires her to develop a database showing all the vegetation types on the ranch. In addition, she needs to know the aboveground biomass (the amount of vegetation by weight that occurs above the ground level) for the grassland vegetation, as well as the locations of large patches of noxious weeds (that might be harmful to her stock) and significant clumps of shrubs (needed by the quail for shelter and foraging). Based on these variables, she produces a cartographic model showing the areas that have the minimal habitat needed to support a reasonable quail population. She also produces a coverage indicating the minimum necessary carrying capacity for the cattle. The model is descriptive in that it shows where the owner could reasonably expect to put her cattle and where quail might survive as well. At this point she could simply herd her cattle into the adequate grazing portions of the ranch and prevent the livestock from moving into the quail habitat by fences or other means.

Suppose, however, that the rancher does not wish her cattle merely to graze successfully but rather wants them to have only the very best grazing lands, to ensure that the animals are fattened as quickly as possible. Such a manipulation of the database could reasonably be called predictive because in effect it is predicting that the areas with the highest carrying capacity will result in the most efficient use of the land for grazing. This model is only slightly more complex than the original, but it carries with it an element of prescription because it "prescribes" the "best" use of the land.

Prescription also can show the model user how manipulating existing attributes can improve an overall situation. For example, some of the best grazing land may contain many shrubs, making it prime quail habitat, as well. If the same location also is found to contain poisonous weeds, the rancher may decide to plant (or transplant) shrubs in areas that are less productive for cattle, while at the same time eliminating the noxious weeds from the most productive grazing sites.

The GIS could be used to "**prescribe**" the best places for moving the shrubs as well as targeting for other manipulation locations with noxious weeds. Thus the GIS is now used to suggest the appropriate manipulation of existing holdings to ensure the best solution to the original problem. The difference between this highly prescriptive model and the preceding example is that the cattle-and-quail model required more predictive capabilities than did the high school scenario. It is the highest form of predictive model because the user has to determine the interactions of many variables through time, and a much higher degree of predictive capability is called for.

You have probably noticed a common theme of predictive modeling, even if an individual case is not purely prescriptive. All these models require, as a prerequisite, a description of existing conditions. That is, when developing prescriptive models, you describe first, then prescribe. There is no need to dwell on the terminology if you remember that the classification of cartographic models as descriptive and prescriptive is, like so many other classifications, arbitrary. If you remember that the change from one to the other is a gradual transition from pure description with no action needed to increased predictive power and increasing prescriptive capabilities, you should have no problem understanding the concept. In any event, the nature and complexity of the problem will dictate the type of model you employ, no matter what you call it.

INDUCTIVE AND DEDUCTIVE MODELING

Whether you are working with purely descriptive models or prescriptive models, or somewhere along the spectrum, there are generally two ways in which models can be formulated. These methods incorporate the same logical approaches used in any scientific research. The first, called the **inductive method** (from inductive reasoning), moves from specific elements to a general statement. That is, you start with a large collection of spatial data for a particular region of study. Or, as occurs often when GIS projects proceed by identifying what data people would like to have without thoroughly researching what the users might actually want to do with the material, you assume that the data can be obtained. As you will see later, such an approach lacks design and often results in failure of the GIS to perform to the users' expectations. Still, it has some validity, especially in a research environment where spatial data have been collected over a long period of time for individual research projects. It is generally assumed that as long as the data are of good quality, we probably can do something with them if we put them into a GIS database.

Under such a setting the empirical spatial data are examined through trial and error. Different coverages are compared and tested for correspondences among the elements of other coverages to determine similar spatial patterns or associations that might indicate an ongoing process within the study area. For a non-GIS example of the inductive method, suppose you go to the grocery store because you are about to prepare a meal. You purchase a large number of food items, choosing each one because you like it or because you think it would be good to have that evening. But upon arriving home, you find yourself in somewhat of a dilemma. Although all the items you purchased are edible, you are hard pressed to convert them into anything resembling a meal. You have purchased enough different foods to come up with a menu, but you wind up preparing a meal that does not include some of the groceries just purchased. These items were unnecessary for the immediate need, although some may prove useful later.

As this example indicates, trial and error can be used to impose order on items acquired without careful planning. The same can be said of using existing GIS databases. If there are enough of the appropriate variables, and if there are resulting correspondences among variables on different coverages, it is entirely possible to create models that will be useful for making some decisions. You will probably also find that coverages you have spent many hundreds of hours creating are never used. Many such projects have contributed to scientific research, but the lack of design in the inductive approach to cartographic modeling is often too inefficient for many commercial applications. While science attempts to explore data, real-world applications are often much more specific in their goals and should therefore be thought out in a more deductive manner.

Deductive models move in the opposite direction from inductive models.

That is, you begin with a specific recipe or formulation that addresses very specific questions. Let's return to our grocery shopping example. In the deductive approach you would select the meal you are going to prepare, determine the specific ingredients needed, and proceed to purchase only those ingredients. You could then combine all your purchases to produce the meal, with no ingredients left over. The advantages of the deductive approach to cartographic modeling should be obvious. You wasted no time or money producing unnecessary coverages. You develop only the coverages you need, and you can concentrate on preparing these coverages with the care needed to answer your questions.

FACTOR SELECTION

As you remember from Chapter 5, one of the difficult problems we face in creating models is the selection of appropriate variables. In some cases it is relatively simple to decide which factors you should use. If we are going to model a transportation net, common sense tells us to include all the available road network. It is also intuitively obvious that we need to know which are dirt roads and which are superhighways, which are one-way streets and which are two-way, which have traffic signals and which do not, and so on. All these factors are accessible in direct data sources that relate explicitly to what we are trying to model. However, not all data are as easily obtained. Nor are all variables explicitly defined.

Let's take another example. Suppose you want to model the potential impact of a hazardous chemical spill along a rail line that runs through your town. Of course, you have explicit information of the location of the rail line, and you might have manifest information about the actual materials being transported. You also have explicit spatial information about the number of housing units located in the hazard zone. Because the population in your area is highly mobile, however, you don't know exactly how many people live in each household, nor do you have information about any special needs populations such as elderly or handicapped people. To develop your model, you will need to use surrogates for these variables. Perhaps you will sample the households in each neighborhood to get an idea of the average number of people living in each unit and the average number of special needs individuals. By generalizing these data, you can obtain an estimate for each neighborhood. It is these estimates that will be used in the development of your potential spill impact model.

MODEL FLOWCHARTING

Whether you are using a deductive or an inductive approach, an extremely useful technique to assist you in formulating your model and determining the appropriate coverages is the **flowchart.** Model flowcharting requires that you isolate each item or element (coverage) that is to be used in your model. Each coverage should have a very specific, unique theme that is representative of a single factor or group of factors in your model. Flowcharting allows you to determine whether you have all the necessary coverages. It also allows you to evaluate whether each coverage is unique. If you have several redundant coverages representing the same theme, you can eliminate one or more of these from your model, and the input time will be reduced.

Figure 13.3 is a simple deductive cartographic model flowchart that attempts to locate the best site for a mountain cabin. As you can see, the result we are seeking is very specific, and at least four major factors need to be considered in making the decision. Let's start with infrastructure. You will need to know about the availability of running water, electricity (no mountain cabin is complete without a television and a VCR), perhaps gas for cooking or heating, roads for access, and other facilities (sewerage, etc.). Then there are political or legal factors. You must, for example, know whether a parcel of land of interest either has not been claimed or is owned by someone willing to sell. You will probably need to obtain a permit for building, or you may be prevented by government mandate from building your mountain cabin on land zoned for nonresidential purposes. If your cabin is to be located in an earthquake-prone area, there may be restrictions as to design or construction methods. Fire-prone areas will require you to use fire-retardant materials. Such restrictions place constraints both on where you can build and how much the effort will cost you.

A third group of factors to be included in your cartographic model are aesthetic. After all, you probably are building a mountain cabin to use as a retreat from your usual environment. You will want the cabin to be out of sight of the city (while still within driving distance). In addition, you want an unobstructed view of the mountains. You may also want the cabin to face in a particular direction. Perhaps you want the back of the cabin facing west so you can view the sunsets from your back porch.

The fourth group of siting factors to be evaluated are physical factors. You will need a slope that is not so steep that construction will require expensive measures to prevent the finished cabin from moving downslope. Also to reduce costs, it would be better to choose a site that does not have dense tree cover. A site with a small clearing surrounded by trees would be ideal. Soil engineering properties will also have to be evaluated, since you do not want to locate your cabin on unstable soils, especially soils with a large amount of clay (which shrinks when it dries and swells when it is wet). Keeping in mind that you may not have access to sewerage facilities, you may have to plan for a septic tank system. This will require you to factor the infiltration capabilities of the soil into your calculations.

Let's move to the actual model now, looking closely at the way in which each coverage was chosen to represent individual unique factors. Later we will examine how these coverages could be combined to complete the model and indicate the best site for our mountain cabin. We can begin by looking at the infrastructure portion of our model. For data regarding the first factor—accessibility to running water—we need a coverage that shows where all the municipal water lines are located. The coverage, "waterline," would be created by digitizing an analog map of water lines. Likewise, we create coverages for electric lines ("electricity"), locations of gas lines ("gaslines"), and all available roads in our study area ("roads").

Next we turn to political and legal factors. First, from cadastral maps (maps showing land ownership) we digitize "ownership," the most current landownership coverage. This coverage will allow us to determine whether there are

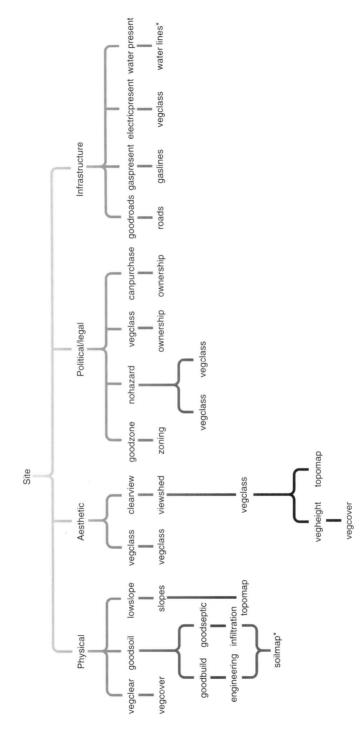

Figure 13.3 Simple mountain cabin model flowchart. This shows each of the elements as well as the interim coverages needed to produce the final solution—the Site.

available unclaimed sites or will tell us whom we need to contact about purchasing the land. Next we obtain a zoning map and digitize "zoning," a coverage that will permit us to determine which areas are not zoned for single-family dwellings. We also needed to know whether there are locations that are earthquake zones or fire hazards. It is likely that data on these two potential problems will likely come from different sources. Going to the state geological survey, we obtain a map of earthquake-prone areas and digitize these ("earthquakes"). The forest service might be the best source for maps of fire hazards, so we use that information to create a coverage called "firezone." Although the last two coverages could easily be considered to be physical parameters rather than political or legal, it is likely that permit restrictions are already in place for such areas. Local government offices might have provided us with maps showing these restrictions. Often, however, a would-be builder is required to make a case for a site's not being in an earthquake-prone area or a fire hazard zone, so we would already have the information if the government offices do not.

An evaluation of aesthetic qualities will require maps showing the locations of nearby municipalities ("urbansites"). Topographic maps would provide information about high topographic forms that could obstruct our view ("topomap"). Maps of forest types, usually available from the forest service offices nearby, tend not to include information about the sizes of trees because of the amount of time required to obtain such information. You might have to create your own forest cover map, either by performing a ground survey of some of your more suitable locations or by using aerial photography of the area and mapping the forest stands yourself to create the needed coverage ("vegcover").

Our final group of factors includes physical parameters. Slope, the degree of change in elevation over horizontal distance, can be obtained by means of a coverage we already have, topomap. While slope is a unique element, it is derived by modifying existing data. Thus we have not violated our rule about selecting unique elements. Another element of our physical parameters that can be obtained from an existing coverage shows areas with little forest vegetation, or with forest vegetation containing an appropriate clearing. The existing coverage called "vegclass" can help us obtain this information.

Our two soil parameters will likely be obtained from a single coverage as well. A detailed soil map will normally have both engineering properties and infiltration properties; either of these will be explicitly encoded to produce separate maps, or they will be associated through a database system and connected to each of the soil-type polygons. We will create these maps by first creating the soil coverage ("soilmap"). The methods we use to keep track of the various soil properties will largely depend on our GIS software. Alternatively, we could simply encode the soil polygons and retain a paper copy of the soil survey from which we obtained the soil polygon information. These survey reports associate a large variety of soil factors with each soil-type polygon (Table 13.1).

Working Through the Model

Look closely at the flowchart we used for the mountain cabin siting model: all the appropriate coverages and elements are included, and they all flow to the

TABLE 13.1 Typical Page from a Soil Survey Report Showing Soil Polygons and Small Portions of Tabular Information Available for Each Soil Type

Soil Name and Map Symbol	Dwellings Without Basements	Dwellings With Basements	Small Commercial Buildings	Local Roads and Streets	Septic Tank Absorption Fields	Sewage Lagoon Areas	Playgrounds
Ev:							
Eudora part	Severe: floods	Severe: floods	Severe: floods	Severe: frost action	Moderate: floods	Moderate: seepage	Moderate: floods
Gravelly land: Ge	Moderate: slope	Moderate: slope	Moderate: slope	Slight	Severe: small stones	Severe: seepage, small stones	Severe: small stones
Judson: Ju	Moderate: shrink/swell	Moderate: shrink/swell	Moderate: shrink/swell	Severe: frost action	Slight	Moderate: seepage	Slight
Kennebec: Kb Kc	Severe: floods	Severe: floods	Severe: floods	Severe: floods, frost action, low strength	Severe: floods, wetness	Severe: floods, wetness	Moderate: floods
Kimo: Km	Severe: floods, shrink/swell	Severe: floods, shrink/swell	Severe: floods, shrink/swell	Severe: shrink/swell, low strength	Severe: percolates slowly	Slight	Severe: wetness
Martin:							
Mb, Mc, Mh	Moderate: shrink/swell	Severe: shrink/swell	Severe: shrink/swell	Severe: low strength, shrink/swell	Severe: percolates slowly	Moderate: slope	Moderate: too clayey, percolates slowly
Morrill:							
Mr	Moderate: shrink/swell, low strength	Moderate: shrink/swell, low strength	Moderate: shrink/swell, low strength	Moderate: shrink/swell	Severe: percolates slowly	Moderate: slope	Moderate: percolates slowly
Ms	Moderate: shrink/swell, low strength	Moderate: shrink/swell, low strength	Severe: slope	Moderate: shrink/swell	Severe: percolates slowly	Severe: slope	Severe: slope
Oska: Oe	Severe: shrink/swell	Severe: shrink/swell, depth to rock	Severe: shrink/swell	Severe: shrink/swell	Severe: depth to rock, percolates slowly	Severe: depth to rock	Moderate: percolates slowly, too clayey, depth to rock
Pawnee: Pb, Pc, Ph	Severe: shrink/swell	Severe: shrink/swell	Severe: shrink/swell	Severe: shrink/swell	Severe: percolates slowly	Moderate: slope	Moderate: percolates slowly
Sharpsburg: Sc, Sd	Severe: shrink/swell	Severe: shrink/swell	Severe: shrink/swell	Severe: shrink/swell, low strength	Severe: percolates slowly	Moderate: slope	Moderate: percolates slowly
Stony steep land: Sx	Severe: depth to rock, slope	Severe: depth to rock, slope	Severe: depth to rock, slope	Severe: depth to rock, slope	Severe: depth to rock, slope	Severe: depth to rock, slope	Severe: depth to rock, slope
Thurman: Tc	Slight	Slight	Moderate: slope	Slight	Slight	Severe: seepage	Severe: slope
Vm:							
Vinland part	Moderate: depth to rock, slope	Moderate: depth to rock	Severe: slope	Moderate: depth to rock	Severe: depth to rock	Severe: depth to rock	Severe: depth to rock
Wabash: Wc Wh	Severe: wetness, floods, shrink/swell	Severe: wetness, floods, shrink/swell	Severe: wetness, floods, shrink/swell	Severe: wetness, floods, shrink/swell	Severe: percolates slowly, floods, wetness	Severe: floods, wetness	Severe: wetness, floods, percolates slowly

Source: U.S. Department of Agriculture.

final site for our cabin. There is still something missing, however. Ask yourself how the intermediate coverages were created from the initial data elements. To complete our flowchart, we need to indicate what functional capabilities of the GIS we use to get from one branch to the next. This step is not only important in modeling, but, as you will see later in the course, it will assist you in determining the needed functionality of any GIS for a specific project. Let's consider a revised version of the flowchart (Figure 13.4).

To avoid using specific commands I will continue to use rather generic functions with mnemonically useful names as I have done so far. To implement the model on your own software you need only to change my names to the specific commands you would use. We begin by examining the infrastructure parameters, namely, roads, gas lines, water lines, and electric utilities. There is a high probability that none of these will directly serve your cabin site. You need to know whether they are close enough—say, within 500 meters—to make it feasible to build extensions for each. To find out, you begin with one of the existing elements. The digitized map of electric utility lines will tell you where they are located, and it will not be too difficult to create a buffer of 500 meters on each side of the existing electric line elements. The result is a map called "electric-present," which shows cabin sites that do or could have access to electric power. Next you might build a buffer around your gas lines coverage extending, say 250 meters, indicating the likely areas where gas lines could be extended to your cabin—a coverage called "gaspresent." Roads are not nearly as important because you have a four-wheel-drive vehicle. You could travel as much as 750 meters off-road to get to your cabin. Therefore, you build your buffer 750 meters around the existing roads coverage to produce a new coverage, "good-roads."

You now have three separate, intermediate coverages, each representing the limits of your requirements for accessibility to roads and electric, water, and gas utilities. Now you must decide how to combine these to produce a final coverage, called "infrastructure," to represent all these factors. Because you need all these elements, rather than just one or two, they must be overlaid in a way that results in a map showing where they intersect. Therefore, the overlay procedure you choose must use some form of "and" search, or intersection, as you learned in Chapter 12. The resulting map should display at least one area that satisfies all three criteria.

At this point you might be wondering, "What happens if there is no area that has all four factors?" This is not an unlikely outcome. If your composite map shows no intersections, you could focus on the factor that is the farthest from intersecting the other two. Let's say that the gas line buffer and the roads buffers intersect, but the electric lines do not. If the distance limits for that factor can be extended—in other words, if the model constraints can be relaxed—you might be able to fulfill the requirements of your infrastructure submodel. For example, you might find that for an additional charge, you could extend the electric lines to 1000 meters. This approach to improving model performance by constraints relaxation is common in cartographic modeling. You use a similar kind of thinking when you decide to buy an automobile you can afford instead of the luxury model you want.

If constraints relaxation is impossible—for example, if the electric company is unwilling to extend lines 1000 meters from their existing facilities—you may have to reduce the importance of one or more of your factors. In creating your

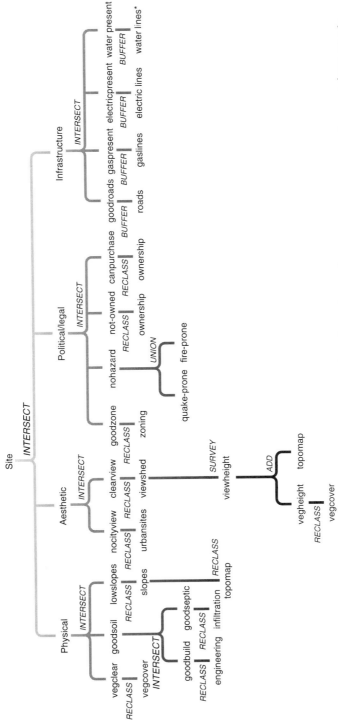

Figure 13.4 Detailed mountain cabin model flowchart. This flowchart shows the processes necessary to move from one interim coverage to the next to obtain the final solution. *Source:* Modified from S.A. Carlson, and H. Fleet. Systems Applications Geographic Information System (SAGIS) and Linked Analytical/Storage Packages. Workshop, Annual meeting of the Association of American Geographers, 1986.

buffers, you based the size of each one on the absolute maximum limits for the corresponding factor. That is, you assumed that all factors were equally important. This is not always the case. In fact, as you remember from Chapter 12, it is rare for all factors in a model to have the same importance. Therefore, unless you are willing to abandon the project, you must weight the four infrastructure factors. Roads are important beyond the 750-meter limit because you will need to get to and from the cabin somewhat regularly to obtain food, medical supplies, and the like. This factor then must retain a high degree of importance. If gas is available, you could use it to heat your home. You could even have natural gas lighting if you wished. What is left is our electricity, which was the problem in the first place. The solution to your problem is to downgrade your perceived need for electricity and give up your television and VCR (or use battery-operated units). Thus you could reduce the importance of electricity to zero, eliminating it as a model constraint.

Now that we have satisfied our infrastructure submodel constraints we can proceed to the political and legal factors. Areas not zoned against single-family dwellings ("goodzone") can be obtained by reclassifying the zoning coverage to eliminate all areas that do not meet this constraint. Likewise, you can create a coverage of property without a registered owner ("unowned") and another of properties whose owners may be willing to sell or have offered the land for sale ("canpurchase"). Both these maps use the same initial coverage ("ownership"), and both are produced by simply reclassifying categories to obtain what you want. Your coverages for the remaining political or legal factors, "quakezone" and "firezone," contain the information you need, but your cabin must be located outside both of these areas. As you remember from Chapter 12, you can combine these coverages using an "or" search or a union overlay operation. The resulting coverage, called "nohazard," shows all areas that do not have seismic and/or fire hazards.

You now have the four intermediate coverages needed to complete your political/legal submodel. Once again, to satisfy all these constraints, you combine them through an overlay operation that intersects the factors to produce a map of political/legal viable locations. Here you must be careful how you intersect the coverages. You must be sure that all four coverages indicate good areas for locating your cabin. If one of your coverages, for example, the nohazard coverage, actually shows locations that have hazards, your cabin will be located in a rather warm and possibly shaky location.

The third submodel, aesthetic factors, has fewer components than the other three but is somewhat more tricky to calculate. Beginning with your coverage of urban sites, you may not be able to eliminate all small towns and villages from your study area. In fact, you probably want a nearby community as a source of supplies. You decide that you want no town larger than, say, 1000 people. By reclassifying your map it is possible to create a coverage called "nourban" that shows small towns and villages, but no large towns or cities. Although this might be appealing, it is not what you want, because you will need to know where the large towns and cities are so you can determine whether they can be viewed from your possible location. So you must reclassify your "urbanzone" map to show where all the cities are located ("cities"). This coverage will then be combined with your next one to eliminate such unsightly anthropogenic artifacts from your view.

The second portion of your aesthetic submodel requires that you be able to see around any peaks and any tall vegetation. Remember that the vegetation is growing on the terrain itself. When we work with maps, we tend to forget that vegetation is not a flat pattern, colored green, on a flat sheet of paper. Instead, it adds elevation to existing topography. To incorporate this feature into our model we must find an approximate height for mature forest stands. (The average height at maturity of the various forest types can be found in standard botanical or forestry references.) Then we reclassify our vegcover based on the vegetation classes at hand and call the map "vegheight." The reclassification will produce a coverage of estimated tree heights and will be stored as ratio data rather than nominal data. To show what the topographic elevations will be like with tree cover on them, we add the topomap data (also in ratio format) with the vegheight data to produce a coverage called "trueheight." This is the intermediate coverage we will need to determine what can and can't be seen. By using a form of "survey" command, or a command that determines the viewshed (see Chapter 10), we produce a viewshed based on our ability to see around peaks of elevation and around the vegetation on those peaks. This coverage can then be reclassified to show areas that offer a clear view of the valley below ("clearview"). Remember, however, we don't want to see cities. The clearview coverage shows areas that are clearly visible from the hillslopes and the cities map shows all communities exceeding 1000 people. By combining these two coverages through intersection, you produce a map that shows both viewable areas with cities and viewable areas without cities. These viewable areas without cities will ultimately be combined with the other three submodels.

Completing the model, we next look at the three major factors of the physical parameters: vegetation, soils, and slope. Returning to our vegcover coverage, it is easy to reclassify the vegetation categories into those with large, mature tree stands and those with other types of vegetation or with relatively small trees that can be cleared. The resulting coverage, called "vegclear," should isolate areas with no mature tree stands. Keep in mind, of course, that the scale of most vegetation maps is rather small, offering little in the way of detail. Thus any decision relying on such data should be checked on the ground. Some mature tree stands on very rugged terrain are not very high, and hence may not be as difficult to clear as a vegetation map seems to imply. This is another area in which it is very easy to modify constraints to accommodate the decision process. In this case, having mature trees may not be particularly negative because of the ready availability of power saws. In addition, a few trees, properly placed, could provide shade without obstructing the view or interfering with the construction of our cabin.

Among the most severe potential physical constraints in siting a mountain cabin is the steepness of the slope. A severe slope will either prohibit the project altogether or radically increase its cost. A slope of 15 degrees or less would be ideal. To find all the areas that fit that constraint, you would begin with the topomap coverage and perform a neighborhood function that separates out the degree of slope throughout the coverage. By reclassifying these slopes so that low slopes are those of 15 degrees or less, you now have a coverage ("lowslope") that fits these constraints.

The two soil characteristics that must be considered are engineering properties and infiltration capacity. The first determines the capacity of the soil to withstand a structure like a mountain cabin, and the second classifies a soil according to whether there is sufficient infiltration capacity to allow the construction of a septic tank. Both these factors come directly from the original soils map; either they will be reclassified to fit the appropriate categories, or they will be selected from an associated attribute table. The coverages produced ("goodbuild" and "goodseptic") can then be intersected to create a more nearly complete coverage called "goodsoil." This coverage completes the three physical factors dictating where we can put our mountain cabin. Through an intersection type of overlay, we can create the final submodel coverage, called "physical."

As you probably have guessed, the final solution is simply to intersect physical, aesthetic, political/legal, and infrastructure maps, because all these constraints must be met. If our model works as planned, we should have at least a single, small portion of land that suits all our needs.

Not so fast. Just when you thought you were finished, yet another problem appears. Earlier we discussed the problem of constraints that were too tight. When we looked at our infrastructure factors, we discovered that insistence on having electricity could limit our choice so severely that the idea of building a mountain cabin would have to be abandoned. There is, however, the opposite possibility. In some regions we may actually have very good soils with few very steep locations, lots of good mountain views, no major zoning or ownership problems, and plenty of close infrastructure. This might seem somewhat unlikely, but let's run with it anyway. If this nearly fairy-tale scenario were true, our model would indicate large portions of the study area in which building conditions were appropriate. In fact, it seems we could build our mountain cabin almost anywhere. Thus an apparently wonderful outcome shows that the model was not very useful in helping us make a decision. Don't give up yet; the GIS is not to blame. Remember, GIS is a tool to assist in making decisions. It isn't meant to replace the decision maker.

Whereas when modeling constraints were too tight we loosened them to make the model more flexible, now we do the opposite. In this case our model constraints are too lax, and we will have to tighten them a little, to give the GIS user the very best mountain cabin location from the model. As before, there are two ways in which to accomplish this. The first would simply be to modify the criteria. For example, we could require soils of the absolute best quality for building the cabin and installing a septic tank. We could also require slopes of less than 7 degrees rather than the 15 degrees originally selected. Another factor that could be modified is the size of town or city that would be acceptable within our viewshed. As you can see, quite a number of changes could be made to the model. In fact, it is generally easier to produce a range of values for each factor, whereupon the answers obtained can be ranked, rather than treated simply as "yes" and "no" answers to the facets of the problem. You might remember this from our earlier discussion of reclassification in Chapter 9.

Again, there is an alternative approach to tightening up our model. Certainly physical, aesthetic, political/legal, and infrastructure parameters are all important. However, they are not all equally important. Physical limitations of the soil and slopes have a substantial impact on your ability to build your cabin. However, if there are no lands zoned for single-family dwellings, and the zoning commission is unwilling to give you a variance, the physical parameters are no longer important. Instead you have one set of factors that can be called absolute limitations, while others might be overcome by modification of the constraints or by a willingness to invest more in the project. A determination of which factors can be weighted and which are absolute or preemptive will help the cartographic modeler adjust the solution to the individual situation. You must be flexible and very aware of the impacts of these weights on the final outcome. There are, however, a few examples in the literature of the use of weighting and reweighting to improve a model and to satisfy the ultimate users (Davis, 1981; DeMers, 1986; Lucky and DeMers, 1987).

Conflict Resolution

It is important to consider factor weights and the importance of different coverages when you produce a cartographic model. However, weighting will not always produce results that satisfy all possible conditions. Consider the problem of producing a cartographic model that involves two essentially competing views of the world—say, selecting timber harvesting versus the preservation of spotted owls to comply with federal regulations. Even without the federal mandates, there are essentially two antagonistic ideas about the potential use of large portions of land. Both those concerned with preserving spotted owls and those trying to use the forests as an industry that will enable them to feed their families have valid standpoints.

How could we approach such a difficult modeling task? Let's look at one readily available technique suggested by Dana Tomlin. Although not the only approach, it shows that the competing portions of a problem can be conceptually separated out and a reasonable compromise arrived at without employing unusual software, weird logics, or high-priced negotiation teams. The approach, called ORPHEUS, involves creating two separate models, one for each competing side. After evaluating all the factors, you begin to put weights on each of the factors for each party (Tomlin and Johnston, 1988). If you continue to reweight the factors on a pair-wise basis, hopefully getting small concessions for each side, your output maps should reveal areas that will allow the competing demands to coexist. The process takes time to perfect, and it seems that there is no substitute for experience. Try this out for yourself by creating a small database, one square mile or so, that includes soils data with associated tables of properties. Then select six to eight possible land uses totaling one full square mile to place in this area. Now use your GIS to place these land uses, keeping in mind the usual needs for size of some uses, locations of others, proximity to pollution, and so on. Even if you don't immediately achieve positive results, the experience will be very instructive.

Some Example Cartographic Models

To finish our examination of flowcharting models, let's look at three basic examples of models and their flowcharts. Our sample cases have separate themes and use quite different approaches both to modeling and to flowcharting. This should give you an idea of some of the options.

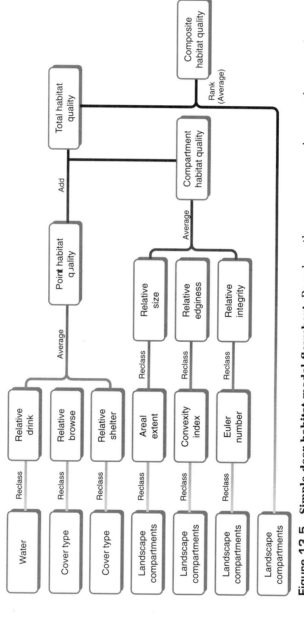

Figure 13.5 Simple deer habitat model flowchart. Based on the presence or absence of some primary survival factors: shelter, food and water, and the spatial configuration of the landscape with respect to size, amount of edge habitat, and the integrity or patchiness of the landscape. *Source:* Adapted from Joseph K. Berry (personal teaching notes). Used with permission.

We begin with a descriptive model of deer habitat quality (Figure 13.5) (Carlson and Fleet, 1986). This simplified model is based on the presence or absence of some factors essential to the survival of the deer: availability of water and vegetation for browsing, and shelter; availability of water is based on a water or hydrology coverage, while browsing and shelter are based on the vegetative cover. Browsers normally obtain their food from woody plants, which have their leaves off the ground. Thus the necessary forage and shelter factors are based on a reclassification of the cover vegetation into woody shrubs and trees as useful and grasses and forbs as nonuseful vegetation types. The model is somewhat vague about the precise types of vegetation needed and whether all types of water can be drunk. Still, it should be easy to imagine how these factors could be obtained from the available coverages with basic zoological/behavioral knowledge (i.e., of the animals' requirements for food, cover, and drink).

A second portion of the model is designed to determine habitat quality based on spatial configuration of the landscape units. The primary considerations here are the size, amount of edge for each landscape compartment or patch (i.e., edginess), and integrity (c.f. Figures 8.2 and 8.3). This portion should remind you of our earlier discussion about the use of landscape ecological units in cartographic modeling. In fact this is an excellent, easy-to-understand example of landscape spatial variables as they are applied to a single-species habitat. The final portion of the model totals the habitat quality components for the separate landscape components and scores or ranks them through, in this case, averaging, to obtain the composite habitat quality. Once again, it should be fairly obvious how simple submodels can be combined to achieve a relatively complex model of deer habitat. By compartmentalizing each set of operations and each landscape compartment, the individual models are easily produced.

Our second model is more of a prescriptive model that attempts to define portions of an area that will receive enough exposure to prevailing winds to operate a wind generator (Figure 13.6) (Carlson and Fleet, 1986). There are only two basic coverages: topographic elevation data and vegetation communities. The topographic coverage exposes the orientation or aspect of the topographic slopes, which have been assigned ordinal categories ranging from best to worst (for wind generators). Level ground will, of course, generally be the most useful because there are no obstructions, while in our example northwest is the next best because the prevailing winds come from that direction. As one moves further away from northwest, usefulness declines.

Another factor we need to consider is the distance to obstructions—in this case, trees and slopes—which is easily obtained by using a functional distance operation; this model uses a command called "Radiate." The required generator height, h', based on the heights of obstructions and the distance just calculated, is a simple difference between the obstruction height and the ratio of the obstruction height minus the generator height divided by the distance. The model assumes that towers will be no higher than 80 feet by subtracting that value from the topography, then ranks the sites independent of the tower heights. Finally the model ranks the sites independent of tower heights, then recombines the aspects to obtain sites available for wind generation. The flowchart for this model is considerably different from that of Figure 13.4; in fact, it is more algorithmic than flowchartlike. Some may feel more comfortable with

374   Chapter 13 • Cartographic Modeling

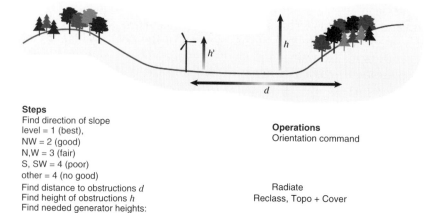

Steps
Find direction of slope
level = 1 (best),
NW = 2 (good)
N,W = 3 (fair)
S, SW = 4 (poor)
other = 4 (no good)

Find distance to obstructions d
Find height of obstructions h
Find needed generator heights:
    ratio $r = (h - h')/d$ for $r = f(d)$
    thus $h' = h - rd$
    select correct r
    find rd
    find h'
Eliminate towers
Display all ranked sites independent of tower heights
Display sites as a function f of (aspect, tower height)

Operations
Orientation command

Radiate
Reclass, Topo + Cover

Subtract $(d - d, f(r))$, Reclass
Multiply
Subtract
Subtract
Reclass, Display
Crosstab, Reclass, Display

Figure 13.6 Wind generator siting flowchart. Model flowchart for the siting of wind generators based on prevailing winds, terrain elevation, and tree heights. This example is designed more as an algorithm than as a flowchart. The mapped data, consisting of elevations and vegetation communities, can be manipulated by means of the commands in the "Operations" column. *Source:* Adapted from C. Dana Tomlin (personal teaching notes). Used with permission.

this recipe approach, although it is relatively easy to transform both the classic flowchart and the algorithmic approaches.

You might want to try the following exercise, which represents a more statistical method of modeling. If you are unfamiliar with regression techniques, however, you may choose to skip this material and proceed to the model implementation section. If you know something about regression, this model is an excellent example of how dasymetric techniques can be applied using the coefficients of a regression equation to modify the values of polygons.

Based on the predictive statistical method known as regression, our last model attempts to predict breakage during timber harvesting (Figure 13.7) (Berry and Tomlin, 1984). Such a value would be useful to the lumber companies as they try to determine their financial bottom line. In this model the independent variables (maps or coverages) used are percent slope, tree diameter, tree height, tree volume, and percent tree defect. Each variable has a measurable impact (indicated by the regression coefficient) on the predicted breakage. By multiplying each coefficient by each associated independent variable, the model produces weighted maps which, when added to each other and a constant (a necessary portion of a regression equation), yield the total predicted breakage occurring during tree harvesting. Because this predictive model uses tested, statistically significant predictions for each independent variable coverage, the results are highly robust and easily tested for validity, a subject we will cover later in the chapter.

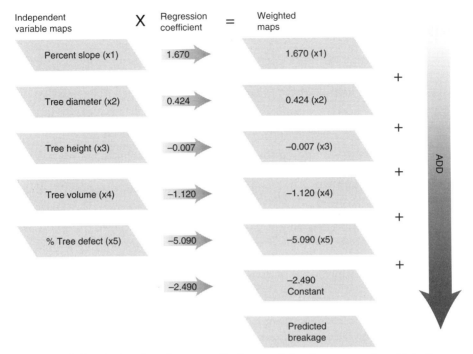

Figure 13.7 Timber breakage predictive model using regression. This flow-chart illustrates the use of dasymetric mapping techniques embedded in following the regression equation for conventional felling practices:

$$Y = -2.490 + [1.670 \times (1)] + [0.424 \times (2)] - [0.007 \times (3)] - [1.120 \times (4)] - [5.090 \times (5)]$$

Source: Adapted from Joseph K. Berry (personal teaching notes). Used with permission.

MODEL IMPLEMENTATION

A discussion of model implementation might seem mundane, given our flow-chart indicating how the model is to be produced and which coverages we need to produce it. But thus far we have assumed that everything is going to work according to plan and that our results will naturally conform to our expectations. Many GIS practitioners have, through trial and error, seen that it is best to take a conservative approach to model implementation. Although input consumes much more of our time than any other phase of GIS, we can nevertheless find ourselves spending many hours correcting mistakes if our implementation fails. For this reason alone it is wise to select a small subset of the study area on which to prototype the model, a technique that will be seen to be very useful in model verification later on.

Using our subset study area and our model flowchart, we begin at the elemental coverage level for each submodel and perform the necessary operations. This activity allows us to examine some of the following basic questions:

1. If we are reclassifying our categories, is our database set up to permit retrieval of the appropriate values? Or, if we are reclassifying categories

in a nondatabase raster environment, does the order of reclassifying make a difference?

2. If there are alternatives to a single technique or alternative combinations of techniques, which ones are the most likely to give us the correct response with the least amount of work?

3. Are we absolutely sure that the operations selected are representative of how the modeling environment really works?

4. How are we going to deal with missing variables, if we have any, in each coverage?

There are, of course, many more questions, but I think this short list makes the point. Be sure you know what it is you are doing before you do it for the entire database. By performing the operations on a sample of the database, you have a better chance to evaluate what is happening at each step than you would if you used the entire database. In addition, a small sample database makes it possible to test some of the results by hand as they are generated, thereby saving a great deal of time during model verification.

Another important aspect of implementation is determining which intermediate coverages will be needed to continue modeling and which should be retained for later model verification. Many GIS practitioners take an extremely conservative approach here, believing that if a coverage is created, it probably should be kept, just in case. If you have very large disk space on your computer, this is certainly a reasonable approach. However, if your study area is very large or if you have mounds of detailed data, or even many coverages, the result can be a model that quickly exceeds your capacity to store the intermediate coverages. This is yet another reason to consider using a small database subset to estimate the amount of data being produced for each intermediate coverage. By extending this information, you will be able to decide whether you can systematically eliminate intermediate coverages that will no longer be used or whether you need a larger storage device.

Here it is appropriate to state another caution, one that was emphasized when we talked about the input and storage/editing subsystems: back up your data. Whatever media you use for backup, the elemental coverages, the intermediate coverages, and the final products should be included. In fact, output or plot files, a subject covered in greater detail in Chapter 14, should also be backed up. Most people in the various branches of computer science use the following simple saying: "Back up your backups." A wise idea. During model implementation you will be creating a large variety of useful output. Some of it will be needed to make refinements to the model during implementation; others you will want to keep in case additional models from the same database are required, and still others will be needed for model verification (our next topic).

Because the implementation is by far the most exciting part of GIS, it is very tempting to try out many different modifications just to see what they do. To help yourself avoid this temptation, keep in mind that the GIS is capable of producing all sorts of meaningless output. Stick to your flowchart unless you have good reason to deviate from it. Continued model tweaking will frequently result in many coverages that cannot be identified later. If you have too large a collection, it can always be purged; because of the sheer numbers, however, you may wind up discarding needed coverages during cleanup operations.

Which brings me to my last cautionary note. If you are implementing a GIS model for yourself, be sure to document every coverage you produce. Keep records of what you have done, and what each coverage represents. Protect your coverages. This will reduce the danger of throwing out good coverages with bad. It will also help to prevent you from unalterably modifying good coverages. If you are implementing a GIS model for a client, the same rules apply, but are to be emphasized. In most consulting work, the GIS is being operated on by numerous analysts. This means the chance of corrupting finished products is increased. Most modern GIS software has built-in security or administration programs or protocols that will prove very useful during the creation of all those intermediate coverages. We will discuss these at greater length in Chapter 15 when we discuss the design issues of GIS.

MODEL VERIFICATION

As we discussed in Chapter 4, the cartographic output is an appealing method of communication. The map has a visual impact that allows people to accept the document as a matter of course without questioning either the data from which it was produced or the cartographic or modeling methods by which the product was derived. It should therefore not be surprising that many, perhaps even most, GIS models are accepted as fact, just as most maps are accepted as fact. This tendency can be very dangerous for the client, who may lose money, prestige, or even an entire business based on decisions made from bad data or invalid combinations of correct data. It can also be dangerous to the GIS consultant, who may spend more time fighting litigation than producing cartographic models. We will look more closely at problems of these kinds later. For the time being let's see if we can find a way to prevent them in the first place.

The solution is **model verification.** There are few examples of model verification either in practical applications of GIS or in the literature. Although this might surprise you, it shouldn't. Remember, the GIS user is not immune to the power of the graphic device. Sometimes a pretty map looks just as good as a correct map. If you attend poster sessions related to GIS applications you will be almost instantly assaulted by a wide range of color and pattern, figure and ground, typography and symbolism. The results can be almost mind-numbing. It is no wonder we spend so much more time on creation than on verification. Besides, it takes a great deal of work to produce a simple cartographic product from a complex model.

Despite the sparse treatment cartographic model verification has received, there are few who would disagree that it is extremely important. So how do we begin? First we need to agree on what we mean by "verification." I define the term loosely to describe not just correct models but useful models. After all, if the results of your analysis are correct but unusable, you have ultimately failed to provide a product. Within this loose definition, three fundamental questions should be asked:

1. Do the data used in the model truly represent the conditions we are attempting to model?

2. Have we combined the model factors correctly, to represent proper fac-

tor interactions, thus correctly describing or prescribing the correct decision-making process?

3. Is the final solution acceptable by the users and/or useful to them as a decision-making tool?

Are the data representative of existing conditions? Our first question might appropriately be asked during the input phase of the project, but this is not always possible. Remember that one important reason for flowcharting a model is to determine missing variables. Another is to determine whether appropriate solutions could be derived with the existing coverages. At times, however, the final models produced strike us as counterintuitive; or as we begin to put the pieces of the model together, we find that the factors we used don't work. Such disappointing results usually mean that something was missing from the original conceptual model. Obviously, it is impossible to evaluate something that wasn't there, even after a flowchart has been produced. Unusual or counterintuitive results might suggest that the model is incomplete, or that the variables used were not adequately representative of model constraints. Therefore it is frequently a good idea to build a small subset of a proposed database and test the model in a small study site that is typical of the overall study area. Such a prototype will allow a quick determination of the ability of the model to characterize the overall study area without having to build the entire database.

If the results are positive, we can invest our time in building the complete database and running the model. If, on the other hand, the results are negative, we must begin to dissect the model to ascertain where the problem lies. This is another reason that flowcharting the model is so important. Most often a problem with a model will appear in one or more of the submodels. In our mountain cabin model, for example, we may experience difficulties with one or more of our physical parameters because very positive mountain cabin site characteristics keep cropping up in areas known to have dense forests. The discrepancy could be due to dated information. That is, our information on vegetation might have been compiled when the stands of trees in question were immature. By updating our information on vegetation stands, and running the model again, we should obtain a more realistic picture.

The foregoing example shows one of the fundamental approaches to evaluating whether data are representative of the important factors—namely, testing them against known conditions. A very good example of this is provided by Duncan et al. (1995), who used bird census data to test whether their Florida scrub-jay habitat model accurately represented the known locations of these birds. The study is particularly useful because it shows that a simple pattern comparison is an insufficient basis for this decision. Instead, these workers used statistical testing to evaluate the results and followed up with a logical discussion of the possible reasons for lack of correspondence.

Another problem that often occurs in determining whether model elements match needed real-world parameters is missing variables. Williams (1985) clearly indicates the possibility of a GIS model with missing variables. Some of these variables may simply be unavailable, and hence can be eliminated from the model. In such a scenario the final model should state which variables are unavailable and acknowledge explicitly that the model is not a complete depiction of the whole. In some cases, the variables are not missing, but rather are either too vague or ill defined to be used, or are by definition aspatial variables that cannot properly be placed in an explicitly spatial modeling context in the absence of a spatial surrogate (DeMers, 1995).

Our next verification parameters involve the twofold question of whether the factors have been properly combined. The first part asks whether the GIS analysis functions are producing the proper results based on the input parameters. Many GIS systems operate somewhat blindly in their analysis functions because the software vendor fails to provide specific information about how the results of, for example, a functional distance operation, are obtained. Research on a variety of GIS software has examined the differences in performance using the same functionality on an identical database (Fisher, 1993). The results suggest that the user might want to test these different software systems in much the same way. This is seldom possible in an operational setting, however, because of the high cost of purchasing and maintaining multiple GIS software and hardware configurations. An alternative is to produce an artificial database with known parameters—a sort of control database—against which to test each analysis operation for validity of results. This approach, like the former, is time-consuming, and you should consider carefully the importance of embarking on it. If your results, especially for highly complex operations, are extremely sensitive, it might prove to be a useful endeavor. There are few if any guidelines for producing control databases. You must rely on your own knowledge of the available data and your notion of how you would expect the software to perform. This is one more reason for the necessity to acquire conceptual knowledge of GIS functionality.

Another approach assumes that you are comfortable with the software algorithms but are not entirely sure that all the correct steps were performed in just the right sequence. To test the validity of the modeling in this case, you can employ the approach you used in elementary school when you were learning how to divide—namely, reversing the process. In the arithmetic case, you multiplied the answer you wanted to check by the divisor, hoping to come out with the dividend. In the cartographic modeling context, you might subset a small portion of the model and its elemental database. Backtracking the model results by following the flowchart from the trunk to the branches should produce the same coverages you started with (Tomlin, 1990). Any step that does not produce the same coverages will immediately become obvious, and you will also know which portions of the overall model need to be recalculated. If you have kept your intermediate coverages (a good practice until the model results are confirmed), you can use them instead of starting the model at the very beginning.

The second part of our model verification task of determining whether the factors are combined correctly is much more difficult. Now we must ask whether the factors themselves interact in a manner that actually simulates or models reality. Because of the nearly infinite number of combinations of factors and model types, there is no simple recipe. Here you must rely on your knowledge of the environment you are attempting to model and your intuition. In most cases the modeler becomes so familiar with both the environment and the data representing the environment that any discrepancies between the data and the model are immediately apparent. This is identical to an evaluation of whether we are using the correct parameters. If the model seems to be counterintuitive, there is a good chance that something is wrong. If you have eliminated bad data as a possibility, and you know that the algorithms were operating correctly, you are left with the possibility that the model itself is incorrect. Rather than being distressed, incorrect or counterintuitive results should, as in statistical analysis, result in a further investigation of how the environment actually operates. In this case, the GIS becomes a useful tool in advancing science as well as implementing it.

Finally, the third major category of model verification concerns the usefulness of the model as a decision tool. The final output format can have an important impact on whether the results of modeling are useful. We will discuss this in more detail in Chapter 15, but a few words of caution are important within the context of verification. As you know, many users of maps are both extremely impressed with the cartographic form and easily swayed by what is presented. By extension, then, the cartographic design process can easily sway decision makers in the wrong direction. This may happen when a model uses colors that usually imply high importance or danger (e.g., red) to indicate relatively benign or unimportant map features. Even if there was no intent to assign heavy weight to a factor that was misinterpreted, the result was a GIS analysis that led to incorrect decision making. For a simple example, consider a choropleth map produced with a classification technique that isolates one small area with very high potential impact from radioactive waste disposal, ignoring several areas with values nearly as high that should be shown as well. If it is relatively easy to lie with one map, it is even easier to lie with many maps.

Another important factor to be examined is whether the format of the output itself is useful to the client. As we will see in more detail in Chapter 14 there are many forms of output, not just the cartographic output. In some cases a tabulation of the results might be more useful to the client than a map. Say, for example, that the client is the publisher of a newspaper that is trying to increase circulation. The analysis might produce a list of local residents who are not current subscribers, together with their telephone numbers. The tabulated results could, of course, be used for telemarketing purposes. In this case a list would be far more useful to the client than a map showing a distribution of circulation.

Or, suppose that your analysis yielded a range of numbers of people infected with a virus for each county of a state or region. Some clients would prefer to see a tabulation of these results from high to low rather than a choropleth map. There are at least two reasons why this might be so. First, users who are uncomfortable with map output, or unfamiliar with choropleth maps, may find the results difficult to interpret. Or, users may want to see the raw data, at least prior to cartographic classification, with a view to determining the critical values for classification.

Still other forms of output are possible. In a few communities, GIS is being used for emergency services. In addition to producing a map showing the shortest route to a fire or other emergency, the system can send an electronic alarm signal directly to the nearest emergency service, therefore shortening response time and hopefully saving lives or property. Although this is perhaps an unusual example of GIS output today, as the technology changes, and as more and different clients begin using GIS, such output will become commonplace.

One other aspect of model verification that needs to be included is the acceptability of the results to the user. This is especially important if your model is not for your personal use. Clients may be searching for particular results to coincide with their intuitive or preconceived idea as to what a correct solution is or to serve their political or legal mandates. If a client has provided data to be included in the model, and has indicated how the model should perform, it might be assumed that the model will meet these expectations. If, for example, you have included values from the users, such as the quality of a particular parcel of land for a preselected land use, you are going to be limited to using those values. The client will probably also expect that the handpicked values will produce model output that will either favor or discourage using the land for the designated purposes. In such a case the GIS is really being used to validate the client's existing attitude and to document its derivation. While you may have difficulty with this from the standpoint of honesty, it is a common practice. Because the client is paying for the result, it is important that you reconfirm that the model output matches the desired outcome. This can be done by presenting the results to the client, explaining them fully, and asking for verbal confirmation. If the results are not what the client wanted, you may have to backtrack through the modeling steps, as before, to see where the model may have departed from expectations. In some cases you will be asked to modify how you combined variables (often a form of constraints relaxation) and in others you may be able to teach the client that the initial view was incorrect (DeMers, 1985).

In another situation, clients may require you to manipulate the model to obtain a particular answer to support a legal mandate to, for example, protect a particular piece of land. Again, the client is looking for a means of validating a decision that has already been made. In any event, you should remember that, regardless of whether your model exactly coincides with the client's view, the client is not obliged to accept the results you have produced. These realities, although not always pleasant, should be considered when you decide on your clientele. You are also encouraged to keep abreast of the ever increasing literature devoted to ethical questions.

Terms

cartographic modeling	descriptive models	deductive models
isolated state model	predictive models	flowchart
location–allocation models	prescriptive models	model verification
gravity models	prescribe	
innovation diffusion models	inductive method	

Review Questions

1. With so many commands and so many options to each command available in a GIS, how can anyone possibly decide how to build a model?

2. What do we mean when we say that the process of cartographic modeling

is cyclical? Within your own area of interest, create a scenario that illustrates this process.

3. What advantage does deductive cartographic modeling have over inductive cartographic modeling? When might you find inductive cartographic modeling most useful?

4. What is the difference between explicit and implicit variables? What is the difference between implicitly spatial and explicitly spatial variables? What are surrogate variables, and how are they used in cartographic modeling?

5. What is the purpose of flowcharting cartographic models? What advantages does this process have over creating cartographic models without them?

6. What do you do if your cartographic modeling constraints are so severe that the model cannot reach a reasonable solution? What are two possible ways around this problem?

7. What do you do if your cartographic modeling constraints are so loose they give results that are too broad to help you make your decision?

8. Why are weights so important to the creation of cartographic models? Why are some factors preemptive rather than simply weighted higher than others? Can you give examples?

9. What are the three basic areas of cartographic model verification? Suggest some methods of model verification for each.

10. Give some examples of cartographic output that might not be as useful as alternative methods or output from cartographic analysis.

11. Why is it important to prototype models before they are implemented on the entire database?

12. What should we keep in mind as we begin model implementation? Why are experimentation and reliance on serendipity not good practices when we are implementing a model?

References

Abler, R., J.S. Adams, and P. Gould, 1971. *Spatial Organization: The Geographer's View of the World.* Englewood Cliffs, NJ: Prentice-Hall.

Baker, W.L., and Y. Cai, 1992. "The r. le Programs for Multiscale Analysis of Landscape Structure Using the GRASS Geographical Information System." *Landscape Ecology,* 7(4):291–301.

Berry, J.K., and C.D. Tomlin, 1984. Geographic Information Analysis Workshop Workbook. New Haven, CT: Yale School of Forestry.

Brush, J.E., 1953. "The Hierarchy of Central Places in Southwestern Wisconsin." *Geographical Review,* 43(3):380–402.

Carlson, S.A., and H. Fleet, 1986. Systems Applications Geographic Information Systems (SAGIS) and Linked Analytical/Storage Packages including Map Analysis Package (MAP), Image Processing Package (IPP) and Manage, A Relational Database Management System. Workshop Materials, Annual Meeting, Association of American Geographers, Minneapolis, MN.

Casetti, E., 1969. Alternate Population Density Models. In A. Scott, ed., Studies in Regional Science. London: Pion Press, pp. 105–116.

Christaller, W., 1966. *Central Places in Southern Germany,* translated by C.W. Baskin. Englewood Cliffs, NJ: Prentice-Hall.

Davis, J.R., 1981. Weighting and Reweighting in SIRO-PLAN. Canberra: CSIRO, Institute of Earth Resources, Division of Land Use Research, Technical Memorandum 81/2.

DeMers, M.N., 1988. "Policy Implications of LESA Factor and Weight Determination in Douglas County, Kansas." *Land Use Policy,* 5(4):408–418.

DeMers, M.N., 1989. "Knowledge Acquisition for GIS Automation of the SCS LESA Model: An Empirical Study." *AI Applications in Natural Resource Management,* 3(4):12–22.

DeMers, M.N., 1995. "Requirements Analysis for GIS LESA Modeling." In *A Decade with LESA: The Evolution of Land Evaluation and Site Assessment,* F.R. Steiner, J.R. Pease, and R.E. Coughlin, eds. Ankney, IA: Soil and Water Conservation Society, pp. 243–259.

Duncan, B.W., D.R. Breininger, P.A. Schmalzer, and V.L. Larson, 1995. "Validating a Florida Scrub Jay Habitat Suitability Model, Using Demography Data on Kennedy Space Center." *Photogrammetric Engineering and Remote Sensing,* 61(11):1361–1370.

Fisher, P.F., 1993. "Algorithm and Implementation Uncertainty in Viewshed Analysis." *International Journal of Geographical Information Systems,* 7(4):331–347.

Forman, R.T.T., and M. Godron, 1987. *Landscape Ecology.* New York: John Wiley & Sons.

Hägerstrand, T., 1967. *Innovation as a Spatial Process,* translated by Allan Pred. Chicago: University of Chicago Press.

Isard, W., 1956. *Location and Space Economy.* Cambridge, MA: MIT Press.

Lucky, D., and M.N. DeMers, 1987. "A Comparative Analysis of Land Evaluation Systems for Douglas County." *Journal of Environmental Systems,* 16(4):259–277.

McGarigal, K., and B.J. Marks, 1994. FRAGSTATS, Spatial Pattern Analysis Program for Quantifying Landscape Structure, Version Two. Corvallis, OR: Forest Science Department, Oregon State University.

Monmonier, M., 1991. *How to Lie With Maps.* Chicago, IL: University of Chicago Press.

Tomlin, C.D., 1990. *Geographic Information Systems and Cartographic Modeling.* Englewood Cliffs, NJ:, Prentice-Hall.

Tomlin, C.D., and J.K. Berry, 1979. "A Mathematical Structure for Cartographic Modelling in Environmental Analysis." In *Proceedings of the 39th Symposium of the American Conference on Surveying and Mapping,* pp. 269–283.

Tomlin, C.D., and K.M. Johnston, 1988. "An Experiment in Land-Use Allocation with a Geographic Information System." In *Technical Papers, ACSM-ASPRS,* St. Louis, MO, pp. 23–34.

Weber, A., 1909. *Theory of Location of Industries,* translated by C.J. Freidrich. Chicago, IL: University of Chicago Press.

Williams, T.H. Lee, 1985. "Implementing LESA on a Geographic Information System—A Case Study." *Photogrammetric Engineering and Remote Sensing,* 51(12):1923–1932.

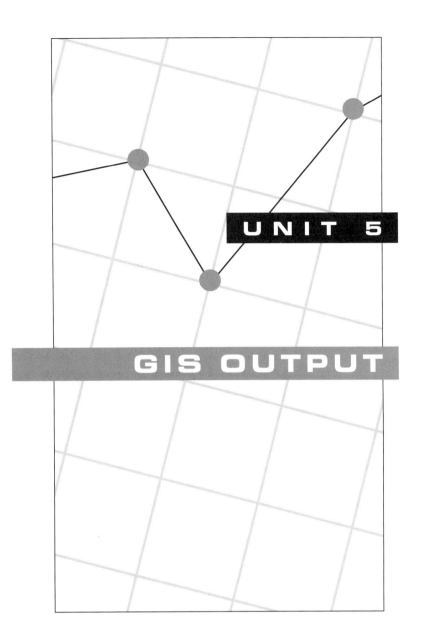

UNIT 5

GIS OUTPUT

CHAPTER 14

The Output
from Analysis

Because analysis is the strength of a geographic information system and because we all enjoy the ability to manipulate spatial data, we often forget that when we are done with our job, others have to be able to make sense out of what the analysis says. Output is the final product of any analysis. If our output is unintelligible, we have failed in our mission. After all, our purpose is not just to perform an analysis, but to communicate the results. In this chapter we will examine the technology of GIS output as well as some basic design criteria for producing good quality, readable output. Both these subjects are important, because both affect the way in which we communicate to the user. The technological aspects of GIS output are important because the limits of technology impose physical constraints on our ability to produce the kind of output we need. Beyond the technology, however, we must be aware that users of GIS output have psychological and biological limits and biases that affect the way they interpret the results.

There is relatively little discussion about the design of GIS output in the current GIS literature, but design literature in the field of cartography is abundant and robust. At this point in your course you have probably seen enough bad maps to know that there has to be a better way. I'll introduce you to some of the basic design criteria in this chapter, but you should try to expand this limited discussion by reading more about graphic and cartographic design. There are many sources for this literature, but a good cartography book is as good a place to start as any. Although you may not see a major need for this knowledge now, as you continue in your automated geography career you will find that well-designed output produces happy users. And happy users are repeat customers.

I have tried to keep the technical jargon down to a minimum, but it is impossible to avoid it completely. As you read on, keep the central theme of producing usable output uppermost in your mind. As the technology changes through the next few years, the terms may change the methods of output themselves will certainly change, but the value of good output will always remain an important aspect of GIS. Thus while knowing how the systems work may prove useful, the relationship of technology to the caliber of your output is fundamental.

To reproduce dozens of good and bad examples would have radically increased the cost of this text. Therefore, I've tried to keep the graphics to an absolute minimum, while still providing enough examples for you to understand what is being discussed. In addition, however, I've included an electronic home page addresse in Appendix B at the end of this book that has been set up in part for you to view the output from GIS analysis. This service, provided by the publisher and some GIS vendors, should give you ample opportunity to view both good and bad output. In addition, you will have a chance to see the wide variety of possible applications and their output. If you don't yet surf the Net, or are still dog-paddling, there is an alternative. Many vendors provide hard-copy examples of output from their projects. Often there is no change for this material. You might ask your instructor for names and addresses of these vendors. In addition, there are numerous conferences devoted to GIS in general and to specific applications of GIS. These are also excellent sources of information and opportunities to view the results of GIS analysis up close. If you attend such conferences, look critically at the output based on what you have learned here. Try to explain verbally what you like and what you don't like about each GIS output you see. Not only will this exercise help you learn what you should and should not do but it will allow you to practice using the design terminology. When you become fluent in the language of the discipline, you will be able to communicate exactly what you need to make your GIS output the best possible.

LEARNING OBJECTIVES

When you have completed this chapter you should be able to:

1. Understand the primary concern for map design in GIS work and why it is important.

2. Describe the conflict between intellectual and aesthetic map design considerations.

3. Understand the use of the three basic design principles and be able to use them in designing and organizing maps.

4. Show examples of the possible map design constraints and considerations for two- and three-dimensional GIS output.

5. Suggest possible uses of cartograms as GIS output.

6. Identify some types of noncartographic output and understand the design considerations for them as GIS output from analysis.

7. Explain the impact of modern technology on GIS output design.

8. Define the parameters for GIS hard-copy output devices desirable in various work settings.

OUTPUT: THE DISPLAY OF ANALYSIS

As the section title implies, the final step in analysis is to display the results so that someone, perhaps even the cartographer, can see what has been done. In Chapter 1 we learned that computer-assisted cartography emphasizes this far more than does GIS. Of course, the primary purpose of a GIS is to analyze spatial data for decision making, but the end product or output is the mode of communication. This topic has been relatively neglected, but we will explore it in some detail. You, in turn, will be a better GIS analyst because you know how to present your results in a clear and understandable fashion.

Output from GIS analysis is generally produced as either permanent or ephemeral products, largely depending on the nature of the output devices. "Permanent" implies that the output is printed or plotted on a tangible medium such as paper or Mylar, or even saved on magnetic or laser disk media for storage. These media are used to preserve the output for an extended period of time, and the form selected generally depends on whether the results are to be displayed or archived and stored. "Ephemeral" is applied to text and image output that most often will not be displayed for long periods of time, but rather will be viewed, perhaps to demonstrate the results of analysis or, alternatively, as a method of reviewing files in the course of deciding whether they will be used for further analysis or permanent output.

GIS output also can be classified according to whether it is produced in human- or machine-compatible form (Burrough, 1986). Machine-compatible forms of output are most often used as a means of storing material—in which case we move from the output subsystem back to the storage and editing subsystem. Human-compatible output is most often meant to be more permanent and, more importantly, to be viewed and understood by an audience. While machine-compatible forms of output are important and will require decisions about media, data structure, and volume compatibility between and among computer systems and their storage devices, the human-compatible forms most often cause difficulties. This is primarily because the types of software and hardware are relatively simple in comparison to the variety of human users, with their dissimilar backgrounds and unequal levels of graphic understanding. Therefore, we will concentrate primarily on human-compatible forms of output. We will include permanent and ephemeral types of output because both can be used as devices of communicating results to human users.

At the outset, you should be reminded that GIS output is primarily but not exclusively cartographic. Because the traditional input to a GIS is mapped or mappable data, it is not unexpected that the primary output consists of maps. Indeed, for the next few years, or even decades, this trend is likely to continue. Thus we will spend a large portion of this chapter on cartographic output, but we'll also look at some alternative methods, and we'll discuss reasons for using these alternative output methods instead of or in addition to the map.

CARTOGRAPHIC OUTPUT

As we have seen many times, the map is still the most compact form of communicating spatial information. When we discussed the paradigm shift from pure communication to a combination of communication and analysis, I made it clear that analysis without communication is a serious mistake. The GIS allows us to create many maps in many different forms based on a wide variety of analyses and based on an increasingly large set of spatial data. In fact, the advent of readily available databases is in large part a response to the increasing availability of GIS software. This power makes an understanding of map design more important than ever, especially because the majority of GIS analysts today have limited experience with cartographic production and design. More important, because there are now far more users of maps of many different types, the requirements for effective communication become even more necessary.

The object of creating a map is to elicit from viewers a response that allows them to understand what the mapped environment looks like (Robinson et al., 1995). This objective is in large part a function of the purpose of the map and is linked closely to the intended audience. The vast numbers of data types, map styles, and available symbols, scales, typography, and other iconographic devices are meant to operate together. Because most GIS output maps are not of the **general reference map** type, which strives to display, on a single map, a wide variety of different geographical phenomena, we will focus on the second basic category of maps—**thematic maps**. These maps usually concentrate on the representation of the structural relationships of a selected theme or subject. In the case of GIS output, we could easily replace "theme" with "solution" because the map output will most often be an answer to a predefined question or the key to the completion of a decision-making process.

The general process of thematic map design involves the selection, creation, and placement of appropriate symbols and graphic objects to show explicitly the important features and spatial relationships of the objects under study. Most often, however, the thematic map will also need a reference system as a comparative framework for locating these thematic objects in geographic space (Robinson et al., 1995). In some cases—for example, when satellite data are the primary graphic objects—a framework becomes especially important because the reader may be relatively unfamiliar with the color representation of earth features through this medium and will not be able to interpret the data without a reference system that includes notations of well-known objects. In most other cases, the reference system provides a method of demonstrating areal extent of the thematic objects and their relationships to other spatial locations.

Beyond the inclusion of a reference system, perhaps the first rule you should remember about designing your GIS thematic map output is that despite the inherent beauty of the cartographic form, you are designing the map first and foremost to be read, analyzed, and interpreted (Muehrcke and Muehrcke, 1992). Simply put: "Eliminate needless objects." Any object placed on the map should say something about the spatial distributions and arrangements. An object should not be placed on the map if it has any objective beyond serving the function of the documents. There should be none of the extraneous objects or fancy flourishes typical of maps showing the location of hidden pirate treasure.

Even if you eliminate extraneous objects from your thematic map design, many decisions remain to be made. Map design involves selecting, processing, and generalizing your output data purposefully and with forethought, then using appropriate symbols to portray them in a way the user can understand easily (Robinson et al., 1995). The decision process entails a sometimes difficult mix of art and science, impression and logic. To best make use of these components, you should produce a number of preliminary sketches of how your map will look before beginning the actual design. Although you may have access to map-composing software, either separately or as part of the GIS, it is a good idea to begin with pencil and paper, outlining which objects might go where and how they will be represented. This manual operation will save time and effort later, when you are using your software, since sketching is generally easier and very often faster than placing, moving, and relocating complex graphic objects on a screen. Doing the work by hand also focuses your attention on the choices that have to be made and the look you are trying to achieve for the map, rather than on the commands or operations needed to perform the various tasks.

As you might have gathered, map design can be very complex, especially with the large number of options available using the computer. You will also find that the logical or intellectual objectives and the graphic or visual objectives of map design will sometimes be in conflict. These conflicting demands on map production indicate quite clearly that design considerations are nearly always resolved by compromise (Robinson et al., 1995). For example, if you are trying to place house symbols on your map, both logic and aesthetics dictate a consistent approach. For example, you might want to place each house symbol in its exact (analog) position. But if you have a line symbol representing a major stream or river that runs very near the house symbols, either the stream symbol or the house symbols must be adjusted. Under the circumstances, you might prefer to move all the house symbols as well as the road symbol to the right of their actual locations to changing the location of the stream (Figure 14.1).

The choice made in Figure 14.1 reflects the fundamental belief that the location of a physical landscape object takes precedence over a typographical object that does not precisely indicate the actual location of a physical object. Over the generations, cartographers have dealt with many hundreds of thousands of similar conflicts and have, therefore, established a rather exhaustive set of **conventions** and traditions. They are a result of trial and error and, perhaps more importantly, testing among map users. Their results show the most effective means of compromise. These should act as guidelines for your own cartographic representations. Deviations from established norms will almost always result in a map that is less effective than it could be at communi-

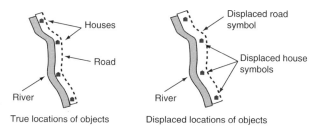

True locations of objects   Displaced locations of objects

Figure 14.1   Symbol placement compromise. Close-up of a portion of a map that illustrates the compromise necessary to allow placement of the stream symbol adjacent to house symbols. One set of cartographic objects must be physically moved to accommodate the other.

cating the results of your analysis. We will now look at the basic processes, perceptual considerations, and design principles needed to produce effective map products. You are also encouraged to consult frequently with textbooks on general cartography or more specific texts on map design.

The Design Process

A time-honored approach to any graphic design process is to begin with a knowledge of the basic cartographic elements and symbology as discussed in Chapter 3, combined with a mental image of maps you have seen that communicated the kinds of information with which you are currently working. The first step is to visualize the type of map you are going to create, the objects to be placed on the map, and a basic layout. This initial stage is an intuitive one and results in a rather general plan for your map. As stated earlier, it is probably best to sketch this out freehand on a piece of paper (Figure 14.2a) rather than using the computer, but this is not absolutely necessary.

As your plan develops, you begin to consider which symbols will be used to represent the objects, and you decide on class limits, colors, line weights, and other graphic elements. Again, it might be best to note these ideas on a pencil sketch before you begin using your software, or at least while you are placing objects on screen. Remember, many changes are likely during this second stage, and having notes available frequently proves useful. Even experienced GIS practitioners generally have a pencil sketch nearby at this stage. It is customary to note measurements among objects on the sketch to ensure that when the software is used for their placement, exact numbers can be input to the software (Figure 14.2b). Even if the objects are later moved, the original numbers will likely be referred to while adjustments are being made.

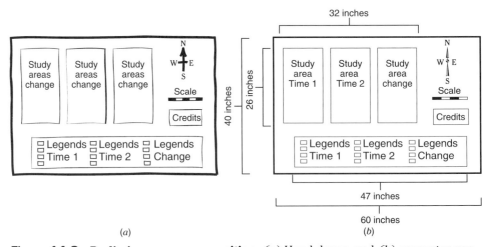

(a)   (b)

Figure 14.2 **Preliminary map composition.** (a) Hand-drawn and (b) computer-generated preliminary sketches showing the general object types, their approximate placement, and a general view of how the map will be composed. The computer-generated sketch includes measurements that indicate distances provided for objects; these notations are useful when the design is to be finalized with the GIS software.

The final stage in the design process consists of fine-tuning your work in the preceding stage. This stage should employ only minor modifications to the graphic plan developed earlier. Among the more important considerations here, however, is the development of design prototypes on your monitor prior to hard-copy output. Printers and plotters are often many times slower than the graphic monitor and proportionately slow the process of prototyping. Another, frequently ignored factor to be kept in mind in the final stage is the visual relationship between your monitor and your output device. Most hard-copy output devices have their own graphic language that can result in a map that looks somewhat different from what you see on your monitor. It is not uncommon to find that a text font used on your monitor is not available on either the printer or the plotter. The plotter may give you text that looks totally different from what you had anticipated or runs one set of text over another. Colors that appear easily separable on your monitor may look nearly identical when output on the hard-copy device. Or fonts that you selected with your software may produce unexpected results because the codes for fonts available for the hard-copy output device are different (Figure 14.3). Most software and hardware vendors provide printed versions of symbols, fonts, and color palettes. You should consult these before you finalize your design; but remember that the mix of sizes, shapes, and proximities, not to mention output material responses to inks and wax output material, will all have an impact on the appearance of your map. When you are finished with your design prototype, it is wise to test the output by plotting or printing a small version, or perhaps a small portion of the map, to examine it for these possible pitfalls. Final adjustments can then be made before you print the entire map.

The Role of Symbols in Design

Ultimately, to produce a map that illustrates the output from your GIS analysis, you must manipulate the appearance of the graphic symbols that represent the

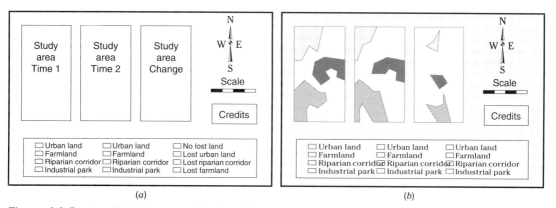

Figure 14.3 Detailed map compilation. (*a*) The map as it might have appeared on the graphics monitor. (*b*) Initial plotter output with text overprinting and a font shift. These problems are most often caused by differences in graphic language between the software for the GIS and for the output devices.

point, line, and area elements so that they can be discriminated. The primary visual variables that can be manipulated for these symbols are **shape, size, orientation, hue** (color), **value** (lightness or darkness), and **chroma** (a complex color variable; other color variables are saturation, intensity, richness, lightness, and purity) (Robinson et al., 1995). Secondary or related variables called **pattern** are achieved by means of manipulating combinations of individual graphic elements to achieve areal graphic patterns. These variables include **arrangement**, or the random or systematic grouping of the elements, **texture**, or the spacing of graphic elements to change the lightness or darkness of the shading patterns, and **orientation**, or directional positioning of the graphic elements (Robinson et al., 1995).

All these elements are manipulated to improve the graphic communication of collections of objects. Because maps are perceived synoptically (i.e., all at once, rather than serially, as we process language), the interactions of the graphic elements require us to pay particular attention to some basic design principles. For cartographic output these are **legibility, visual contrast, figure–ground,** and **hierarchical structure.**

Graphic symbols must first be legible. If the marks on the page are unclear or indistinguishable like messy handwriting, the reader will be unable to understand what the map is trying to communicate. The lines must be easily separable; patterns, shapes, colors, and shadings should be distinct; and the shapes must be clear and identifiable (Robinson et al., 1995). The sizes of the objects must also be appropriate for the distance at which the map is to be viewed. Because GIS analysis is frequently employed for relatively large portions of the earth, this latter point should be especially noted. You will choose different sizes for the text and the symbols of a poster-sized output that will be viewed from a few feet and for an output that will be placed on a 10-inch × 10-inch sheet to be viewed at close distances. In addition, as we will see later, the graphic devices themselves have physical limitations: some graphic elements are not distinguishable when produced in smaller sizes.

Other legibility factors concern the inherent visibility of the symbols themselves. Line symbols, for example, are easily seen, so their widths need not be large. Some color combinations modify the visibility of the symbols as well—consider the impact of black lettering on a bright yellow sheet of paper, as opposed to black lettering on dark blue. Finally, easily recognizable symbols and symbol combinations increase the likelihood of legibility. A classic illustration is the set of shapes used for traffic signs that allow motorists to easily receive the message without having to read any text.

The second principle of graphic design, visual contrast, is also necessary to allow symbols and text to be seen and recognized easily. Robinson et al. (1995) show that the way in which a graphic element compares with its background or with the adjacent elements is the most important of all graphic design factors. The use of visual contrast in map output entails consideration of two issues. First, a clean and uncluttered map appearance depends largely on whether the objects themselves appear distinctly different from their background or adjacent objects. However, the second part of visual contrast deals with the differences in appearance inherent in the graphic elements. Thus if some elements look essentially the same because of their sizes, shapes, patterns, or other factors, you might find it useful to vary the symbols used to set them apart. In addition, too much contrast may result in an unpleasant monot-

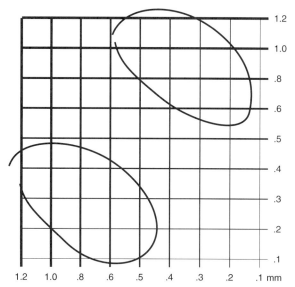

Figure 14.4   Line width and boredom. Diagram
illustrating the concept of boredom due to nonvary-
ing line width. The circled areas tend to be crisper
and more interesting than those with less contrast.
Source: A. H. Robinson et al., *Elements of Cartography,* 6th
ed., John Wiley & Sons, Inc., New York, © 1995. Adapted
from Figure 18.10, page 330. Used with permission.

ony that can be avoided by variations in color, shape, and size to add visual
interest—still keeping the need for legibility uppermost (Figure 14.4).

Graphic Design Principles

As you begin the first stage in map composition and design, you will notice that
among your earliest decisions is the placement of objects in the limited space.
Many maps are uninteresting because there is either too much white background or too little. You will probably find such choices to be quite intuitive as
you contemplate that annoying space in the bottom left-hand corner produced
by the diagonal orientation of your primary map object. Look at a map of
California (Figure 14.5). Because the state is shaped somewhat like a boomer-
ang, you have a large space in the upper right and another in the lower left. It
is easy enough to fill them with the scale in the lower left, a north arrow in the
upper right, and other conventional map components, all of which can be
adjusted in size and line weight to fill the spaces. What you are manipulating is
the "figure–ground"—that is, the ratio of figure to ground or background. A
map that is all figure is less desirable than one that has at least some amorphous
background to isolate the figure and give a sense of belonging to the earth in
some respect (Figure 14.6). The background also adds visual interest and contrast. A map that has too much background and too little figure diminishes the
importance of the figure and leaves the viewer suspecting that the map is
somehow incomplete.

California base map

Legend

Scale

Credits

Figure 14.5 Map balance. Preliminary sketch of components of a map of California. The odd shape of the state produces large blank spaces on the upper right and lower left portions of the map. These can be filled with titles, north arrows, legends, scale bars, and other ancillary cartographic information. The result is a more balanced map product, and a better figure—ground ratio, as well.

But the figure–ground issue goes beyond simply deciding how much object and how much background should be drawn. If our map of California has the same white internal color as the background, the viewer can easily become confused as to which areas are ocean and which are land. Addition of names, familiar boundary symbols, graticules or grid lines, and shading patterns can be used to permit the reader to easily identify the important study area as well as the background or nonanalyzed portions of the map (Figure 14.7).

A number of graphic mechanisms can be employed to enhance the separation of figure from ground and allow the viewer to focus on the important portions of the map. The use of different patterns and colors will make the important figure of the map appear visually more homogeneous than the surrounding background. This does not mean that the map figure should be devoid of symbols. In fact, the detail contained in the figure adds internal homogeneity and helps to promote it as important, while the map margins are filled with a heterogeneous mix of north arrow, scale, legend, graticule, and other necessary, but widely differing objects. If you have a study area that can be well defined by outside boundaries, such as the island of Tasmania, it is best to show the

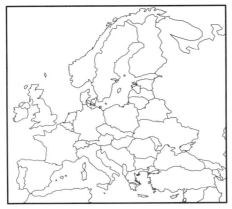
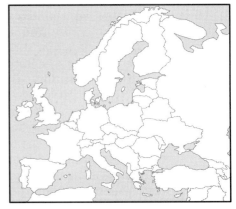

Figure 14.6 Stereogrammic hierarchy. Two maps of the same area showing how a distinct background can effectively separate the figure from the ground. The map that uses the same background throughout results in confusion over which portions are land and which are water.

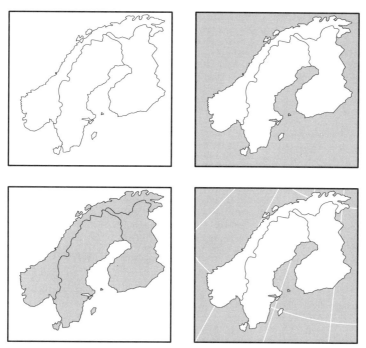

Figure 14.7 Figure—ground relationships. The use of a grati-cule and shading demonstrate the impression of depth known as stereogrammic hierarchy.

entire island because a single closed form imposed on a background is likely to be viewed as the area of interest, whereas if the figure's boundaries ran off the map, the focus would be hard to determine.

Other approaches to improved figure–ground relationships involve using familiar shapes (because we are drawn to what we know). If, for example, you were analyzing the distribution of Dutch elm disease in Minnesota, isolating that state from neighboring states will allow people to instantly identify the familiar shape of Minnesota's borders as the figure under study. As you map the area you will also find that a variation in lightness or darkness between the figure and its surroundings will help isolate the figure.

A more complex aspect of figure–ground is referred to as "good contour." Essentially this means that the elements on your map give a logical impression of what is on the map. Good contour can mean simply that objects appear to belong as they would if this were a real portion of the earth. You may find, for example, that to make space for a place name you will have to break the line of a land area. This does not violate logical expectations, for even though the boundary line has been broken, the map reader perceives an "implied line" and is still able to read the name. Other examples include logical differentiation of land from water, borders from roads, and vegetation from urban. Even size can be used to improve good contour because we tend to identify smaller areas, for example, an island surrounded by ocean, as the figure rather than larger figures. As many of these relationships as possible should be considered when designing your figure–ground relationships, but there will be constraints in many mapping situations, especially when the nature of your study area limits the use of one or more of these components. If your map area is not a closed form, use the other figure–ground aspects remaining to you. The goal, remember, is to make a useful map, not just to employ as many design principles as possible.

One last principle of graphic design is hierarchical organization. The myriad graphic elements present in a map must be organized to make the important elements stand out. This principle is poorly developed in general reference maps because their purpose is to allow all the elements to be of equal value, permitting a series of many individual readers to focus on what is important to each one at a given time. However, in thematic maps, the usual output from GIS analysis, our purpose is to highlight specific features or results from analysis; thus it is vital that we be able to prominently display the most important elements. This can be done by developing a hierarchical organization or separating the elements into levels of visual importance.

There are three primary methods of obtaining a hierarchical organization. The **stereogrammic** method requires the modification of graphic devices to allow certain essential elements to appear to be physically higher on the map than unimportant elements. This is a useful technique for improving figure–ground. Depth clues include the use of three–dimensional objects, or differences in line width, color, shade, or size. We can even obtain the impression of depth by placing all or portions of some objects physically on top of others.

A second method of hierarchical organization is called **extensional** and is most often used for ranking line networks or point symbols. Here the purpose is not to make the elements appear to move up or down on the page, but rather to force the most important elements to stand out. For example, freeways should appear more prominently than single-lane highways, interstate highways more than state roads, and city streets more than unimproved roads. From this example it is clear that we can modify the sizes of the lines, their values or textures, or some combination to show these distinctions in road importance ranging from a high with the freeways down to a low with the unimproved roads. This can be difficult with current GIS technology that often limits your choices of line widths, but there are usually ways to achieve the desired results. You will have to experiment with your own software to see what works best.

The final method of producing a hierarchical graphic structure, called **subdivisional hierarchy,** is chosen primarily when you want to show differences in internal structure. Here your main concerns are areas and area symbols (Figure 14.8). Distinctions can be made by subdividing larger categories, such as dividing pastureland into heavily grazed, moderately grazed, and lightly grazed. This subdivision is not unlike the convention seen on political maps, where nations are outlined with heavy, dark, solid lines, and lighter colors enclosed in broken lines indicate internal subdivisions. As you can see, the latter two examples were fundamentally different in that the extensional system develops a hierarchy based on ordinal differences in regions, whereas a subdivisional hierarchy could more appropriately be defined in terms of nominal differences. All scales of geographic data measurement can be associated with a graphic hierarchy. The primary factor to be remembered, no matter what the scale of data measurement, is that the distinctions should be more obvious for larger categories than for subcategories.

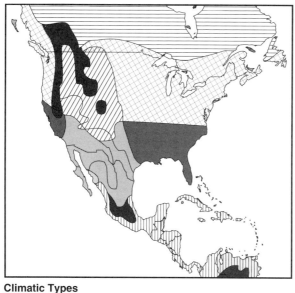

Climatic Types

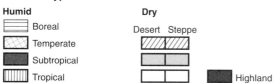

Figure 14.8 Subdivisional hierarchy. The use of
subdivision hierarchy to separate out area patterns.
Note how the highest level of the hierarchy uses the
heaviest line weight to separate it from its subdivisions.   *Source:* A. H. Robinson et al., *Elements of Cartography.* 6th ed., John Wiley & Sons, Inc., New York, © 1995.
Used with permission.

MAP DESIGN CONTROLS

All the design principles we have examined are dependent on or influenced by
a number of external forces known as **controls**. These controls determine the
nature of the map produced, the types of graphic element used, and the design
principles that will come into play as you draft your final output. To prevent
design problems and conflicts later on, you need to keep all these considerations in mind from the outset.

The first and most essential control on map design is the **purpose** for which
the map is being prepared. The **substantive objective** relates to the nature of
the data and information you are attempting to display. Thematic maps produced by GIS analysis are meant to present the distributional form of a specific
set of analytical outcomes. Here simplicity is very important. The more focused
your graphic output, the more easily understood. With the lower cost of producing maps from a GIS as opposed to traditional manual cartography, the
tendency to produce highly complex maps is considerably less prevalent. Still,
the large volume of data, combined with the possibility of multiple analyses
from a single database, might create a tendency to put too much information

into a single map. Your results will be much easier to understand if you present the results of one analysis on each map you produce. If multiple analyses have been performed, you can always combine several maps on a single output, thus maintaining the purposeful identity of each map.

The other aspect of map purpose, called **affective objective**, relates less to what is to be presented than to how the data are presented. Once you have decided what each map is to represent, you must decide on a form of presentation that will convey the appropriate message. If, for example, the map output is meant to illustrate potential hazards, your map design should make it clear that the hazardous areas are important. The output should be stated boldly and decisively in a way that does not disguise the dangers at hand—pleasant appearance might result in an artistically beautiful map product, but it would send the wrong message to the map reader.

A second control on map design is **reality**, meaning that every area mapped has its own character that will place limits on how well you can employ your design criteria. Selected study areas for GIS analysis are frequently locations with highly complex physical, transportation, social, or economic structures. Indeed, these are the areas that most need spatial data analysis. The complexity of soils, road networks, distribution of socioeconomic phenomena, or even the shape of the study area may place serious constraints on name placement, symbol sizes and styles, shading patterns, and the like. Important objects may not always appear in prominent locations on the map, or the numerical differences among variable measurements may be very subtle. Because of the inevitability of such restrictions on the display of your graphical hierarchy, you should become familiar with the physical layout of your study area as well as the nature of the data before you begin the design process. A knowledge of the restrictions placed on your design capabilities will allow you to plan ahead and anticipate solutions before you reach the final stage, when change is difficult.

The nature of the **available data** is a particularly difficult problem in GIS output because the GIS is a spatial analysis tool rather than merely an aid for cartographic output. Most GIS databases produce very large data sets, often with equally large ranges of variables. There are at least two reasons for this enormity of output: first, the purpose of GIS is to analyze data; and second, such analysis is frequently performed for large territories. It is not uncommon to see a single map of land use that contains over 100 categories, even though the sheer amount of data creates difficulties. Selecting 100 color categories that are separable to the map reader is a particularly nasty design task. You may have to modify your design procedure to include a subdivisional hierarchy with different color or shading pattern sets for each subdivision (Figure 14.9).

The opposite problem may be encountered with available data, as well. That is, data either are not sufficient to be portrayed (or analyzed) easily, or they are old and were not collected according to rigorous protocols, or they were collected in such a coarse sampling scheme that their positions may not accurately portray where they actually occur on the ground. Take the example of a temporal comparison of vegetation maps from century-old surveyor's notes versus a 1990s vegetation map compiled from *LANDSAT* Thematic Mapper (TM) data. The historical surveyor's maps are likely to contain a few broad categories of vegetation associations, frequently biased by the experience of the surveyor. In addition, these associations often are based on a combination of the structure of the vegetation and species type. By contrast, the TM data will reflect a high

SPATIO-TEMPORAL ANALYSIS OF LANDCOVER PATTERNS

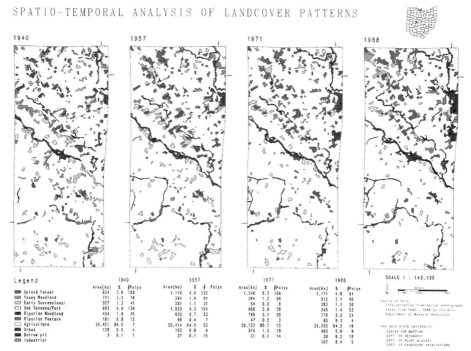

Legend	1940 Area(ha)	%	#Polys	1957 Area(ha)	%	# Polys	1971 Area(ha)	%	#Polys	1988 Area(ha)	%	#Polys
Upland Forest	924	3.8	100	1,110	4.6	122	1,248	5.2	109	1,171	4.8	94
Young Woodland	771	3.2	76	394	1.6	61	284	1.2	58	512	2.1	80
Early Successional	327	1.3	41	261	1.1	31	54	0.2	8	283	1.1	26
Oak Savanna/Park	962	4.0	130	1,033	4.3	101	698	2.9	78	345	1.4	53
Riparian Woodland	434	1.8	35	652	2.7	23	760	3.1	25	775	3.2	21
Riparian Pasture	191	0.8	12	86	0.4	7	47	0.2	3	85	0.3	4
Agriculture	20,421	84.5	7	20,414	84.5	23	20,723	85.7	12	20,353	84.2	19
Urban	128	0.5	4	183	0.8	9	315	1.3	10	492	2.0	9
Borrow pit	3	0.1	1	27	0.1	15	31	0.1	14	38	0.2	18
Industrial										107	0.4	5

SCALE 1 : 143,130

Source of Data
Interpretation from aerial photographs
taken from 1940 - 1988 by the Ohio
Department of Natural Resources.

THE OHIO STATE UNIVERSITY
CENTER FOR MAPPING
DEPT. OF GEOGRAPHY
DEPT. OF PLANT BIOLOGY
DEPT. OF LANDSCAPE ARCHITECTURE

Figure 14.9   Map category confusion. GIS output showing a large number of land use categories. The visual confusion results from a lack of hierarchy separating the categories. Because people can seldom visually separate more than eight or ten shades in a given color, a map like this prevents users from determining what category name belongs to which polygon.

number of classes developed through digital classification algorithms and based mostly on vegetation structure as observed by the satellite. Performing an analysis of vegetation change between the two is difficult enough, but displaying the results in map form presents an entirely new set of problems. Because TM data are limited to 30 meter grid cells, the spatial accuracy will be limited to what the satellite is capable of viewing. In the case of the surveyor, the spatial accuracy may be quite precise, but the classification of the vegetation will likely be much coarser than that obtained from satellite data interpretation. In addition, because the TM data are in the form of raster grid cells, the final map of vegetation change is going to be rather blocky, as well. Awareness of such limitations before you begin to design will prove invaluable.

But presentation is also, in part, dictated by the **scale** at which the map is to be produced. As map scale decreases, so too will the amount of detail you can place on the map. Decreasing map scale does not simply mean that you make all the objects proportionately smaller. You will soon reach a point at which none of the symbols will be observable to the reader. Instead you will be required to select the data you will represent in accordance with the scale of output. Selection, simplification, and generalization will become important as you design a map for differences in scale of output.

While scale is a factor in allowing the viewer to perceive the map distributions, the **audience** for the map is also important. Many users of GIS output are unsophisticated both geographically and cartographically (Robinson et al., 1995). Under such circumstances the map should be as easy to understand as possible, with an emphasis on important shapes, names, and pictographic symbols. Alternatively, more sophisticated readers may obtain more information through a higher density representation with more abstract symbols (Robinson et al., 1995). Age is also a factor in presentation because older people often have more difficulty viewing small symbols and text. Some younger readers, too, have partial sight impairment and need larger symbols and text, and people who are color blind can process black-and-white formats best.

Conditions of use will play an important role in how the map is designed. This category is more far-reaching than whether the map will be displayed as a poster-sized document or put into service as a small, desktop piece, and the full set of viewing conditions anticipated should be accommodated before the design is finalized. If your map is to be used in conditions of poor lighting, for example, the shades, colors, and symbol sizes should be more prominent than would be necessary for good lighting conditions. In addition, maps that will be used in the field, where outdoor conditions would destroy traditional paper, must be laminated with a plastic coating. During the Gulf War of 1993, for example, the National Geographic Society provided the military with hundreds of maps of the region in laminated form. Each set of viewing conditions should be considered before the final design is developed.

Finally, **technical limits** will affect the approach to map design. In GIS output, the manner in which the map is to be produced most often is controlled by hardware and software capabilities. For example, if you are using a low-end color ink-jet printer, your color combinations may be limited to as few as 16 colors. Or perhaps you have only black-and-white output devices, restricting your design possibilities to the gray-shade environment. In addition, the resolution capabilities, often measured in dots per inch or lines per inch, will affect the size of symbols and text alike. All these factors should be considered prior to the commencement of map design. We will look in more detail at some of the technical aspects of computer map output in a later section.

NONTRADITIONAL CARTOGRAPHIC OUTPUT

Traditionally, thematic maps have been seen as orthographic forms that present the display as if the map reader were observing the map from directly overhead. These maps have also concentrated on the production of graphic forms based on their relative shapes and sizes as seen in the real world. While these forms of thematic cartography are still among the most frequently used, they are far from the only ones available, especially with the computer graphics devices available with most commercial GIS.

Among the first vector graphics display options used for presenting somewhat nontraditional thematic maps were **fishnet maps**. A computer program called SYMVU, one of the earlier computer mapping programs, specialized in producing these maps, also known as **wire-frame diagrams**. This now prevalent form of cartographic output produces an impression of three dimensions by changing the viewer's perspective from vertical to oblique. While these outputs are not often useful as analytical devices, they are extremely powerful at communicating the output of analysis, especially where topography is involved. Three additional design factors must be kept in mind when designing these oblique views: **angle of view, viewing azimuth,** and **viewing distance.** There is ample literature from which to obtain details concerning these three aspects (Imhof, 1982; Kraak, 1993), as well as other factors such as sun angle (Moellering and Kimerling, 1990) and the impact of DEM resolution. A detailed examination of this aspect of GIS output is beyond the scope of this text. However, in general, the importance of each factor depends on the perception of size, vertical relief, and the portion of the study area that is most important to the map reader. In addition, today's graphic software allows us to visually overlay other coregistered map data on these apparent three-dimensional maps. Thus the viewer can readily identify relationships between topographic surfaces and other factors such as land use or vegetation.

Today, some vendors also offer an alternative to static viewing of these and many other types of maps. For two-dimensional maps, animation allows the viewer to inspect dynamic systems as they are seen in real life (Moellering, n.d.) or to illustrate phenomena that would otherwise not be perceived because their natural motion is too slow (Tobler, 1970). In addition, displays can be stopped, slowed, even reversed to obtain additional insights. Three-dimensional maps can now be viewed in animation as well (Moellering, 1980). The effect is one of an airplane flying over terrain that is covered with both natural and anthropogenic patterns, giving the viewer an added dimension from which to analyze possible data relationships. The power of this technique of visualization should not be diminished. Because the human eye was designed to notice moving objects more readily than static objects, the use of computer animation of all kinds can only enhance a viewer's ability to see pattern interactions and to formulate both hypotheses and answers about the phenomena studied.

If your system lacks the computing resources to manage computer animation, there are still a number of useful variations on thematic maps that can enhance the viewer's understanding of your results. Shaded relief maps are particularly good tools for visualizing topographic surfaces, without the requirement of creating three-dimensional maps. The shading patterns themselves give an impression of shadow produced by the interaction of the sun with the topographic surface, thus making the map look more like a photograph. Digital orthophotographs can also be used as background to illustrate the relationships of mapped variables to the easily perceived aerial photographic image. For topographic as well as subsurface information, cross sections can be produced along any transect the user wishes. Even block diagrams are easily produced and provide a visually interesting output, while displaying both surface and subsurface features simultaneously. The list of options and combinations of these approaches to nontraditional cartographic output is nearly endless.

Finally, the GIS analyst should consider a set of nontraditional cartographic forms that has been relatively neglected among the community of users. The graphic devices called **cartograms** have the appearance of maps, but the spatial arrangements of the depicted objects have been modified not by their actual locations in space, but rather by the changes in character of the variable being measured. Thus the distances, directions, and other spatial relationships are relative rather than absolute. Cartograms are seen frequently and give evidence of their usefulness. For example, many urban mass-transit networks are posted inside the municipal buses and trains. These maps plot the relative sequential locations of each connecting stop on a straight line, as opposed to showing the actual route taken. In addition, the spacing between stops is not necessarily a true depiction of the scaled distance between each stop. These conventions simplify the map and emphasize the most important aspect of the spatial relationship. While riding in a bus or subway car, your most important concern is where you want to get off, perhaps followed by where the transit line you are traveling on will connect with the next (Figure 14.10). Such **routed line cartograms** are also found in many road atlases that indicate, by using straight lines rather than road networks, both the distances between major cities on the map and the estimated travel times.

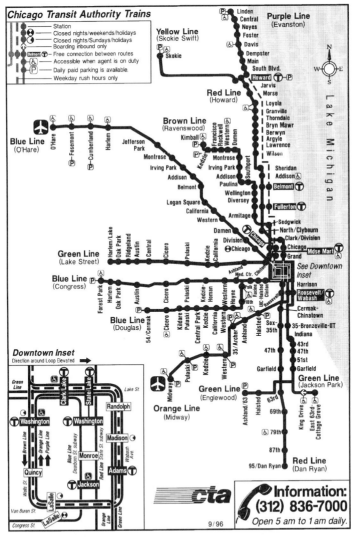

Figure 14.10 Linear cartogram. Linear cartogram for the Chicago transit system. *Source:* Chicago Transit Authority.

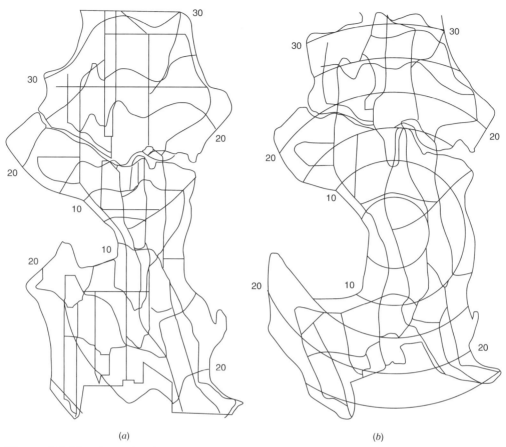

(a)   (b)

Figure 14.11   Central point cartogram. (From Bunge, 1962.) *(a)* The illustration shows
a conventional representation of travel times from central Seattle in 5-minute intervals.
(b) This is a cartogram showing functional distances in 5-minute intervals. *Source:* Adapted
from W. Bunge, in *Theoretical Geography,* 1962. Lund Studies in Geography. Series C. General and
Mathematical Geography, No. 1 (Lund, Sweden: C.W.K. Gleerup, Publishers for the Royal University of
Lund, Sweden, Dept. of Geography).

As you can see, the cartogram modifies geographic space to produce an
easily understandable model of reality. Another form of cartogram, the **central
point linear cartogram,** could easily be used to modify the output from a
functional distance model used by your GIS (Bunge, 1962) (Figure 14.11). And
among the most used forms of cartogram are **area cartograms,** which vary the
sizes of each mapped study area based on the value being examined. A classic
example of the use of this technique was presented by de Blij and Muller (1994),
who modified the physical size of each nation to illustrate the amount of population, rather than the physical land area. An area cartogram can be either
contiguous, where all the areas are touching, or noncontiguous, where the areas
are not touching (Campbell, 1991; Figure 14.12). The latter type is sometimes
called an exploded cartogram and has the advantage of not requiring the lines
to be connecting, thus simplifying the production of the output.

The interpretation of cartograms may require a minor rethinking by users

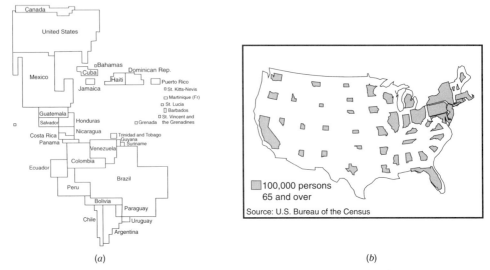

(a)   (b)

Figure 14.12 Contiguous versus noncontiguous cartograms. *(a)* Contiguous and *(b)* noncontiguous value by area cartograms. (The requirement for contiguity applies only to areas on the same landmoss; hence, the Caribbean islands are shown separate from each other and from the mainland of the Americas.) *Source:* Part (b), J.M. Olson, "Noncontiguous Area Cartograms," *The Professional Geographer* 28(4) p. 377. Blackwell Publishers. Used with permission.

unfamiliar with this form of output. Still, the time spent should be minimal. And once users become fluent in the reading of cartograms, the results generally are well received. However, many users will want to have traditional cartographic forms against which to compare this radically different device of graphic communication.

You may not see many examples of cartograms as you look over the output from GIS analysis at conferences and meetings. This is partly because most of the members of the GIS community either are not familiar with the form or are not used to producing it. A second, and perhaps more important reason for the relative lack of cartogram output is that few vendors offer software that readily performs this task in an automated environment. Tubler (1974) developed a number of FORTRAN programs designed to automate the process. As you become more familiar with your software, and as you gain confidence in your ability to create macro programs for unique tasks, you may want to refer to Tobler's programs, modifying them to suit your own needs or the macro language used by your software. In any event, the use of cartograms as an alternative to traditional map output is sure to increase as both GIS operators and users become more familiar with them.

NONCARTOGRAPHIC OUTPUT

Despite the prevalence of maps as a form of GIS output, certain data are better presented in noncartographic form or in addition to the map product. There are two general reasons for this relatively frequent occurrence. Either the map provides output that is less immediately understood because of the audience, or the map is not the desired output from analysis. In the former case, the map usually is supplemented with other forms of output, while in the latter, it most often is replaced. A few examples of enhanced or alternative output should prove useful here.

Interactive Output

Perhaps the classic example of replacing map output with alternative types has already been alluded to: namely, the enhancement of emergency services such as the 911 system by means of GIS. In the standard implementation of this system, residents of large metropolitan areas may not get a speedy response when they dial the 911 network to report a fire because there are many fire stations throughout the city that could respond. The 911 operator wants to make certain that the closest station will always be the one to take the call. In the enhanced implementation, information about, say, the location of a fire, is transmitted electronically to an automated GIS system while the operator is still on the phone. The GIS matches the address of the actual road network inside the database. In turn, the program determines which fire station should respond and reacts by immediately sending an alarm signal to that unit, alerting the dispatcher to the current emergency. In this case the result of the GIS analysis is an electronic response rather than a map.

As indicated earlier, the program also could be designed to determine the quickest path from the fire station to the fire, and this information could be electronically transmitted to an output device at the appropriate fire station. The output could be in the form of a text-based route, perhaps including a map. Although computer networking and GIS implementation today are seldom so sophisticated, the scenario is not beyond current capabilities. It is much more likely that multiple output would be available under less pressing (i.e., non-emergency) circumstances.

Suppose that you are using the network functions of your GIS to route a moving van from one location to another. This is an analytical task commonly performed by auto clubs for individual motorists as well. If you have used these services, you know that there are generally two types of output. The first is a map that highlights the roads that lead most efficiently from one place to another. The second is a text-based road log that shows, in miles, the distance traveled on each road, the names of the roads, the intersections at which you are to turn and the direction of the turn, as well as the number of interchanges when available. While the correct route could be discerned by careful attention to any of the individual components, together they provide additional information that makes navigating along the highways a simpler task. While many of us are comfortable with a regular map, having the road log along makes us more confident, particularly when the available map is highly generalized. Thus road logs offer a good example of how cartographic and noncartographic output can complement each other to provide a more comprehensive solution to a problem.

Tables and Charts

There are, of course, many other forms of alternative output for GIS (Garson and Biggs, 1992), but among the most common are tables and charts. Earlier we considered a GIS client in the newspaper business and saw that the most useful output for this publisher would be a tabulation of potential customers. Here a simple, noncartographic output that could be used to increase circulation could be produced without the use of expensive equipment.

In other situatons, the use of tables and charts could greatly improve the understanding of cartographic results. Assume that you are analyzing land use patterns as they change through time in an area that is losing natural habitat and agricultural lands as a result of population pressures. A series of maps shows where the changes from wildlife habitat and farming to urban land uses are most severe. The map legend may include tabulations of total land uses by type for each time period. However, there is much more specific information available in the database. The tabulation could be enhanced simply by producing a bar graph under each map to show the total amounts or percentages of each land class. In addition you might produce a bar graph showing the percentage change for each land class for each subsequent time period. You could improve this simple graphic technique by creating a matrix showing "from" and "to" changes for each time period. Such a table would prove very useful in understanding which types of land use were responsible for the loss of the largest proportion of natural habitat and which were most responsible for the loss of agricultural lands.

The foregoing example illustrates that for any given set of GIS analysis data, there are many options for noncartographic output, both in type and in form. Just as in cartographic display there are design issues, there are also design issues for noncartographic output. One simple example is encountered very often among users of computerized spreadsheets. Nearly all these software programs allow the user to produce a wide array of different types of tabular output, graphs, and diagrams showing the results of analysis. The number of design questions that arise are numerous and can encompass the complexities associated with color perceptions, personal preferences, and the capabilities of output devices. We'll look at just a few of the more important considerations.

Tabular output from GIS analysis is of two general types. The first is contained in the textual material of the map legend and can often be treated as part of the map design process because it is used to connect the attribute data to the entities presented on the map. A second type of tabular output employs lists of attribute data, tables showing relationships between raw attribute data and map entities, text showing extended-data dictionary entities for improved understanding of what the legend means, correspondence matrices showing relationships among sets of attributes, and so on. These outputs can be placed inside the exterior map boundary (the neat line) if they are to be presented as a combined finished product, or they can be output separately for the user to peruse at will (i.e., alone or in conjunction with a separate map).

Design considerations for text-based output involve the basic concepts of purpose, readability, and audience, familiar from our introduction to map design. Before producing text and tabular data, consider the needs of the user. First, will the text-based output effectively replace the map as a communication device? Or will it sufficiently enhance the readability of the map output to warrant its production or inclusion? In our example of the newspaper publisher needing lists of potential subscribers together with their telephone numbers, the answer to the first question is an obvious yes. This is exactly what would serve this client best. Many other situations will be far more subtle, and you will have to determine the value of the tables themselves or the value added to the map by inclusion of tables and text. Because each GIS project is fundamentally different from its predecessors, you will need to consider each situation separately, most often during the design of the GIS (see Chapter 15). If you are going to include tables and text, however, the next item you should consider is readability.

Text and tables are fairly commonplace, and more easily understood by the general public than maps; hence they tend to be easier to design. Still, a few things should be kept in mind to ensure basic readability. First, the font should be simple in style and large enough to be seen at the designated combination of viewing distance and lighting conditions. Fancy fonts may by themselves seem particularly attractive, but simpler is usually better. Keep in mind the need for contrast when you select the color of your text output. If the tables or additional text are to be included inside the map boundary, be sure that their colors contrast sufficiently with the map background. For further assurance of separation, add a border, taking care that it is not so heavy that it conflicts with graphics devices on the map proper. Avoid acronyms and abbreviations whenever possible. The less explanation your tables require, the more useful they will be. For large tables you may decide to outline major categories, or even all the cells in the table, since these options are familiar to many from word processing or spreadsheet programs. Text or tables lying inside the map boundaries should be separated with a border, which should be functional but not so heavy that it conflicts with other graphics devices on the map. Finally, employ the same common sense in depicting your data that you used in designing your map.

As you produce text and tables you should also keep in mind the audience you are trying to reach, using terms that will be familiar to the client. In addition, the type of table you create should be useful to the client, as well as understandable. Returning to our example of land use change, a set of tables indicating the amount of change in each category for each time period may not contain as much information as a change matrix table or to–from matrix. However, if the client is interested only in the amount of increase or decrease in a select few land use types, the additional information may be not only uninteresting, but generally confusing. The same principle that we followed in cartography applies here. The text or table should say something about the phenomena studied, but it should say no more than what the reader needs to see. Alternatively, if you or your client really need to know both the amounts and the directions of the land use changes, a change matrix might be entirely appropriate. Before you produce your tabular output, be specific about what you or your client needs to know and how best to present it.

Tables can easily become too large or cumbersome to be useful. Monstrosities with 40 rows and 40 columns are not likely to be useful because they require far too much time to examine. An easy alternative is to present the results not as tables, but rather as graphic representations of tabular data. Most common microcomputer spreadsheet programs illustrate a wide variety of traditional business graphics devices for the presentation of data: line graphs, pie charts, bar graphs, and the like. There are, however, far more options available to represent the phenomena that can be output from GIS analysis. Among these are triangular graphs representative of, for example, soil textures; climographs indicating the relationships among soil moisture, temperature, and precipitation variables for a region; and wind roses or star diagrams indicating predominant directions of wind, glacial flow, or other phenomena for selected points (Monkhouse and Wilkinson, 1971). It would be futile to try to examine each in detail, or even to consider all the possibilities. Instead we'll look again at some basic considerations for the more general types of line graph, trusting in your own background and experience to select the appropriate specimens for the data with which you are working.

The most common line graphs employ a Cartesian coordinate system with the standard ordinate (vertical scale) and abscissa (horizontal scale). Another, less common type of graph is the polar or circular graph, where the points are plotted based on a combination of angular bearing (the vectorial angle) and distance from a point of origin (the radius vector). In one variation, the points of the line graph are plotted on a triangular grid. Such curves, as mentioned earlier, are common when showing the relationships of three variables; in a soil texture triangle, for example, the percentages of sand, silt, and clay are represented on the same graph. For all line graphs, in general, the observed points for the variables, usually classified as dependent and independent, are connected either by individual straight-line segments between pairs of observations or by a smooth curve indicating the general trend along the points. A modification of the standard Cartesian line graph is the histogram or bar graph, which displays the response of each observation by means of a wide band extending either vertically from the abscissa or horizontally from the ordinate. The histogram is more often used when the number of observations is low; it is more generally representative of the noncontinuous nature of many observations. For data that represent continuous variables, such as changes in barometric pressure, the lines themselves replace the points (Monkhouse and Wilkinson, 1971). You are not likely to encounter these true line graphs in GIS analysis, where the spatial dimension is normally stronger than the temporal dimension.

Line graphs can be simple line graphs, displaying a single series of values connected by a line, or they can resemble a polygraph, producing multiple lines, each representing a selected variable. Simple histograms may have a single series of values rising like ribbons, or multiple values, each represented by a ribbon of unique color or pattern. These multiple line graphs are produced primarily to allow the direct comparison of numerous variables under the same circumstances. Multiple line graphs can be produced in two general fashions: as multiple sets of adjacent lines or bar symbols, or as stacked graphics. For stacked graphics versions, the values of each observation are physically combined with those of the preceding values, to give an impression of area. Such plots are especially useful if the values themselves are meant to indicate an additive relationship. In the case of stacked histograms, for example, the height of the bar is based on a combination of all the different variables at a single observation point. Under this circumstance, if the bar was meant to represent the variables %sand, %silt, and %clay, the total bar would be at 100% because each of the values is additive.

Designing line graphs has become relatively simple with the power of the computer. Some software, however, reduces our options for output or preselects the ways in which the output will appear. These default settings can, and often should, be overridden by the user.

Good line graph design choices for GIS output can be indicated in general with a few simple comments. While most software offers very colorful and stylish options such as the appearance of three dimensions for histograms, and ribbons for lines, these tend to be unnecessary and generally are less readable. Try to avoid using large, garish symbols as data points because these outsized symbols often interfere with the lines. Unwieldy data points can easily be replaced by numerical values, which are more informative and less confusing. Avoid radical differences in direction of the internal marks used to produce shading patterns, such as diagonal lines next to crossed lines next to diagonal lines in the other direction. Such inelegant graphic design will often result in unnecessary eyestrain. (The same can be said for adjacent choropleth patterns on maps.)

Keep the important lines prominent. Although most software allows you to place grids on your graph, the fewer you use and the lighter the line shade and line width, the more prominently the important lines will show. Make the size of the vertical and horizontal scale extents as close to the maximum value as possible. This is not always easy with available software. And you may want to ignore this rule if you have multiple graphs that show relationships between or among different places. In this case, all the graphs should carry the same maximum sizes to avoid confusion, or the false impression that small values are larger than they really are. Finally, remember what you are seeking: simplicity and legibility. Simple, clear fonts, and lines and symbols with good contrast should remain the most important criteria.

The sheer numbers of graph types and options can be somewhat overwhelming. There are a number of good references, primarily based on business graphics and graphics presentations. You should consult these or others like them when you intend to include business-type graphics in your work (Cleveland and McGill, 1988; Holmes, 1984; Meilach, 1990; Robertson, 1988; Sutton, 1988; Tufte, 1983; Zelazny, 1985).

A final note should be made about another form of noncartographic output that is becoming more prevalent in GIS output. Many commercial vendors now include the option of incorporating digital photographs showing, for example, buildings, site characteristics, and specimens of plants and animals found in certain areas. These digital images, often explicitly linked to the attribute database, can add a great deal to an understanding of what is represented on the map by giving the reader a set of familiar examples. The images can either be scanned into appropriate software or photographed directly with the use of digital still cameras or even videotape. More of these images are likely to appear as output from a GIS. Like all graphic and nongraphic devices, they should be used sparingly and only to achieve a desired effect. Because of the graphic complexity typical of these images, it is easy to overwhelm the associated map. And, as always, make sure the images are appropriate, useful, and sure to add value to the output.

TECHNOLOGY AND GIS OUTPUT

As I've indicated from time to time in the preceding discussion, the technology of modern electronic GIS has been a mixed blessing. It has produced a wide range of options for map design within an environment that allows relative ease of movement of the cartographic objects. But it also has created some limitations on the nature of the output by restricting color sets, symbol types, and the like. Whether your software and hardware offer you limits or options is largely dependent on the degree of sophistication of the installation.

Ephemeral output is most often produced on computer display devices referred to as monitors. These were only recently limited to the cathode ray tube (CRT) technology akin to black-and-white (monochrome) or color televisions, though usually resolution was higher. New monitor displays based on thin-screen technologies include gas discharge, plasma panels, light-emitting diodes (LED), and liquid crystal displays (LCD) (Robinson et al., 1995). The latter technologies, although becoming more prevalent and expected to continue to expand, especially for laptop computers, still form a relatively minor part of the GIS market. The CRT display is by far the dominant method of display and is considered here in more detail because of its present status.

The CRT monitor can operate in either raster or vector. In the raster mode, a beam of electrons is directed against a phosphor-coated screen and the image is built up horizontally, line by line. The phosphor coating on the screen reacts to the electron beams by glowing. On monochrome monitors the effect of gray shading is modified by changes in beam intensity. Color monitors have three beams and three phosphors: red, green, and blue. Each primary color phosphor can be modified in intensity and combined with one or both of the others to produce a wide range of hues and intensities (Figure 14.13).

In vector or random-scan display devices, only the points and lines necessary to create the image are scanned by the electron gun(s). This approach is an

Figure 14.13 Raster display monitor. The monitor shows ephemeral output of GIS analysis.

expensive alternative to the normal raster CRT output and is useful for enhanced image precision, as might be necessary for detailed CAD graphics. As
image complexity increases, however, the monitor needed to produce the image
also increases (Robinson et al., 1995). Our remaining discussion focuses on
raster CRT monitors, again because of their current domination of the market.

The phosphors on the screens of CRT monitors glow only briefly and then
have to be "refreshed": that is, to keep the image visible to the viewer, the
electron beam must continue to scan across the phosphor screen to prevent
screen flicker. Prevention of screen flicker requires the phosphor to be excited
by the electron beam approximately 72 times every second. This is called the
refresh rate. Some monitors, usually less expensive ones, reduce the number
of scans and simulate the same refresh rate by scanning every other line with
each pass of the electron beam. In other words, only half the lines are glowing
at any given time. This is called interlacing. Monitors that use interlacing will
reduce the costs of GIS operations because the equipment is less expensive,
but the interlacing process often produces results that are not as crisp and may
not completely eliminate display flicker.

Unlike random-scan monitors, raster CRTs must store information for each
of the picture elements (**pixels**) on the screen, regardless of whether it is turned
on. As a result, there is a rapid increase with screen resolution of the number
of data processes, and therefore the amount of memory required to process
them. The more pixels your monitor uses, the more memory will be required
just to display images. In terms of being able to view the detail of a map on
screen, this means that the output from printing or plotting devices is likely to
be far superior. For example, if you have a 20-inch diagonal monochrome monitor with 1024×768-pixel resolution (786,432 pixels), each of these nearly three-
quarter million pixels must be excited every 1/72 of a second just to avoid
flicker. A monitor of this size yields approximately 70 dots per inch (dpi), which
is slightly less than the resolution available from newspapers (Robinson et al.,
1995). As a rule, 70 dpi is not acceptable quality for final output. And this is
monochrome. To achieve color at the same resolution you must operate on
three times the number of electron guns with additional variation added to
allow for the photographic quality of 256 intensity levels (8 bits per gun). You
will soon see that the CRT monitors allowing more colors and more dots per
inch call for very large amounts of graphics memory.

The graphics monitor, then, is the source of some serious limitations on the
quality of graphic output. But it has the advantage of allowing users to interact
with the map, enabling them to vary colors, shadings, and intensities at will.
And, again, it allows for the physical manipulation of the graphic objects for
layout purposes. Thus the monitor facilitates improvements in the quality of
the final, permanent output.

Permanent, hard-copy output from GIS also relies on the variety of output
devices available for production. The capabilities of these devices depend on
their ability to use on-board software to handle and translate the large data
files produced through analysis. They are also limited by the size of the output
required, color handling capabilities, available resolutions, and necessary output speed. To a great degree, the capabilities of these devices, as with the
monitors, depend on the amount of money that can be invested (Robinson et
al., 1995).

As with ephemeral devices, hard-copy output devices come in two general forms—raster and vector. In raster devices, such as printers and imagesetters, the output is built up by evenly spaced rows, much as the raster CRT builds an image line by line. As a result, the speed of these devices is not affected by the complexity of the image. The output from raster devices is produced through a number of possible means. Electrostatic printers use electrostatically charged dots placed on a constantly moving, specially coated paper or film, yielding resolutions of up to 400 dpi. Both monochrome and color capabilities exist, but monochrome is prevalent because full color is more expensive to produce, because the computer files must be color-separated and the paper must pass through four separate ink arrays to be printed. Ink-jet printers, including those commonly available as peripherals for personal computer output, spray streams of ink droplets through tiny holes. The droplets are electrically charged by passing through charged plates, and droplets are either attracted to or deflected away from the paper. These printers are commonly available in both monochrome and color and can, through the use of resolution enhancement technology, achieve resolutions that appear to the user to be near 600 dpi. A major advantage of ink-jet printing is the relatively low cost, but the devices are often slow.

Laser printers use technology similar to that for xerography (Figure 14.14). Based on the digital information provided, a laser beam scans back and forth along a photosensitive drum, and individual dot positions on the drum are

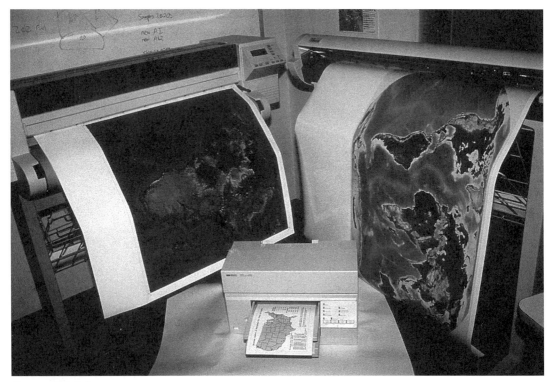

Figure 14.14   Laser printer. The photo shows a printer with 600 dpi resolution illustrating the output from a GIS.

given a positive charge by the laser. This positively charged drum is then dusted with negatively charged toner particles that adhere to the charged drum locations. Next, a sheet of positively charged paper or plastic is rolled over the drum, and the charged toner particles are attracted to it. Finally, heat and pressure are added to fuse the toner particles to the paper or plastic, making the image more permanent. Laser printers are available in both monochrome and color, although the color versions are considerably more expensive. Normal resolutions for these devices range between 300 and 600 dpi, but enhancements with ink-jet printers will give the semblance of up to 1200 dpi. These 1200-dpi devices produce excellent quality output, while 300-dpi quality usually is not sufficient for large format output.

Thermal printers use a page-sized inking ribbon coated with soft wax or plastic containing color pigments (or temperature-sensitive dye in the case of dye-sublimation printers). The pigments are temperature sensitive, and the on-board software controls the 256 levels of heat each color pigment receives, thereby determining the amount of pigment applied for each dot on the page. In general, thermal printers, which are capable of producing over 16 million colors, achieve quality superior to ink-jet printers, with resolutions as high as 300 dpi and 256 levels per color. The primary drawback is the cost of hardware.

Imagesetters, the highest quality output devices, produce output by plotting in raster fashion directly on photographic film. These devices use a laser beam to write directly to the photographic film, achieving resolutions of 1200 lines per inch for paper and 2400 lines per inch for film output. The very high cost of these devices often prohibits their general use in the GIS environment.

A relatively inexpensive alternative to raster devices is the vector plotter that simulates the actions of the human draftsperson. These devices come as flatbed surface models or drum plotters. In the former case, a sheet of paper is laid on the bed and the plotting device does all the moving. In the drum plotter, the pens or scribing points can be moved back and forth in one direction, while the paper moves at 90 degrees to it. Generally, but not always, these devices are incremental, in that the pens or scribing tools move in short increments, dependent on the capabilities of the hardware. This so-called **step rate** determines the resolution of the graphic output. Nonincremental plotters, which produce considerably better resolution output without the "zigzag" effect seen along diagonal lines made by the incremental plotters, are also available.

Raster and vector devices, alike are driven by software and control languages that convert the graphic images from your GIS database into commands to reproduce the output. Commands for plotters, for example, can be summarized as a set of "pen up, pen down, move x steps, turn right x degrees," and the like. These commands are different for each output language, and they must be compatible with the language of your GIS software. A detailed description of all these languages would not prove useful in an introduction to GIS, but you will find that each vendor supplies information about language compatibility. You are also directed to see Appendix B in Robinson et al. (1995) for a well-organized description of the common graphics languages. As you become acquainted with your own GIS configuration, you may want to become more familiar with these languages. It is essential that your GIS communicate properly with your output device; otherwise, your well-designed map output will not reflect what you designed. Font sets, symbol sets, and color sets frequently differ for each type

of hardware and for each software system. A test of the output at lower resolution (and usually faster output times) can give you an easy indication of
whether your GIS and its peripherals are literally talking the same language.

Terms

general reference map	orientation	hue
size	chroma	pattern
value	texture	orientation
arrangement	visual contrast	figure–ground
legibility	stereogrammic hierarchy	extensional hierarchy
hierarchical structure	purpose	substantive objective
subdivisional hierarchy	reality	available data
controls	audience	conditions of use
affective objective	fishnet maps	wire-frame diagrams
scale	viewing azimuth	viewing distance
technical limits	routed line cartograms	step rate
angle of view	area cartograms	
cartograms	pixels	
central point linear cartograms	thematic maps	
subdivisional hierarchy	conventions	

Review Questions

1. Why is it important to be familiar with cartographic design in GIS? Why is it even more important with GIS than with traditional, hand-drawn cartography?

2. What is our primary concern when producing a map from GIS analysis? How does this matter relate to the map as an art form?

3. Give a concrete example, other than what is in your text, of how intellectual and aesthetic design considerations can conflict. Show how the problem you describe can be solved.

4. Given the ready availability of computer software for designing and organizing graphics, why should you begin the map design process with pencil and paper?

5. Indicate, through concrete examples, the three basic design principles.

6. Show, with examples, what some of the possible design constraints might be in GIS map output.

7. Give an example of how too many data can produce problems in map design. Give an example of map design problems due to insufficient data.

8. What are some design considerations involved in producing good quality, three-dimensional map output?

9. Give as many examples as you can think of to describe how you might find animation useful as a form of cartographic output.

10. What are cartograms? Why are they not in general use in GIS today? When might they prove useful as output from GIS analysis? Give some examples, perhaps from your own laboratory exercises.

11. List some types of noncartographic output. What are some of the more unique types? Think about how you might decide whether these should replace or augment an existing map.

12. What are the basic design considerations for the production and use of line graphs as GIS output?

13. How does technology impact the design considerations for GIS output?

14. What are some of the factors you would consider in selecting the type of hard-copy output device for a GIS? When would you be more likely to purchase a vector plotter as opposed to a raster printer?

References

Bunge, William, 1962. *Theoretical Geography.* No. 1 in Lund Studies in Geography, Series C, General and Mathematical Geography. Lund, Sweden: C.W.K. Gleerup, Publishers, for the Royal University of Lund, Sweden, Department of Geography.

Burrough, P.A., 1986. *Geographical Information Systems for Land Resources Assessment.* New York: Oxford University Press.

Campbell, John, 1991. *Map Use and Analysis.* Dubuque, IA: Wm. C. Brown, Publishers.

Cleveland, W.S., and M.E. McGill, 1988. *Dynamic Graphics for Statistics.* Belmont, CA: Wadsworth Publishers.

de Blij, H.J, and P.O. Muller, 1994. *Geography: Realms, Regions and Concepts,* 7th ed. New York: John Wiley & Sons.

Garson, G.D., and R.S. Biggs, 1992. *Quantitative Applications in the Social Sciences.* No. 87 in Analytic Mapping and Geographic Databases Series: Newbury Park, CA: Sage Publications.

Holmes, N.,1984. *Designer's Guide to Creative Charts and Diagrams.* New York: Watson-Guptill.

Imhof, E., 1982. *Cartographic Relief Presentation,* edited by H.J. Steward. Berlin and New York: Walter de Gruyter, 1982.

Kraak, M.J., 1993. "Cartographic Terrain Modeling in a Three-Dimensional GIS Environment." *Cartography and Geographic Information Systems,* 20:13–18.

Meilach, D.Z., 1990. *Dynamics of Presentation Graphics,* 2nd ed. Homewood, IL: Dow Jones-Irwin.

Moellering, H., 1980. "The Real-Time Animation of Three-Dimensional Maps." *American Cartographer,* 7:67–75.

Moellering, H., n.d., Traffic Crashes in Washtenaw County "Michigan, 1968–1970," Highway Safety Research Institute, University of Michigan, Ann Arbor.

Moellering, H., and A.J. Kimerling, 1990. "A New Digital Slope–Aspect Display Process." *Cartography and Geographic Information Systems,* 17:151–159.

Monkhouse, F.J., and H.R. Wilkinson, 1971. *Maps and Diagrams: Their Compilation and Construction,* 3rd ed. London: Methuen & Co. Ltd.

Muehrcke, P.C., and J.O. Muehrcke, 1992. *Map Use: Reading, Analysis, Interpretation.* Madison, WI: J.P. Publications.

Robertson, B., 1988. *How to Draw Charts and Diagrams.* Cincinnati, OH: North Light.

Robinson, A.H., J.L. Morrison, P.C. Muehrcke, A.J. Kimerling, and S.C. Guptill, 1995. *Elements of Cartography,* 6th ed. New York: John Wiley & Sons.

Sutton, J., 1988. *Lotus Focus on Graphics* (5 vols.). Cambridge, MA: Lotus Development Corporation.

Tobler, W.A., 1967. Automated Cartograms. Department of Geography, University of Michigan, Ann Arbor.

Tobler, W., 1970. "A Computer Movie Simulating Urban Growth in the Detroit Region." *Economic Geography,* 46:234–240.

Tufte, E.R., 1983. *The Visual Display of Quantitative Information.* Cheshire, CT: Graphics.

Zelazny, G., 1985. *Say It with Charts: The Executive's Guide to Successful Presentations.* Homewood, IL, Dow Jones-Irwin.

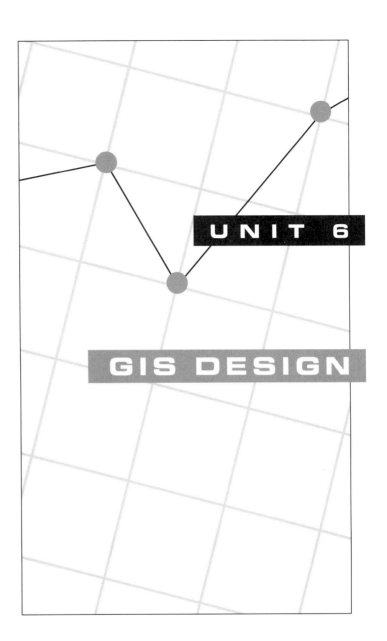

UNIT 6

GIS DESIGN

CHAPTER 15

GIS Design and Implementation

Our journey is now complete. We have gone from abstracting geographic objects to representing them cartographically to converting them to digital form. We've learned how we might use both the objects we encountered and the intervening spaces to answer geographic questions. And we've seen how different coverages can work together to give us still more insights, and how it is possible to combine all the analytical capabilities of the GIS into larger, more complex, and more meaningful models of the lands we have seen. Our geographic filter has expanded beyond its simple beginnings—now we recognize far more possibilities for conceptualizing, recording, and analyzing our environment, and for presenting the results of analysis. But as our journey ends, new ones will surely begin. We will use the tools we have honed and sharpened in many new and exciting adventures. Perhaps even more importantly, however, we are now in a position to be able to show others the utility of automated geography for a wide array of investigations. If our journey is to have been worth our while, we need to be able to use our knowledge in different settings, both for ourselves and for others who ask for our services. In this next chapter we will both look back at what we have learned and look forward to tasks that remain.

As we worked through the first 14 chapters of this book, our knowledge accumulated incrementally. We will now learn how we can combine what we know for future journeys. Our experiences will guide us in the design and implementation of GIS databases and analysis for large-scale, real-world applications. In this chapter we will see how every project must be carefully planned to ensure the success of the new implementations.

The idea of design will be a unifying theme throughout this chapter. As with our initial journey through the world of automated geography, we will need to select the right tools, define the applicable geographic objects and their relationships, identify the correct study area, and examine the availability of data to build an operational GIS for ourselves or for a client. The importance of good design has been indicated. Poor design has been responsible for the vast majority of GIS failures in the past and contributes to the problems associated with many of today's poorly functioning systems.

As you examine more GIS operations through meetings and shared experiences, you will begin to see how system flaws have contributed to system failures. Some GIS systems are incapable of producing the desired results because of poorly selected or badly organized data, incorrect data models, or software with limited capabilities; in some cases, system managers have underestimated the time involved in developing the databases. On the other hand, you also see well-designed GIS systems that are underutilized, either because system complexity prohibits easy use, or because of lack of understanding of the overall capabilities of the software. In still other cases, systems that were once highly useful have lost that utility because of aging software and hardware configurations, aging data, or other shortcomings that rob the system of flexibility required to adapt to changing demands. Finally, some failed systems will be seen as direct reflections of personnel problems that arise when organizational structures are changed forever by the introduction of new technologies into previously manual settings. Much of what we know of successful systems can be learned by looking at failed systems, just as successful businesses have learned from disasters experienced by colleagues.

Although some of you may be learning GIS with the goal of performing your own analyses, for other classes, honors theses, or personal curiosity, many will embark on careers as GIS consultants, systems analysts, applications specialists, and the like. We will focus more on the business side of GIS for two reasons. First, the vast majority of GIS clients are institutional: environmental consulting firms; economic analysis firms; transportation consulting companies; fire, police, and emergency services; mining, utility, and public service operations; and governmental organizations. In other words, most of the users are organizations that provide a service requiring the use of GIS. Our second reason for concentrating on the business side of GIS is that the success or failure of the system encountered by those who plan to use GIS for scientific research projects depends on much the same set of criteria used in the business community. If you can make a system work in the budget-conscious business world, it is likely to be successful for other investigations.

As you read this chapter, remember that your business is the business of automated geography. While the subject matter may seem to deviate considerably from what you have learned in this text, success depends on being able to combine business-oriented systems theory and operations research with the basic data models, abstractions, and analytical capabilities of the practicing geographer. Whether you choose a career directly in geography or elect some other application of these geographic concepts, remember that the utility of geography is what makes GIS a successful enabling technology. Companies use GIS because they need to answer geographic questions. They employ people who can interact as liaisons between the organization that is in place and the geography they need to answer spatial queries. So rather than wishing you a fond farewell, I invite you, as you finish this last chapter, to continue in your study of geography through the application of GIS to the many tasks for which it is uniquely designed.

LEARNING OBJECTIVES

When you are finished with this chapter you should be able to:

1. Define what we mean by GIS design as it applies to the success or failure of modern systems, as opposed to the GIS setups of the 1960s.

2. Define system design and discuss why it is important for GIS.

3. Understand the concepts of the system life cycle and the waterfall model, their objectives, and the limitations they impose on GIS design.

4. Understand why designing a system correctly the first time will save a great deal of effort later.

5. List and explain the function of the internal and external players in the GIS institutional setting.

6. Understand the concepts of conceptual design and cost/benefit analysis as they apply to GIS, and have a feel for how people problems are a primary concern in this area.

7. Describe the advantage of the Marble model for GIS design as opposed to the waterfall model.

8. Define SIPs, and discuss how they can drive the design of a GIS and how they are integrated from local views to global views of the GIS in an organization.

9. Understand the major database design considerations, especially with a view toward system verification and validation.

WHAT IS GIS DESIGN?

When GIS first came on the scene in the 1960s, there was a rapid proliferation of raster and vector GIS software. Some of these systems were developed for experimental academic use in colleges and universities, while others were created as operational systems much as most of today's GIS software is targeted. Unfortunately, most of these systems failed, as evidenced by the limited number that survive as complete systems. When we say that these systems failed, we mean that, by and large, they did not function. In some cases they didn't work to any degree at all as analytical tools. In other cases they produced erroneous results, and some systems tended to stop functioning entirely, leaving the computer hung up, trying to figure out what to do next.

These systems did not survive because they were poorly designed as software systems. Today, however, there are more and more successful implementations of the spatial analytical software we generally refer to as GIS. And while many of these systems can only marginally be called GIS because of the limited subset of automated geographic principles they employ, they are most often capable of performing the tasks for which they were designed relatively efficiently and at a reasonable cost. So when we talk about GIS design we are not necessarily talking about actual software design, although that is also an important aspect of the process.

Most of the problems related to GIS implementation today are nontechnical. In fact, much that is available in the way of software capabilities far outstrips the requirements of the user community, especially in most commercial applications of the technology. The primary problem is that there is often a mismatch between the capabilities of the software and the needs of the users: data and analytical needs, as well as personnel needs, including training and user acceptance. Unfortunately little attention has been given to the actual implementational problems. In the rest of this chapter we will see why we need to examine the design of GIS implementations and how structured design processes, primarily modified from the field of software engineering, can be useful for ensuring GIS success.

THE NEED FOR GIS DESIGN

Because of the enormous costs and time involved in developing operational GIS systems, whether for ourselves or for a third-party client, our desire is to avoid, as often as possible, the consequences of failure (Marble, 1994). Design is no guarantee of success of course—the Edsel automobile, perhaps the most designed automobile in United States history, was an absolute failure in terms of sales. Still, careful design will most often shift the odds in favor of success.

When we think of success or failure in any type of computer system, we need to be aware that the failures are most often not as dramatic as the rocket failures in the days of the space industry that resulted in explosions and months of setbacks. Computer software failures are far more subtle. Poorly written software sometimes results in obscure computational errors that are not detected by the users, or input or formatting errors may cause the software to malfunction and hang up your computing session, resulting in lost data. In the first case the error may not be found until thousands of operations have been performed on erroneous results for intermediate coverages. This discouraging scenario suggests that the more complex the system, and the more transactions it performs, the higher the degree of failure. Because GIS uses some of the more complex software available in the commercial marketplace, often requiring millions of computations in a single session, even minor errors due to the software will have a cumulative effect, eventually becoming highly accentuated.

But, even if we assume that the computational accuracy of the GIS is sufficient, and that care was taken to eliminate as much error as possible in the input phase of an operation, we must concern ourselves with even basic functions: namely, serving the proper users and serving the users properly.

Regarding the first function, we need to understand that especially in a multiuser **institutional setting,** there is the possibility of excluding potential users during the implementation phase. For example, systems in the public sector are devised to answer questions about fairly large areas such as counties, or even states or regions. Nevertheless, you might be surprised to find that many county and state organizations, perhaps having offices within easy walking distance, begin developing large databases without much consultation or interaction with other obviously interested parties. Such a myopic view of GIS implementation results in duplication of effort, which wastes much time and money (DeMers and Fisher, 1991). The problem is one of blindness to the needs of different institutional users for some of the same databases. Indeed, GIS systems have been developed without prior assessment of the potential utility of the system or without any indication at all of a potential clientele. An example was provided by the satellite data-driven Regional Environmental Assessment Program (REAP), started in North Dakota during the 1970s, when the usefulness of the *LANDSAT* satellite's data in evaluating environmental conditions became apparent. While the idea behind the operation was well founded scientifically, a clientele that might have been able to use the products of such a program did not exist. Lack of participation by the legislative constituency resulted in lack of sponsorship by legislators, who finally ended state funding for the program.

Regarding the second function—the requirement that the GIS meet the needs of the users properly—we are primarily concerned with software capabilities, data needs, and institutional fit. The first two concerns are strong indications of why a knowledge of both system functional capabilities and proper database construction are so important to the long-term functioning of a GIS. Before a GIS system is purchased, with its own data model and analytical capabilities, the spatial data handling needs of the user must be determined. As an example, the Ohio Environmental Protection Agency commissioned a system that emphasized a TIN-like vector data model, which was intended to handle the many operations dealing with surfaces that were needed for the agency's studies of the spread of pollution (DeMers and Fisher, 1991). For other organizations, needing constant updates of their study areas using digital satellite data, software that allows the performance of image enhancement and image classification procedures would be more appropriate. In both cases, the spatial data handling needs dictate the type of system and its analytical capabilities.

But it is not enough simply to have software and data that meet the spatial data handling needs of the organization. The system must fit well with the institutional organization and its personnel. Just as GIS systems differ in what they do and how they do it, the organizations that will employ GIS differ in their operations. A university research setting is likely to be the site of complex experiments that push the GIS to its limits. In addition, the research agenda changes quickly, and the GIS data and analytical needs are often changing or ill defined (Burrough, 1986). Rapid turnaround and widely varied skills of university personnel, not to mention the restrictions of budgets and deadlines for the research program, also impose constraints on a university-based GIS. Alternatively, a business may have more money, a more consistent workforce, and a better defined, more narrowly focused applications agenda. In addition, the analytical uses for which the GIS is employed may be somewhat limited compared to those one might encounter in a university. For an organization funded by federal, state, or county government, analytical and data requirements will be predominantly dictated by mandates from the governing body.

Burrough (1987) gives an excellent set of general concepts concerning the operational considerations for these different settings in which GIS might be employed. Experience indicates, however, that while the generalizations provide a nice framework for decision making, few operations completely fit within a given structure. Here we will assume that each organization's structure is unique and needs to be treated as such to provide the best GIS for individual uses. This perspective also allows us to view the organization in terms of our private usage as well, especially if we have a team of people working on a class project, independent study, or grant.

INTERNAL AND EXTERNAL GIS DESIGN QUESTIONS

Before we pursue the approach to GIS design we need to pause here to define our terminology. A brief look at Figure 15.1 shows that the overall process of

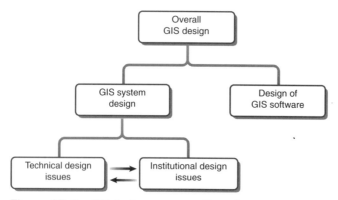

Figure 15.1   GIS design process. Note the separation be-
tween software design and system design. Under system
design also note the separation between technical design
(our major concern) and institutional design.   *Source:* D. F.
Marble, 1994. *An Introduction to the Structured Design of Geographic
Information Systems,* Association for Geographic Information
Source Book, New York, John Wiley & Sons, © 1994. Adapted from
Figure 1.

GIS design can be broken down into two major components—GIS **software
design** and GIS **system design**. The design of GIS software requires extensive
technical knowledge of data structures, data models, and computer programming. This work commands among the highest salaries in the profession, and
it requires the equivalent of a computer science or computer engineering degree
in addition to a knowledge of GIS basics. Those who are so inclined may find
this field very rewarding, especially because of the rapid changes in the industry. Our primary focus here, however, is with the second major issue of system
design.

System design is concerned with the interactions of individual people, groups
of people, and computers as they function within organizations (Yourdon, 1989).
This aspect of GIS study should allow you to broaden your understanding and
even your definition of GIS. GIS is not just about computing. It is also about how
incorporating the system into an organization affects people, how it changes
the way they do things, and how that, in turn, changes the functioning of the
organization itself. If you think about how the automobile industry has changed,
from one-by-one manufacturing through the development of the assembly line
approach to the introduction of robots, you will begin to see the potential
impact of GIS as an enabling technology on the performance of businesses,
government agencies, and university research groups. Just as word processors
and computer referencing services in libraries have drastically changed the
way we write term papers, the GIS fundamentally changes the way organizations
specializing in the analysis of spatial data go about their business. The introduction of the new technology requires different or additional training for people doing the work, requires that funds be allocated for software and hardware,
changes the flow of information within the organization and, therefore, modifies
the organization's structure. And now that we have large databases to work
with, we must be more concerned about the integrity and quality of the data
that are accessible by many different individuals. Wide accessibility, in turn,

increases the need for security and quality control measures within the organization.

GIS system design can be subdivided into two highly interactive parts: **technical design** (internal) issues and **institutional design** (external) issues. The internal issues deal most often with the system functionality and the database. Will the system work the way we need it to? Can we answer the questions we need answered? Have we got the correct data in the right format? Do we have people with the right training to run the system? Can our system adapt to changing demands? These are some of the general questions that accompany the process of internal or technical design. But while we need to be sure that our system functions properly, we also need to understand the relationship between the GIS operation and the organizational setting. Do we have the funding necessary to permit continued operations? Can we obtain data at reasonable cost? Do we need to employ applications programmers to customize the software? Will we have adequate software support from our GIS vendor? Will we be legally responsible for errors made through our analyses? And are we meeting the larger goals of the organization beyond the immediate end of performing GIS analyses? All these questions are important institutional considerations.

As you can probably see, the technical design issues cannot be separated from the institutional issues. Even GIS operations that are brilliantly successful from the technical design standpoint will fail if we lose the support of our organization or external sponsor. For example, even if your GIS-based efforts to enhance delivery of newspapers within the *New York Times* organization are successful, if upper management sees the added expense of operating the GIS as excessive, your operation will likely be suspended. Or, if your state fails to get continued federal funding for the use of GIS by your Natural Resources Department, you will have to close up shop. And, of course, if your system fails to provide the needed answers to the questions at hand, you are far more likely to lose your institutional support. Internal and external forces impact the overall system in which the GIS operates. And, as with any system, proper planning can make all the parts work together more effectively. We will look at a modern structured approach to successful GIS system design in the following pages.

THE SOFTWARE ENGINEERING APPROACH

A relatively recent subfield of computer science called **software engineering** has developed primarily in response to the problems of software that failed to achieve its intended goals. The problems encountered in the development of software are often analogous to those typical of the implementation phase of a large-scale GIS operation. Software engineering deals with the creation of software that successfully automates tasks that had been performed manually. For example, your effort might be to produce a word processing system that allows you to write, edit, check spelling and grammar, use a thesaurus, import graphics, typeset, and print documents easily and efficiently. The original word processors, while they worked reasonably well, provided only a few of these features, which are now standard in sophisticated word processing programs. As this example indicates, the fundamental problem is not simply to write good code, but rather to know what code to write in the first place. As with the implementation of a GIS, we need to understand the user's needs before we begin the process. In GIS systems design, the relevant user needs involve the kinds of spatial analytical technique to be applied, training, and the capacity of the analytical techniques to meet the overall organizational goals. Rather than writing code, we as GIS systems designers provide information about data structures and models, the software that provides the needed analytics at the most reasonable cost, the types of training that will be called for, and the system that provides the best overall fit to the organization.

System Design Principles

Among the first concepts developed in the field of system design was the **project life cycle**. In most operational GIS settings, many projects are going on simultaneously. For each project we must make basic decisions about what needs to be done, when it needs to be done, and who will be responsible for doing it. Because each project has a beginning, a middle, and an end (we hope), we can say that it has a life cycle, which in turn dictates the operations and organizational structure that will bring the project to a successful conclusion. If you are working on a single project, the project life cycle will focus on that one undertaking. The principles remain the same, but the operational details will differ.

When the output of a final product signals the end of the group activity that resulted in product creation, a substantially different approach will be used. Management styles differ, goals and objectives differ, and the personnel involved differ. Such project life cycles are developed in an ad hoc manner, often in a very flexible setting. This is particularly true of research projects, where the goals and objectives may change based on interim information. Larger, more business-oriented organizations tend to adopt a more formal, structured approach to the life cycle of a project. Where projects are of longer duration and personnel change frequently, it is often more useful for the life cycle structure to accommodate the need of new employees to discern their precise role in the larger project.

Whether your project is to be performed by yourself, within a small group, or within a larger organizational structure, there must be methodology to give a framework for finishing the job. You might think of an assigned term paper as a simple example of a project having a life cycle. The paper may require field work, computations and analysis, review of the literature, compilation of results, and writing. You don't (I hope) begin writing the paper until the preparatory operations are complete. There is a need for a structure that enables you to perform the correct tasks in the right order to get to the final result. Because you are the only one involved, defining the tasks and establishing the proper order of execution probably will suffice. In a setting characterized by many interconnected projects designed to meet the needs of management, however, two other objectives of the project life cycle must be met. There is a need to assure consistency among all the different projects so that all will have the same level of reliability. In addition, there must be built-in points at which management can determine whether it is time for the project to go forward. We can formalize these three objectives of the project life cycle as follows:

1. Define the activities of the project and in what order they are to be performed.

2. Assure consistency among many projects in the same organization.

3. Provide points for management decision making regarding starting and stopping individual phases of the larger project.

It is important to note, however, that the project life cycle is only a guide to management, a framework for the project; the manager still makes the fundamental decisions. Many important aspects of the organization (supporting the workers, fighting political battles, assuring good morale, etc.) are still fundamentally up to the manager. The life cycle provides guidelines to facilitate the making of the right decisions at the correct time (Yourdon, 1989).

System Development Waterfall Model

Among the first approaches to the project life cycle was a technique called **bottom-up implementation**. This method, although flexible in terms of the number of steps and the exact steps taken, is highly structured and progresses in linear fashion. The steps cover the definition of user requirements, the specification of functional needs, systems analysis, detailed design, the testing of individual modules, the testing of subsystems, and finally system testing. In the computer industry this approach is often called the **waterfall model** of system design (Boehm, 1981; Royce, 1970) (Figure 15.2).

The waterfall model was developed to provide a structure for the systematic movement from requirements analysis through testing and final operation of an information system. The waterfall model generally cascades from conceptual design through program detail design, code creation, and code testing to program implementation. There are several problems with the waterfall model as a tool for developing a GIS. First, because the model requires each step to be completed properly before the next proceeds, any delay in a single step will slow the entire system (Yourdon, 1989). Within a GIS context, then, if we know most but not all of the user requirements, we may not proceed to the input phase until the rest of the requirements have been received. While this approach may not seem to be unreasonable, new requirements frequently are discovered late in the requirements phase. Many days or weeks of data input could be lost by waiting to be absolutely sure that all requirements are known.

Another problem with the waterfall model is its linear structure. While we may want our GIS system to progress in a linear fashion, nearly every implementation is going to encounter problems. And, as it turns out, people are better at making improvements to an imperfect system they have watched in operation than at trying to anticipate all possible problems before the initial implementation. Think about how you write term papers: after you have your first draft, it is relatively easy to edit the material, identifying incomplete ideas, poor grammar, or out-of-place paragraphs. In addition, many times clients neglect important details at the outset, or discover new uses for the GIS as they see the system implemented. Or a client may find the economic situation chang-

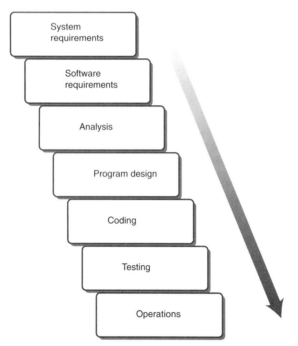

Figure 15.2 **Waterfall model of system life cycle.**

ing (perhaps through shortfalls of cash or a sudden influx of funding for a particular project), resulting in either more uses being included or costly portions of the system being cut. If we followed the waterfall approach to system development, we are likely to have a completed or nearly completed system before we discover that other things need to be added or mistakes need to be fixed.

The Mythical Man-Month

Mistakes revealed after a system is complete might seem to be relatively unimportant. But let's say that we have built a functional GIS for a third-party client and have proceeded to another project. A few months later we get a call from the client indicating that the GIS won't answer queries related to converting road log miles to and from map distance miles. It turns out that somewhere along the line, we forgot to provide the software that performs this operation. In the meantime, some of the people who worked on the original design, in particular our transportation experts, have obtained employment elsewhere. Yet we are legally responsible for providing the missing service. Think about the possible consequences of this scenario, which easily fits under the heading of the **mythical man-month** concept coined by Brooks (1975).

In simplest terms, the mythical man-month states that a $1 error in the first step will cost you $400 to $4000 in the fifth step. In other words, mistakes that might have been easily corrected early on are going to cause substantial problems in a later phase. A common response to such disasters is to "throw money at the problem." Thus our response to the missed transportation module for our client's GIS might be to assign a large group of people to the tasks. As Brooks (1975) predicted, however, this simply won't work. Although we may have many people to assign to the problem, they will not necessarily be familiar with the project, and none of them will have experience with transportation modules. Therefore much time must be spent in training. This is not only a particularly inefficient way of dealing with people, it also puts other projects on hold. And while we are trying to quick-grow the expertise to develop the appropriate module, the client is unable to perform some of the tasks the GIS was bought to handle.

The mythical man-month can also apply to groups of researchers trying to build a GIS for a scientific application. Because the goals of scientific applications are often ill defined, and because of the rapid turnaround in student workers, a strictly modular, linear approach to project design often turns out to be inefficient in these systems. An alternative approach might be advisable. We will see such an approach a little later in the chapter, but first we need to consider some basic principles about systems and look at the institutional setting.

Some General Systems Characteristics

Whether we are dealing with a software system or a GIS project, the more specialized it is, the less adaptable it is for new tasks. If, for example, you are developing a GIS system specifically oriented toward digital remote sensing applications, you are most likely to incorporate data models and techniques designed to enhance and classify the pixels obtained from a satellite system. If you decide later to incorporate a gravity model designed for use with a vector data structure, your tools will likely be inadequate for the new tasks. You will have to add vector data modeling capabilities to the system. Then you will have to train the users. This is nearly like creating a new system and placing it on top of the existing one.

The size of the system has much to do with the amount of resources that must be applied. In general, the larger the system, the more resources needed to run it. For example, if you are creating a GIS for a very large region, with many possible tasks, and hundreds of coverages, the input alone is going to consume more time for digitizing, more costs for obtaining data, more digitizing tablets, and probably more copies of the software than in a smaller system. Once the system is implemented, you will need more people to analyze data, you will need people to keep the data secure, you will need more output devices, and you will have to hire more managerial staff to keep track of the many parts of the analysis.

Fortunately, systems can be broken into component parts, each of which can be managed separately. This is another basic principle of systems, and one that is of paramount importance for larger systems. Each of the separate analyses can be envisioned as a subsystem that can be operated much like a single project. Strongly related to the ability to break GIS tasks into smaller components is the concept that many systems, perhaps even most, have a tendency to grow. So even if your original GIS operations performed only a few analytical tasks, as the knowledge and skill of the personnel grow, and as the utility of the system becomes better known both to management and to the client base, the system will have to grow to adapt. In fact, the more successful the operation, the more likely it is to grow. But success is often partly a function of communicating success to the institution and its sponsors. Most commercial operations serve many people. Therefore a knowledge of the institutional setting of the operation becomes important.

THE INSTITUTIONAL SETTING FOR GIS

Whether a GIS is operating in a university environment or in a business or government organization, it is not in a vacuum. If you are doing your own work, it is likely that you will need access to the hardware and software, but you should be sensitive to the requirements of other users so that your project does not interfere with theirs. If you are working on a larger project, you will need to coordinate your activities with the other users, while still keeping the goals and objectives of the overall project in mind. In turn, a project funded by external sources will require the system manager to meet the goals and objectives of that outside sponsor. The hardware and software most often will come from outside sources as well. As you can plainly see, there are many players in most GIS operations, and these players are both internal and external.

The Relation Between the System and the Outside World

Figure 15.3 shows a simplified model of the relation between most GIS operations (the system) and the world at large. Inside the system there are interactions with the people responsible for day-to-day operations of the GIS, with those who operate the system, and with those responsible for project management and oversight. This is illustrated by the two-way internal arrow in Figure 15.3. The second set of arrows shows inputs and outputs to and from the outside world. This is the larger framework within which the GIS must operate. Now let's identify the internal (system) players and the external (world) players and see how they interact.

Internal Players   The system comprises three basic sets of players, each with different objectives. The **system users**—those who will use the GIS to solve spatial problems—are most often people who are well trained in GIS, perhaps in a specific GIS. Digitizing, error checking, editing, analysis of the raw data, and output of the final solutions to queries are the primary tasks of the system users. Whether trained in university courses or on the job, the system users will require additional training because of the constant changes both in software and in the demands of the system for new analytical techniques. In many GIS operations, this group is likely to be the most transitory, with new employees replacing those who have moved on to different positions.

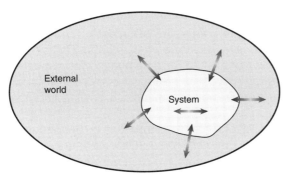

Figure 15.3 Internal and external players. The GIS system in relation to the outside world.

System operators are responsible for the day-to-day operations of the system, more often performing tasks that allow the system users to function efficiently. Their work involves troubleshooting when programs hang up or the complexity of the analysis requires additional insights. They are also responsible, in most cases, for training the users. This suggests that in many cases system operators have experience as system users. They also know enough about the hardware and software configurations to be able to make adjustments for upgrades. In addition to acting as system administrators, they frequently act as database administrators, keeping track of the security and integrity of the database to prevent possible loss or corruption of data.

Both system users and system operators operate within a larger organization often called the **system sponsor**. The system sponsor is the institutional parent that provides the funding for the software, hardware, and salaries. This entity also provides the political viability of the system. In other words, if there is no larger organization, there probably is no GIS. The system sponsor can be a university research group that obtained grant funding for a project, a government body whose work requires the use of GIS, or a commercial organization that performs GIS operations for paying clients. Funding and political viability of a GIS system also require that the system sponsor engage in the long-term planning needed to assure a steady supply of projects and continued funding.

External Players Because most GIS operations today do not use homemade software, they rely on the first of our external players, the **GIS supplier,** a company that develops and markets GIS software systems of one or more types. Suppliers also are responsible for providing software support and updates of the software as new and improved methods are put into the system; they may work in concert with hardware development companies to provide bundled packages of software and hardware. In many cases, training is provided for GIS users, either on site or at this supplier's facility, through contacts with the system administrator.

The data suppliers, the second external player in the GIS world, tend to be either private or public. The private company may provide internally generated data or data obtained from public agencies, modified to better fit needs expressed by the user community. Public agencies, primarily federal government agencies, provide data for large portions of the country. In many cases these data are designed for use by the agencies themselves, but the data are available to external users as well. Private companies are most often for-profit organizations that supply data at a cost-plus profit level. In some cases the data can be provided for specific regions and in the formats needed for a specific use. Some public agencies offer data at cost, or in some cases at no cost at all, provided the use is for nonprofit GIS operations. Appendix A gives a list of some public sources of data.

Another group of external players, a group that is becoming increasingly important as GIS software becomes more sophisticated, consists of the **application developers**. Application developers are generally trained programmers who will provide user interfaces to reduce the reliance on specialized GIS professionals to perform common tasks. In many cases the programming is done in macro languages provided by the GIS supplier to support applications development without the need for interfacing with traditional computer languages. While in the past most application developers were internal personnel who had years of experience with the analytical capabilities of the GIS software, today there are many small consulting firms specializing in the development of applications for third-party clients. Application programmers can limit the training costs incurred as new GIS systems users enter the organization. They provide a "point and click" environment that allows the system to be used by analysts having only a limited amount of knowledge of the internal functionality of the GIS software.

Finally, we have the **GIS systems analysts**. Members of this group of exernal players specialize in the study of systems design, the primary focus of this chapter. Most often systems analysts are part of a team of professionals responsible for determining the goals and objectives of the GIS system within an organization, fine-tuning the system so that it provides the right analytical techniques, and assuring the successful integration of the system into the organizational framework. In short, the systems analysts act as navigators for organizations using GIS. Commonly the systems analysts are part of a larger consulting firm specializing in the implementation of GIS for third-party users, but in some cases they are employees of the GIS suppliers.

A STRUCTURED DESIGN MODEL

Technical Design

In our overall view, we broke the GIS design process down into software design and system design (Figure 15.1). System design can be further divided into technical and institutional design categories. We will discuss institutional design later in the chapter. For the time being, we concentrate on the technical design.

Technical design has two basic parts: system functionality and system database. System functionality relates primarily to the ability of the GIS to perform the analyses it is expected to be able to handle. The system database is the set of coverages and their respective components on which the personnel and the software interactively operate to fulfill the goals of the GIS project. Any project

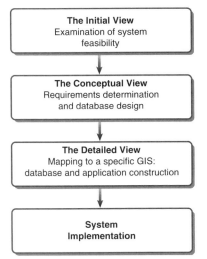

Figure 15.4   Simplified GIS design model. General system design model indicating level of detail encompassed at each step.   *Source:* Duane F. Marble, Department of Geography, The Ohio State University, Columbus, Ohio. Used with permission.

entails movement from the conceptual to the detailed to the implementational levels (Figure 15.4) (Marble, 1994). The level of detailed knowledge about the functionality and the database necessarily increases from concept to implementation. Before we can begin to understand the details and to finalize the implementation, we must develop a concept of system performance and an idea of how it might be obtained. A good conceptual design is an important first step to any good GIS.

The Reasons for Conceptual Design

We begin our project by making very general recommendations about what the system is meant to do, what problems it needs to solve, the general types of data, and so on. The existence of a good conceptual view allows us to plan future modifications on a continuing basis. Unlike the waterfall approach, the concept needs to be flexible enough to account for ongoing changes in goals, data availability, personnel, and management requirements. Thus independence of any particular GIS is essential. Once a specific GIS has been selected, the software may force us to limit our objectives based on the limitations of its data model(s) and functional capabilities.

Our approach in using a conceptual design as a first step is to create a global view of the whole system and its data requirements. If the requirements change for any reason, we can modify the global view at such time. Instead of using the conceptual view as a hard and fast structure, we use it as a starting point to provide an initial direction and to allow us to ask the right questions along the way. In addition to being software independent, the conceptual view allows us to view the availability of data from all different sources (called **federated databases**). While the GIS should not be exclusively data driven, its functionality and even its viability may very well be determined by the availability, cost, and quality of the data for a particular set of tasks. This means that we may find ourselves deciding to use different software; or we may request budgetary changes that will allow the purchase of additional data, or even suggest that the GIS is not a viable solution for the organization.

Conceptual Design Shortcomings

While the conceptual design is an improvement over the structured (waterfall) approach, it is not complete as a design formula for GIS operations. Even the best conceptual designs will fail if they are not effectively employed by the organization. Let us say that we are new consultants for a GIS project in a government organization tasked with monitoring the long-term availability of natural resources for a large state or province. The organization may have as many as 100 employees, all with different views of what the GIS may or may not offer by way of empowering them to perform their mandates. Because each individual is responsible for a portion of the larger mission, each will, by definition, have an interest in how he or she might interact with the GIS. In other words, the individuals have a stake in the implementation of GIS. The stakeholders are a vital part of the organization: trying to get people to use a system they don't need, don't find useful, or even see as threatening is a quick way to assure that the system will never work properly. Even the most powerful management level commands cannot make a system function if those who must use it feel that the decision to implement was made without their input. Perhaps a classic example is the unpopularity of the Susan B. Anthony dollar in the United States; the coin remains largely unused because the average person—that is, the intended user—found it too similar in size to the quarter. Or, take the congressional decision to designate the metric system of measurement as the standard for use by the federal government. The designation not withstanding, most private citizens still measure things in feet and miles, ounces and pounds, rather than meters and kilometers, and kilograms.

People Problems of GIS

The primary problem with the conceptual design is that it cannot ignore the players. The largest single hurdle to the implementation of any information technology, whether it be word processors, accounting systems, or GIS, is people. People problems more often than not explain the lack of return on large organizational investments in technology. One fundamental reason for such disappointing performance is that the introduction of technology fundamentally changes the way business is done. Take the simple example of the introduction of word processors into an office setting where everyone uses a typewriter. Some employees will welcome the change as a blessing, while others may be terrified at the thought of having to learn how to use a computer to type a letter. Some will become very proficient at the use of the software, while others may resist every inducement to use computers for even simple tasks. This problem is also seen when computerized accounting systems designed by outside consultants are introduced without regard for the people who must use the software daily. While the software per se may be able to handle the tasks needed by management, it must be operated by people. Those who expect the computer to perform all the essential tasks precisely as they were done before automation may revert to the manual approach until either the system is changed or they find jobs elsewhere.

But even software prepared with the needs of the individual users in mind cannot guarantee that employees will instantly adopt the technology. For starters, it will take time for people to learn and to use the systems put in place. As you might guess, if a system is implemented before the people are trained, there is likely to be a lengthy period showing an overall drop in organizational productivity. So the problem regarding training is twofold: the personnel must be willing and able to learn how to use the system, and the organization must provide the opportunity for the appropriate training.

But while many employees will welcome the introduction of a new technology (e.g., GIS) that will benefit them by improving efficiency, there is another problem, as well. It is not unusual to find individuals, groups of individuals, or even whole divisions whose knowledge and experience is their guarantee of continued employment. Because GIS, like any information technology, essentially opens the doors to the free flow of information and empowers many people to perform tasks that were once the province of single divisions, those whose power is about to be shared are likely to resist every effort to implement the new system. This often explains why GIS is not easily incorporated into many organizations.

Cost/Benefit Issues

Even if we assume that the employees are interested in the use of the technology because they see that it will make their tasks easier to perform and might even assure them of a more secure employment situation, we need to convince the system sponsor that the benefits of introducing GIS into the organization will outweigh the costs. By and large corporate CEOs feel that there is generally no totally reliable measure of what the return on investment is. Cost/benefit analysis, developed shortly after World War II, is a common technique for examining the return in real dollars based on the dollar amounts invested in a system. Little is known about the exact costs of implementing a GIS, however, and even less about the economic benefits from such an implementation. This is partly because for each GIS implementation, the numbers and complexity of the maps that are input to the GIS change radically. In addition, because of the long time periods necessary to develop the initial databases, short-term returns on this investment frequently are minimal. Even to begin a GIS project an organization's management must be convinced that there will be long-term benefits at all. Among the more successful ways of demonstrating this is to cite the positive experiences of similar organizations' efforts at implementing GIS. An alternative method requires the sponsoring organization to announce the cost/benefit ratio they are seeking and try to match that (Tomlinson, n.d.). In this rather conservative approach, the costs are inflated to ensure that cost overruns are avoided during the implementation phase.

Data and Applications Requirements Models

While the idea of structured technical design may sound like a single approach, it can take two separate forms. The **data requirements model** bases its methodology on the idea that the availability of data largely determines what can be done analytically. It begins, as always, with a conceptual design, proceeding to logical design (how the data are logically related to one another), and finally to the physical design of the database. The second approach, called the **application requirements model,** is based on the idea that the system is driven by the analysis the system is to perform. This model proceeds from a general functional analysis to higher level application designs and finally to the development of the specific details of how the analyses can be carried out. But while both these models represent logical approaches, the GIS is more easily seen as being driven by data *and* applications, because the two are interrelated. The basic design model we will now see is based on this concept.

FORMAL GIS DESIGN METHODOLOGY

There have been several attempts to develop and refine GIS design models, beginning with a model that was developed after painstakingly evaluating the actual implementation of a system by Calkins (1982). Other approaches include the extensive evaluation of the manual implementation of the GIS by the USGS (Guptill, 1977). Later, Johnson (1981) showed how improvements could be made to Calkins's (1982) original design in view of the increased management information systems (MIS) literature. By that time a number of computer tools were being developed to assist with software engineering. These tools might have been immediately useful to the GIS community except that GIS design focuses not on software, but rather on databases to operate within software, with both operating in an institutional context.

The Spiral Model: Rapid Prototyping

Duane Marble (1994) conceptualized and developed a flexible, multilevel design process modified from a software engineering approach pioneered by Boehm (1981, 1986, 1988). This spiral model for GIS design separates the three tasks of acquiring, organizing, and analyzing information and imposes three levels of detail. The first, called the initial view or initial model, is the most general and is the basis for discussion of the feasibility of the GIS system. The initial view is actually a precursor to the conceptual view or conceptual design process, which represents the commencement of requirements analysis and carries over into the first discussions of the database design. With time, the system works its way to the detailed design, which begins to answer questions about specific software to implement the system. Figure 15.5 (Marble, 1994) shows the spiral model and its component parts.

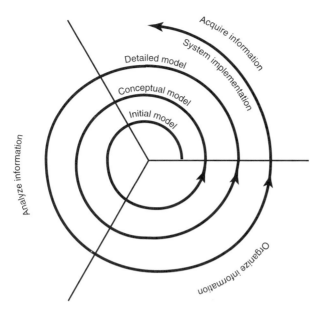

Figure 15.5   Marble spiral GIS design model. The
labels indicate the levels of design.   *Source:* D. F. Marble, 1994. *An Introduction to the Structured Design of Geographic Information Systems,* Association for Geographic Information Source Book, New York, John Wiley & Sons, Inc. ©
1994. Adapted from Figure 2.

Overview of the Initial GIS Design Model (Level 1)

In the initial GIS design view or model, Marble has created a flowchart that
details the individual tasks involved in the process. Each block of the model
has several levels, each revealing more details. In addition, the conceptual and
detailed models have several levels of detail that show how the process moves
forward. A detailed discussion of these three models and their levels is best
left for advanced GIS students and perhaps best described by Marble himself.
We will look only at the initial model and discuss some of the specifics to give
you a general feel of how the spiral works.

Figure 15.6 shows the general flow of the decision-making process as it proceeds from step 1.1.1 to step 1.2.6. In the first step, the client's input is used to
decide on the general goals of the organization. A GIS implementation that does
not reflect an understanding of the overall organization goals, perhaps expressed as a statement of the services or products normally generated, will not
be accepted by management. The next task (step 1.1.2) is to determine what
the GIS is meant to do. This is often difficult for potential users, whose interest
in GIS stems from the high profile of this glamorous technology in some business
circles. Many users get interested in GIS because someone else is using it. At
this point we need to ask potential users if they are truly interested in analysis.
Perhaps they see the GIS as a convenient archive for their data, or they might
be better served by a CAD or CAC system that requires less extensive analysis.

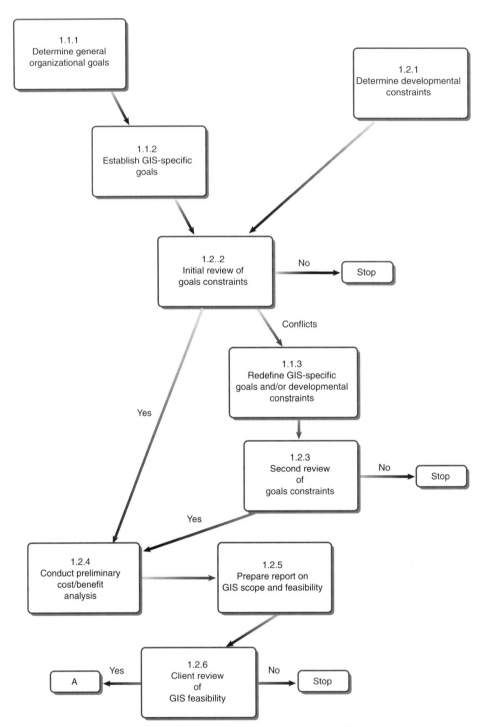

Figure 15.6 Steps in the design process. Decision model showing how the initial model moves from step to step, based on a series of decisions to continue or to stop the introduction of GIS into an organization. *Source:* D. F. Marble, 1994. *An Introduction to the Structured Design of Geographic Information Systems,* Association for Geographic Information Source Book, New York, John Wiley & Sons, Inc., © 1994. Adapted from Figure 3.

All the potential users in the organization should be part of the GIS goals decision, which means of course that we must know who they are. The users' requirements can then be identified individually. One of the best approaches is to find out what final products the users might want. These final products, called **spatial information products (SIP),** are covered in more detail as a separate topic. Here we present the general idea of ascertaining from the client what the final products might generally look like. We need to define the relationships between the SIPs that each user in the organization might be required to produce and the goals that each is required to fulfill. As we proceed, we shall keep in mind that each SIP will call for specific data coverages. Also, as we proceed from user to user, we begin the initial integration of each of the **local views** of what the GIS is to do into a larger **global view,** more often carried by management to meet the organization's larger goals. After completion of this step, the normal procedure is to produce for the potential client a final, product level requirements report to verify that our understandings thus far are accurate.

Concurrent with determining goals is an analysis of the limitations of development (step 1.2.1 of Figure 15.6). These limitations might include the budget allocated for start-up and continued support, the amount of time available to produce the final result, and the accessibility and costs of project data. It is important to separate out each constraint according to whether it is monetary, time related, or related to the modeling needs. Some special restrictions might entail the available hardware (perhaps the organization is limited to working with PCs). We should also consider work-arounds for each constraint as we encounter it rather than waiting until later. And, of course, we will need to consider the availability of data based on the constraints we have already seen. If, for example, there is a need for multiple coverages at different times, but the cost of satellite data is prohibitively high, we must investigate either alternative data sources or alternative methods of analysis. When step 1.2.1 is complete, we move to 1.2.2, the comparison of general user needs with the limitations. In other words, we decide whether it is appropriate to continue to pursue GIS. A good systems consultant will not suggest a $500 answer to a $5 problem. If the products needed and the budget, time, and data availability suggest the use of GIS, then and only then should we continue.

The next steps (1.1.3 and 1.2.4) are in direct response to the results of the comparison of requirements and constraints. The redefinition of GIS-specific goals and/or developmental constraints (step 1.1.3) may call for a certain amount of conflict resolution as the requirements of the system are balanced against the constraints to determine the feasibility of a GIS implementation. This balance is often a difficult one, and you may need to develop hierarchical views of the system requirements and the constraints. It may also be necessary to prioritize the requirements as to level of importance; the system costs will be subject to this ranking as well. When very important GIS requirements are also among the most costly implementational features of the GIS, a compromise may have to be reached, either by finding an alternative, cheaper solution or by simplifying the requirements to reduce their cost. In some cases the decision to halt a GIS activity that appears to be too costly will be delayed to allow for the possibility that the constraints can be reduced or the applications simplified. If the decision is to continue the development of a GIS for the organization, we can proceed to the next phase.

While affordability of the implementation is a major consideration, it may not be controlling, especially in the commercial environment, where profit is a driving force. In step 1.2.4 we can proceed with a cost/benefit analysis. First, the costs are evaluated: data purchase; hardware, software, and maintenance; personnel; training; input (among the most difficult to determine); space; and many more (Aronoff, 1989). The costs should also be separated into start-up and long-term operational costs. Benefits are much more difficult to determine but can generally be categorized as follows (Aronoff, 1989):

1. Increased efficiency of new methods over old.

2. New nonmarketable products and services, generally related to better quality of the products currently produced.

3. New marketable products and services, including GIS expertise sold to other organizations.

4. Better decisions (a major driving force for many GIS systems but very difficult to quantify).

5. Intangibles, including improved communication, streamlined organization, better outside image, and improved morale.

As we have seen, performing a cost/benefit analysis on a GIS is not an exact science, but there is an increasing literature that can be consulted (Goodchild and Rizzo, 1986; Green and Moyer, 1985; Kenney and Hamilton, 1985, 1986; Kevany, 1986; Laroche and Hamilton, 1986). But while the use of cost/benefit analysis may provide some financial basis for a decision, it is rarely the final determinant (Aronoff, 1989). Frequently when management has a clear understanding of the potential of GIS, it is the products of a GIS that drive both the decision to consider GIS and the implementation plan itself.

GIS INFORMATION PRODUCTS

In large part, GIS information products are the result of analysis within the software. What the specific information products are depends on the nature of the organization, its goals, and its experience with the system. While many organizations have a fairly structured set of products—for example, Rand McNally Corporation makes a relatively definable number of products (Calkins and Marble, 1987)—others, especially scientific research organizations, have a large, and ever changing set of potential products that may be difficult to define (Rhind and Green, 1988; Wellar, 1994). But even in scientific organizations there are most often general ideas of what the GIS is capable of doing and what the technololgy might offer the individual institution.

How Information Products Drive the GIS

Organizations that have a particular goal in mind when they consider implementing a GIS are quite likely to have some ideas about what the output of analysis might look like, even if they have only a vague sense of how it might be derived. In addition, they most likely know roughly what general data might be part of the system. In fact, many organizations recognize the value of the data in an automated form even before they are able to see the potential utility for analysis. To extract particular spatial information products from the users, we must recognize the close linkage between what goes into the system (data availability) and what comes out.

But, because members of the user community cannot be expected to be instant GIS experts, it is our job to help them define their needs. You should note that I did not say it is our job to define their needs! We are not present to tell them what they need. Instead, we are present to help them tell us what they need. Thus the systems analyst must play the role of educator to elicit descriptions of needed products from the users. Perhaps this will clarify my insistence on the need to understand the analytical capabilities of the GIS as well as the database creation and management tasks.

To act as educators we first spend a great deal of time determining how a GIS could best serve each user. This is best done in person, rather than through a review of questionnaires, because interactive question-and-answer sessions are most likely to lead to an understanding of the tasks each user normally performs, and to enhance explanations of functions the GIS might be able to perform to match those tasks. Any documentation indicating what the user produces (sample maps, written responses to questions or reports of decisions, applicable legal mandates, etc.) will prove useful in making a good match between the existing products and the SIPs obtainable through GIS. As you perform this review you should keep in mind that the users will be more interested in what can be obtained than in how it is derived. As the process moves toward implementation, the how questions can be addressed.

Organizing the Local Views

Because we most often will be working with many potential users of differing individual needs, it can be useful to be able to keep track of the relationships between each user and each SIP. Tomlinson (n.d.) has suggested the use of a **decision system matrix,** listing the users along the side of and the products along the top; Figure 15-7 shows a highly simplified version. Not only will this

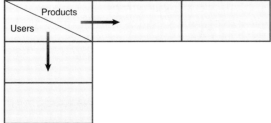

Figure 15.7   Decision system matrix. Simplified decision system matrix used to organize individual user views of the GIS.

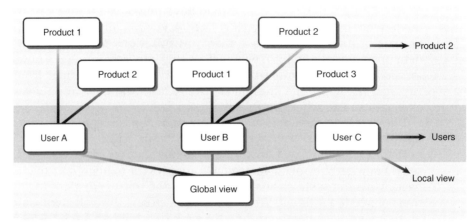

Figure 15.8   View integration. Organizational chart used to display individual user views of the GIS.   *Source:* Duane F. Marble, Department of Geography, The Ohio State University, Columbus, Ohio. Used with permission.

device allow us to keep track of the SIPs for each user, but later, when we must integrate each of the local (individual user) views into a larger global view, data from this matrix will provide much of what we need. An alternative to the decision matrix is an **organizational diagram** such as that shown in Figure 15.8. The global view at the bottom of the diagram illustrates the overall organization's needs; the next tier isolates the local views of individual users, and the products, differentiated by user, appear at the top. This type of diagram, like the matrix format, is useful for integrating the local views into a more general global view. Both can also be used to decide which products are the most important, based on which are most often called for by individual users.

Avoiding Design Creep

Design creep, which occurs in the absence of organizational learning, can be defined in terms of its results: at one extreme is a system with more functionality than is necessary, and at the other is a system having an incorrect functionality. Figure 15.9*a* shows a model for a rather structured approach to GIS design, proceeding from the feasibility study at the left through the design phase and finally into the implementation phase. The two curves indicate the relationship between the technical design process (light curve) and organizational learning. As you can see, organizational learning begins late in the process, when the system is almost completely designed and is nearly ready for implementation. Thus users are forced to learn and operate a system that may not meet their needs.

Figure 15.9*b* shows the more flexible spiral model (Marble, 1994), in which organizational learning drives system development. As users become familiar with what a GIS is capable of, they can describe their system needs at a point well in advance of the run-up to implementation. You can also see that the more sophisticated the organizational learning, the more sophisticated the system

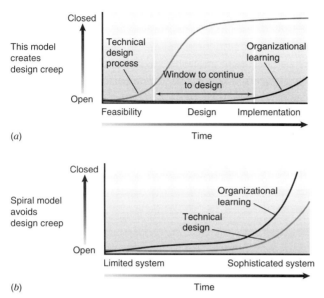

(a)

(b)

Figure 15.9 Avoiding design creep. Two approaches to system design: *(a)* a structured, linear approach that leads to design creep and *(b)* the Marble spiral model, which avoids design creep.

becomes, but the learning curve is always above the design curve. In short, the users are driving the design of the GIS, rather than being driven to learn by an in-place system. The latter approach avoids design creep and allows the system to grow in complexity as the organization's needs increase.

VIEW INTEGRATION

As we have seen, most organizations will have many GIS users, and it is vital that the needs of each be ascertained and addressed. The individual sets of requirements are called local views (of what the GIS is going to do and how it is to do it), and as a preintegration task we often find it useful to list a set of preferences for each of these local views, based on management needs. We also need to decide on a methodology—perhaps a decision matrix or an organizational chart—for later integration of the local views (called **view integration**). One useful technique is to produce a pair-wise (dyadic) grouping of similar individuals or groups. This can be performed easily with either the decision matrix or the organizational chart approach. The most common needs, those being met for the most people, will provide us with one method of identifying a set of priorities that can be compared to the larger organizational goals.

When conflicts exist between and among local views, additional discussion will be necessary to pinpoint the problems. Conflict resolution may require arbitration, modification, and even redefinition of the local views to achieve correspondence to the overall organizational goals. If isolated needs are found to be incompatible with the global view, several iterations of intermediate views may be required to combine individual local views into larger groups. This merger of local views will assure that the vast majority of individual user requirements are being met. Final decisions for determining the global view will often rest with management, where individual user needs are considered as part of the overall needs.

DATABASE DESIGN: GENERAL CONSIDERATIONS

Study Area

Beyond the basic idea of spatial information products there are other facets of the development of a GIS for an organization. When there are many different projects, it is necessary to consider the respective study areas separately. Study areas are most often driven by a formal decision as to why the area is under study. Thus for a county level study, the entire county will be the study area. For a study based on a natural phenomenon, such as pollution in a stream, the entire stream drainage basin may be selected as the study area because the pollution will come from all upland areas that drain into all the stream's tributaries. For a study for a lumber company contemplating the use of GIS to keep track of its logging operations and landholdings, the landownership will be the deciding factor for selecting the study area. In short, the study area may be selected based on political boundaries, physical boundaries, or landownership, or it may reflect primarily the economic and data constraints that are found in the early stages of the design process; in any event, its choice is project driven.

If there are detailed data for a small portion of a prospective study area, but more general data for the rest, the detailed data need not dictate the extent of the study area. Instead, the larger area of interest can be considered to be the overall study area, and the subarea for which there is more detailed information can act as a prototype for a detailed analysis of the larger study when data become available or affordable. In addition, the detailed data for the subarea might be useful for improving the knowledge of the larger area through the use of density of parts or another dasymetric mapping technique.

Whenever any form of interpolation is scheduled, the study area boundary, at least for the elevational coverage, must be extended sufficiently to ensure that the interpolation will produce the correct results. Finally, the larger the study area, the more expensive in time and money will be the production of the database. Often cost awareness is a major constraint in defining a study area in the first place. In addition, the larger the study area, the greater the chance that there will be a lack of data for all the coverages needed to perform the analysis. Selecting a smaller study as an initial prototype is a good choice if it will allow the organization to demonstrate the utility of the system. Once that has been effectively done, the system sponsor is much more likely to supply more resources to increase the size of the study area.

Scale, Resolution, and Level of Detail

Related to the size of the study area is the scale at which you want your data input. This is seldom a straightforward matter because of the disparate scales of cartographic data available for building cartographic databases. As you remember, the smaller the scale of the maps, the larger the amount of error, and the greater the generalization. A good rule of thumb is to take the largest scale maps available for the task. The importance of the coverage to the models being built will also dictate what scales will be acceptable. While there are no formal guidelines for determining an acceptable scale, most GIS professionals apply the "best available data" approach, recognizing that more detail is better than less detail.

When working with raster data models you will also need to know the acceptable size of grid cells within the larger map area. Again, smaller provides better detail, but also drastically increases the data volume (i.e., the smaller the grid cells, the more you will need a single coverage). While storage may not be a problem for most modern computer systems, functions that require extensive neighborhood searches will be slowed by massive increases in the grid array. In some cases the grid cell size may be dictated by the smallest item you wish to represent, in others by the requirements of the model (DeMers, 1992), and in still others by issues of compatibility with other digital data such as input from a satellite.

Classification

Now we return to the same old problem we encountered in Chapter 9: when designing the GIS, we need to consider the available data sources, as well as the classification system that fits our modeling needs. It might prove useful to examine the kinds of input data that are available prior to deciding on the classification system. The use of a more detailed classification is often preferred for two reasons. First, it gives the user the largest amount of data; and second, if one coverage can be compared to another using a low level of detail, it is relatively easy to aggregate classes, whereas disaggregation may be difficult or even impossible. Of course, the most detailed classifications also present two possible problems. First, the classifications may be designed for a particular task that has little to do with the project at hand; and second, the more detail, the greater the chance for errors in classification in the first place.

But classification is more than just selecting the correct level of classification detail. It is important to remember that in GIS we frequently are making comparisons among coverages. This means that we need to be able to correctly compare the classifications from one coverage to another. We also need to be able to produce consistent classifications within a single coverage, often from different sources. Suppose, for example, that you require a regional coverage of soil classes based on different county-level soil surveys. When surveys have been developed in widely differing time periods, it will be necessary to "**cross-walk**" them to ensure a uniform classification. Burrough (1986) provides a more detailed discussion of the problems of classification, especially as they related to potential error in the GIS.

Coordinate System and Projection

We have seen that while GIS is capable of converting from and to different projections and coordinate systems, the more often this is done, the greater the probability of error. The decision of which should be used is often dictated by the regional extent of the study area and by the availability of data. Again, there are no hard and fast rules, just some basic common sense guidelines that can be applied to the problem. For study areas covering many states, it is generally advisable to avoid the use of a state plane coordinate system, or any other coordinate system that has a limited range of accuracy. As for projections, we have seen that the properties needing to be preserved (i.e., the properties that will be most important to analysis) are those that will dictate the correct projection. In general, for both coordinate system and projection, spatial and temporal compatibility are vital to the decision-making process.

Selecting a System

It is always difficult to select the correct software system for an organization. In many cases, the decision is made by the system vendor even before the design process takes place. While this is a backward approach, it is a reality for many organizations and individuals. Burrough (1986) provides a number of guidelines, some of which are outdated because of advances in software and hardware. As he points out, the choice of data model is often based on the needed analyses, which, in turn can be identified by the SIPs for the project. But many modern GIS systems provide more than one data model, or offer a suite of analytical modules that can be addressed, menu style, depending on style and individual user needs. The choice of hardware platforms and peripherals is dictated by economic constraints, accuracy requirements, learning curves, and even personal preference.

So how do we decide for ourselves or for our clients which system to purchase? A reasonable approach is to prepare a document indicating analytical needs, cost limitations, accuracy needs, and training needs. Then these specifications are submitted to a large number of vendors for competitive bidding. You may wish to append a request that each bidder provide a list of current customers, both satisfied users and clients who have experienced problems. The responses you receive will tell you about the kinds of organization that use the system (giving you an idea of their modeling needs) and will allow you to learn from firsthand responses about various users' personal satisfaction levels. The final selection of the GIS should be among the last decisions to be made.

VERIFICATION AND VALIDATION

Perhaps from the operational standpoint a design methodology is only as good as the results obtained from the implementation. The primary question that needs to be asked is, Will the GIS perform the tasks the organization needs in a timely and correct manner? If the design process has been performed properly, it should yield a product description that incorporates all the organizational needs based on the individual user needs. These needs must be specific and detailed enough for the vendor to be able to determine whether the system under consideration can meet those needs. As is well known, "right the first time" is better than having to make expensive fixes later. Many an organization has contracted with a second vendor after the original vendor's system failed to meet the stated needs. Often such disappointments are not the fault of the vendor but are more likely a result of a failure of the organization, or its system analyst, to provide the complete specifications. The costs of fixing a broken system are often far higher than those of developing the correct system through a systematic, well-organized, and complete analysis of the needs of the organization.

Terms

institutional setting
technical design
project life cycle
mythical man-month
system sponsor
GIS systems analysts
application requirements model
spatial information products (SIP)
global view
decision system matrix

crosswalk
software design
institutional design
bottom-up implementation
system users
GIS supplier
federated databases
Marble model
organizational diagram

system design
software engineering
waterfall model
system operators
application developers
data requirements model
local views
design creep
view
integration

Review Questions

1. What accounted for most system failures in GIS in the 1960s? What causes most GIS system failures today?

2. Why do we need to consider system design for GIS? Why aren't some users' needs met? Why is it important to know what GIS analytical capabilities are when we are designing a system for third-party users?

3. What is a system? How does this design construct expand our definition of GIS? What is the difference between technical design and institutional design?

4. What is the concept of a system life cycle? What are its objectives? Why is the waterfall model of the system life cycle inappropriate for designing and implementing GIS in organizations?

5. What is the mythical man-month? What does this tell us about the cost-effectiveness of doing a job right the first time?

6. Who are the internal players in the institutional setting? What important tasks do they perform?

7. Who are the external players in the institutional setting? What role do they play in organizational implementations of GIS?

8. What is conceptual GIS design? Why is it not enough for designing a GIS for an organization? What are some of the people problems of GIS? What is a cost/benefit analysis and what role does it play in designing a GIS?

9. Describe and diagram the Marble spiral design model for GIS design. What advantages does it have over the life cycle model?

10. Describe the steps involved in the initial GIS design using the Marble spiral design model.

11. List and describe the potential benefits of GIS in an organizational framework. Create a fictitious organization and discuss the potential benefits for it.

12. What are SIPs? How are they related to the input to a GIS? How do they drive the implementation of a GIS?

13. Describe and diagram the decision system matrix approach to organizing local views of a GIS. Do the same for the organizational chart approach. Suggest how these methods might be used for later view integration.

14. What is design creep? Diagram and describe how the organization of the spiral model of GIS design functions to prevent design creep.

15. What major database design considerations do we need to look at for GIS design? Define some general rules for each.

16. What do "verification" and "validation" mean in the context of GIS design rather than GIS modeling?

References

Aronoff, S., 1989. *Geographic Information Systems: A Management Perspective,* Ottawa, Canada: WDL Publications.

Boehm, Barry, 1981. *Software Engineering Economics.* Englewood Cliffs, NJ: Prentice-Hall.

Boehm, B.W., 1986. "A Spiral Model of Software Development and Enhancement," *ACM Software Engineering Notes* Vol. 11, No. 4.

Boehm, B., and P.N. Papccio, 1988. "Understanding and Controlling Software Costs." *IEEE Transactions on Software Engineering,* SE-M(10):1462–1477.

Brooks, F.P., 1975. *The Mythical Man-Month.* Reading, MA: Addison-Wesley.

Burrough, P.A., 1986. *Geographical Information Systems for Natural Resources Assessment.* New York: Oxford University Press.

Calkins, H.W., 1982. "A Pragmatic Approach to Geographic Information System Design." In *Proceedings of the U.S./Australia Workshop on Design and Implementation of Computer-Based Geographic Information Systems,* D. Peuguet and J. O'Callaghan, Eds. Amherst, MA: IGU Commission on Geographical Data Sensing and Processing.

Calkins, H.W., and D.F. Marble, 1987. "The Transition to Automated Production Cartography: Design of the Master Cartographic Database." *American Cartographer,* 14(2):105–119.

DeMers, M.D., 1992. "Resolution Tolerance in a Forest Land Evaluation System." *Computers, Environment and Urban Systems,* 16:389–401.

DeMers, M.D., and P.F. Fisher, 1991. "Comparative Evolution of Statewide Geographic Information Systems in Ohio," *International Journal of Geographical Information Systems* 5(4):469–485.

Goodchild, M.F., and B. R., 1986. "Performance Evaluation and Workload Estimation for Geographic Information Systems." In *Proceedings of the Second International Symposium on Spatial Data Handling,* pp. 497–509.

Green, J., and D.D. Moyer, 1985. "Implementation Costs of a Multipurpose County Land Information System." In *Proceedings of the URISA 1985 Conference,* Vol. 1. Washington, DC: Urban and Regional Information Systems Association, pp. 145–151.

Guptill, S.C. (ed.), 1977. A Process for Evaluating Geographic Information Systems, US Geological Survey Open-File Report 88–105.

Johnson, T.R., 1981. "Evaluation and Improvement of the Geographic Information System Design Model." Unpublished M.A. thesis, Department of Geography, State University of New York, Buffalo.

Kenney, H., and A. Hamilton, 1985. "Unit Costs for Property Mapping in Northern New Brunswick." In *Proceedings of the URISA 1985 Conference,* Vol. 1. Washington, DC: Urban and Regional Information Systems Association, pp. 132–144.

Kenney, H., and A. Hamilton, 1986. "Unit Costs for Parcel Indexing and Related Activities in Northern New Brunswick." In *Proceedings of the URISA 1985 Conference,* Vol. 1. Washington, DC: Urban and Regional Information Systems Association, pp. 141–149.

Kevany, M.J., 1986. "Assessing Productivity Gains in Advance: Feasibility Studies." In *Proceedings of the URISA 1986 Conference,* Vol. 2. Washington, DC: Urban and Regional Information Systems Association, pp. 40–46.

LaRoche, S., and A.C. Hamilton, 1986. "Unit Costs for Topographic Mapping." In *Proceedings of the URISA 1985 Conference,* Vol. 1. Washington, DC: Urban and Regional Information Systems Association, pp. 150–158.

Marble, D.F. 1994. *An Introduction to the Structured Design of Geographic Information Systems* (Source Book, Association for Geographic Information). London: John Wiley & Sons.

Rhind, D.W., and N.P.A. Green, 1988. "Design of a Geographical Information System for a Heterogeneous Scientific Community." *International Journal of Geographical Information Systems,* 2(2):171–189.

Royce, W.W., 1970. "Managing the Development of Large Software Systems," *Proceedings, IEEE Wescon,* pp. 1–9.

Tomlinson, R.F., n.d. *The GIS Planning Process.* Tomlinson Associates Ltd., Consulting Geographers, Ottawa, Canada.

Wellar, B., 1994. "Progress in Building Linkages Between GIS and Methods and Techniques of Scientific Inquiry." *Computers, Environment and Urban Systems,* 18(2):67–80.

Yourdon, E., 1989. *Modern Structured Analysis.* Englewood Cliffs, NJ: Prentice-Hall,

Software and Data Sources

This appendix is a sample of available GIS software and sources of data for use with the software. Although this list is not complete, it should be enough to get you started either in a learning setting or a commercial operation. Additional data sources and vendors can be found by searching through some of the new GIS, remote sensing, and surveying trade journals, as well as through your instructor.

GIS SOFTWARE VENDORS

Arc/Info, Arc/View, MapObjects
Environmental Systems Research Institute
380 New York Street
Redlands, CA 92372
(714) 793-2853

Spans GIS
Intera TYDAC Technologies Inc.
1500 Carling Ave, Ottawa, Ontario K1Z 8R7
Canada
(613) 722-7508

MicroStation GIS Environment (MGE)
Integraph Corporation
Huntsville, AL 35894
(205) 730-2000

ERDAS IMAGINE
ERDAS, Inc.
2801 Buford Highway
Atlanta, GA 30329
(404) 248-9000

GRASS
   GRASS Information Center
   USACERL
   ATTN: CECER-ECA
   P.O. Box 9005
   Champaign, IL 61826-9005

Atlas GIS
   Strategic Mapping Inc.
   4030 Moorpark Avenue
   San Jose, CA 95117-4103
   (408) 985-7400

GisPlus
   Caliper Corporation
   1172 Beacon Street
   Newton, MA 02161
   (617) 527-4700

MapInfo
   MapInfo Corporation
   200 Broadway
   Troy, NY 12180-3289
   (518) 274-6000

IDRISI
   The Clark Labs for Cartographic Technology and Geographic Analysis
   Clark University
   950 Main Street
   Worcester, MA 01610
   (508) 793-7526

MapGrafix
   ComGraphix, Inc.
   620 E Street
   Clearwater, FL 34616
   (813) 443-6807

OSU Map-for-the-PC
   Duane Marble
   Department of Geography
   The Ohio State University
   Columbus, OH 43210

Appendix B contains some of these vendors as well as a few others as Internet home pages that you can contact.

D A T A   S O U R C E S

U.S. DEPARTMENT OF AGRICULTURE (USDA)

Soil Conservation Service (SCS)

The following information is available from:

National Cartographic and Geographic Information Systems Center
USDA-Soil Conservation Service
P.O. Box 6567
Fort Worth, TX 76115
(817) 334-5559
FAX: (817) 334-5290

Soil Survey Geographic Database (SSURGO) This is the most detailed soil mapping for the United States, ranging in scale from 1:12,000 to 1:31,680. It duplicates the original soil survey maps found in county soil survey documents.

State Soil Geographic Database (STATSGO) Generalized SSURGO data at a scale of 1:250,000, which covers state, regional, and multi-county areas.

National Soil Geographic Database (NATSGO) Data based on generalized state-level soils maps and formed from the major land resource area (MLRA) and land resource region (LRR) boundaries. Digitized at 1:7,500,000.

U.S. DEPARTMENT OF COMMERCE

Bureau of the Census

Census data in the form of TIGER files are available from:

Customer Services
Bureau of the Census
Washington Plaza, Room 326
Washington, DC 20233
(301) 763-4100
FAX: (301) 763-4794

National Oceanic and Atmospheric Administration (NOAA)

National Environmental Satellite, Data, and Information Service (NESDIS)

United States weather records collected by the National Climatic Data Center (NCDC), and international weather records collected by the World Data Center A (WDC-A) are available from:

NOAA
National Climatic Data Center
Climate Services Division
NOAA/NESDIS E/CC3
Federal Building
Asheville NC 28801-2696
(704) 259-0682
FAX: (704) 259-2876

National Geophysical Data Center (NGDC)

Solid earth, solar-terrestrial physics, and snow and ice data from worldwide sources are available from:

National Geophysical Data Center
Information Services Division
NOAA/NESDIS E/GC4
325 Broadway
Boulder CO 80303-3328
(303) 497-6958
FAX: (303) 497-6513

National Oceanic Data Center (NODC)

Oceanographic station data, bathythermograph, current, biological, sea surface, and GEOSAT data are available from:

National Oceanographic Data Center
User Services Branch
NOAA/NESDIS E/OC21
1825 Connecticut Avenue, NW
Washington DC 20235
(202) 606-4549
FAX: (202) 606-4586

National Ocean Service (NOS)

Sailing charts, coast charts, and harbor charts are available from NOS. For information contact:

Coast and Geodetic Survey
National Ocean Service, NOAA
1315 East-West Highway, Station 8620
Silver Spring MD 20910-3282
(301) 713-2780

U.S. DEPARTMENT OF THE INTERIOR

U.S. Geological Survey (USGS)

Digital line graphs, digital elevation models, side-looking airborne radar (SLAR), advanced very high resolution radiometer (AVHRR), digital orthophoto quad, land satellite (LANDSAT) data, systeme probatoire d'observation de la terre (SPOT) data, defense mapping agency (DMA) data, national wetlands inventory (NWI) data, and national biological service (NBS) data are all accessible from:

Earth Science Information Center
U.S. Geological Survey
507 National Center
Reston VA 22092
(703) 648-5920
FAX: (703) 648-5548
Toll Free: 1-800-USA-MAPs

or

Sioux Falls ESIC
U.S. Geological Survey
EROS Data Center
Sioux Falls SD 57189
(605) 594-6151
FAX: (605) 594-6589

CANADIAN DATA

National Digital Topographic Data are available from:

> Topographic Surveys Division
> Surveys and Mapping Branch
> Energy, Mines and Resources Canada
> 615 Booth Street
> Ottawa, Ontario K1A 0E9
> (613) 992-0924

Data from the Canada Land Data System (including the Canada Geographic Information System) are available from:

> Environmental Information Systems Division
> State of the Environment Reporting Branch
> Environment Canada
> Ottawa, Ontario K1A 0H3
> (613) 997-2510

Canadian soils information system data are available from:

> CanSIS Project Leader
> Land Resource Research Centre
> Agriculture Canada, Research Branch
> K.W. Neatby Building
> Ottawa, Ontario K1A 0C6
> (613) 995-5011

Landsat and SPOT data in Canada are available from:

> Canada Centre for Remote Sensing
> 2464 Sheffield Road
> Ottawa, Ontario K1A 0Y7
> (613) 952-2171

or

> Canada Centre for Remote Sensing
> Prince Albert Receiving Station
> Prince Albert, Saskatchewan S6V 5S7
> (306) 764-3602

Census data for Canada are available from:

> User Summary Tapes
> Electronic Data Dissemination Division
> Statistics Canada
> 9th Floor, R.H. Coates Building
> Ottawa, Ontario K1A 0T6
> (613) 951-8200

OTHER SOURCES OF DATA

The preceding sources of data are limited to resources for Canada and the United States, but there are many sources of data for many other nations. To list them all here would add considerably to the overall size of this book, and would also require frequent updates. As an alternative, I have provided information about Internet sources in Appendix B.

Using the Wiley World Wide Web to Find Data and GIS Examples

In Appendix A, I describe a number of locations for data available through the mail. Many of these sources of data are now accessible using the World Wide Web (WWW). In fact, even a cursory perusal of the net using the keywords "Geographic Information System" will give you a wide array of data sources, listservers, interest groups, class materials, active GIS research agendas, GIS output, sources for GIS employment, software and hardware vendors, and the like. A major problem with the web is that you can spend an enormous amount of time surfing before you find the object of your search. I have compiled a list of sources that have come to my attention while searching. My intent is to shorten the amount of time you spend surfing and increase your productivity. This list is available from John Wiley & Sons home page through links established to my book. Instructions for access are given below.

In addition to providing you with existing sources of materials through the web, I have also created what I view as an active manual for students and faculty alike. Rather than being a manual per se, it is a home page that will allow for active participation among students, faculty, and myself. Initially this home page contains a response form so that relevant questions about material in the text can be asked. While I don't expect to be able to answer everyone's questions immediately, I will make every attempt to respond directly to questions posed to me. In addition, I will create a discussion group based on the conerns of those of you using my text. As users send material to me through my home page, you will be asked if you are interested in being included on the mailing list. Those who choose to do so will be informed of materials that others would like to make available, for instance, suggestions for course exercises and ideas. This is meant as a forum for sharing exercises, insights, questions, and ideas. Hopefully this will be an active, growing, dynamic forum for dissemination of materials for learning GIS.

To find the home page for your first step is to use your browser software to contact John Wiley & Sons home page. Their URL is

http://www.wiley.com

Once you have arrived you will have a number of options where you can search for books, software, personnel, and many other offerings by Wiley. For my home page your next stop is the textbooks icon. This in turn provides you with a number of options. You should next select the Geography Newsletter. That newsletter shows you a number of offerings of geography textbooks, including this one. Select the highlighted title "Fundamentals of Geographic Information Systems," which will provide you with another menu. From this select the highlighted "course materials" text. Within that page you will find a wide offering of information plus the reply form. From there the choice is yours as to what material you wish to view.

Absolute barriers: Barriers that prevent movement through them.

Absolute location: Location on the earth or on a map that has associated with it a specific set of locational coordinates.

Accessibility: A measure of arrangement that focuses on the connectedness of an area object to other area objects.

Address matching: The process of defining exact addresses and linking them to specific locations along linear objects.

Affective objective: One of two parts of the purpose of a map, this one relating to the overall appearance of the map. Controls how information contained in a map is to be portrayed.

Aggregation: One of a number of numerical processes that place individual data elements into classes.

Alber's equal area projections: Widely used equal area conic projections that contain two standard parallels. Good for midlatitude areas of greater east-west than north-south extent.

Allocate: The process of selecting a portion of geographic space to satisfy a predefined set of locational criteria. This process is usually performed on a vector data structure for the purpose of selecting a portion of a network.

Alpha index: A measure of network complexity that examines network circuitry.

AM/FM: Automated mapping/facilities management. Computer assisted cartography, particularly applied to the display and subsequent analysis of facilities associated with the function of urban and rural areas.

Analytical paradigm: A conceptual model, sometimes called the holistic paradigm, where the map is viewed both as a means of graphic communication and as a means of numerical spatial analysis.

Angle of view: The angular distance above the ground from which a perspective map can be shown.

Angular conformity: The property of some map projections that retain all angular relations on the map.

Application developers: GIS professionals specializing in the development of specific databases, analytical functional capabilities, and appropriate graphical user interfaces to allow specific applications to be performed by non-GIS specialists.

Application requirements model: A database design model focusing on the application needs as a means of properly completing the design.

Applications development: The process employed by a group of GIS professionals to allow non-GIS users easy access to the technology for specific tasks.

Arbitrary buffers: Buffers whose distances are selected without regard for any known criteria.

Area cartograms: Cartograms that vary the size of the polygons based on the statistical value represented, rather than on their actual areal dimensions.

Arrangements: 1. A number of measurements indicating how spatial data are organized through such attributes as nearness and connectivity, as well as others. 2. The spatial distribution of internal markings for cartographic symbols to allow visual separation.

Aspect: The azimuthal direction of surface features.

Association: Spatial relationship that exists between different elements of the earth that occur at the same locations. An example would be a particular vegetation type occurring on north-facing slopes.

Attribute error: Incorrect or missing attributes.

Attribute pseudo nodes: Spatial pseudo nodes that are a result of explicit attribute changes along a line.

Attributes: Nongraphic descriptors of point, line, and area entities in a GIS.

461

Audience: Map design consideration based on the background and map reading skills of the potential map user.

Available data: Map design limitation based on the quality, quantity, and reliability of the data used for mapping.

Azimuthal equidistant projections: Types of azimuthal map projections in which the linear scale is uniform along the radiating straight lines through the center. Allows for the entire sphere to be shown on a single projection.

Azimuthal projections: A family of map projections resulting from conceptually transferring the earth's coordinates onto a flat surface placed perpendicular to the sphere.

Barrier: Some object whose attributes either stop or impede movement through the rest of the coverage.

Binary maps: Maps that contain only two possible values for the same characteristic or attribute. The attribute is either present or absent.

Biophysical mapping: A traditional, analog technique used by environmental planners to combine relevant mapped biological and physical data into meaningful combinations for decision making.

Block codes: A compact method of storing raster data as multifaceted blocks of homogeneous grid cells.

Boolean overlay: A type of overlay operation that relies on Boolean algebra.

Borders: Higher order groups of line objects that exemplify some functional or political demarcation from one region to another.

Bottom-up implementation: One of the major weaknesses of the classical (waterfall model) project life cycle design methodology where final system testing is left until last, thus requiring large chunks of uninterrupted computer time. Because such large chunks of computer time are often hard to obtain, the development often falls behind schedule.

Buffering: The process of creating areas of calculated distance from a point, line, or area object.

CAD: See computer assisted drafting.

Cadastral: Having to do with Cadastre, which involves interests in land ownership and management.

Cartesian coordinate system: A mathematical construct defined by an origin and a unit distance in the X and Y direction from that origin.

Cartograms: Maps that emphasize the communication paradigm to a degree that often modifies actual geographic space in deference to the phenomena being displayed.

Cartographic database: A digital database developed from an existing cartographic document. This process results in the appearance of increased accuracy where it does not necessarily exist.

Cartographic modeling: The process of combining individual GIS analytics to create complex models used for decision making.

Cartographic process: The steps involved in producing a map, beginning with data collection and ending with the final map product.

Causative buffers: Buffers whose distances are selected based on some physical phenomena that dictate their location.

Census: A method of collecting data that normally involves tabulating data about an entire population rather than a sample of the population.

Center of gravity: Center of a polygon based on the arrangements of objects within it rather than simply as a measure of the area geometry.

Central place: A node that exhibits a high degree of network linkage intensity.

Central point linear cartograms: Cartograms that show distances from a central point as concentric lines, thus requiring the underlying graphic objects to be modified to conform to those lines.

Centroid-of-cell method: A method of raster encoding where a grid cell is encoded based on whether or not the object in question is located at the exact centroid of the geographic space occupied by each cell.

Centroids: For graphical objects, the exact geometric center.

Chains: In the POLYVRT vector data model, collections of line segments that begin and end with specific nodes that indicate topological information such as to and from directions as well as left and right polygons.

Changes of dimensionality: The process of either increasing or decreasing the dimension of cartographic objects based often on the scale at which the map is produced. For example, a line (1 dimension) at one scale may be represented as an area at a substantially larger scale.

Chi-square: A statistical test that examines the relationship between observed and expected distributions of objects.

Chroma: The perceived amount of white in a hue compared to a gray tone of the same value level.

Circuitry: The degree to which nodes are connected by circuits of alternative routes. Defined by the Alpha index.

Circuits: Type of network that allows movement to and from the initial point.

Circular variance: The circular analog to standard deviation. A measure of the degree of variation from a mean resultant length.

Classed choropleth map: The traditional method of value by area mapping that first aggregates the data into classes before the polygons are assigned color or shading patterns.

Class interval selection: The process of using a variety of statistical techniques to group cartographic data into classes for cartographic display. This data aggregation results in loss of data for analysis, but is essential for many map production purposes to allow readability of the map.

Classless choropleth mapping: A proposed mapping technique that assigns a color or shading pattern to each polygon based on its unit value rather than grouping polygons into classes.

Closed cartographic form: Polygonal object whose entire extent is contained within a single map.

Clustered: Spatial arrangement of objects where they occur as groups located close to one another, leaving large empty spaces between. This type of arrangement indicates that the processes operating on the objects are different near the clusters rather than in the intervening space.

Clustering: The process that produces a spatial arrangement where some objects are located very close to one another while others are widely separated.

Color video: A passive remote sensing device designed to produce color images of the ground surface and storing it on videotape for later retrieval and analysis.

Communication paradigm: A conceptual model where the map is viewed primarily as a means of graphic communication.

Communities: Higher order groups of objects that either occur in a particular pattern that differs from their surrounding objects or that are linked as a functional region.

Computer assisted cartographic systems (CAC): A computerized system designed primarily to assist in the production of maps.

Computer assisted drafting (CAD): A set of computer programs designed to assist in the process of drafting. It is normally used for architectural purposes but can also be used for drafting maps and as an input to GIS.

Concavity: A measure of the degree to which an area object produces concave shapes along its perimeter.

Conditions of use: Map design consideration based on such factors as viewing distance and lighting that limit the readability of the document as it is used.

Conflation: The computational equivalent of stretching a map until its internal components can be rectified. See also rubber sheeting.

Conformal projection: A type of map projection that maintains all angular relations norally found on the reference globe.

Conformal stereographic projections: Azimuthal projections that exhibit symmetrical distortion around the center point. Useful when the shape of the area to be mapped is relatively compact.

Conical projections: A family of map projections resulting from conceptually transferring the earth's coordinates onto a cone.

Connectivity: The degree to which lines in a network are linked to one another.

Constrained math: Any of a number of mathematical functions applied to cartographic coverages to reclassify their attributes.

Contiguity: A measure of the degree of wholeness within a region or of the degree to which polygons are in contact with one another.

Contiguous regions: Regions that are defined based both on category homogeneity but also on all parts being in direct contact with one another.

Continuous: Data that occur everywhere on the surface of the earth. An example would be temperature.

Continuous surface: Any statistical surface

whose values occur at an infinite number of possible locations.

Contour interval: Class interval illustrating the difference in elevation between contour lines.

Contour line: An isarithm connecting points of identical elevation.

Control points: Any points on a cartographic document for which the geographic coordinates are well known and reasonably accurate. These points are often used during the process of co-registration of two or more coverages or for adjusting the locations of other objects on the same coverage through conflation.

Controls: The limitations on the map design process.

Conventions: For maps, agreed upon rules that have been carefully selected and are generally accepted by the mapping community.

Convexity: A measure of the degree to which an area object produces convex shapes along its perimeter.

Coregistered: The result of the process of precisely locating the coordinates for two or more GIS coverages so that their spatial locations match each other.

Coverage: The common terminology signifying a single thematic map in a multimap GIS database. Sometimes referred to as a data layer or an overlay.

Cross-sectional profile: A method of visualizing statistical surfaces by examining one transect through the surface viewed from the side.

Crosswalk: The process of matching disparate categories between maps or GIS coverages.

Cumulative distance: Usually associated with cost surfaces. The accumulation of cost in terms of time, energy, etc., as one travels from place to place.

Cylindrical projections: A family of map projections resulting from conceptually transferring the earth's coordinates onto a cylinder.

Dangling node: A node located at the end of an undershoot.

Dasymetric mapping: A variety of techniques designed to improve area homogeneity for choropleth maps.

Dasymetry: See Dasymetric mapping.

Database: A collection of many files collectively associated with a single general category.

Database management system: Any of a variety of computer organizational structures that allow search and retrieval of individual files or items in the database. These systems are generally of three primary types: hierarchical, network, and relational.

Database structures: A set of approaches to organizing large collections of computer files commonly relating to a single major subject. Database structures are designed to enable the user to store, edit, search, and retrieve files or individual pieces of data within those files.

Data dictionary: Detailed description of the data contents of a database, with particular attention being paid to explanations of categories.

Data requirements model: A database design model focusing on the required data needs as a means of properly completing the design.

Dead reckoning: A method of surveying that measures distance and direction from one point to the next.

Decision system matrix: An aid to determining system design requirements, especially for linking spatial information products to necessary data elements.

Deductive models: Cartographic models that move from a general goal to the selection of individual components needed to achieve it.

DEMs: See digital elevation models.

Density: Measure of the number of objects per unit area.

Density of parts: A type of dasymetric mapping that uses improved quantitative information about subarea density on maps to calculate improved information about the remaining polygons.

Density zone outlining: A type of dasymetric mapping that creates neighborhoods of uniform density of cartographic objects that are then mapped as individual polygons. Also known as density of parts.

Descriptive: A type of cartographic model that describes an outcome based on a set of conditions.

Design creep: Process that occurs when the design process exceeds the organizational learning curve. Usually results in a system that is over designed and offers more than the client needs.

DD: Decimal degrees method of designation for geocoded objects.

Differential rectification: Process used to adjust aerial photographs for planimetric errors due to changes in aircraft altitude and elevational features on the ground.

Diffusion: A spatio-temporal process wherein objects move from one area to another through time, or where additional objects appear where they had not previously existed.

Digital elevation models: Digital model of landform data represented as point elevation values. Also called digital terrain models (DTM).

Digital line graphs: Digital representations of the graphics contained on USGS topographic maps. These do not include topographic data as one might find in a Digital Elevation Model.

Digital orthophotoquads: Digital version of aerial photographs that are constructed to eliminate image displacement due to changes in aircraft tilt and topographic relief.

Digital photography: A passive remote sensing device that produces still digital images similar to that from video, but as single frames.

Digitizers: Electronic devices designed to transfer analog cartographic data to digital form.

Directed network: Type of network that places restrictions on the direction of movement (e.g., a one-way street).

Direct files: Within an indexed file structure a condition in which the data items themselves provide the primary order of the files.

Directional filter: A high-pass filter that enhances linear objects that lie in a particular direction (e.g., NE-SW).

Dirichlet diagrams: See Thiessen polygons.

Discrete: Data that only occur in selected places on the earth's surface. An example is human population.

Discrete altitude matrix: A lattice of point values that represents elevation, used to model topography in a GIS.

Discrete surface: Any statistical surface whose values are limited to selected locations.

Dispersed: A spatial arrangement that exhibits the greatest possible distance between objects in a confined space.

Dispersion: A measure of arrangement that focuses on the relationship between one area and its neighbors, especially regarding the average distances and average density of patches in the map.

DMS: Degrees, minutes, seconds method of designation for geocoded objects.

Dominant type method: A method of raster encoding where a grid cell is encoded based on whether or not the object in question occupies greater than 50% of the geographic space occupied by each cell.

Dot distribution map: See dot mapping.

Dot mapping: The cartographic technique of representing one or more objects as points on a map, and where the exact geographical locations are not precisely recorded.

Doughnut buffer: A series of buffers of varying distance produced one inside the other.

Drift: The random, spatially correlated elevational component for a kriging model. The general trend in elevational values.

Dumpy level: A telescopic survey device that has an elongated base with a moveable joint that allows vertical movement so that elevations can be measured easily in steep terrain.

Easting: On some grid systems, a measurement of distance east of a preselected standard starting meridian.

Edge enhancement: A type of filter, often called a high-pass filter, that enhances values that change rapidly from place to place so that these changes can easily be observed.

Edge matching: The process of aligning the edges of two or more cartographic documents.

Edginess: Measure of the amount of edge in a polygonal cartographic form.

Encode: To place analog graphic data into a form that is compatible with computer cartographic data structures.

Entities: Points, lines, and polygons as they are represented by computerized cartographic data structures.

Entity-attribute agreement error: Related to attribute error. A mismatch between entity objects and the attributes assigned to them.

Entity error: Error of position for cartographic objects.

Equal area projections: A group of map projections that maintain the property of equal area for cartographic objects. These projections are useful for small-scale general reference and instructional maps.

Equidistant projections: A group of map projections that maintain the property of equal distance along straight lines radiating through a central starting point from which the projection was made.

Equivalent projections: See equal area projections.

Euclidean distance: Distance measured simply as a function of Euclidean geometric space.

Euler function: Mathematical measure of spatial integrity that compares the number of regions to the number of perforations within all the regions.

Euler number: Number resulting from the implementation of the Euler function.

Exclusionary variables: Coverage variables whose importance outweighs the results of normal Boolean overlay operations for the purposes of decision making.

Extended neighborhoods: Regions that extend beyond those immediately adjacent to the focal point of analysis.

Extensional hierarchy: The ranking of groups of cartographic features to achieve a graphical hierarchy.

Extrapolation: The process of numerically predicting missing values by using existing values that occur on only one side of the point in question.

False eastings: On some coordinate systems, arbitrary, large values given to the Y axis of the false origin to allow only positive distance measured east of that point.

False northings: On some coordinate systems, arbitrary, large values given to the X axis of the false origin to allow only positive distance measured north of that point.

False origins: Arbitrary starting points on a rectangular coordinate system designed to allow only positive distances east and north.

Federated databases: Databases whose component parts are derived from disparate sources, each with its own data quality standards.

Figure-ground: The spatial relationship between the amount of area occupied by cartographic objects and the amount of space occupied by the background.

Filter: In raster GIS and digital remote sensing, a matrix of numbers used to modify the grid cell or pixel values of the original through a variety of mathematical procedures.

First normal form: Based on the theory of normal forms, the first requirement that all tables must contain rows and columns and that the column values cannot contain repeating groups of data.

Fishnet maps: A term often applied to wireframe diagrams, but representing cartographic statistical surfaces.

Flowchart: A graphical device that illustrates the exact coverages and data elements, the operations to be applied to each, and the order in which the operations are applied to produce a cartographic model.

Foreign key: In relational database management systems, a column in a secondary table linked by a primary key in the primary table that is being used to join two tables.

Fragmented regions: Regions based on category homogeneity but with parts spatially separate from each other.

Freeman-Hoffman chaincodes: Compact raster data models that use eight unique directional vectors to indicate the directional orientation or change of linear features.

Free sampling: In the join count statistic, the method of testing that assumes that we can determine the expected frequency of within and between category joins based on theory or known patterns.

Friction surface: Some assigned attribute on a surface representation of a portion of the earth that acts to impede movement.

Functional distance: Distance measured as a function of difficulty rather than as a simple geometric distance.

Fuzzy tolerance: A user defined distance of error to allow for minute digitizing mistakes.

Gamma index: A measure of complexity of a network that compares the number of links on a given network to the maximum possible number of links.

GBF/DIME: Geographic base file/dual independent map encoding system. A topological vector data model created by the U.S. Bureau of the Census, and based on graph theory.

General reference map: A map, usually at small scale, whose primary objective is to show locations of different features on a single document.

Geocoding: The process inputting spatial data into a GIS database by assigning geographical coordinates to each point, line, and area entity.

Geodetic framework: A carefully measured system of ground-based coordinates designed to assure accurate locations for cartographic documents.

Geographic database: A digital database developed from field-based spatial data. This process results in decreased accuracy and increased generalization because of the sampling necessary to produce it.

Geographic data measurement: The processes and data levels that combine to characterize data observed on the surface of the earth.

GIS systems analysts: GIS professionals specializing in the proper design of the overall GIS operations.

Global Positioning System: A satellite-based device that records locational (X,Y,Z) and ancillary data for portions of the earth.

Global view: Organizational view of a GIS database requiring the integration of multiple local views. Spatial Information Products (SIP)

Gnomonic projections: Azimuthal projections where all great-circles are represented as straight lines. Useful for marine navigation.

GPS: See global positioning system.

Graphicacy: The level of understanding of graphic devices of communication, especially maps, charts, and diagrams.

Graphic data structure: A method of storage of analog graphical data into a computer that enables the user to reconstruct a close approximation of the analog graphic through some output procedure.

Graphic scale: One method of representing map scale where a preselected, real-world distance is shown as a bar or line on the map document. Sometimes called a bar scale.

Graph theory: A mathematical theory that examines the relationships among linear graphs.

Gravity model: A measure of the interaction of nodes based on their distance and some functional measure of their individual importance.

Grid cells: Raster data structures in which the geographical space is quantized into rectangular shapes of equal size and shape.

GRID/LUNR/MAGI: Raster GIS data model where each grid cell is referenced or addressed individually and organized in columnar fashion.

Grid system: A system of horizontal and vertical lines on a globe or projected map that allow for locating objects in geographic space and making measurements among objects.

Ground control points (GCPs): Points of known geographic location used to register satellite imagery and other coverage data to the geodetic framework.

Heterogeneity: Opposite of homogeneity. The amount of diversity within a selected region.

Hierarchical data structures: Computer database structures employing parent/child or one-to-many relationship that requires direct linkages among items for a search to be successful.

Hierarchical structure: The graphic appearance of some cartographic objects appearing to be more important than others, resulting from one or more cartographic techniques to produce this result.

High-pass filter: See edge enhancement.

Holistic paradigm: See analytical paradigm.

Homogeneity: The degree to which attributes in a region are similar. If all attributes are identical within a selected region they are said to be homogeneous.

Hue: The color of a cartographic symbol based on the wavelength of electromagnetic radiation reaching the viewer.

Hybrid systems: GIS systems whose entity and attribute tables are separate and are linked through a series of pointers and identification codes.

Identity overlay: The use of tabulations of data from two or more coverages to make decisions about how attribute values are to be combined. This technique can be employed to enhance a number of other methods of overlay by providing a decision tool for their use.

IMGRID: Raster GIS data model where grid cells are referenced as part of a two-dimensional array and where the thematic attributes are coded in Boolean fashion (i.e., 1 and 0 only).

Immediate neighborhoods: Regions limited to those immediately adjacent to the focal point of analysis.

Impedance value: A user specified numerical value used to simulate the effects of barriers, or friction surfaces. These are also employed in the same way for network modeling where the impedance values indicate the degree to which a network allows travel.

Inductive method: A method of cartographic

model that begins with individual observations (data elements) and proceeds to develop general patterns.

Innovation diffusion: A set of geographic models that examine the movement of ideas, innovations, or strategies through geographic space.

Institutional design: Design issues concerning the role of institutional considerations such as training and personnel, data security, organizational functioning, and many others.

Integrated terrain unit: A unit of geographic space defined by the explicit collection of various attributes at one time, usually from aerial photography.

Interactions: A measure of arrangement that focuses on the proximity, sizes, and amount of edge between neighboring areas, especially regarding the edges of these areas.

Intervisibility: Analytical technique that allows for the determination of visibility from one object to another and back.

Incorrect attribute values: Attribute error resulting from the assignment of incorrect or nonexistent attribute values to entities.

Incremental distance: The simple addition of distance at each successive step.

Indexed files: Computer file structures that allow faster search than ordered sequential because each entry is assigned an index location through the use of software pointers.

Information theory: A field of study that deals with information content. It is used in GIS primarily as an estimate of the amount of sampling necessary to encode a graphic object.

Institutional setting: The size, type, organizational hierarchy, goals, and overall objectives of each organization likely to adopt GIS.

Integrated systems: GIS systems whose entity and attribute tables are shared.

International date line: Meridian drawn at 180 degrees east and west of the prime meridian.

Interpolation: A process of predicting unknown elevational values by using known values occurring on multiple locations around the unknown value.

Interval: Comparative data with a relatively high degree of accuracy, but with an arbitrary starting point (e.g., degrees Centigrade).

Inverse map projection: The process of converting from two-dimensional (projected) map coordinates to geographical coordinates.

Inverted files: Within an indexed file structure a condition in which the data items are organized by a second or topic file that provides the primary order of the files. In such a system the topic or inverted file is searched rather than the data items themselves.

Irregular lattice: A series of point locations whose interspatial distances are not identical.

Isarithm: General term for a line drawn on a map to connect points of known continuous statistical surface values.

Isarithmic mapping: The process of using line symbols to estimate and display continuous statistical surfaces. **Isarithmic map:** The result of the process of isarithmic mapping.

Island pseudo nodes: Spatial pseudo nodes that result when a single line connects with itself.

Isolated State Model: An early model to explain the arrangement of agricultural activities based on their distance to a single market.

Isolation: A measure of arrangement that focuses on the distance between objects in geographic space.

Isoline: A line connecting points of known or predicted equal statistical surface value.

Isometric map: A map composed of isolines whose known Z values are sampled at point locations.

Isoplethic map: A map composed of isolines whose known Z values are recorded for polygonal areas rather than at specific point locations.

Isotropic surface: A measure of simple Euclidean distance from some central point outward throughout a coverage where no obstructions or frictional changes exist.

ITUs: Integrated terrain units.

Join count statistic: A method of analyzing contiguity in vector by comparing the contacts between like polygons to unlike polygons.

Kriging: An exact interpolation routine that depends on the probabilistic nature of surface changes with distance.

Lag: On a semivariogram, the distance between sample locations, plotted on the horizontal axis.

Lambert's conformal conic projections: Conformal map projections that have concentric parallels and equally spaced, straight meridians meeting the parallels at right angles. Provide good directional and shape relationships for east-west midlatitudinal zones.

Lambert's equal area projections: Azimuthal equivalent projections with symmetrical distortion around a central point. Useful for areas that have nearly equal east-west and north-south dimensions.

Land information systems: Subset of geographic information systems that pay special attention to land related data.

Latitude: Angular measurement north and south from the equator.

LCGU: Least common geographic unit. A computer graphics construct that provides topological information for vector overlay operations to determine the ability or lack of ability of polygons to be further divided.

Least convex hull: Straight-line segments connecting all outside points within a distribution in such a fashion that the smallest possible polygon is produced.

Least cost distance: A distance measure based on the minimization of friction or cost for a single path.

Least cost surface: A distance measure based on the minimization of friction or cost for an entire surface.

Legibility: The ability of a cartographic object to be visually observed and recognized by the map user.

Limiting variables: Variables that result in the absence of one or more attributes because of how they interact with one another. A method of dasymetric mapping.

Linear interpolation: A type of interpolation linearity: Measure of the ability of a digitizer to be within a specified distance or tolerance of the correct value as the puck is moved over large distances.

Line dissolve: In vector GIS the process of eliminating line segments between polygons so that they become larger polygons with identical attributes.

Line-in-polygon: A method of overlay that combines a coverage with linear objects with a polygon coverage to determine the numbers and extents of linear objects that fall within selected polygons.

Line intersect methods: A group of techniques for analyzing the spatial distribution of linear objects by drawing one or more sample lines at random and noting where they intersect the coverage lines.

Local operator models: Methods of interpolation that rely heavily on existing elevational data that are located within close proximity to the points being predicted.

Local views: Individual users' views of a GIS database.

Locate: Determining the best location for some form of geographic activity, most often used for economic activities.

Locational information: Any information about the absolute or relative coordinates of points, lines, or areas. This information is often used to reclassify coverage values.

Location-allocation: A group of models designed to determine the best locations for activities in geographic space and to assign a portion of geographic space to existing facilities based on demand and location.

Longitude: Angular measurement east and west from the prime meridian.

Low-pass filter: See smoothing.

Mandated buffer: Buffers whose distances are dictated by legal mandate (e.g., frontage along homes dictated by zoning ordinances).

MAP: Raster GIS data model where grid cells are referenced as part of a two-dimensional array and where the thematic attributes for each coverage are referenced by separate number codes or labels, thus allowing a range of values for each category in a single theme.

Map legend: The portion of the map document that describes the symbols. Conceptually, the map legend ties the entities and attributes together in an analog map document.

Map projections: Any of a number of approaches to transfer the spherical earth onto a two dimensional surface. Each projection is an approximation and imposes its own limitations on the utility of the map.

Marble model: A spiral structured GIS design model that uses an increasing level of detail to assure design flexibility.

Mathematical overlay: A method of overlay, usually associated with raster GIS, that uses mathematical and algebraic expressions to create new coverage attribute variables.

Mean center: Center of a group of point objects located in space derived by calculating the average X and Y coordinate distances.

Mean resultant length: The result of averaging the resultant length by the number of observed component vectors.

Measurable buffers: Buffers whose distances are selected based on exact measurements of existing phenomena as they occur in space.

Mercator projections: Conformal map projections introduced specifically for nautical navigation. It maintains the property that rhumb lines always appear as straight lines.

Meridians: Lines of longitude drawn north and south on the globe and converging at the poles.

Metadata: Data about data. An overall description of the contents of a database.

Method of ordinates: A simplified method for calculating volumes.

Missing attributes: Attribute error resulting from a failure to assign labels to point, line, or polygonal features or to grid cells.

Missing labels: Entity-attribute agreement error resulting from failure to label a polygon during the input process.

Mixed pixels: See mixels.

Mixels: Mixed pixels. Pixels whose spatial extent is coincident with more than one category of object on the earth. Mixels often result in category confusion during the classification process.

Model verification: The process of comparing expected results to observed results of a cartographic model to assure it is performing as it should.

Mythical man-month: The erroneous concept that if a problem arises the simple addition of more workers to the problem will result in its solution. This idea ignores the necessary learning curve for people asked to perform tasks with which they are unfamiliar.

Nearest neighbor analysis: A statistical test to compare the distance between each point object and its nearest neighbor to an average "between neighbor" distance.

Neighborhood functions: GIS analytical functions that operate on regions of the database within proximity of some starting point or grid cell.

Network complexity: Overall network patterning that combines the number of links and the degree of connectivity.

Networks: 1. Two or more interconnected computers often used to allow communications between them. 2. Higher order linear objects. Interconnected lines (arcs, chains, strings, etc.) defining the boundaries of polygons or a linear object that allows movement (e.g., road network).

Network systems: Computer database structures employing a series of software pointers from one data item to another. Unlike hierarchical data structures, network systems are not restricted to paths up and down hierarchical pathways.

Node: In vector data structures, a point that acts as the intersection of two or more lines (links) and explicitly identifies either the beginning or ending of each line to which it is attached.

Nominal: Level of data measurement that is noncomparative, usually representing a description or name.

Non-free sampling: In the join-count statistic, the sampling procedure that compares the number of joins of a random pattern of joins to those measured.

Nonlinear interpolation: A method of interpolation that accounts for the nonlinear nature of elevational change with distance.

Normal forms: In relational database management systems, a theory of table design that specifies what types of values columns may contain and how the table columns are to be dependent on the primary key.

Northing: On some grid systems, a measurement of distance north of a preselected standard starting parallel.

Nugget variance: Spatially uncorrelated noise associated with a Kriging model.

Object-oriented database management systems: GIS systems based on object-oriented programming methods and demonstrating object inheritance.

Ordered sequential files: Computer file structures that are ordered based on some form of alphanumeric scheme much like alphabetizing a mailing list. This file structure makes it more difficult for data input, but allows searches and retrieval with greater speed than with simple lists.

Ordinal: Ranked data (e.g., good, better, best) that are comparable only within a given spectrum.

Organizational diagram: Hierarchical diagram

showing the structure of an organization from management to workers.

Orientation: 1. The azimuthal directions in which linear objects are placed. For area objects this can also be applied to the major axis of the object. 2. The azimuthal directions of internal symbol markings used to separate one category from another.

Orthogonal: The normal point of view for viewing a cartographic document directly from above.

Orthographic projections: Azimuthal projections that look like perspective views of the globe at a distance. Useful for illustrations where the sphericity of the globe must be maintained.

Orthomorphic projections: See conformal projection.

Orthophotoquads: Aerial photographs that are constructed to eliminate image displacement due to changes in aircraft tilt and topographic relief.

Overlay: The operation of comparing variables among multiple coverages.

Overshoot: An arc that extends beyond the arc with which it was meant to connect.

Parallels: Lines of latitude drawn east and west around the globe parallel to the equator.

Passive remote sensing: A method of remote sensing that uses ambient energy from the object being sensed, rather than sending out a signal for later sensing.

Pattern: The regular arrangement of cartographic objects or internal symbol markings.

Percent occurrence: A method of raster encoding where a grid cell is encoded so that it includes all categories falling within the geographic space occupied by each cell as well as the percentage of each cell it occupies.

Perforated regions: Regions that demonstrate category homogeneity, but within which other dissimilar polygonal forms exist (e.g., islands within a lake).

Perimeter-area ratio: Method of measuring polygons that compares the perimeter of each polygon to its area as a ratio.

Pixels: In remote sensing, picture elements, each of which contains a preselected amount of geographic space and a set of electromagnetic measurements assigned to each. Pixels most often resemble grid cells when displayed at small scale.

Plane coordinates: See rectangular coordinates.

Plane table and alidade: A common traditional field device used for surveying and consisting of a table upon which a draft survey document would be placed, a telescopic device for determining line of site at a distance, and a tripod upon which they both rest.

Planimeters: Mechanical devices designed to measure lengths and areas on analog maps.

Point-in-polygon: The process of overlaying a point coverage with a polygon coverage to determine which polygons contain which points.

Polygon: A multifaceted vector graphic figure that represents an area.

Polynomials: Mathematical equations with two or more terms, used for approximating surface trends. The more complex the polynomial, the more complex the surface estimate.

POLYVRT: Polygon converter model. A vector data model that uses explicit topological data stored separately for each type of entity.

Position measures: Analytical techniques that characterize cartographic objects based on their absolute positions or on their positions and arrangements relative to other cartographic objects.

Predictive models: A group of cartographic models (either descriptive or prescriptive) that rely on the prediction of outcomes given a set of conditions.

Presence/absence method: A method of raster encoding where a grid cell is encoded based on whether or not the object in question is either present or absent from that portion of geographic space occupied by each cell.

Prescribe: To recommend the best solution.

Prescriptive: A type of cartographic model that determines the best possible solution from a set of conditions.

Primary key: A set of attributes in a relational database that are designed as the primary search criteria for the database. For example, a database of students may have as a primary key the student name.

Prime meridian: Arbitrary starting point for lines of longitude located at Greenwich, England.

Principal scale: A given representative fraction for a reference globe derived by dividing the earth's radius by the radius of the globe.

Project life cycle: A highly structured, project life cycle model, developed to assure proper software development that proceeds from module testing to subsystem testing and finally system testing.

Projection family: A group of map projections based on a conceptual model of transferring the spherical earth onto a cylinder, a cone, or a flat surface. The conceptual model is visualized as if the spherical coordinates are physically transferred onto these individual surfaces by projecting a light source located at the center of the sphere toward the projected surface.

Proximal region: The area of influence defined for each point object when subjected to Thiessen polygonal.

Pseudo nodes: A node where two, and only two, arcs intersect, or where a single arc connects with itself.

Public land survey system (PLSS) grid: Common grid system used in the United States to divide the land into square mile sections, each equal to 640 acres.

Purpose: A primary limitation on the design of a cartographic document relating to what the map is to be used for.

Quadrat analysis: A standard method of testing point distribution patterns by comparing the numbers of objects from one subarea (quadrat) to another.

Quadrats: Uniform subareas designed for sampling point objects.

Quadtrees: Compact data structures that quantize geographical space into variably sized quarters, each of which exhibits attribute homogeneity.

Quantize: The process of dividing data into quanta or packets. In GIS the process is most commonly associated with the encoding of geographic space into some form of raster data structure.

Radio telemetry: A method of tracking large animals with the use of radio-collars that transmit a signal that can be received either on the ground or through low-flying aircraft.

Random: Spatial arrangement of objects where their locations relative to one another are unpredictable. This type of arrangement is indicative of processes that operate through pure chance occurrences.

Random noise: The nonspatially correlated error or residual component for a Kriging model.

Random walk: A line intersect method for analyzing the distribution of linear objects by drawing a zigzag path and recording its points of intersection with lines on the coverage.

Range: On a semivariogram, the critical value where the variance levels off or stays flat.

Range graded: The result of a cartographic classing technique that groups ranges of numerical values into single classes.

Raster: A form of GIS graphic data structure that quantizes space into a series of uniformly shaped cells.

Raster chain codes: A compact method of storing raster data as chains or grid cells bordering homogeneous polygonal areas.

Ratio: The highest level of data measurement that includes an absolute starting point and allows ratios of values to be produced (e.g.. salaries).

Ray: A line based on optical geometry representing the uppermost bound of visible objects from a viewer location.

Ray tracing: Analysis of locations on a map that are visible from a viewer location by using a series of rays.

Reality: A map design constraint resulting from the often irregular shapes of real geographic objects.

Rectangular coordinates: Coordinate system based on a two-dimensional planar surface, using X and Y coordinate locations from an origin or starting point. The origin is normally based on an arbitrary Cartesian coordinate system. This is the common coordinate system used by digitizer tablets.

Reductions: In mapping, the idea that the cartographic document is normally smaller than the area it represents.

Reference globe: A hypothetical reduction of the earth and its coordinates mapped onto a sphere that is reduced in size to that chosen for the scale of the flat map being produced.

Region growing: In remote sensing the procedure that is most often used in unsupervised classification as it attempts to define groups of pixels (regions) with similar spectral signatures.

Regions: Areas of relative uniformity of content within a coverage.

Registration points: Digitizer locations that locate the corners of the map document so that the user can digitize different parts of the document at different sessions.

Regular: Spatialarrangement of objects where all objects are the same distance away from their immediate neighbors. Such an arrangement implies some form of process that places these objects this way.

Regular grid: A series of locational lines whose interspatial distances are identical.

Regular lattice: A series of point locations whose interspatial distances are identical.

Related variables: Variables from two coverages whose interactions can be defined by mathematical equations. A method of dasymetric mapping.

Relational database structures: Computer database structures employing an ordered set of attribute values or records known as tuples grouped into two-dimensional tables called relations.

Relational join: In relational database management systems, the process of creating functional links between two or more tables sharing some common attribute.

Relations: Two-dimensional tables of relational database attribute records.

Relative barriers: Barriers that, like friction surfaces but at discrete locations, act to impede movement.

Relative location: Location determined in relation to a second object rather than by its own coordinates.

Remote sensing: The observation of objects or groups of objects, normally at a distance, most often with the use of some form of mechanical or electronic device. The data can either produce an image or be stored for later retrieval.

Repeatability: Synonym for repeatability. Deals with how close the digitizer will read to an original value if it is recorded more than once.

Representational fraction (RF): One method of representing map scale where both map units and earth units are represented by the same measurement units and are shown as a fraction (e.g., 1:24,000).

Resolution: 1. The amount of earth surface represented by a single grid cell in a raster GIS. 2. The ability of a digitizer to record increments in space. The smaller the units it can handle the better its resolution.

Resultant length: The length of the resultant vector calculated through the use of the Pythagorean theorem on component vectors.

Rose diagram: A circular graph that displays the orientation of linear objects or phenomena by starting at the center of a circle and drawing each observation as a single line outward from that point.

Rotation: The process of moving part or all of a cartographic object in some azimuthal direction around some focal point.

Rough surface: A statistical surface whose values change rapidly and often unpredictably with changes in horizontal distance.

Routed line cartograms: Cartograms that show the order of objects located along a line, but do not show the actual distances between them.

Routing: The process of using networks to define travel paths from one place to another.

Roving window neighborhood functions: A group of analytical functions that employ a moving filter to reclassify the values of the coverage.

Rubber sheeting: A method to adjust coverage features in a nonuniform manner. Used to rectify objects within a coverage.

Rules of combination overlay: A method of overlay that allows the user to define which categories will be combined, which will take precedence, and which may be ignored.

Run length codes: A compact method of storing raster data as strings of homogeneous grid cells.

Runs test: A simple statistical technique for analyzing the pattern of sequences of data such as those derived through line intersect methods.

Saddle-point problem: For interpolation, the problem that arises when one pair of diagonally opposite Z values forming the corners of a rectangle is located above and the second pair below the value the algorithm is attempting to solve.

Sampled area: The portion of a larger area that is to be sampled to draw inferences about the whole.

Sampling frame: A complete list of all members of a population to be sampled, or the entire area within which a sample is to be selected.

Scale: In mapping, the size relationship or ratio

between the map document and the portion of the earth it represents.

Scale change: The process of changing the size of part or all of a cartographic object.

Scale dependent error: Spatial data error that is primarily a function of the scale of the input map document.

Scale factor (SF): On a reference globe, the actual scale divided by the principal scale. By definition the scale factor for a reference globe should be 1.0 because the actual scale and the principal scale should be the same everywhere.

Scan lines: Lines of pixels resulting from the process of scanning used by the remote sensing device.

Scanning radiometers: Passive remote sensing devices that receive electromagnetic radiation in groups of wavelengths.

Second normal form: Based on the theory of normal forms, the second requirement that every column that is not part of the primary key must be completely dependent on the primary key.

Selective overlay: See rules of combination overlay.

Semivariance: One half the square of the standard deviation between each elevational value and its neighbors.

Semivariogram: A graphical device used in the process of Kriging that compares the variance of the difference in value of elevational points to their distances.

Shape measures: Analytical techniques that characterize cartographic objects based on comparisons of their own dimensional attributes (e.g., perimeter-area ratios), comparisons to known geometric shapes (e.g., circles), or to alternative geometry's (e.g., fractal geometry).

Shortest path: A coverage created by determining the shortest Euclidean distance from one place to another. This can be determined for a surface or for networks.

Shortest path surface: A coverage created by determining the distance from a point, line, or area to all other locations on the coverage. This coverage includes the shortest path as part of the result.

Side-looking airborne radar (SLAR): A system mounted at the base of an aircraft that sends out and receives radar signals in a sweeping motion perpendicular to the flight path of the aircraft.

Simple list: The simplest computer file structure that stores data as it is input, but does not organize it in any sequence or order. This type of file structure allows very easy input, but makes searches and retrieval operations very difficult.

Sinuosity: The relationship between straight-line distance and actual distance of linear objects.

Sinuous: Wandering wildly in a lateral direction along its length.

Size measures: Analytical techniques that characterize cartographic objects based on their absolute sizes or comparisons of their sizes with other cartographic objects.

Skew: Measure of the squareness of the results of digitizing four corners of a document on a digitizing table.

Slicing: A group of GIS functions that selectively change the class interval for statistical surfaces.

Sliver polygons: Small polygons, often without attributes, that result either from digitizing the same line twice or following the overlay of two or more coverages.

Slope: Measure of the amount of rise divided by the amount of horizontal distance traveled.

Smoothing: A type of filter, often called a low-pass filter, that reduces the value of extraordinarily high cell or pixel values through some process of averaging.

Smooth surface: A statistical surface whose values change at a relatively constant rate with changes in horizontal distance.

Software design: The structured process of transforming software functional requirements to specific code.

Software engineering: A sub-field of computer science specializing in the systematic approaches to software code generation.

Spacing: The attribute of spatial objects that shows their distances from one another.

Spaghetti model: The simplest vector data structure representing a one-for-one translation of the graphic image. Often used as a data structure for vector input.

Spatial: Anything dealing with the concept of space. In the geographic context, primarily dealing with the distribution of things on the surface of the earth.

Spatial arrangement: The placement, ordering,

concentration, connectedness, or dispersion of multiple objects within a confined geographic space.

Spatial information product: The desired output from a set of GIS analysis operations.

Spatial integrity: The degree of perforation of a perforated region.

Spatially correlated: Sets of objects that are spatially associated may also show statistically significant relationships that may indicate some connection between the processes acting on both objects or even the processes acting on one set of objects influencing those of another set of objects in the same area.

Spatial pseudo nodes: Pseudo nodes that represent island polygons or where some attribute along a line changes attributes. Both of these are acceptable pseudo nodes.

Spatial surrogates: Spatially explicit data used to replace data that do not exhibit spatial properties.

Splines: Mathematical methods of smoothing linear cartographic objects. Often used in interpolation methods.

Stability: The property of a digitizer that deals with the propensity of the device to change the readings it provides as it warms up.

Standard parallels: Lines on a reference globe that maintain a scale factor of 1.0.

Static neighborhood functions: Analytical functions that modify the attributes of neighboring cells without the use of a moving filter.

Statistical surface: Any ordinal, interval, or ratio data that can be represented as and operated on as a surface.

Step rate: The incremental distance covered by an output device during the process of drawing lines. The smaller the step rate, the finer the resolution of the output document.

Stereographic projections: See conformal stereographic projections.

Stereogrammic hierarchy: Giving mapped objects the impression of different visual levels to direct the map reader to different types of information.

Structure: See drift.

Subdivisional hierarchy: A method of using categorical subdivision to portray the internal relationships of a map in a hierarchical fashion.

Substantive objective: One of two parts of the purpose of a map, this one relating to the infor-

mation the map must include. Controls what is to be mapped.

Supervised classification: In remote sensing the process of classifying digital data with the interaction of a user. The user normally selects pixels that are of known category and uses these to train the software about the electromagnetic properties likely to be associated with each category.

Supplier for GIS (GIS Supplier): Third party vendor for GIS software, and sometimes hardware, used by the GIS community.

Surface fitting models: Interpolation methods that attempt to fit the existing data points into a mathematical surface equation.

System design: The process whereby the output from system analysis is transformed from technology independent statements of user requirements to the system that is going to be used to implement the GIS.

System operators: GIS professionals responsible for assuring that the system functions correctly, and for maintaining the hardware and software in a condition that allows the users to perform their tasks.

System sponsor: The organizational leadership responsible for making funding and logistical decisions and for selecting priorities for resource allocation within the organization.

System users: GIS professionals that interact most closely with the GIS through data input, storage, editing, and performing the actual analytical functions necessary to provide answers to queries.

Targeted analysis: An analysis that is directed toward reclassifying a target cell or target groups of cells as a function of neighboring values.

Target population: That portion of a whole population that is likely to contain attributes for which a sample is being taken.

Technical design: GIS design issues that deal with system requirements, hardware and software needs, and other noninstitutional issues.

Technical limits: A set of limitations on map design based on the physical limitations of the equipment to produce readable results.

Template: A user-selected area on a map that is employed as the outline for subsequent coverages to compare their attributes.

Texture: The visual appearance of a symbol resulting from the spacing and placement of internal markings.

Thematic maps: Maps whose primary purpose is to display the locations of a single attribute or the relationships among several selected attributes.

Theodolite: A modern optical survey device that includes digital readouts and improved accuracy over its predecessors. This device is still limited in its ability to provide survey data for large areas.

Thiessen polygons: A method of creating polygons or proximal regions around point objects by defining them mathematically, dividing the space between each point, and connecting these distances with straight lines.

Third normal form: Based on the theory of normal forms, the requirement that every non-primary key column must be nontransitively dependent on the primary key.

Tic marks: Digitizer locations of known latitude and longitude, used for the process of geocoding.

Tiling: Process of separating large databases into predefined subsections for archival purposes.

Too many labels: Entity-attribute agreement error resulting from placing more than one label point within a single polygon during input.

Topological models: A vector data structure that incorporates explicit spatial information about the relative locations of objects in the database. These are necessary to allow advanced analysis in a GIS.

Topologically Integrated Geographic Encoding and Referencing System (TIGER): A topological vector data model created by the U.S. Bureau of the Census as an improvement over GBF/DIME. Points, lines, and areas are explicitly addressed, allowing direct retrieval of census block data, and real-world objects are portrayed in their true geographic shape.

Topology: The mathematically explicit rules defining the linkages of geographical elements.

Total analysis of neighborhood: A function that analyzes the entire neighborhood and returns new values for that entire neighborhood.

Translation: The process of moving part or all of a cartographic object to a new location in Cartesian space.

Transverse projections: A type of Mercator projection that does not exhibit the property

that all rhumb lines are straight. As a result, scale exaggeration increases away from the standard meridian, limiting the usefulness of this type of projection to a small zone along the central meridian.

Trapezoidal rule: A methodology for calculating the centroid of a highly irregular polygonal shape.

Trend surface analysis: An interpolation routine that simplifies the surface representation to allow visualization of general or overall changes in elevation value with distance.

Triangulated irregular network (TIN): A vector data model that uses triangular facets as a means of explicitly storing surface information.

Triangulation: A method of surveying that measures distance and direction from a baseline outward to the selected points.

Trilateration: A method of surveying that measures both from and to a baseline and each point to be located.

Tuples: Individual records (rows) in a relational database structure.

Undershoot: An arc that does not extend far enough to connect with another arc. Sometimes called dangling arcs or just dangles.

Undirected networks: Type of network that does not place restrictions on the direction of movement.

Uniform: A distribution of objects that exhibits the same density in one small subarea as that in each additional subarea.

Unit value: The actual numerical value assigned to individual polygons in a choropleth map.

Universal polar stereographic (UPS) grid: A grid system used to cover the polar areas of the globe. Each circular polar zone is divided in half at 0 and 180 degrees meridian.

Unsupervised classification: In remote sensing the process of classifying digital data without the interaction of a user. The process, most often statistical, randomly selects pixels and attempts to find spectral similarities between the selected pixels and the remainder of the pixels in the image.

Value: The lightness or darkness of a cartographic object compared to a standard black and white.

Variance-mean ratio (VMR): An index showing the relationship between the frequency of subarea variability to the average number of points per quadrat.

Vector: A graphic data structure that represents the points, lines, and areas of geographical space by exact X and Y coordinates.

Vector resultant: The result of combining the magnitudes and directions of component vectors.

Verbal scale: One way of representing map scale where both map units and earth units are stated as text (e.g., one inch equals one hundred feet).

Vertical exaggeration: A technique used for production of cross-sectional profiles that increases the vertical dimension for easier viewing of surface changes.

Viewing azimuth: The geographical direction from which a perspective map can be displayed so as to observe its different sides.

Viewing distance: One of a number of factors that can be modified to effectively display perspective maps.

View integration: The process of satisfying multiple user needs for a GIS database and its applications by incorporating each local view into the global view.

Viewshed: A polygonal map resulting from analysis of topographic surfaces that portrays all locations visible from a preselected viewpoint.

Visual contrast: The relationship between cartographic symbols and the underlying map background, often responsible for legibility.

Voronoi diagrams: See Thiessen polygons.

Waterfall model: See project life cycle.

Weighted mean center: Center of a group of point objects located in space and modified by a weighting factor applied to each point. The weight is representative of some physical, economic, or cultural factor.

Weighting methods: Nonlinear interpolation methods that weight the predicted elevation values by the distance to its nearest neighbor elevation values.

Weird polygons: Graphical artifacts resembling real polygons but lacking nodes. Usually the result of digitizing points for polygonal objects out of sequence.

Wire-frame diagrams: A perspective view of a set of statistical data displayed as a skeleton form.

Witness trees: Trees left behind by surveyors during pioneer periods in the United States. These trees were left as witnesses for the metes and bounds survey method and have also been used more recently as indicators of previous forest vegetation.

Zones: On the UTM grid system the earth is divided into columns 6 degrees of longitude wide. Each of these is called a zone and is numbered from 1 to 60 eastward beginning at the 180th meridian.

Zoom transfer scope: A mechanical, optical device designed for accurately transferring information from aerial photographs to projected map documents.

PHOTO CREDITS

Chapter 2

Figure 2.2 (left): Joe Scherschel, Life Magazine © Time Inc. *Figure 2.2 (right):* Courtesy of Del E. Webb Development Company. *Figure 2.11:* Courtesy of NAVSTAR.

Chapter 3

Figure 3.12: Courtesy of Thomas M. Lillesand, University of Wisconsin, Madison.

Chapter 5

Figure 5.1: Courtesy of CalComp Inc.

Chapter 10

Figure 10.1: USGS/Ice & Climate Project.

Chapter 14

Figure 14.13: Photo courtesy of Silicon Graphics, Inc. Landsat imagery courtesy of EOSAT. SPOT imagery courtesy of SPOT Image Corporation. *Figure 14.14:* Courtesy of Environmental Systems Research Institute, Inc.

INDEX

479